ADVANCED CALCULUS

ADVANCED CALCULUS

Gerald B. Folland

Department of Mathematics
University of Washington

PRENTICE HALL, Upper Saddle River, NJ 07458

Library of Congress Cataloging-in-Publication Data

Folland, Gerald B.
  Advanced Calculus / Gerald B. Folland.
     p.   cm.
  Includes bibliographical references and index.
  ISBN: 0-13-065265-2
  1. Calculus. II. Title

QA303.2 F67 2002   2001055359
515--dc21

Acquisition Editor: *George Lobell*
Editor-in-Chief: *Sally Yagan*
Vice-President/Director of Production and Manufacturing: *David W. Riccardi*
Executive Managing Editor: *Kathleen Schiaparelli*
Senior Managing Editor: *Linda Mihatov Behrens*
Assistant Managing Editor: *Bayani Mendoza de Leon*
Production Editor: *Jeanne Audino*
Manufacturing Buyer: *Alan Fischer*
Manufacturing Manager: *Trudy Pisciotti*
Marketing Manager: *Angela Battle*
Marketing Assistant: *Rachel Beckman*
Editorial Assistant: *Melanie Van Benthuysen*
Art Director: *Jayne Conte*
Cover Photo: *László Moholy-Nagy, "Komposition A XXI" 1925. Oil and Mixed Media on Canvas,* 96 x 77 *cm,*
          Westfälisches Landesmuseum fürKunst und Kulturgeschichte, München, Germany

Pearson Education Ltd.
Pearson Education Australia Pty., Limited
Pearson Education Singapore, Pte. Ltd
Pearson Education North Asia Ltd.
Pearson Education Canada, Ltd.
Pearson Education de Mexico, S.A. de C.V.
Pearson Education—Japan
Pearson Education Malaysia, Pte. Ltd

CONTENTS

PREFACE

This is a book about the theory and applications of derivatives (mostly partial), integrals (mostly multiple or improper), and infinite series (mostly of functions rather than of numbers), at a deeper level than is found in the standard calculus books.

In recent years there has been a tendency for the courses that were once called "advanced calculus" to turn into courses on the foundations of analysis. Students typically start with a year and a half of calculus that emphasizes computations and applications, then proceed (perhaps by way of a "bridge course" on mathematical reasoning) to a course of an entirely theoretical nature that covers such things as the topology of Euclidean space, the theory of the Riemann integral, and proofs of some theorems that have been taken on faith before.

I am not persuaded that such a divorce of the practical from the theoretical aspects of the subject is a good idea. On the one hand, the study of theoretical underpinnings of ideas with which one is already familiar tends to be dry and tedious, and the development of unfamiliar ideas can be rather daunting unless it is accompanied by some hands-on experience with concrete examples and applications. On the other hand, relegation of the computations and applications to the elementary courses means that students are not exposed to these matters on a more sophisticated level. (How many students recognize that Taylor polynomials should be part of one's everyday tool kit? How many know that the integral test gives an effective way of approximating the sum of a series?)

This book is an attempt to present a unified view of calculus in which theory and practice can reinforce each other. On the theoretical side, it is reasonably complete and self-contained. Accordingly, it contains a certain amount of "foundations of analysis," but I have kept this material to the bare minimum needed for the main topics of the book. I also place a higher premium on intuitive understanding than on formal proofs and technical definitions. Along with the latter, therefore, I often offer informal arguments and ideas, sometimes involving infinitesimals, that may provide more enlightenment than the strictly rigorous approach. The worked-out examples and exercises run the gamut from routine calculations to theoretical arguments; many of them involve a mixture of the two. The reader whose interest in the theory is limited should be able to benefit from the book by skipping many of the proofs.

The essential prerequisite for this book is a sound knowledge of the mechanics of one-variable calculus. The theory of differentiation and integration on the real line is presented, rather tersely, in Sections 2.1 and 4.1, but I assume that the reader is thoroughly familiar with the standard techniques for calculating derivatives and integrals. Some previous experience with infinite series, partial derivatives, and multiple integrals might be helpful but is not really necessary. And, of course, for a full appreciation of the theory one needs a certain level of comfort with mathematical reasoning, but that is best acquired with practice and experience.

An acquaintance with linear algebra is needed in a few places, particularly §2.8 (classification of critical points), §2.10 (differentiation of vector-valued functions of vector variables), §3.1 and §§3.4–5 (the implicit function theorem for systems of equations, the inverse mapping theorem, and functional dependence), and §4.4 (change of variables for multiple integrals). However, most of this material can be done in the two- and three-dimensional cases (perhaps by eliding parts of some proofs) with vector algebra and a little ad hoc discussion of matrices and determinants. In any case, Appendix A provides a brief summary of the necessary concepts and results from linear algebra.

A few of the more formidable proofs have been exiled to Appendix B. In some of them, the ratio of the amount of work required to the amount of understanding gained is especially high. Others involve ideas such as the Heine-Borel theorem or partitions of unity that are best appreciated at a more advanced level. Of course, the decisions on what to put into Appendix B reflect my personal tastes; instructors will have to make their own choices of what to include or omit.

In this book a single numeration system is used for theorems, lemmas, corollaries, propositions, *and* displayed formulas. Thus, for each m and n there is only one item of any of these types labeled $m.n$, and it is guaranteed to follow $m.(n-1)$ and precede $m.(n+1)$. This procedure minimizes the effort needed to locate referenced items.

In a few places I offer glimpses into the world of more advanced analysis. Chapters 4 and 5 end with brief, informal sketches of the Lebesgue integral and the theory of differential forms; Chapter 8 leads to the point where the realm of eigenfunction expansions and spectral theory is visible on the horizon. I hope that many of my readers will accept the invitation to explore further.

Acknowledgments. This book has benefited from the comments and suggestions of a number of people: my colleague James Morrow, the students in the advanced calculus classes that he and I have taught over the past three years in which preliminary versions of this book were used, and several reviewers, especially Jeffrey Fox. I am also grateful to my editor, George Lobell, for his support and enthusiasm.

Errata. Responsibility for errors in this book, of course, remains with me. Responsibility for informing me of these errors, however, rests with my readers. Anyone who finds misprints, mistakes, or obscurities is urged to write to me at the address below. I will post such things on a web site that will be accessible from `www.math.washington.edu`.

Gerald B. Folland
Department of Mathematics
University of Washington
Seattle, WA 98195-4350
`folland@math.washington.edu`

ADVANCED CALCULUS

Chapter 1

SETTING THE STAGE

The first half of this chapter (§§1.1–4) presents basic facts and concepts concerning geometry, vectors, limits, continuity, and sequences; the material in it is used throughout the later chapters. The second half (§§1.5–8) deals with some of the more technical topological results that underlie calculus. It is quite concise and includes nothing but what is needed in this book. The reader who wishes to proceed quickly to the study of differentiation and integration may scan it quickly and refer back to it as necessary; on the other hand, the reader who wishes to see a more extensive development of this material is referred to books on the foundations of analysis such as DePree and Swartz [5], Krantz [12], or Rudin [18].[1]

At the outset, let us review some standard notation and terminology for future reference:

- *Sums:* If $a_1, a_2, \ldots, a_k$ are numbers, their sum $a_1 + a_2 + \cdots + a_k$ is denoted by $\sum_1^k a_n$, or by $\sum_{n=1}^k a_n$ if necessary for clarity. The sum need not be started at $n = 1$; more generally, if $j < k$, we have

$$\sum_j^k a_n = a_j + a_{j+1} + \cdots + a_k.$$

  The letters j and k denote the limits of summation; the letter n is analogous to a dummy variable in an integral and may be replaced by any other letter that is not already in use without changing the meaning of the sum. We shall occasionally write simply $\sum a_n$ when the limits of summation are understood.

---

[1]Numbers in brackets refer to the bibliography at the end of the book.

- *Factorials:* If n is a positive integer, $n!$ ("n **factorial**") is the product of all the integers from 1 to n. By convention, $0! = 1$, so that the formula $n! = n \cdot (n-1)!$ remains true even for $n = 1$.

- *Sets:* If S and T are two sets, $S \cup T$ and $S \cap T$ denote their union and intersection, respectively, and $S \setminus T$ denotes the set of all elements of S that are not in T. The expressions "$S \subset T$" and "$T \supset S$" both mean that S is a subset of T, including the possibility that $S = T$, and "$x \in S$" and "$x \notin S$" mean, respectively, that x is or is not an element of S. The set of all objects x satisfying a property $P(x)$ is denoted by $\{x : P(x)\}$, and empty set is denoted by $\varnothing$.

  The union and intersection of a family $S_1, S_2, \ldots, S_k$ of sets are denoted by $\bigcup_1^k S_n$ and $\bigcap_1^k S_n$. The conventions for using the symbols $\bigcup$ and $\bigcap$ are the same as those for the summation sign $\sum$ described above.

- *Real numbers:* The set of real numbers is denoted by $\mathbb{R}$. The following notations are used for intervals in $\mathbb{R}$:

$$(a, b) = \{x : a < x < b\}, \qquad [a, b] = \{x : a \le x \le b\}$$
$$(a, b] = \{x : a < x \le b\}, \qquad [a, b) = \{x : a \le x < b\}.$$

  Intervals of the form (a, b) are called **open**; intervals of the form $[a, b]$ are called **closed**; and intervals of the forms $(a, b]$ and $[a, b)$ are called **half-open**. (Of course, the symbol (a, b) is also used to denote the ordered pair whose first and second members are a and b, respectively; remarkably enough, this rarely causes any confusion.)

  If $\{x_1, \ldots, x_k\}$ is a finite set of real numbers, its largest and smallest elements are denoted by $\max(x_1, \ldots, x_k)$ and $\min(x_1, \ldots, x_k)$, respectively.

- *Infinity.* In discussing limits it is often convenient to add two "points at infinity" ∞ (also called $+\infty$) and $-\infty$ to the real number system. These are *not* real numbers, and one can perform arithmetical operations on them only with great caution, but there is no harm in thinking of them as actual mathematical objects. The points $\pm\infty$ may be used as endpoints of intervals; for example, $(a, \infty) = \{x : x > a\}$. Intervals of the form $[a, \infty)$ and $(-\infty, b]$ are classified as closed intervals; (a, ∞) and $(-\infty, a)$ are open.

- *Complex numbers:* The imaginary unit $\sqrt{-1}$ is denoted by i, although the letter i may be used for other purposes when complex numbers are not under discussion. The set of complex numbers, that is, numbers of the form $x + iy$

where $x, y \in \mathbb{R}$, is denoted by $\mathbb{C}$. As a set, $\mathbb{C}$ may be identified with the Cartesian plane by the correspondence $x + iy \longleftrightarrow (x, y)$, and we speak of "the complex plane $\mathbb{C}$." If $z = x + iy$ is a complex number, x and y are called its **real** and **imaginary parts**, respectively, and are denoted by $\operatorname{Re} z$ and $\operatorname{Im} z$. The number $x - iy$ is called the **complex conjugate** of z and is denoted by $\bar{z}$, and the number $\sqrt{z\bar{z}} = \sqrt{x^2 + y^2}$ (the distance from (x, y) to the origin in the plane) is called the **absolute value** of z and is denoted by $|z|$.

- *Mappings and functions:* A **mapping**, or **map**, is a rule f that assigns to each element of some set A an element of some other set B (possibly equal to A). We write $f : A \to B$ to display all these ingredients together. If $x \in A$, the element of B assigned to x by f is called the **value** of f at x and is denoted by $f(x)$. If S is a subset of A, the set of values $\{f(x) : x \in S\}$ is denoted by $f(S)$. The set A is called the **domain** of f, and the set $f(A)$ (a subset of B) is called the **range** of f. The mapping $f : A \to B$ is called **one-to-one** if $f(x) = f(y)$ only when $x = y$, and f is said to map A **onto** B if $f(A) = B$.

  If $f : A \to B$ and $g : B \to C$ are mappings, their **composition** is the mapping $g \circ f : A \to C$ defined by $(g \circ f)(x) = g(f(x))$.

  A mapping $f : A \to B$ is said to be **invertible** if there is another mapping $g : B \to A$ such that $g(f(x)) = x$ for all $x \in A$ and $f(g(y)) = y$ for all $y \in B$. The equation $g(f(x)) = x$ can be valid for all $x \in A$ only if f is one-to-one, and the equation $f(g(y)) = y$ can be valid for all $y \in B$ only if f maps A onto B. Conversely, if these two conditions are satisfied, it is easy to verify that f is invertible. In this case, the mapping g is called the **inverse** of f and is commonly denoted by f^{-1}.

  Mappings are sometimes called "functions," but we shall reserve the term **function** for mappings whose values are real numbers, complex numbers, or vectors. Mappings of a set A into itself ($B = A$) are sometimes called **transformations**.

- *Special functions:* In this book, we denote the natural logarithm by log rather than ln, this being the common usage in advanced mathematics. Also, we denote the principal branches of the inverse trig functions by arcsin, arccos, and arctan; arcsin and arccos map $[-1, 1]$ onto $[-\frac{1}{2}\pi, \frac{1}{2}\pi]$ and $[0, \pi]$, respectively, and arctan maps $\mathbb{R}$ onto $(-\frac{1}{2}\pi, \frac{1}{2}\pi)$.

- *Logical symbols:* We shall sometimes use the symbols $\implies$ and $\iff$ to denote logical implication and equivalence, respectively. That is, if A and B

are mathematical statements, "$A \implies B$" is read "A implies B" or "If A, then B," and "$A \iff B$" is read "A is equivalent to B" or "A if and only if B." We point out that "$A \implies B$" and "not $B \implies$ not A" are logically equivalent; that is, in order to prove that hypothesis A implies conclusion B, one may assume that B is false and show that A is false.

1.1   Euclidean Spaces and Vectors

We shall be studying functions of several real variables, say $f(x_1, x_2, \ldots, x_n)$. In elementary treatments of the subject one usually focuses on the cases $n = 2$ and $n = 3$, because these are the ones where ordered n-tuples of numbers can represent points in physical space. However, most of the ideas work equally well for any number of variables, and it is helpful to continue using geometric language in this more general setting even though "n-dimensional space" doesn't correspond directly to a physical object that can be visualized.

The set of all ordered n-tuples of real numbers is called n-**dimensional Euclidean space** and is denoted by $\mathbb{R}^n$. We will denote such n-tuples either by writing out the components or by single boldface letters:

$$\mathbf{x} = (x_1, x_2, \ldots, x_n).$$

The n-tuple whose components are all zero is denoted by $\mathbf{0}$:

$$\mathbf{0} = (0, 0, \ldots, 0).$$

When $n = 2$ or 3, we shall often write (x, y) or (x, y, z) instead of (x_1, x_2) or (x_1, x_2, x_3), but we shall still use $\mathbf{x}$ as a single symbol to denote the ordered pair or triple.

Ordered n-tuples of numbers lead a double life. We usually think of the n-tuple $(x_1, \ldots, x_n)$ as representing the Cartesian coordinates of a point in the n-dimensional space $\mathbb{R}^n$. However, sometimes we think of it as representing a "quantity with magnitude and direction" such as a force or velocity and visualize it as an arrow. There is some virtue in maintaining a notational distinction between these two concepts, but we shall not attempt to do so.

To express the basic ideas of n-dimensional geometry it is convenient to use the language of vector algebra. Most of the vector operations work equally well in any number of dimensions:

$$\begin{aligned}
\text{Addition}: \quad & \mathbf{x} + \mathbf{y} = (x_1 + y_1, \ldots, x_n + y_n), \\
\text{Scalar multiplication}: \quad & c\mathbf{x} = (cx_1, \ldots, cx_n), \\
\text{Dot product}: \quad & \mathbf{x} \cdot \mathbf{y} = x_1 y_1 + \cdots + x_n y_n.
\end{aligned}$$

The exception is the cross product, which is peculiar to 3 dimensions; we shall discuss it at the end of this section. If $\mathbf{x} \in \mathbb{R}^n$, the **norm** of $\mathbf{x}$ is defined to be

$$|\mathbf{x}| = \sqrt{x_1^2 + \cdots + x_n^2} = \sqrt{\mathbf{x} \cdot \mathbf{x}}.$$

Some people denote norms by double vertical bars, thus: $\|\mathbf{x}\|$.

There are two fundamental inequalities involving the dot product and norm, Cauchy's inequality and the triangle inequality. The reader is probably familiar with them in dimensions 2 and 3, and the ideas are exactly the same in higher dimensions.

1.1 Proposition (Cauchy's Inequality). *For any* $\mathbf{a}, \mathbf{b} \in \mathbb{R}^n$,

$$|\mathbf{a} \cdot \mathbf{b}| \leq |\mathbf{a}| \, |\mathbf{b}|.$$

Proof. If $\mathbf{b} = \mathbf{0}$ then both sides of the inequality are 0. Otherwise, we introduce a real variable t and consider the function

$$f(t) = |\mathbf{a} - t\mathbf{b}|^2 = (\mathbf{a} - t\mathbf{b}) \cdot (\mathbf{a} - t\mathbf{b}) = |\mathbf{a}|^2 - 2t\mathbf{a} \cdot \mathbf{b} + t^2|\mathbf{b}|^2.$$

This is a quadratic function of t. Its minimum value occurs at $t = (\mathbf{a} \cdot \mathbf{b})/|\mathbf{b}|^2$, and that minimum value is

$$f((\mathbf{a} \cdot \mathbf{b})/|\mathbf{b}|^2) = |\mathbf{a}|^2 - \frac{(\mathbf{a} \cdot \mathbf{b})^2}{|\mathbf{b}|^2}.$$

On the other hand, clearly $f(t) \geq 0$ for all t, so

$$|\mathbf{a}|^2 - \frac{(\mathbf{a} \cdot \mathbf{b})^2}{|\mathbf{b}|^2} \geq 0.$$

Multiplying through by $|\mathbf{b}|^2$, we obtain the desired result: $|\mathbf{a}|^2|\mathbf{b}|^2 \geq (\mathbf{a} \cdot \mathbf{b})^2$. □

Note. Cauchy's inequality is also called *Schwarz's inequality*, the *Cauchy-Schwarz inequality*, or *Buniakovsky's inequality*. (Schwarz and Buniakovsky independently discovered the corresponding result for integrals of functions, namely,

$$\left| \int_a^b f(x)g(x)\,dx \right| \leq \left[\int_a^b |f(x)|^2\,dx \right]^{1/2} \left[\int_a^b |g(x)|^2\,dx \right]^{1/2},$$

which can be proved in much the same way.)

1.2 Proposition (The Triangle Inequality). *For any* $\mathbf{a}, \mathbf{b} \in \mathbb{R}^n$,

$$|\mathbf{a} + \mathbf{b}| \le |\mathbf{a}| + |\mathbf{b}|.$$

Proof. We have $|\mathbf{a} + \mathbf{b}|^2 = (\mathbf{a} + \mathbf{b}) \cdot (\mathbf{a} + \mathbf{b}) = |\mathbf{a}|^2 + 2\mathbf{a} \cdot \mathbf{b} + |\mathbf{b}|^2$. By Cauchy's inequality, this last sum is at most $|\mathbf{a}|^2 + 2|\mathbf{a}|\,|\mathbf{b}| + |\mathbf{b}|^2 = (|\mathbf{a}| + |\mathbf{b}|)^2$, so the result follows by taking square roots. $\qquad\square$

The **distance** between two points $\mathbf{x}$ and $\mathbf{y}$ in 3-space is given by

$$\sqrt{(x_1 - y_1)^2 + (x_2 - y_2)^2 + (x_3 - y_3)^2} = |\mathbf{x} - \mathbf{y}|,$$

and similarly for points in the plane. We shall take this as a *definition* of distance in n-space for any n:

$$\text{Distance from } \mathbf{x} \text{ to } \mathbf{y} = |\mathbf{x} - \mathbf{y}|.$$

By taking $\mathbf{a} = \mathbf{x} - \mathbf{y}$ and $\mathbf{b} = \mathbf{y} - \mathbf{z}$ in the triangle inequality, we see that

$$|\mathbf{x} - \mathbf{z}| \le |\mathbf{x} - \mathbf{y}| + |\mathbf{y} - \mathbf{z}|$$

for any $\mathbf{x}, \mathbf{y}, \mathbf{z} \in \mathbb{R}^n$. That is, the distance from $\mathbf{x}$ to $\mathbf{z}$ is at most the sum of the distances from $\mathbf{x}$ to $\mathbf{y}$ and from $\mathbf{y}$ to $\mathbf{z}$, for any intermediate point $\mathbf{y}$. Hence the name "triangle inequality": One side of a triangle is at most the sum of the other two sides.

If we think of two vectors $\mathbf{x}$ and $\mathbf{y}$ as arrows emanating from the same point, we can speak of the **angle** θ between them. The familiar formula for θ in dimensions 2 and 3 remains valid in higher dimensions:

$$\theta = \arccos\left(\frac{\mathbf{x} \cdot \mathbf{y}}{|\mathbf{x}|\,|\mathbf{y}|}\right).$$

Cauchy's inequality says that the quotient in parentheses always lies in the interval $[-1, 1]$, so it is indeed the cosine of some number $\theta \in [0, \pi]$.

In particular, the directions of two vectors $\mathbf{x}$ and $\mathbf{y}$ are perpendicular to each other if and only if $\mathbf{x} \cdot \mathbf{y} = 0$. In this case the vectors are said to be **orthogonal** to each other.

In many situations we need to control the magnitude, i.e., the norm, of a vector $\mathbf{x} = (x_1, \ldots, x_n)$, but it is often more convenient to work with the magnitudes of the components x_j of $\mathbf{x}$. In such cases the following inequalities are useful. Let M be the largest of the numbers $|x_1|, \ldots, |x_n|$. Then $M^2 \le x_1^2 + \cdots + x_n^2$ (because M^2 is one of the numbers on the right), and $x_1^2 + \cdots + x_n^2 \le nM^2$ (because each number on the left is at most M^2). In other words,

$$(1.3) \qquad \max(|x_1|, \ldots, |x_n|) \le |\mathbf{x}| \le \sqrt{n}\,\max(|x_1|, \ldots, |x_n|).$$

Cross Products. Let $\mathbf{i} = (1, 0, 0)$, $\mathbf{j} = (0, 1, 0)$, and $\mathbf{k} = (0, 0, 1)$ be the standard basis vectors for $\mathbb{R}^3$; then an arbitrary vector $\mathbf{a} \in \mathbb{R}^3$ can be written as

$$\mathbf{a} = (a_1, a_2, a_3) = a_1\mathbf{i} + a_2\mathbf{j} + a_3\mathbf{k}.$$

The **cross product** of two vectors $\mathbf{a}, \mathbf{b} \in \mathbb{R}^3$ is defined by

$$\mathbf{a} \times \mathbf{b} = \det \begin{pmatrix} \mathbf{i} & \mathbf{j} & \mathbf{k} \\ a_1 & a_2 & a_3 \\ b_1 & b_2 & b_3 \end{pmatrix} = (a_2b_3 - a_3b_2)\mathbf{i} + (a_3b_1 - a_1b_3)\mathbf{j} + (a_1b_2 - a_2b_1)\mathbf{k}.$$

(For a review of determinants, see Appendix A, (A.24)–(A.33).) It is easily verified that cross products distribute over addition and scalar multiplication in the usual way:

$$(c_1\mathbf{a}_1 + c_2\mathbf{a}_2) \times \mathbf{b} = c_1(\mathbf{a}_1 \times \mathbf{b}) + c_2(\mathbf{a}_2 \times \mathbf{b}),$$
$$\mathbf{a} \times (c_1\mathbf{b}_1 + c_2\mathbf{b}_2) = c_1(\mathbf{a} \times \mathbf{b}_1) + c_2(\mathbf{a} \times \mathbf{b}_2).$$

The cross product is anticommutative:

$$\mathbf{a} \times \mathbf{b} = -\mathbf{b} \times \mathbf{a}.$$

It is not associative; that is, $\mathbf{a} \times (\mathbf{b} \times \mathbf{c}) \neq (\mathbf{a} \times \mathbf{b}) \times \mathbf{c}$ in general. Instead, it satisfies a quasi-associative law called the *Jacobi identity*:

$$\mathbf{a} \times (\mathbf{b} \times \mathbf{c}) + \mathbf{b} \times (\mathbf{c} \times \mathbf{a}) + \mathbf{c} \times (\mathbf{a} \times \mathbf{b}) = \mathbf{0}.$$

A messy but straightforward calculation shows that

$$|\mathbf{a} \times \mathbf{b}|^2 = |\mathbf{a}|^2|\mathbf{b}|^2 - (\mathbf{a} \cdot \mathbf{b})^2.$$

($|\mathbf{a} \times \mathbf{b}|^2$ is the sum of the squares of the components of $\mathbf{a} \times \mathbf{b}$. Multiply it out and rearrange the terms to get $|\mathbf{a}|^2|\mathbf{b}|^2 - (\mathbf{a} \cdot \mathbf{b})^2$.) If θ is the angle between $\mathbf{a}$ and $\mathbf{b}$ ($0 \leq \theta \leq \pi$), we know that $\mathbf{a} \cdot \mathbf{b} = |\mathbf{a}|\,|\mathbf{b}| \cos\theta$, so

$$|\mathbf{a} \times \mathbf{b}|^2 = |\mathbf{a}|^2|\mathbf{b}|^2(1 - \cos^2\theta), \text{ or } |\mathbf{a} \times \mathbf{b}| = |\mathbf{a}|\,|\mathbf{b}| \sin\theta.$$

If $\mathbf{a}$ and $\mathbf{b}$ represent two sides of a parallelogram and we take $\mathbf{a}$ to be the "base," then $|\mathbf{b}| \sin\theta$ is the "height"; hence, $|\mathbf{a} \times \mathbf{b}|$ *is the area of the parallelogram generated by* $\mathbf{a}$ *and* $\mathbf{b}$. Another easy calculation shows that

$$\mathbf{a} \cdot (\mathbf{a} \times \mathbf{b}) = \mathbf{b} \cdot (\mathbf{a} \times \mathbf{b}) = 0;$$

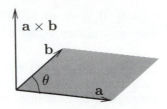

FIGURE 1.1: The geometry of the cross product.

in other words, $\mathbf{a} \times \mathbf{b}$ *is orthogonal to both* $\mathbf{a}$ *and* $\mathbf{b}$. See Figure 1.1.

The two italicized statements specify the magnitude and direction of $\mathbf{a} \times \mathbf{b}$ in purely geometric terms and show that $\mathbf{a} \times \mathbf{b}$ has an intrinsic geometric meaning, independent of the choice of coordinate axes. Well, almost: The fact that $\mathbf{a} \times \mathbf{b}$ is orthogonal to both $\mathbf{a}$ and $\mathbf{b}$ specifies its direction only up to a factor of ± 1, and this last bit of information is provided by the "right hand rule": If you point the thumb and first finger of your right hand in the directions of $\mathbf{a}$ and $\mathbf{b}$, respectively, and bend the middle finger so that it is perpendicular to both of them, the middle finger points in the direction of $\mathbf{a} \times \mathbf{b}$. Thus the definition of cross product is tied to the convention of using "right-handed" coordinate systems. If we were to switch to "left-handed" ones, all cross products would be multiplied by -1.

EXERCISES

1. Let $\mathbf{x} = (3, -1, -1, 1)$ and $\mathbf{y} = (-2, 2, 1, 0)$. Compute the norms of $\mathbf{x}$ and $\mathbf{y}$ and the angle between them.

2. Given $\mathbf{x}, \mathbf{y} \in \mathbb{R}^n$, show that
   a. $|\mathbf{x} + \mathbf{y}|^2 = |\mathbf{x}|^2 + 2\mathbf{x} \cdot \mathbf{y} + |\mathbf{y}|^2$.
   b. $|\mathbf{x} + \mathbf{y}|^2 + |\mathbf{x} - \mathbf{y}|^2 = 2(|\mathbf{x}|^2 + |\mathbf{y}|^2)$.

3. Suppose $\mathbf{x}_1, \dots, \mathbf{x}_k \in \mathbb{R}^n$.
   a. Generalize Exercise 2a to obtain a formula for $|\mathbf{x}_1 + \cdots + \mathbf{x}_k|^2$.
   b. (*The Pythagorean Theorem*) Suppose the vectors $\mathbf{x}_j$ are mutually orthogonal, i.e., that $\mathbf{x}_i \cdot \mathbf{x}_j = 0$ for $i \neq j$. Show that $|\mathbf{x}_1 + \cdots + \mathbf{x}_k|^2 = |\mathbf{x}_1|^2 + \cdots + |\mathbf{x}_k|^2$.

4. Under what conditions on $\mathbf{a}$ and $\mathbf{b}$ is Cauchy's inequality an equality? (Examine the proof.)

5. Under what conditions on $\mathbf{a}$ and $\mathbf{b}$ is the triangle inequality an equality?

6. Show that $\big| |\mathbf{a}| - |\mathbf{b}| \big| \le |\mathbf{a} - \mathbf{b}|$ for every $\mathbf{a}, \mathbf{b} \in \mathbb{R}^n$.

7. Suppose $\mathbf{a}, \mathbf{b} \in \mathbb{R}^3$.

    a. Show that if $\mathbf{a} \cdot \mathbf{b} = 0$ and $\mathbf{a} \times \mathbf{b} = \mathbf{0}$, then either $\mathbf{a} = \mathbf{0}$ or $\mathbf{b} = \mathbf{0}$.

    b. Show that if $\mathbf{a} \cdot \mathbf{c} = \mathbf{b} \cdot \mathbf{c}$ and $\mathbf{a} \times \mathbf{c} = \mathbf{b} \times \mathbf{c}$ for some nonzero $\mathbf{c} \in \mathbb{R}^3$, then $\mathbf{a} = \mathbf{b}$.

    c. Show that $(\mathbf{a} \times \mathbf{a}) \times \mathbf{b} = \mathbf{a} \times (\mathbf{a} \times \mathbf{b})$ if and only if $\mathbf{a}$ and $\mathbf{b}$ are proportional (i.e., one is a scalar multiple of the other).

8. Show that $\mathbf{a} \cdot (\mathbf{b} \times \mathbf{c})$ is the determinant of the matrix whose rows are $\mathbf{a}$, $\mathbf{b}$, and $\mathbf{c}$ (if these vectors are considered as row vectors) or the matrix whose columns are $\mathbf{a}$, $\mathbf{b}$, and $\mathbf{c}$ (if they are considered as column vectors).

1.2   Subsets of Euclidean Space

In this section we introduce some standard terminology for sets in $\mathbb{R}^n$.

First, the set of all points whose distance from a fixed point $\mathbf{a}$ is equal to some number r is called the **sphere** of radius r about $\mathbf{a}$, and the set of points whose distance from $\mathbf{a}$ is less than r is called the (open) **ball** of radius r about $\mathbf{a}$. (In ordinary English the word "sphere" is often used for both these purposes, but mathematicians have found it helpful to reserve the word "sphere" for the spherical *surface* and to use "ball" to denote the solid body.) We shall use the notation $B(r, \mathbf{a})$ for the ball of radius r about $\mathbf{a}$:

$$B(r, \mathbf{a}) = \big\{ \mathbf{x} \in \mathbb{R}^n : |\mathbf{x} - \mathbf{a}| < r \big\}.$$

Of course, when in dimension 1, a ball is just an open interval, and in dimension 2, the words "disc" and "circle" may be used in place of "ball" and "sphere."

A set $S \subset \mathbb{R}^n$ is called **bounded** if it is contained in some ball about the origin, that is, if there is a constant C such that $|\mathbf{x}| < C$ for every $\mathbf{x} \in S$.

When one studies functions of a single variable, one frequently considers *intervals* in the real line, and it is often necessary to distinguish between *open* intervals (with the endpoints excluded) and *closed* intervals (with the endpoints included). When $n > 1$, there is a much greater variety of interesting subsets of $\mathbb{R}^n$ to be considered, but the notions of "open" and "closed" are still fundamental. Here are the definitions.

Let S be a subset of $\mathbb{R}^n$.

- The **complement** of S is the set of all points in $\mathbb{R}^n$ that are *not* in S; we denote it by $\mathbb{R}^n \setminus S$ or by S^c:

$$S^c = \mathbb{R}^n \setminus S = \big\{ \mathbf{x} \in \mathbb{R}^n : \mathbf{x} \notin S \big\}.$$

- A point $\mathbf{x} \in \mathbb{R}^n$ is called an **interior point** of S if all points sufficiently close to $\mathbf{x}$ (including $\mathbf{x}$ itself) are also in S, that is, if S contains some ball centered at $\mathbf{x}$. The set of all interior points of S is called the **interior** of S and is denoted by S^{int}:

$$S^{\mathrm{int}} = \{\mathbf{x} \in S : B(r, \mathbf{x}) \subset S \text{ for some } r > 0\}.$$

- A point $\mathbf{x} \in \mathbb{R}^n$ is called a **boundary point** of S if every ball centered at $\mathbf{x}$ contains both points in S and points in S^c. (Note that if $\mathbf{x}$ is a boundary point of S, $\mathbf{x}$ may belong to either S or S^c.) The set of all boundary points of S is called the **boundary** of S and is denoted by ∂S:

$$\partial S = \{\mathbf{x} \in \mathbb{R}^n : B(r, \mathbf{x}) \cap S \neq \varnothing \text{ and } B(r, \mathbf{x}) \cap S^c \neq \varnothing \text{ for every } r > 0\}.$$

(*Remark.* We shall use the term "boundary" slightly differently in §5.7 in connection with Stokes's theorem, in the context of surfaces in $\mathbb{R}^3$ being "bounded" by curves. But the present definition is the general-purpose one.)

- S is called **open** if it contains none of its boundary points.

- S is called **closed** if it contains all of its boundary points.

- The **closure** of S is the union of S and all its boundary points. It is denoted by $\overline{S}$:

$$\overline{S} = S \cup \partial S.$$

- Finally, a **neighborhood** of a point $\mathbf{x} \in \mathbb{R}^n$ is a set of which $\mathbf{x}$ is an interior point. That is, S is a neighborhood of $\mathbf{x}$ if and only if $\mathbf{x}$ is an interior point of S.

Let us examine these ideas a little more closely. First, notice that the boundary points of S are the same as the boundary points of S^c; the definition of boundary point remains unchanged if S and S^c are switched. Moreover, if $\mathbf{x}$ is neither an interior point of S nor an interior point of S^c, then $\mathbf{x}$ must be a boundary point of S. In other words, given $S \subset \mathbb{R}^n$ and $\mathbf{x} \in \mathbb{R}^n$, there are exactly three possibilities: $\mathbf{x}$ is an interior point of S, or $\mathbf{x}$ is an interior point of S^c, or $\mathbf{x}$ is a boundary point of S.

1.4 Proposition. *Suppose $S \subset \mathbb{R}^n$.*

a. *S is open $\iff$ every point of S is an interior point.*

b. *S is closed $\iff$ S^c is open.*

Proof. Every point of S is either an interior point or a boundary point; thus S is open $\Longleftrightarrow$ every point of S is an interior point. On the other hand, S is closed $\Longleftrightarrow$ it contains all of ∂S, which is the same as $\partial(S^c)$; this happens precisely when S^c contains none of its boundary points, i.e., when S^c is open. $\qquad\square$

EXAMPLE 1. Let S be $B(\rho, \mathbf{0})$, the ball of radius ρ about the origin. First, given $\mathbf{x} \in S$, let $r = \rho - |\mathbf{x}|$. If $|\mathbf{y} - \mathbf{x}| < r$, then by the triangle inequality we have $|\mathbf{y}| \le |\mathbf{y} - \mathbf{x}| + |\mathbf{x}| < \rho$, so that $B(r, \mathbf{x}) \subset S$. Therefore, every $\mathbf{x} \in S$ is an interior point of S, so S is open. Second, a similar calculation shows that if $|\mathbf{x}| > \rho$ then $B(r, \mathbf{x}) \subset S^c$ where $r = |\mathbf{x}| - \rho$, so every point with $|\mathbf{x}| > \rho$ is an interior point of S^c. On the other hand, if $|\mathbf{x}| = \rho$, then $c\mathbf{x} \in S$ for $0 < c < 1$ and $c\mathbf{x} \in S^c$ for $c \ge 1$, and $|c\mathbf{x} - \mathbf{x}| = |c - 1|\rho$ can be as small as we please, so $\mathbf{x}$ is a boundary point. In other words, the boundary of S is the sphere of radius ρ about the origin, and the closure of S is the *closed* ball $\{\mathbf{x} : |\mathbf{x}| \le \rho\}$.

EXAMPLE 2. Now let S be the ball of radius ρ about the origin together with the "upper hemisphere" of its boundary:

$$S = B(\rho, \mathbf{0}) \cup \{\mathbf{x} \in \mathbb{R}^n : |\mathbf{x}| = \rho \text{ and } x_n > 0\}.$$

The calculations in Example 1 show that S^{int} is the open ball $B(\rho, \mathbf{0})$; ∂S is the sphere $\{\mathbf{x} : |\mathbf{x}| = \rho\}$, and $\overline{S}$ is the closed ball $\{\mathbf{x} : |\mathbf{x}| \le \rho\}$. The set S is neither open nor closed.

EXAMPLE 3. In the real line (i.e., $n = 1$), let S be the set of all rational numbers. Since every ball in $\mathbb{R}$ — that is, every interval — contains both rational and irrational numbers, *every* point of $\mathbb{R}$ is a boundary point of S. The set S is neither open nor closed; its interior is empty; and its closure is $\mathbb{R}$.

Subsets of $\mathbb{R}^n$ are often specified in terms of equations or inequalities — for example, by an expression of the form

(1.5) $$S = \{\mathbf{x} \in \mathbb{R}^n : f(\mathbf{x}) \,\square\, 0\},$$

where $\square$ denotes one of the relations $=, <, >, \le, \ge$. (Taking the quantity on the right of $\square$ to be 0 is no restriction; just move all the terms over to the left side.) We anticipate some results from §1.3 in giving the following rule of thumb: *Sets defined by strict inequalities are open; sets defined by equalities or weak inequalities are closed. More precisely, if S is given by* (1.5) *where the function f is continuous, then S is open if $\square$ denotes $<$ or $>$, and S is closed if $\square$ denotes $=, \le,$ or $\ge$.* The reader may feel free to use this rule in doing the exercises.

EXERCISES

1. For each of the following sets S in the plane $\mathbb{R}^2$, do the following: (i) Draw a sketch of S. (ii) Tell whether S is open, closed, or neither. (iii) Describe S^{int}, $\overline{S}$, and ∂S. (These descriptions should be in the same set-theoretic language as the description of S itself given here.)
   a. $S = \{(x, y) : 0 < x^2 + y^2 \leq 4\}$.
   b. $S = \{(x, y) : x^2 - x \leq y \leq 0\}$.
   c. $S = \{(x, y) : x > 0,\ y > 0,\ \text{and } x + y > 1\}$.
   d. $S = \{(x, y) : y = x^3\}$.
   e. $S = \{(x, y) : x > 0 \text{ and } y = \sin(1/x)\}$.
   f. $S = \{(x, y) : x^2 + y^2 < 1\} \setminus \{(x, 0) : x < 0\}$.
   g. $S = \{(x, y) : x \text{ and } y \text{ are rational numbers in } [0, 1]\}$.

2. Show that for any $S \subset \mathbb{R}^n$, S^{int} is open and ∂S and $\overline{S}$ are both closed. (*Hint:* Use the fact that balls are open, proved in Example 1.)

3. Show that if S_1 and S_2 are open, so are $S_1 \cup S_2$ and $S_1 \cap S_2$.

4. Show that if S_1 and S_2 are closed, so are $S_1 \cup S_2$ and $S_1 \cap S_2$. (One way is to use Exercise 3 and Proposition 1.4b.)

5. Show that the boundary of S is the intersection of the closures of S and S^c.

6. Give an example of an infinite collection $S_1, S_2, \ldots$ of closed sets whose union $\bigcup_{j=1}^{\infty} S_j$ is not closed.

7. There are precisely two subsets of $\mathbb{R}^n$ that are both open and closed. What are they?

8. Give an example of a set S such that the interior of S is unequal to the interior of the closure of S.

9. Show that the ball of radius r about $\mathbf{a}$ is contained in the ball of radius $r + |\mathbf{a}|$ about the origin. Conclude that a set $S \subset \mathbb{R}^n$ is bounded if it is contained in some ball (whose center can be anywhere in $\mathbb{R}^n$).

1.3   Limits and Continuity

We now commence our study of functions defined on $\mathbb{R}^n$ or subsets of $\mathbb{R}^n$. For the most part we shall be dealing with real-valued functions, but in many situations we shall deal with vector-valued or complex-valued functions, that is, functions whose values lie in $\mathbb{R}^k$ or $\mathbb{C}$. For our present purposes we can regard $\mathbb{C}$ as $\mathbb{R}^2$ by identifying the complex number $u + iv$ with the ordered pair (u, v), so it is enough to consider vector-valued functions. But we begin with the real-valued case.

Suppose f is a real-valued function defined on $\mathbb{R}^n$. We say that

$$\lim_{\mathbf{x} \to \mathbf{a}} f(\mathbf{x}) = L,$$

and call L the **limit** of $f(\mathbf{x})$ as $\mathbf{x}$ approaches $\mathbf{a}$, if $f(\mathbf{x})$ becomes as close as we wish to L provided $\mathbf{x}$ is sufficiently close to, but not equal to, $\mathbf{a}$. More formally, the statement $\lim_{\mathbf{x} \to \mathbf{a}} f(\mathbf{x}) = L$ means that for any positive number ϵ there is a positive number δ so that

(1.6) $$|f(\mathbf{x}) - L| < \epsilon \text{ whenever } 0 < |\mathbf{x} - \mathbf{a}| < \delta.$$

This condition can be rephrased in terms of the individual components $x_j - a_j$ of $\mathbf{x} - \mathbf{a}$, as follows: $\lim_{\mathbf{x} \to \mathbf{a}} f(\mathbf{x}) = L$ if and only if for every positive number ϵ there is a positive number δ' so that

(1.7) $$|f(\mathbf{x}) - L| < \epsilon \text{ whenever } 0 < \max\big(|x_1 - a_1|, \ldots, |x_n - a_n|\big) < \delta'.$$

The equivalence of (1.6) and (1.7) follows from (1.3): If (1.6) is satisfied, then (1.7) is satisfied with $\delta' = \delta/\sqrt{n}$; and if (1.7) is satisfied, then (1.6) is satisfied with $\delta = \delta'$.

More generally, we can consider functions f that are only defined on a subset S of $\mathbb{R}^n$ and points $\mathbf{a}$ that lie in the closure of S. The definition of $\lim_{\mathbf{x} \to \mathbf{a}} f(\mathbf{x})$ is the same as before except that $\mathbf{x}$ is restricted to lie in the set S. It may be necessary, for the sake of clarity, to specify this restriction explicitly; for this purpose we use the notation

$$\lim_{\mathbf{x} \to \mathbf{a},\ \mathbf{x} \in S} f(\mathbf{x}).$$

In particular, for a function f on the real line we often need to consider the **one-sided limits**

$$\lim_{x \to a+} f(x) = \lim_{x \to a,\ x > a} f(x) \quad \text{and} \quad \lim_{x \to a-} f(x) = \lim_{x \to a,\ x < a} f(x).$$

For example, let $f : \mathbb{R} \to \mathbb{R}$ be the function defined by $f(x) = x + 1$ for $|x| \le 1$ and $f(x) = 0$ for $|x| > 1$. Then $\lim_{x \to 1} f(x)$ does not exist, but $\lim_{x \to 1-} f(x) = 2$ and $\lim_{x \to 1+} f(x) = 0$.

Notice that the definition of $\lim_{\mathbf{x} \to \mathbf{a}} f(\mathbf{x})$ does not involve the value $f(\mathbf{a})$ at all; only the values of f at points near $\mathbf{a}$ but unequal to $\mathbf{a}$ are relevant. Indeed, f need not even be defined at $\mathbf{a}$ — a situation that arises, for example, in the limits that define derivatives. On the other hand, if $\lim_{\mathbf{x} \to \mathbf{a}} f(\mathbf{x})$ and $f(\mathbf{a})$ both exist and are equal, that is, if

$$\lim_{\mathbf{x} \to \mathbf{a}} f(\mathbf{x}) = f(\mathbf{a}),$$

then f is said to be **continuous at a**.

If f is continuous at every point of a set $U \subset \mathbb{R}^n$, f is said to be **continuous on** U. Going back to the condition (1.6) that defines limits, we see that the continuity of f on U is equivalent to the following condition: For every positive number ϵ and every $\mathbf{a} \in U$ there is a positive number δ so that

$$(1.8) \qquad |f(\mathbf{x}) - f(\mathbf{a})| < \epsilon \text{ whenever } |\mathbf{x} - \mathbf{a}| < \delta.$$

Informally speaking, f is continuous if changing the input values by a small amount changes the output values by only a small amount.

The same definitions apply equally well to vector-valued functions, that is, functions $\mathbf{f}$ with values in $\mathbb{R}^k$ for some $k > 1$. In this case the limit $\mathbf{L}$ is an element of $\mathbb{R}^k$, and $|\mathbf{f}(\mathbf{x}) - \mathbf{L}|$ is the norm of the vector $\mathbf{f}(\mathbf{x}) - \mathbf{L}$. In view of (1.3), it is clear that

$$\lim_{\mathbf{x} \to \mathbf{a}} \mathbf{f}(\mathbf{x}) = \mathbf{L} \quad \Longleftrightarrow \quad \lim_{\mathbf{x} \to \mathbf{a}} f_j(\mathbf{x}) = L_j \text{ for } j = 1, \ldots, k.$$

Thus the study of limits and continuity of vector-valued functions is easily reduced to the scalar case, to which we now return out attention.

We often express the relation $\lim_{\mathbf{x} \to \mathbf{a}} f(\mathbf{x}) = L$ informally by saying that $f(\mathbf{x})$ approaches L as $\mathbf{x}$ approaches $\mathbf{a}$. In one dimension this works quite well; we can envision $\mathbf{x}$ as the location of a particle that moves toward $\mathbf{a}$ from the right or the left. But in higher dimensions there are infinitely many different paths along which a particle might move toward $\mathbf{a}$, and for the limit to exist one must get the same result no matter which path is chosen. It is safer to abandon the "dynamic" picture of a particle moving toward $\mathbf{a}$; we should simply think in terms of $f(\mathbf{x})$ being close to L provided that $\mathbf{x}$ is close to $\mathbf{a}$, without reference to any motion.

EXAMPLE 1.   Let $f(x,y) = \dfrac{xy}{x^2 + y^2}$ if $(x,y) \neq (0,0)$, and let $f(0,0) = 0$. Show that $\lim_{(x,y) \to (0,0)} f(x,y)$ does not exist — and, in particular, f is discontinuous at $(0,0)$.

Solution. First, note that $f(x,0) = f(0,y) = 0$ for all x and y, so $f(x,y) \to 0$ as (x,y) approaches $(0,0)$ along the x-axis or the y-axis. But if we consider other straight lines passing through the origin, say $y = cx$, we have $f(x,cx) = cx^2/(x^2 + c^2 x^2) = c/(1 + c^2)$, so the limit as (x,y) approaches $(0,0)$ along the line $y = cx$ is $c/(1 + c^2)$. Depending on the value of c, this can be anything between $-\frac{1}{2}$ and $\frac{1}{2}$ (these two extreme values being achieved when $c = -1$ or $c = 1$). So there is no limit as (x,y) approaches $(0,0)$ unrestrictedly.

The argument just given suggests the following line of thought. We wish to know if $\lim_{\mathbf{x}\to\mathbf{a}} f(\mathbf{x})$ exists. We look at all the straight lines passing through $\mathbf{a}$ and evaluate the limit of $f(\mathbf{x})$ as $\mathbf{x}$ approaches $\mathbf{a}$ along each of those lines by one-variable techniques; if we always get the same answer L, then we should have $\lim_{\mathbf{x}\to\mathbf{a}} f(\mathbf{x}) = L$, right? Unfortunately, this doesn't work:

EXAMPLE 2. Let $g(x, y) = \dfrac{x^2 y}{x^4 + y^2}$ if $(x, y) \neq (0, 0)$ and $g(0, 0) = 0$. Again we have $g(x, 0) = g(0, y) = 0$, so the limit as $(x, y) \to (0, 0)$ along the coordinate axes is 0. Moreover, if $c \neq 0$,

$$g(x, cx) = \frac{cx^3}{x^4 + c^2 x^2} = \frac{cx}{c^2 + x^2} \to 0 \text{ as } x \to 0,$$

so the limit as $(x, y) \to (0, 0)$ along any other straight line is also 0. But if we approach along a *parabola* $y = cx^2$, we get

$$g(x, cx^2) = \frac{cx^3}{x^4 + c^2 x^4} = \frac{c}{1 + c^2},$$

which can be anything between $-\frac{1}{2}$ and $\frac{1}{2}$ as before, so the limit does not exist. (The similarity with Example 1 is not accidental: If f is the function in Example 1 we have $g(x, y) = f(x^2, y)$.)

After looking at examples like this one, one might become discouraged about the possibility of ever proving that limits *do* exist! But things are not so bad. If f is a *continuous* function, $\lim_{\mathbf{x}\to\mathbf{a}} f(\mathbf{x})$ is simply $f(\mathbf{a})$. Moreover, most of the functions of several variables that one can easily write down are built up from continuous functions of one variable by using the arithmetic operations plus composition, and these operations all preserve continuity (except for division when the denominator vanishes).

Here are the precise statements and proofs of the fundamental results. (The reader may wish to skip the proofs; they are of some value as illustrations of the sort of formal arguments involving limits that are important in more advanced analysis, but they contribute little to an intuitive understanding of the results.)

1.9 Theorem. *Suppose* $\mathbf{f} : \mathbb{R}^n \to \mathbb{R}^m$ *is continuous on* $U \subset \mathbb{R}^n$ *and* $\mathbf{g} : \mathbb{R}^m \to \mathbb{R}^k$ *is continuous on* $\mathbf{f}(U) \subset \mathbb{R}^m$. *Then the composite function* $\mathbf{g} \circ \mathbf{f} : \mathbb{R}^n \to \mathbb{R}^k$ *is continuous on* U.

Proof. Let $\epsilon > 0$ and $\mathbf{a} \in U$ be given, and let $\mathbf{b} = \mathbf{f}(\mathbf{a})$. Since $\mathbf{g}$ is continuous on $\mathbf{f}(U)$, we can choose $\eta > 0$ so that $|\mathbf{g}(\mathbf{y}) - \mathbf{g}(\mathbf{b})| < \epsilon$ whenever $|\mathbf{y} - \mathbf{b}| < \eta$. Having chosen this η, since $\mathbf{f}$ is continuous on U we can find $\delta > 0$ so that $|\mathbf{f}(\mathbf{x}) - \mathbf{b}| < \eta$ whenever $|\mathbf{x} - \mathbf{a}| < \delta$. Thus,

$$|\mathbf{x} - \mathbf{a}| < \delta \quad \Longrightarrow \quad |\mathbf{f}(\mathbf{x}) - \mathbf{f}(\mathbf{a})| < \eta \quad \Longrightarrow \quad |\mathbf{g}(\mathbf{f}(\mathbf{x})) - \mathbf{g}(\mathbf{f}(\mathbf{a}))| < \epsilon,$$

which says that $\mathbf{g} \circ \mathbf{f}$ is continuous on U. $\qquad\qquad\square$

1.10 Theorem. *Let $f_1(x, y) = x + y$, $f_2(x, y) = xy$, and $g(x) = 1/x$. Then f_1 and f_2 are continuous on $\mathbb{R}^2$ and g is continuous on $\mathbb{R} \setminus \{0\}$.*

Proof. To prove continuity of f_1 and f_2, we need to show that $\lim_{(x,y)\to(a,b)} x+y = a + b$ and $\lim_{(x,y)\to(a,b)} xy = ab$ for every $a, b \in \mathbb{R}$. That is, given $\epsilon > 0$ and $a, b \in \mathbb{R}$, we need to find $\delta > 0$ so that if $|x - a| < \delta$ and $|y - b| < \delta$, then (i) $|(x + y) - (a + b)| < \epsilon$ or (ii) $|xy - ab| < \epsilon$. For (i) we can simply take $\delta = \frac{1}{2}\epsilon$, for if $|x - a| < \frac{1}{2}\epsilon$ and $|y - b| < \frac{1}{2}\epsilon$, then

$$|(x + y) - (a + b)| = |(x - a) + (y - b)| \le |x - a| + |y - b| < \tfrac{1}{2}\epsilon + \tfrac{1}{2}\epsilon = \epsilon.$$

For (ii) we observe that $xy - ab = (x - a)y + a(y - b)$, so we can make $xy - ab$ small by making the two terms on the right small. Indeed, let

$$\delta = \min\left(1, \frac{\epsilon}{2(|a| + 1)}, \frac{\epsilon}{2(|b| + 1)}\right).$$

If $|x - a| < \delta$ and $|y - b| < \delta$, then $|y| < |b| + \delta \le |b| + 1$, so

$$|xy - ab| \le |x - a||y| + |a||y - b|$$
$$\le \frac{\epsilon}{2(|b| + 1)}(|b| + 1) + |a|\frac{\epsilon}{2(|a| + 1)} < \frac{\epsilon}{2} + \frac{\epsilon}{2} = \epsilon.$$

This proves the continuity of f_1 and f_2. As for g, to show that $\lim_{x\to a} 1/x = 1/a$ for $a \ne 0$, we observe that

$$\frac{1}{x} - \frac{1}{a} = \frac{a - x}{ax}.$$

Given $\epsilon > 0$, let δ be the smaller of the numbers $\frac{1}{2}|a|$ and $\frac{1}{2}\epsilon a^2$. If $|x - a| < \delta$, then $|a| \le |a - x| + |x| < \frac{1}{2}|a| + |x|$ and hence $|x| > \frac{1}{2}|a|$, so

$$\left|\frac{x - a}{ax}\right| < \left|\frac{\epsilon a^2}{2ax}\right| = \epsilon\left|\frac{a}{2x}\right| < \epsilon,$$

as desired. $\qquad\qquad\square$

1.11 Corollary. *The function $f_3(x, y) = x - y$ is continuous on $\mathbb{R}^2$, and the function $f_4(x, y) = x/y$ is continuous on $\{(x, y) : y \ne 0\}$.*

Proof. With notation as in Theorem 1.10, we have $f_4(x, y) = f_2(x, g(y))$, so f_4 is the composition of continuous mappings and hence is continuous on the set where $y \neq 0$. Likewise, $f_3(x, y) = f_1(x, f_2(-1, y))$, so f_3 is continuous. (Alternatively, continuity for f_3 may be proved in exactly the same way as for f_1.) $\qquad \square$

1.12 Corollary. *The sum, product, or difference of two continuous functions is continuous; the quotient of two continuous functions is continuous on the set where the denominator is nonzero.*

Proof. Combine Theorem 1.10 and Corollary 1.11 with Theorem 1.9. For example, if f and g are continuous functions on $U \subset \mathbb{R}^n$, then $f + g$ is continuous because it is the composition of the continuous map (f, g) from U to $\mathbb{R}^2$ and the continuous map $(x, y) \mapsto x+y$ from $\mathbb{R}^2$ to $\mathbb{R}$. Likewise for the other arithmetic operations. $\qquad \square$

The elementary functions of a single variable (polynomials, trig functions, exponential functions, etc.) are all continuous on their domains of definition, and elementary functions of several variables are generally built up out of functions of one variable by the arithmetic operations and composition. The preceding results therefore allow the continuity of such functions to be established almost immediately in most cases. For example, the function $\varphi(x, y) = \dfrac{\sin(3x + 2y)}{x^2 - y}$ is continuous everywhere except along the parabola $y = x^2$, because it is built up from the continuous functions of one variable $3x$, $2y$, x^2, and $-y$ by taking sums ($3x + 2y$ and $x^2 - y$), composing with the sine function ($\sin(3x + 2y)$), and then taking a quotient. For another example, the function $\psi(x, y) = x^y$, defined on the region where $x > 0$, is continuous there, because it can be rewritten as $\psi(x, y) = e^{y \log x}$, which is assembled from the (continuous) exponential and logarithmic functions and the operation of multiplication ($y \cdot \log x$). Similarly, the functions in Examples 1 and 2 are continuous everywhere except at the origin.

Let us look at one more example:

EXAMPLE 3. Let $h(x, y) = \dfrac{xy(x^2 - y^2)}{x^2 + y^2}$ for $(x, y) \neq (0, 0)$ and $h(0, 0) = 0$. Evaluate $\lim_{(x,y)\to(2,3)} h(x, y)$ and $\lim_{(x,y)\to(0,0)} h(x, y)$. Is h continuous at $(0, 0)$?

Solution. The first limit is easy: Clearly h is continuous everywhere except at the origin, so $\lim_{(x,y)\to(2,3)} h(x, y) = h(2, 3) = 6(4 - 9)/(4 + 9) = -\frac{30}{13}$. The behavior of h at the origin requires a closer examination. Since $h(x, 0) = 0$ for all x, if the limit exists it must equal 0. Experimentation with lines and parabolas as in Examples 1 and 2 fails to yield any evidence to the contrary.

In fact, the limit *is* 0, and this can be established with a little ad hoc estimating. Clearly $|x^2 - y^2| \leq x^2 + y^2$, so $|h(x, y)| \leq |xy|$. But $xy \rightarrow 0$ as $(x, y) \rightarrow (0, 0)$, so $h(x, y)$, being even smaller in absolute value than xy, must also approach 0. Thus $\lim_{(x,y) \rightarrow (0,0)} h(x, y) = 0$ and h is continuous at $(0, 0)$.

We now establish the relation between inequalities on continuous functions and open and closed sets that was mentioned at the end of the preceding section.

1.13 Theorem. *Suppose* $\mathbf{f} : \mathbb{R}^n \rightarrow \mathbb{R}^k$ *is continuous and* U *is a subset of* $\mathbb{R}^k$, *and let* $S = \{\mathbf{x} \in \mathbb{R}^n : \mathbf{f}(\mathbf{x}) \in U\}$. *Then* S *is open if* U *is open, and* S *is closed if* U *is closed.*

Proof. Suppose U is open. We shall show that S is open by showing that every point $\mathbf{a}$ in S is an interior point of S. If $\mathbf{a} \in S$, then $f(\mathbf{a}) \in U$. Since U is open, some ball centered at $f(\mathbf{a})$ is contained in U; that is, there is a positive number ϵ such that every $\mathbf{y} \in \mathbb{R}^k$ such that $|\mathbf{y} - \mathbf{f}(\mathbf{a})| < \epsilon$ is in U. Since f is continuous, there is a positive number δ such that $|\mathbf{f}(\mathbf{x}) - \mathbf{f}(\mathbf{a})| < \epsilon$ whenever $|\mathbf{x} - \mathbf{a}| < \delta$. But this means that $f(\mathbf{x}) \in U$ whenever $|\mathbf{x} - \mathbf{a}| < \delta$, that is, $\mathbf{x} \in S$ whenever $|\mathbf{x} - \mathbf{a}| < \delta$. Thus $\mathbf{a}$ is an interior point of S.

On the other hand, suppose U is closed. Then the complement of U in $\mathbb{R}$ is open by Proposition 1.4b, so the set $S' = \{\mathbf{x} : f(\mathbf{x}) \in U^c\}$ is open by the argument just given. But S' is just the complement of S in $\mathbb{R}^n$, so S is closed by Proposition 1.4b again. $\square$

The result about the openness or closedness of sets defined by inequalities or equations at the end of §1.2 is a corollary of Theorem 1.13. For example, if $f : \mathbb{R}^n \rightarrow \mathbb{R}$ is a continuous function, the set $\{\mathbf{x} : f(\mathbf{x}) > 0\}$ (resp.[2] $\{\mathbf{x} : f(\mathbf{x}) = 0\}$) is of the form $\{\mathbf{x} : f(\mathbf{x}) \in U\}$ where $U = (0, \infty)$ (resp. $U = \{0\}$), and this U is open (resp. closed).

Theorem 1.13 can be generalized to functions that are only defined on subsets of $\mathbb{R}^n$; with notation as above, the correct statement is that if U is open (resp. closed) then S is the intersection of the domain of $\mathbf{f}$ with an open (resp. closed) set. (For example, the set $\{x \in \mathbb{R} : \log x \leq 0\}$, namely $(0, 1]$, is the intersection of the domain of log, namely $(0, \infty)$, with the closed set $[0, 1]$. On the other hand, the set $\{x \in \mathbb{R} : \sqrt{x} < 1\}$, namely $[0, 1)$, is the intersection of the domain of the square root function, namely $[0, \infty)$, with the open set $(-1, 1)$.) In particular, if U and the domain of $\mathbf{f}$ are both open (resp. closed), then so is S.

The converse of Theorem 1.13 is also true; see Exercise 8.

---

[2]"resp." is an abbreviation for "respectively."

EXERCISES

1. For the following functions f, show that $\lim_{(x,y)\to(0,0)} f(x,y)$ does not exist.

   a. $f(x,y) = \dfrac{x^2 + y}{\sqrt{x^2 + y^2}}$

   b. $f(x,y) = \dfrac{x}{x^4 + y^4}$

   c. $f(x,y) = \dfrac{x^4 y^4}{(x^2 + y^4)^3}$

2. For the following functions f, show that $\lim_{(x,y)\to(0,0)} f(x,y) = 0$.

   a. $f(x,y) = \dfrac{x^2 y^2}{x^2 + y^2}$ b. $f(x,y) = \dfrac{3x^5 - xy^4}{x^4 + y^4}$

3. Let $f(x,y) = x^{-1} \sin(xy)$ for $x \neq 0$. How should you define $f(0,y)$ for $y \in \mathbb{R}$ so as to make f a continuous function on all of $\mathbb{R}^2$?

4. Let $f(x,y) = xy/(x^2 + y^2)$ as in Example 1. Show that, although f is discontinuous at $(0,0)$, $f(x,a)$ and $f(a,y)$ are continuous functions of x and y, respectively, for any $a \in \mathbb{R}$ (*including* $a = 0$). We say that f is **separately continuous** in x and y.

5. Let $f(x,y) = y(y - x^2)/x^4$ if $0 < y < x^2$, $f(x,y) = 0$ otherwise. At which point(s) is f discontinuous?

6. Let $f(x) = x$ if x is rational, $f(x) = 0$ if x is irrational. Show that f is continuous at $x = 0$ and nowhere else.

7. Let $f(x) = 1/q$ if $x = p/q$ where p and q are integers with no common factors and $q > 0$, and $f(x) = 0$ if x is irrational. At which points, if any, is f continuous?

8. Suppose $\mathbf{f} : \mathbb{R}^n \to \mathbb{R}^k$ has the following property: For any open set $U \subset \mathbb{R}^k$, $\{\mathbf{x} : \mathbf{f}(\mathbf{x}) \in U\}$ is an open set in $\mathbb{R}^n$. Show that $\mathbf{f}$ is continuous on $\mathbb{R}^n$. Show also that the same result holds if "open" is replaced by "closed."

9. Let U and V be open sets in $\mathbb{R}^n$ and let $\mathbf{f}$ be a one-to-one mapping from U onto V (so that there is an inverse mapping $\mathbf{f}^{-1} : V \to U$). Suppose that $\mathbf{f}$ and $\mathbf{f}^{-1}$ are both continuous. Show that for any set S whose closure is contained in U we have $\mathbf{f}(\partial S) = \partial(\mathbf{f}(S))$.

1.4   Sequences

Generally speaking, a **sequence** is a collection of mathematical objects that is indexed by the positive integers. The objects in question can be of any sort, such as numbers, n-dimensional vectors, sets, etc. If the kth object in the sequence is X_k, the sequence as a whole is usually denoted by $\{X_k\}_{k=1}^{\infty}$, or just by $\{X_k\}_1^{\infty}$ or even $\{X_k\}$ if there is no possibility of confusion. (We shall comment further on this notation below.) Alternatively, we can write out the sequence as $X_1, X_2, X_3, \ldots$. We speak of a sequence **in** a set S if the objects of the sequence all belong to S.

EXAMPLE 1.
a. A sequence of numbers: $1, 4, 9, 16, \ldots$. The kth term in the sequence is k^2, and the sequence as a whole may be written as $\{k^2\}_1^{\infty}$.
b. A sequence of intervals: $(-1, 1)$, $(-\frac{1}{2}, \frac{1}{2})$, $(-\frac{1}{3}, \frac{1}{3})$, $(-\frac{1}{4}, \frac{1}{4})$, $\ldots$. The kth term in the sequence is the interval $(-\frac{1}{k}, \frac{1}{k})$, and the sequence as a whole may be written as $\{(-\frac{1}{k}, \frac{1}{k})\}_1^{\infty}$.

Sequences can be defined by formulas, as in the examples above: $x_k = k^2$, or $I_k = (-\frac{1}{k}, \frac{1}{k})$. They can also be defined by **recursion** (or **induction**), that is, by specifying the first term or the first few terms and then giving a rule that tells how to obtain the kth term from the preceding ones.

EXAMPLE 2. The **Fibonacci sequence** is the sequence

$$1, 1, 2, 3, 5, 8, 13, 21, 34, 55, 89, \ldots,$$

in which the first two terms are equal to 1 and each of the remaining terms is the sum of the two preceding ones (that is, $x_k = x_{k-2} + x_{k-1}$).

EXAMPLE 3. Define a sequence $\{x_k\}$ as follows: x_1 is a given positive integer a. If x_k is odd, then $x_{k+1} = 3x_k + 1$; if x_k is even, then $x_{k+1} = x_k/2$. For example, if $a = 13$, the sequence is

$$13, 40, 20, 10, 5, 16, 8, 4, 2, 1, 4, 2, 1, 4, 2, 1, \ldots,$$

ending in the infinite repetition of $(4, 2, 1)$. It is a famous unsolved problem (as of this writing) to prove or disprove that this sequence eventually ends in the repeating figure $(4, 2, 1)$ no matter what initial number a is chosen. (Try a few values of a to see how it works! For more information, see Lagarias [13].)

It is convenient to make the definition of sequence a little more flexible by allowing the index k to begin with something other than 1. Thus, we may speak of a sequence $\{X_k\}_0^{\infty}$ whose objects are $X_0, X_1, X_2, \ldots$, or a sequence $\{X_k\}_7^{\infty}$, whose objects are $X_7, X_8, X_9, \ldots$. We may also speak of a **finite sequence** whose terms are indexed by a finite collection of integers, such as $\{X_k\}_1^8$ (a finite sequence of eight terms), or a **doubly infinite sequence** whose terms are indexed by the whole set of integers: $\{X_k\}_{-\infty}^{\infty}$.

Strictly speaking, a sequence in a set S is a rule that assigns to each positive integer (or each integer in some other suitable set, as indicated above) an element of S, in other words, a function or mapping from the positive integers to S. The common functional notation would be to write $X(k)$ instead of X_k for the value of this mapping at the integer k, but for sequences it is customary to write the input variable k as a subscript.

It is sometimes necessary to distinguish between the *sequence* $\{X_k\}_1^\infty$ and the *set of values* (i.e., the *range*) of the sequence, because a sequence may assume the same value many times. For example, consider the sequence of numbers $a_k = (-1)^k$. Then the *sequence* $\{a_k\}_1^\infty$ is the function on the positive integers whose values are alternately -1 and $+1$, which may be written out as

$$-1, 1, -1, 1, -1, 1, \ldots,$$

but its *set of values* is just the two-element set $\{-1, 1\}$. Since curly brackets are commonly used to specify sets (as we just did with $\{-1, 1\}$), the notation $\{X_k\}_1^\infty$ for a sequence invites confusion with the set whose elements are the X_k's, and for this reason some authors use other notations such as $\langle X_k \rangle_1^\infty$. However, the notation $\{X_k\}_1^\infty$ is by far the most common one, and in practice it rarely causes problems, so we shall stick with it.

For the remainder of this section we shall be concerned with sequences of numbers or n-dimensional vectors. We reserve the letter n for the dimension and use letters such as k and j for the index on a sequence. Thus, for example, if $\{\mathbf{x}_k\}$ is a sequence in $\mathbb{R}^n$, the components of the vector $\mathbf{x}_k$ are $(x_{k1}, \ldots, x_{kn})$.

A sequence $\{\mathbf{x}_k\}$ in $\mathbb{R}^n$ is said to **converge** to the limit $\mathbf{L}$ if for every $\epsilon > 0$ there is an integer K such that $|\mathbf{x}_k - \mathbf{L}| < \epsilon$ whenever $k > K$; otherwise, $\{\mathbf{x}_k\}$ **diverges**. If $\{\mathbf{x}_k\}$ converges to $\mathbf{L}$, we write $\mathbf{x}_k \to \mathbf{l}$ or $\mathbf{L} = \lim_{k\to\infty} \mathbf{x}_k$.

We say that $\lim_{k\to\infty} \mathbf{x}_k = \infty$ (or $+\infty$) if for every $C > 0$ there is an integer K such that $x_k > C$ whenever $k > K$, and $\lim_{k\to\infty} x_k = -\infty$ if for every $C > 0$ there is an integer K such that $x_k < -C$ whenever $k > K$. (However, a sequence whose limit is $\pm\infty$ is still called divergent.)

It follows easily from the estimates (1.3) that $\mathbf{x}_k \to \mathbf{L}$ if and only if each component of $\mathbf{x}_k$ converges to the corresponding component of $\mathbf{L}$, that is, $x_{km} \to L_m$ for $1 \le m \le n$. The study of convergence of sequences of vectors is thus reducible to the study of convergence of numerical sequences.

EXAMPLE 4.
  a. The sequence $\{1/k\}$ converges to 0, since $|(1/k) - 0| < \epsilon$ whenever $k > (1/\epsilon)$.
  b. The sequence $\{k^2\}$ diverges; more precisely, $\lim_{k\to\infty} k^2 = \infty$.

c. The sequence $\{x_k\} = \{(-1)^k\}$ diverges, but the subsequence $\{y_j\} = \{x_{2j-1}\}$ of odd-numbered terms converges to -1, and the subsequence $\{z_j\} = \{x_{2j}\}$ of even-numbered terms converges to 1.

EXAMPLE 5. If C is any positive number, $C^k/k! \to 0$ as $k \to \infty$ (that is, $k!$ grows faster than exponentially as $k \to \infty$). Indeed, pick an integer $K > 2C$. For $k > K$, we then have

$$0 < \frac{C^k}{k!} = \frac{C^K}{K!} \cdot \frac{C}{K+1} \cdot \frac{C}{K+2} \cdots \frac{C}{k} < \frac{C^K}{K!} \cdot \frac{1}{2} \cdot \frac{1}{2} \cdots \frac{1}{2} = \frac{C^K}{K!} \cdot \frac{1}{2^{k-K}}.$$

But $C^K/K!$ is a fixed number, and $1/2^{k-K} \to 0$ as $k \to \infty$.

Sequential convergence is often a useful tool in studying questions relating to open and closed sets, continuity, and related matters. The fundamental results are the following two theorems.

1.14 Theorem. *Suppose $S \subset \mathbb{R}^n$ and $\mathbf{x} \in \mathbb{R}^n$. Then $\mathbf{x}$ belongs to the closure of S if and only if there is a sequence of points in S that converges to $\mathbf{x}$.*

Proof. If $\{\mathbf{x}_k\}$ is a sequence in S that converges to $\mathbf{x}$, then every neighborhood of $\mathbf{x}$ contains elements of S — namely, $\mathbf{x}_k$ where k is sufficiently large — so $\mathbf{x}$ is in the closure of S. Conversely, suppose $\mathbf{x}$ is in the closure of S. If $\mathbf{x}$ is in S itself, let $\mathbf{x}_k = \mathbf{x}$ for all k. If not, for each k the ball of radius $1/k$ about $\mathbf{x}$ contains points of S; pick one and call it $\mathbf{x}_k$. In either case, $\{\mathbf{x}_k\}$ is a sequence of points in S that converges to $\mathbf{x}$. $\square$

1.15 Theorem. *Given $S \subset \mathbb{R}^n$, $\mathbf{a} \in S$, and $\mathbf{f} : S \to \mathbb{R}^m$, the following are equivalent:*

a. *$\mathbf{f}$ is continuous at $\mathbf{a}$.*

b. *For any $\{\mathbf{x}_k\}$ sequence in S that converges to $\mathbf{a}$, the sequence $\{\mathbf{f}(\mathbf{x}_k)\}$ converges to $\mathbf{f}(\mathbf{a})$.*

Proof. Suppose $\mathbf{f}$ is continuous at $\mathbf{a}$ and $\mathbf{x}_k \to \mathbf{a}$. Given $\epsilon > 0$, we wish to show that $|\mathbf{f}(\mathbf{x}_k) - \mathbf{f}(\mathbf{a})| < \epsilon$ provided k is sufficiently large. But by the continuity of $\mathbf{f}$, there exists $\delta > 0$ such that $|\mathbf{f}(\mathbf{x}_k) - \mathbf{f}(\mathbf{a})| < \epsilon$ when $|\mathbf{x}_k - \mathbf{a}| < \delta$, and since $\mathbf{x}_k \to \mathbf{a}$, there exists an integer K such that $|\mathbf{x}_k - \mathbf{a}| < \delta$ whenever $k > K$. Combining these, we get $|\mathbf{f}(\mathbf{x}_k) - \mathbf{f}(\mathbf{a})| < \epsilon$ whenever $k > K$, as desired.

On the other hand, suppose $\mathbf{f}$ is *not* continuous at $\mathbf{a}$. This means that there exists $\epsilon > 0$ such that for every $\delta > 0$ there is a point $\mathbf{x} \in S$ with $|\mathbf{x} - \mathbf{a}| < \delta$ but $|\mathbf{f}(\mathbf{x}) - \mathbf{f}(\mathbf{a})| \geq \epsilon$. Taking δ equal to $1, \frac{1}{2}, \frac{1}{3}, \ldots$, we see that for each positive integer k there is a point $\mathbf{x}_k \in S$ such that $|\mathbf{x}_k - \mathbf{a}| < k^{-1}$ but $|\mathbf{f}(\mathbf{x}_k) - \mathbf{f}(\mathbf{a})| \geq \epsilon$. The sequence $\{\mathbf{x}_k\}$ then converges to $\mathbf{a}$, but the sequence $\{\mathbf{f}(\mathbf{x}_k)\}$ does not converge to $\mathbf{f}(\mathbf{a})$.

We have shown that if (a) is true then (b) is true, and that if (a) is false then (b) is false, so the proof is complete. $\qquad\square$

EXERCISES

1. For each of the following sequences $\{x_k\}$, find the limit or show that the sequence diverges.

   a. $\quad x_k = \dfrac{\sqrt{2k+1}}{2\sqrt{k}+1}.$
   b. $\quad x_k = \dfrac{\sin k}{k}.$
   c. $\quad x_k = \sin\dfrac{k\pi}{3}.$

2. Let $x_k = \dfrac{3k+4}{k-5}$; then $\lim_{k\to\infty} x_k = 3$. Given $\epsilon > 0$, find an integer K so that $|x_k - 3| < \epsilon$ whenever $k > K$.

3. Define a sequence $\{x_k\}$ recursively by $x_1 = 1$ and $x_{k+1} = kx_k/(k+1)$ for $k \geq 1$. Find an explicit formula for x_k. What is $\lim_{k\to\infty} x_k$?

4. Let $\{x_k\}$ and $\{y_k\}$ be sequences in $\mathbb{R}$ such that $x_k \to a$ and $y_k \to b$. Show that $x_k + y_k \to a + b$ and $x_k y_k \to ab$. (Use Theorems 1.10 and 1.15.)

5. Given $\mathbf{f} : \mathbb{R}^n \to \mathbb{R}^m$; show that $\lim_{\mathbf{x}\to\mathbf{a}} \mathbf{f}(\mathbf{x}) = \mathbf{l}$ if and only if $\mathbf{f}(\mathbf{x}_k) \to \mathbf{l}$ for every sequence $\{\mathbf{x}_k\}$ that converges to $\mathbf{a}$. (Adapt the proof of Theorem 1.15.)

A point $\mathbf{a} \in \mathbb{R}^n$ is called an **accumulation point** of a set $S \subset \mathbb{R}^n$ if every neighborhood of $\mathbf{a}$ contains infinitely many points of S. (The point $\mathbf{a}$ itself may or may not belong to S. Some people use the terms "limit point" or "cluster point" instead of "accumulation point.") For example, the accumulation points of the interval $(-1, 1)$ in $\mathbb{R}$ are the points in the closed interval $[-1, 1]$, and the only accumulation point of the set $\{1, \frac{1}{2}, \frac{1}{3}, \frac{1}{4}, \dots\}$ is 0.

6. Show that $\mathbf{a}$ is an accumulation point of S if and only if there is a sequence $\{\mathbf{x}_k\}$ of points in S, none of which are equal to $\mathbf{a}$, such that $\mathbf{x}_k \to \mathbf{a}$. (Adapt the proof of Theorem 1.14.)

7. Show that the closure of S is the union of S and the set of all its accumulation points.

1.5 Completeness

The essential properties of the real number system that underlie all the theorems of calculus are summarized by saying that $\mathbb{R}$ is a *complete ordered field*. We explain the meaning of these terms one by one:

A *field* is a set on which the operations of addition, subtraction, multiplication, and division (by any nonzero number) are defined, subject to all the usual laws of arithmetic: commutativity, associativity, etc. Besides the real numbers, examples of fields include the rational numbers and the complex numbers, and there are many others. (For more precise definitions and more examples, consult a textbook on abstract algebra such as Birkhoff and Mac Lane [4] or Hungerford [8].)

An *ordered field* is a field equipped with a binary relation $<$ that is transitive (if $a < b$ and $b < c$, then $a < c$) and antisymmetric (if $a \neq b$, then either $a < b$ or $b < a$, but not both), and interacts with the arithmetic operations in the usual way (if $a < b$ then $a + c < b + c$ for any c, and also $ac < bc$ if $c > 0$). The real number and rational number systems are ordered fields (with the usual meaning of "$<$"), but the complex number system is not.

Finally, *completeness* is what distinguishes the real numbers from the smaller ordered fields such as the rational numbers and makes possible the transition from algebra to calculus; it means that there are "no holes" in the real number line. There are several equivalent ways of stating the completeness property precisely. The one we shall use as a starting point is the existence of least upper bounds.

If S is a subset of $\mathbb{R}$, an **upper bound** for S is a number u such that $x \leq u$ for all $x \in S$, and a **lower bound** for S is a number l such that $x \geq l$ for all $x \in S$.

The Completeness Axiom. *Let S be a nonempty set of real numbers. If S has an upper bound, then S has a least upper bound, called the* **supremum** *of S and denoted by* $\sup S$. *If S has a lower bound, then S has a greatest lower bound, called the* **infimum** *of S and denoted by* $\inf S$.

If S has no upper bound, we shall define $\sup S$ to be $+\infty$, and if S has no lower bound, we shall define $\inf S$ to be $-\infty$.

EXAMPLE 1.
a. If S is the interval $(0, 1]$, then $\sup S = 1$ and $\inf S = 0$.
b. If $S = \{1, \frac{1}{2}, \frac{1}{3}, \frac{1}{4}, \ldots\}$, then $\sup S = 1$ and $\inf S = 0$.
c. If $S = \{1, 2, 3, 4, \ldots\}$, then $\sup S = \infty$ and $\inf S = 1$.
d. If S is the single point a, then $\sup S = \inf S = a$.
e. If $S = \{x : x \text{ is rational and } x^2 < 2\}$, then $\sup S = \sqrt{2}$ and $\inf S = -\sqrt{2}$. This is an example of a set of rational numbers that has no supremum or infimum within the set of rational numbers.

If S has an upper bound, the number $a = \sup S$ is the unique number such that

i. $x \le a$ for every $x \in S$ and

ii. for every $\epsilon > 0$ there exists $x \in S$ with $x > a - \epsilon$.

(i) expresses the fact that a is an upper bound, whereas (ii) expresses the fact that there is no smaller upper bound. In particular, while $\sup S$ may or may not belong to S itself, it always belongs to the closure of S. Similarly for $\inf S$ if S is bounded below.

The completeness of the real number system plays a crucial role in establishing the convergence of numerical sequences. The most basic result along these lines is the following. First, some terminology: A sequence $\{x_k\}$ is called **bounded** if all the numbers x_n are contained in some bounded interval. A sequence $\{x_n\}$ is called **increasing** if $x_n \le x_m$ whenever $n \le m$, and **decreasing** if $x_n \ge x_m$ whenever $n \le m$. A sequence that is either increasing or decreasing is called **monotone** (or monotonic).

1.16. Theorem (The Monotone Sequence Theorem). *Every bounded monotone sequence in $\mathbb{R}$ is convergent. More precisely, the limit of an increasing (resp. decreasing) sequence is the supremum (resp. infimum) of its set of values.*

Proof. Suppose $\{x_k\}$ is a bounded increasing sequence. Let l be the supremum of the set of values $\{x_1, x_2, \ldots\}$; I claim that $x_k \to l$. Since l is an upper bound, we have $x_k \le l$ for all k. On the other hand, since l is the *least* upper bound, for any $\epsilon > 0$ there is some K for which $x_K > l - \epsilon$. Since the x_k's increase with k, we also have $x_k > l - \epsilon$ for all $k > K$. Therefore, $l - \epsilon < x_k \le l$ for all $k > K$, and this shows that $x_k \to l$.

Similarly, if $\{x_k\}$ is decreasing, it converges to $\inf\{x_1, x_2, \ldots\}$. $\qquad\square$

EXAMPLE 2. Given a positive real number a, define a sequence $\{x_k\}$ recursively as follows. x_1 is some fixed positive real number, and for $k \ge 2$,

$$x_k = \frac{1}{2}\left(x_{k-1} + \frac{a}{x_{k-1}}\right).$$

Observe that if $x_{k-1} > 0$ then $x_k > 0$ too; since we assume that $x_1 > 0$, every term of this sequence is positive. (In particular, division by zero is never a problem.) We claim that $x_k \to \sqrt{a}$, no matter what initial x_1 is chosen. Indeed, if we *assume* that the sequence converges to a nonzero limit L, by letting $k \to \infty$ in the recursion formula we see that

$$L = \frac{1}{2}\left(L + \frac{a}{L}\right), \quad \text{or} \quad L^2 = \tfrac{1}{2}L^2 + \tfrac{1}{2}a,$$

so that $L^2 = a$. Since $x_k > 0$ for every k, we must have $L > 0$, and hence $L = \sqrt{a}$. But this argument is without force until we know that $\{x_k\}$ converges to a nonzero limit.

To verify this, observe that for $k \geq 2$,

$$x_k^2 = \tfrac{1}{4}(x_{k-1}^2 + 2a + a^2 x_{k-1}^{-2}) = a + \tfrac{1}{4}(x_{k-1}^2 - 2a + a^2 x_{k-1}^{-2})$$
$$= a + (x_{k-1} - a x_{k-1}^{-1})^2 > a.$$

Thus, starting with the second term, the sequence $\{x_k\}$ is bounded below by $\sqrt{a} > 0$, and it is decreasing:

$$x_{k+1} - x_k = \tfrac{1}{2}(a x_k^{-1} - x_k) < \tfrac{1}{2}(x_k - x_k) = 0.$$

The convergence to a limit $L \geq \sqrt{a}$ now follows from the monotone sequence theorem. (The verification that $\{x_k\}$ converges is not just a formality; see Exercise 4.)

The sequence $\{x_k\}$ gives a computationally efficient recursive algorithm for computing square roots.

The following consequence of the monotone sequence theorem is also a useful technical tool.

1.17 Theorem (The Nested Interval Theorem). *Let* $I_1 = [a_1, b_1]$, $I_2 = [a_2, b_2]$, *... be a sequence of closed, bounded intervals in* $\mathbb{R}$. *Suppose that (a)* $I_1 \supset I_2 \supset I_3 \supset \cdots$, *and (b) the length* $b_k - a_k$ *of* I_k *tends to 0 as* $k \to \infty$. *Then there is exactly one point contained in all of the intervals* I_k.

Proof. The condition $I_1 \supset I_2 \supset I_3 \supset \cdots$ means that $a_1 \leq a_2 \leq a_3 \leq \cdots$ and $b_1 \geq b_2 \geq b_3 \geq \cdots$, so the sequences $\{a_k\}$ and $\{b_k\}$ are monotone. They are also bounded, since all a_k and b_k are contained in I_1; hence, by the monotone sequence theorem, they are both convergent. Moreover, since $b_k - a_k \to 0$, their limits are equal. Call their common limit l. Then $a_k \leq l \leq b_k$ for all k, so $l \in I_k$ for all n. No other point l' can be common to all I_k, for the length of I_k is less than the distance $|l - l'|$ when k is sufficiently large. $\qquad\square$

It should be emphasized that the real point of the nested interval theorem is that the intersection $\bigcap_1^\infty I_n$ is nonempty; the fact that it can contain no more than one point is pretty obvious from the assumption that the length of I_n tends to zero.

If $\{x_k\}$ is a sequence (in any set, not necessarily $\mathbb{R}$), we may form a *subsequence* of $\{x_k\}$ by deleting some of the terms and keeping the rest in their original order. More precisely, a **subsequence** of $\{x_k\}$ is a sequence $\{x_{k_j}\}_{j=1}^\infty$ specified by a one-to-one, increasing map $j \to k_j$ from the set of positive integers into itself. For example, by taking $k_j = 2j$ we obtain the subsequence of even-numbered terms; by taking $k_j = j^2$ we obtain the subsequence of those terms whose index is a perfect square, and so on.

The following theorem is one of the most useful results in the foundations of analysis; it is one version of the *Bolzano-Weierstrass theorem,* whose general form will be found in Theorem 1.21.

1.18 Theorem. *Every bounded sequence in $\mathbb{R}$ has a convergent subsequence.*

Proof. Let $\{x_k\}$ be a bounded sequence, say $x_k \in [a, b]$ for all k. Bisect the interval $[a, b]$ — that is, consider the two intervals $[a, \frac{1}{2}(a + b)]$ and $[\frac{1}{2}(a + b), b]$. At least one of these subintervals must contain x_k for infinitely many k; call that subinterval I_1. (If both of them contain x_k for infinitely many k, pick the one on the left.) Now bisect I_1. Again, one of the two halves must contain x_k for infinitely many k; call that half I_2. Proceeding inductively, we obtain a sequence of intervals I_j, each one contained in the preceding one, each one half as long as the preceding one, and each one containing x_k for infinitely many k. By the nested interval theorem, there is exactly one point l contained in every I_j.

It is now easy to construct a subsequence of $\{x_k\}$ that converges to l, as follows. Pick an integer k_1 such that $x_{k_1} \in I_1$, then pick $k_2 > k_1$ such that $x_{k_2} \in I_2$, then pick $k_3 > k_2$ such that $x_{k_3} \in I_3$, and so forth. By construction of the I_j's, this process can be continued indefinitely. Since x_{k_j} and l are both in I_j, and the length of I_j is $2^{-j}(b - a)$, we have $|x_{k_j} - l| \le 2^{-j}(b - a)$, which tends to 0 as $j \to \infty$; that is, $x_{k_j} \to l$. $\square$

Theorem 1.18 generalizes easily to higher dimensions:

1.19 Theorem. *Every bounded sequence in $\mathbb{R}^n$ has a convergent subsequence.*

Proof. If $|\mathbf{x}_k| \le C$ for all k, then the components $x_{k1}, \ldots, x_{kn}$ all lie in the interval $[-C, C]$. Hence, for each $m = 1, \ldots, n$ we can extract a convergent subsequence from the sequence of mth components, $\{x_{km}\}_{k=1}^{\infty}$. The trouble is that the indices on these subsequences might all be different, so we can't put them together. (We might have chosen the odd-numbered terms for $m = 1$ and the even-numbered terms for $m = 2$, for example.) Instead, we have to proceed inductively. First we choose a subsequence $\{\mathbf{x}_{k_j}\}$ such that the first components converge; then we choose a sub-subsequence $\{\mathbf{x}_{k_{j_i}}\}$ whose second components also converge, and so on until we find a $(\text{sub})^n$sequence whose components all converge. $\square$

Another way to express the completeness of the real number system is to say that every sequence whose terms get closer and closer to each other actually converges. To be more precise, a sequence $\{\mathbf{x}_k\}$ in $\mathbb{R}^n$ is called a **Cauchy sequence** if

$\mathbf{x}_k - \mathbf{x}_j \to 0$ as $k, j \to \infty$, that is, if for every $\epsilon > 0$ there exists an integer K such that $|\mathbf{x}_k - \mathbf{x}_j| < \epsilon$ whenever $k > K$ and $j > K$.

1.20 Theorem. *A sequence $\{\mathbf{x}_k\}$ in $\mathbb{R}^n$ is convergent if and only if it is Cauchy.*

Proof. Suppose $\mathbf{x}_k \to \mathbf{l}$. Since $\mathbf{x}_k - \mathbf{x}_j = (\mathbf{x}_k - \mathbf{l}) - (\mathbf{x}_j - \mathbf{l})$, we have $0 \leq |\mathbf{x}_k - \mathbf{x}_j| \leq |\mathbf{x}_k - \mathbf{l}| + |\mathbf{x}_j - \mathbf{l}|$. Both terms on the right tend to zero as $k, j \to \infty$; hence so does $\mathbf{x}_k - \mathbf{x}_j$. Thus $\{\mathbf{x}_k\}$ is Cauchy.

Now suppose $\{\mathbf{x}_k\}$ is Cauchy. Taking $\epsilon = 1$ in the definition of "Cauchy," we see that there is an integer K such that $|\mathbf{x}_k - \mathbf{x}_j| < 1$ if $k, j > K$. Then $|\mathbf{x}_k| < |\mathbf{x}_{K+1}| + 1$ for all $k > K$, and it follows that the sequence $\{\mathbf{x}_k\}$ is bounded. By Theorem 1.18, there is a subsequence $\{\mathbf{x}_{k_j}\}$ that converges to a limit $\mathbf{l}$. But then since $\{\mathbf{x}_k\}$ is Cauchy, the whole sequence must also converge to $\mathbf{l}$. Indeed, given $\epsilon > 0$, there is an integer J such that $|\mathbf{x}_{k_j} - \mathbf{l}| < \frac{1}{2}\epsilon$ if $j > J$, and there is an integer K such that $|\mathbf{x}_k - \mathbf{x}_m| < \frac{1}{2}\epsilon$ if $k, m > K$. Pick an integer $j > J$ such that $k_j > K$; then for $k > K$ we have

$$|\mathbf{x}_k - \mathbf{l}| \leq |\mathbf{x}_k - \mathbf{x}_{k_j}| + |\mathbf{x}_{k_j} - \mathbf{l}| < \tfrac{1}{2}\epsilon + \tfrac{1}{2}\epsilon = \epsilon.$$

Therefore, $\mathbf{x}_k \to \mathbf{l}$. $\square$

EXERCISES

1. Find $\sup S$ and $\inf S$ for the following sets S. Do these numbers belong to S or not?
   a. $S = \{x : (2x^2 - 1)(x^2 - 1) < 0\}$.
   b. $S = \{(-1)^k + 2^{-k} : k \geq 0\}$.
   c. $S = \{x : \arctan x \geq 1\}$.

2. Construct a sequence $\{x_k\}$ that has subsequences converging to three different limits.

3. Consider the sequence $\frac{1}{2}, \frac{1}{3}, \frac{2}{3}, \frac{1}{4}, \frac{2}{4}, \frac{3}{4}, \frac{1}{5}, \frac{2}{5}, \frac{3}{5}, \frac{4}{5}, \ldots$, obtained by listing the rational numbers in $(0, 1)$ with denominator n in increasing order, for n successively equal to $2, 3, 4, \ldots$. Show that for any $a \in [0, 1]$, there is a subsequence that converges to a. (*Hint:* Consider the decimal expansion of a.)

4. Given a real number a, define a sequence $\{x_k\}$ recursively by $x_1 = a$, $x_{k+1} = x_k^2$.
   a. Show, as in Example 2, that *if* $\{x_k\}$ converges, its limit must be 0 or 1.
   b. For which a is the limit equal to 0? equal to 1? nonexistent?

5. Define a sequence $\{x_k\}$ recursively by $x_1 = \sqrt{2}$, $x_{k+1} = \sqrt{2 + x_k}$. Show by induction that (a) $x_k < 2$ and (b) $x_k < x_{k+1}$ for all k. Then show that $\lim x_k$ exists and evaluate it.

6. Let r_k be the ratio of the $(k + 1)$th term to the kth term of the Fibonacci sequence (Example 2, §1.4). (Thus the first few r_k's are $1, 2, \frac{3}{2}, \frac{5}{3}, \ldots$) Our object is to show that $\lim_{k \to \infty} r_k$ is the "golden ratio" $\varphi = \frac{1}{2}(1 + \sqrt{5})$, the positive root of the equation $x^2 = x + 1$.

   a. Show that
$$r_{k+1} = \frac{r_k + 1}{r_k}, \qquad r_{k+2} = \frac{2r_k + 1}{r_k + 1}.$$

   b. Show that $r_k < \varphi$ if k is odd and $r_k > \varphi$ if k is even. Then show that $r_{k+2} - r_k$ is positive if k is odd and negative if k is even. (*Hint:* For $x > 0$ we have $x^2 < x + 1$ if $x < \varphi$ and $x^2 > x + 1$ if $x > \varphi$.)

   c. Show that the subsequences $\{r_{2j-1}\}$ and $\{r_{2j}\}$ of odd- and even-numbered terms both converge to φ.

7. Let $\{\mathbf{x}_k\}$ be a sequence in $\mathbb{R}^n$ and $\mathbf{a}$ a point in $\mathbb{R}^n$. Show that some subsequence of $\{\mathbf{x}_k\}$ converges to $\mathbf{x}$ if and only if every ball centered at $\mathbf{x}$ contains $\mathbf{x}_k$ for infinitely many values of k.

8. Show that every infinite bounded set in $\mathbb{R}^n$ has an accumulation point. (See Exercises 6–7 in §1.4.)

Let $\{x_k\}_1^\infty$ be a bounded sequence in $\mathbb{R}$. For $m = 1, 2, 3, \ldots$, let

$$Y_m = \sup\{x_m, x_{m+1}, x_{m+2}, \ldots\}, \qquad y_m = \inf\{x_m, x_{m+1}, x_{m+2}, \ldots\}.$$

Then the sequence $\{Y_m\}$ is bounded and decreasing, and $\{y_m\}$ is bounded and increasing (because the sup and inf are being taken over fewer and fewer numbers as m increases), so they both converge. The limits $\lim Y_m$ and $\lim y_m$ are called the **limit superior** and **limit inferior** of the sequence $\{x_k\}$, respectively; they are denoted by $\lim \sup_{k \to \infty} x_k$ and $\lim \inf_{k \to \infty} x_k$:

$$\lim_{k \to \infty} \sup x_k = \lim_{m \to \infty} \left(\sup\{x_k : k \geq m\}\right), \quad \lim_{k \to \infty} \inf x_k = \lim_{m \to \infty} \left(\inf\{x_k : k \geq m\}\right).$$

The following exercises pertain to these ideas.

9. Show that $\lim \sup x_k$ is the number a uniquely specified by the following property: For any $\epsilon > 0$, there are infinitely many k for which $x_k > a - \epsilon$ but only finitely many for which $x_k > a + \epsilon$. What is the corresponding condition for $\lim \inf x_k$?

10. Show that there is a subsequence of $\{x_k\}$ that converges to $\limsup x_k$, and one that converges to $\liminf x_k$.

11. Show that if $a \in \mathbb{R}$ is the limit of some subsequence of $\{x_k\}$, then $\liminf x_k \leq a \leq \limsup x_k$.

12. Show that $\{x_k\}$ converges if and only if $\limsup x_k = \liminf x_k$, in which case this common value is equal to $\lim x_k$.

1.6   Compactness

A subset of $\mathbb{R}^n$ is called **compact** if it is both closed and bounded. (*Note:* The notion of compactness can be extended to settings other than $\mathbb{R}^n$, but a different definition must be adopted; see the concluding paragraph of this section.) Compactness is an important property, principally because it yields existence theorems for limits in many situations. The fundamental result is the following theorem.

1.21 Theorem (The Bolzano-Weierstrass Theorem). *If S is a subset of $\mathbb{R}^n$, the following are equivalent:*

  a. *S is compact.*
  b. *Every sequence of points in S has a convergent subsequence whose limit lies in S.*

Proof. Suppose S is compact. If $\{\mathbf{x}_k\}$ is a sequence in S, it has a convergent subsequence by Theorem 1.19 since S is bounded, and the limit lies in S by Theorem 1.14 since S is closed; thus (b) holds.

On the other hand, suppose S is not compact, i.e., S is either not closed or not bounded. If S is not bounded, there is a sequence of points $\{\mathbf{x}_k\}$ in S such that $|\mathbf{x}_k| \to \infty$. But then $\{\mathbf{x}_k\}$ has no convergent subsequence, as any subsequence must also satisfy $|\mathbf{x}_{k_j}| \to \infty$. If S is not closed, there is a point $\mathbf{x}$ that lies in $\overline{S}$ but not in S. By Theorem 1.14 there is a sequence $\{\mathbf{x}_k\}$ in S that converges to $\mathbf{x}$. Every subsequence also converges to $\mathbf{x}$, which is not in S. Thus (b) is false if S is either not closed or not bounded. $\square$

Remark. Every finite subset of $\mathbb{R}^n$ is obviously compact. If S is finite, (b) is true because if $\{\mathbf{x}_k\}$ is a sequence in S, then there must be a single point $\mathbf{x} \in S$ such that $\mathbf{x}_k = \mathbf{x}$ for infinitely many k; the subsequence consisting of those $\mathbf{x}_k$'s trivially converges to $\mathbf{x}$.

The Bolzano-Weierstrass theorem paves the way to the fundamental connection between continuity and compactness:

1.22 Theorem. *Continuous functions map compact sets to compact sets. That is, suppose that S is a compact subset of $\mathbb{R}^n$ and $\mathbf{f} : S \to \mathbb{R}^m$ is continuous at every point of S. Then the set*

$$\mathbf{f}(S) = \{\mathbf{f}(\mathbf{x}) : \mathbf{x} \in S\}$$

is also compact.

Proof. Suppose $\{\mathbf{y}_k\}$ is a sequence in the image $\mathbf{f}(S)$. For each k there is a point $\mathbf{x}_k \in V$ such that $\mathbf{y}_k = \mathbf{f}(\mathbf{x}_k)$. Since S is compact, by the Bolzano-Weierstrass theorem the sequence $\{\mathbf{x}_k\}$ has a convergent subsequence $\{\mathbf{x}_{k_j}\}$ whose limit $\mathbf{a}$ lies in S. Since $\mathbf{f}$ is continuous at $\mathbf{a}$, by Theorem 1.15 the sequence $\{\mathbf{y}_{k_j}\} = \{\mathbf{f}(\mathbf{x}_{k_j})\}$ converges to the point $\mathbf{f}(\mathbf{a}) \in \mathbf{f}(S)$. Thus, every sequence in $\mathbf{f}(S)$ has a subsequence whose limit lies in $\mathbf{f}(S)$. By the Bolzano-Weierstrass theorem again, $\mathbf{f}(S)$ is compact. $\qquad\square$

It is not true, in general, that continuous functions map closed sets to closed sets, or bounded sets to bounded sets. (See Exercises 1–2.) Only the *combination* of closedness and boundedness is preserved.

An immediate consequence of Theorem 1.22 is the fundamental existence theorem for maxima and minima of real-valued functions.

1.23 Corollary (The Extreme Value Theorem). *Suppose $S \subset \mathbb{R}^n$ is compact and $f : S \to \mathbb{R}$ is continuous. Then f has an absolute minimum value and an absolute maximum value on S; that is, there exist points $\mathbf{a}, \mathbf{b} \in S$ such that $f(\mathbf{a}) \leq f(\mathbf{x}) \leq f(\mathbf{b})$ for all $\mathbf{x} \in S$.*

Proof. By Theorem 1.22, the set $f(S)$ is a compact subset of $\mathbb{R}$. Thus, it is *bounded,* so $\inf f(S)$ and $\sup f(S)$ exist, and *closed,* so $\inf f(S)$ and $\sup f(S)$ actually belong to $f(S)$. But this says precisely that the set of values of f on V has a smallest and a largest element, as desired. $\qquad\square$

The assumption that S is compact is necessary. If S is not closed or not bounded, the function f might be unbounded, or its extreme values might occur at points on the boundary of S that are not in S or "at infinity." Here are a few simple counterexamples with $n = 1$:

- $f(x) = x$, $S = (0, 1)$. (The extreme values occur on the boundary.)

- $f(x) = \cot \pi x$, $S = (0, 1)$. (The values of f range from $-\infty$ to ∞.)

- $f(x) = \arctan x$, $S = \mathbb{R}$. (f approaches but does not achieve the extreme values $\pm \frac{1}{2}\pi$.)

- $f(x) = 3x - x^3$, $S = \mathbb{R}$. (f has a local maximum at $x = 1$ and a local minimum at $x = -1$, but no absolute maximum or minimum.)

Compactness also has another consequence that turns out to be extremely useful in more advanced mathematical analysis, although its significance may not be very clear at first sight. (It will not be used elsewhere in this book except in some of the technical arguments in Appendix B, so it may be regarded as an optional topic.) Suppose S is a subset of $\mathbb{R}^n$. A collection $\mathcal{U}$ of subsets of $\mathbb{R}^n$ is called a **covering** of S if S is contained in the union of the sets in $\mathcal{U}$. For example, for each $\mathbf{x} \in S$ we could pick an open ball $B_\mathbf{x}$ centered at $\mathbf{x}$; then $\mathcal{U} = \{B_\mathbf{x} : \mathbf{x} \in S\}$ is a covering of S.

1.24 Theorem (The Heine-Borel Theorem). *If S is a subset of $\mathbb{R}^n$, the following are equivalent:*

a. *S is compact.*
b. *If $\mathcal{U}$ is any covering of S by open sets, there is a finite subcollection of $\mathcal{U}$ that still forms a covering of S. (In brief: Every open covering of S has a finite subcovering.)*

Proof. The proof is given in Appendix B.1 (Theorem B.1). □

Much of what we have done in this section and the preceding ones can be generalized from subsets of $\mathbb{R}^n$ to subsets of more general spaces equipped with a "distance function" that behaves more or less like the Euclidean distance $d(\mathbf{x}, \mathbf{y}) = |\mathbf{x} - \mathbf{y}|$. (Such spaces are known as *metric spaces*; see DePree and Swartz [5], Krantz [12], or Rudin [18].) For example, in studying the geometry of a surface S in $\mathbb{R}^3$, one might want to take the "distance" between two points $\mathbf{x}, \mathbf{y} \in S$ to be not the straight-line distance $|\mathbf{x} - \mathbf{y}|$ but the length of the shortest curve on S that joins $\mathbf{x}$ to $\mathbf{y}$. Another class of examples is provided by spaces of functions, where the "distance" between two functions f and g can be measured in a number of different ways; we shall say more about this in Chapter 8. In this general setting, the Bolzano-Weierstrass and Heine-Borel theorems are no longer completely valid. The conditions on a set S in Theorem 1.21b and Theorem 1.24b still imply that S is closed and bounded, *but not conversely.* These conditions are still very important, however, so a shift in terminology is called for. The condition in Theorem 1.24b — that every open cover of S has a finite subcover — is usually taken as the *definition* of compactness in the general setting, and the condition in Theorem 1.21b — that every sequence in S has a subsequence that converges in S — is called *sequential compactness.*

EXERCISES

1. Give an example of
   a. a closed set $S \subset \mathbb{R}$ and a continuous function $f : \mathbb{R} \to \mathbb{R}$ such that $f(S)$ is not closed;
   b. an open set $U \subset \mathbb{R}$ and a continuous function $f : \mathbb{R} \to \mathbb{R}$ such that $f(U)$ is not open.

2. a. Give an example of a bounded set $S \subset \mathbb{R} \setminus \{0\}$ and a real-valued function f that is defined and continuous on $\mathbb{R} \setminus \{0\}$ such that $f(S)$ is not bounded.
   b. However, show that if $f : \mathbb{R}^n \to \mathbb{R}^m$ is continuous everywhere and $S \subset \mathbb{R}^n$ is bounded, then $\mathbf{f}(S)$ is bounded.

3. Show that an infinite set $S \subset \mathbb{R}^n$ is compact if and only if every infinite subset of S has an accumulation point that lies in S. (See Exercises 6–7 in §1.4 and Exercise 8 in §1.5.)

4. Suppose $S \subset \mathbb{R}^n$ is compact, $f : S \to \mathbb{R}$ is continuous, and $f(\mathbf{x}) > 0$ for every $\mathbf{x} \in S$. Show that there is a number $c > 0$ such that $f(\mathbf{x}) \geq c$ for every $\mathbf{x} \in S$.

5. (A generalization of the nested interval theorem) Suppose $\{S_k\}$ is a sequence of nonempty compact subsets of $\mathbb{R}^n$ such that $S_1 \supset S_2 \supset S_3 \supset \dots$. Show that there is at least one point contained in all of the S_k's (that is, $\bigcap_1^\infty S_k \neq \varnothing$). (This can be done using either the Bolzano-Weierstrass theorem or the Heine-Borel theorem. Can you find both proofs?)

6. The **distance** between two sets $U, V \subset \mathbb{R}^n$ is defined to be

$$d(U, V) = \inf\big\{ |\mathbf{x} - \mathbf{y}| : \mathbf{x} \in U, \, \mathbf{y} \in V \big\}.$$

   a. Show that $d(U, V) = 0$ if either of the sets U, V contains a point in the closure of the other one.
   b. Show that if U is compact, V is closed, and $U \cap V = \varnothing$, then $d(U, V) > 0$.
   c. Give an example of two closed sets U and V in $\mathbb{R}^2$ that have no point in common but satisfy $d(U, V) = 0$.

1.7 Connectedness

A set in $\mathbb{R}^n$ is said to be connected if it is "all in one piece," that is, if it is not the union of two nonempty subsets that do not touch each other. The formal definition is as follows: A set $S \subset \mathbb{R}^n$ is **disconnected** if it is the union of two nonempty subsets S_1 and S_2, neither of which intersects the closure of the other one; in this

FIGURE 1.2: The sets S and T in Example 1.

case we shall call the pair (S_1, S_2) a **disconnection** of S. The set S is **connected** if it is not disconnected.

EXAMPLE 1. Let

$$S_1 = \{(x,y) : (x+1)^2 + y^2 < 1\}, \qquad S_2 = \{(x,y) : (x-1)^2 + y^2 < 1\},$$
$$\overline{S}_2 = \{(x,y) : (x-1)^2 + y^2 \leq 1\}.$$

Then the set $S = S_1 \cup S_2$ is disconnected, for the only point common to the closures of S_1 and S_2 is $(0,0)$, which belongs to neither S_1 nor S_2. However, the set $T = S_1 \cup \overline{S}_2$ is connected, for $(0,0)$ belongs both to $\overline{S}_2$ and the closure of S_1; this point "connects" the two pieces of T. See Figure 1.2.

The connected subsets of the real line are easy to describe.

1.25 Theorem. *The connected subsets of $\mathbb{R}$ are precisely the intervals (open, half-open, or closed; bounded or unbounded).*

Proof. If $S \subset \mathbb{R}$ is not an interval, there exist $a, b \in S$ and $c \notin S$ such that $a < c < b$. Let $S_1 = S \cap (-\infty, c)$ and $S_2 = S \cap (c, \infty)$. Then $S = S_1 \cup S_2$ (since $c \notin S$), and S_1 and S_2 are nonempty since $a \in S_1$ and $b \in S_2$. The closures of S_1 and S_2 are contained in $(-\infty, c]$ and $[c, \infty)$, so the only point where they can intersect is c, which is not in either S_1 or S_2. Thus S is disconnected.

Conversely, suppose S is an interval. We shall suppose that S is disconnected and derive a contradiction.

We first consider the case where S is compact, say $S = [a, b]$. Suppose (S_1, S_2) is a disconnection of S. By relabeling if necessary, we take S_2 to be the set that contains b. Let $c = \sup S_1$. Then c belongs to the closure of S_1, so it cannot be in S_2; hence $c \in S_1$. In particular, $c \neq b$. But then the interval $(c, b]$ is included in S_2, and c is in the closure of this interval; so c is in the closure of S_2 and so cannot belong to S_1. This contradiction shows that S must be connected.

Finally, suppose S is a noncompact interval and (S_1, S_2) is a disconnection of S. Pick $a \in S_1$ and $b \in S_2$; then $[a, b] \subset S$ since S is an interval. But then $[a, b] = T_1 \cup T_2$ where $T_1 = [a, b] \cap S_1$ and $T_2 = [a, b] \cap S_2$. The sets T_1 and T_2 are nonempty ($a \in T_1$ and $b \in T_2$), and they are contained in S_1 and S_2, so neither one can intersect the closure of the other. But this means that $[a, b]$ is disconnected, which we have just proved to be false. Therefore, S is connected. □

The following result, a cousin of Theorem 1.22, gives the basic relation between continuity and connectedness:

1.26 Theorem. *Continuous functions map connected sets to connected sets. That is, suppose $\mathbf{f} : S \to \mathbb{R}^m$ is continuous at every point of S and S is connected. Then the set*

$$\mathbf{f}(S) = \{\mathbf{f}(\mathbf{x}) : \mathbf{x} \in S\}$$

is also connected.

Proof. We proceed by contraposition; that is, we assume that $\mathbf{f}(S)$ is disconnected and deduce that S is disconnected. Thus, suppose that (U_1, U_2) is a disconnection of $\mathbf{f}(S)$. Let

$$S_1 = \{\mathbf{x} \in S : \mathbf{f}(\mathbf{x}) \in U_1\}, \qquad S_2 = \{\mathbf{x} \in S : \mathbf{f}(\mathbf{x}) \in U_2\}.$$

Then S_1 and S_2 are nonempty, and their union is S. If there were a point $\mathbf{x} \in S_1$ belonging to the closure of S_2, $\mathbf{x}$ would be the limit of a sequence $\{\mathbf{x}_k\}$ in S_2 by Theorem 1.14. But then $\mathbf{f}(\mathbf{x}) \in U_1$ and $\mathbf{f}(\mathbf{x}_k) \in U_2$, so $\mathbf{f}(\mathbf{x}) = \lim \mathbf{f}(\mathbf{x}_k)$ would be in the closure of U_2 by Theorem 1.14 again. This is impossible; hence S_1 does not intersect the closure of S_2, and likewise, S_2 does not intersect the closure of S_1. Thus $S = S_1 \cup S_2$ is disconnected. □

1.27 Corollary (The Intermediate Value Theorem). *Suppose $f : S \to \mathbb{R}$ is continuous at every point of S and $V \subset S$ is connected. If $\mathbf{a}, \mathbf{b} \in V$ and $f(\mathbf{a}) < t < f(\mathbf{b})$ or $f(\mathbf{b}) < t < f(\mathbf{a})$, there is a point $\mathbf{c} \in V$ such that $f(\mathbf{c}) = t$.*

Proof. By Theorems 1.25 and 1.26, $f(V)$ is an interval. It contains $f(\mathbf{a})$ and $f(\mathbf{b})$ and hence contains the entire interval between them. □

There is another notion of connectedness that is important in many situations. A set $S \subset \mathbb{R}^n$ is called **arcwise connected** (or **pathwise connected**) if any two points in S can be joined by a continuous curve in S, that is, if for any $\mathbf{a}, \mathbf{b}$ in S there is a continuous map $\mathbf{f} : [0, 1] \to \mathbb{R}^n$ such that $\mathbf{f}(0) = \mathbf{a}$, $\mathbf{f}(1) = \mathbf{b}$, and $\mathbf{f}(t) \in S$ for all $t \in [0, 1]$.

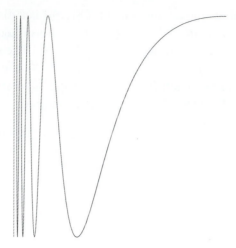

FIGURE 1.3: The set defined in (1.29).

It is useful to observe that the relation of being joined by a continuous curve is transitive; that is, if there is a continuous curve in S from $\mathbf{a}$ to $\mathbf{b}$, and one from $\mathbf{b}$ to $\mathbf{c}$, then there is one from $\mathbf{a}$ to $\mathbf{c}$. Namely, if $\mathbf{f} : [0, 1] \to S$ and $\mathbf{g} : [0, 1] \to S$ are continuous maps with $\mathbf{f}(0) = \mathbf{a}$, $\mathbf{f}(1) = \mathbf{g}(0) = \mathbf{b}$, and $\mathbf{g}(1) = \mathbf{c}$, we obtain a continuous $\mathbf{h} : [0, 1] \to S$ by gluing $\mathbf{f}$ and $\mathbf{g}$ together:

$$\mathbf{h}(t) = \begin{cases} \mathbf{f}(2t) & \text{if } 0 \le t \le \frac{1}{2}, \\ \mathbf{g}(2t - 1) & \text{if } \frac{1}{2} \le t \le 1. \end{cases}$$

The following results explain the relation between connectedness and arcwise connectedness.

1.28 Theorem. *If $S \subset \mathbb{R}^n$ is arcwise connected, then S is connected.*

Proof. We shall assume that S is disconnected and show that it is not arcwise connected. Accordingly, suppose (S_1, S_2) is a disconnection of S. Pick $\mathbf{a} \in S_1$ and $\mathbf{b} \in S_2$; we claim that there is no continuous $\mathbf{g} : [0, 1] \to S$ such that $\mathbf{g}(0) = \mathbf{a}$ and $\mathbf{g}(1) = \mathbf{b}$. If there were, the set $V = \mathbf{g}([0, 1])$ would be connected by Theorems 1.25 and 1.26. But this cannot be so: V is the union of $V \cap S_1$ and $V \cap S_2$; these sets are nonempty since $\mathbf{a} \in V \cap S_1$ and $\mathbf{b} \in V \cap S_2$, and neither of them intersects the closure of the other. Hence S is not arcwise connected.   □

The converse of Theorem 1.28 is false: A set can be connected without being arcwise connected. A typical example is

$$(1.29) \quad S = \big\{(x, y) : 0 < x \leq 2 \text{ and } y = \sin(\pi/x)\big\} \cup \big\{(0, y) : y \in [-1, 1]\big\},$$

pictured in Figure 1.3. S consists of two pieces, the graph of $\sin(\pi/x)$ and the vertical line segment. These two sets do not form a disconnection of S, as the line segment is included in the closure of the graph, but a point on the line segment cannot be connected to a point on the graph by a continuous curve. The details are sketched in Exercise 11.

However, *open* connected sets are arcwise connected:

1.30 Theorem. *If $S \subset \mathbb{R}^n$ is open and connected, then S is arcwise connected.*

Proof. Fix a point $\mathbf{a} \in S$. Let S_1 be the set of points in S that can be joined to $\mathbf{a}$ by a continuous curve in S, and let S_2 be the set of points in S that cannot; thus S_1 and S_2 are disjoint and $S = S_1 \cup S_2$. We shall show that

a. if $\mathbf{x} \in S_1$, then all points sufficiently close to $\mathbf{x}$ are in S_1;

b. if $\mathbf{x} \in S$ is in the closure of S_1, then $\mathbf{x} \in S_1$.

(a) shows that no point of S_1 can be in the closure of S_2, and (b) shows that no point in the closure of S_1 can be in S_2. Thus (S_1, S_2) will form a disconnection of S, contrary to the assumption that S is connected, *unless* S_2 is empty — which means that S is arcwise connected.

To prove (a) and (b), we use the fact that S is open, so that if $\mathbf{x} \in S$, there is a ball B centered at $\mathbf{x}$ that is included in S. If $\mathbf{x} \in S_1$, then every $\mathbf{y} \in B$ is also in S_1, for $\mathbf{y}$ can be joined to $\mathbf{a}$ by first joining $\mathbf{x}$ to $\mathbf{a}$ and then joining $\mathbf{y}$ to $\mathbf{x}$ by the straight line segment from $\mathbf{x}$ to $\mathbf{y}$, which lies in B and hence in S. Similarly, if $\mathbf{x}$ is in the closure of S_1, by Theorem 1.14 there is a sequence $\{\mathbf{x}_k\}$ of points in S_1 that converges to $\mathbf{x}$. We have $\mathbf{x}_k \in B$ for k sufficiently large, so again, $\mathbf{x}$ can be joined to $\mathbf{a}$ by joining $\mathbf{x}_k$ to $\mathbf{a}$ and then joining $\mathbf{x}$ to $\mathbf{x}_k$ by a line segment in B; hence $\mathbf{x} \in S_1$. This completes the proof. $\square$

EXERCISES

1. Show directly from the definition that the following sets are disconnected. (That is, produce a disconnection for each of them.)

   a. The hyperbola $\{(x, y) \in \mathbb{R}^2 : x^2 - y^2 = 1\}$.

   b. Any finite set in $\mathbb{R}^n$ with at least two elements.

   c. $\{(x, y, z) \in \mathbb{R}^3 : xyz > 0\}$.

2. Show that the unit sphere $\{(x, y, z) : x^2 + y^2 + z^2 = 1\}$ in $\mathbb{R}^3$ is arcwise connected. Can you generalize your argument to show that the unit sphere in $\mathbb{R}^n$ is arcwise connected for all $n > 1$?

3. Suppose I is an interval in $\mathbb{R}$ and $f : I \to \mathbb{R}$ is continuous and one-to-one (i.e., $f(x_1) \neq f(x_2)$ unless $x_1 = x_2$). Show that f must be strictly increasing or strictly decreasing on I.

4. Suppose S_1 and S_2 are connected sets in $\mathbb{R}^n$ that contain at least one point in common. Show that $S_1 \cup S_2$ is connected. Is it true that $S_1 \cap S_2$ must be connected?

5. Show that an open set in $\mathbb{R}^n$ is disconnected if and only if it is the union of two disjoint nonempty open subsets.

6. Show that a closed set in $\mathbb{R}^n$ is disconnected if and only if it is the union of two disjoint nonempty closed subsets.

7. Show that $S \subset \mathbb{R}^n$ is disconnected if and only if there is a continuous function $f : S \to \mathbb{R}$ such that $f(S)$ consists of the two points 0 and 1.

8. Show that the closure of a connected set is connected.

9. Let $S = \{\mathbf{x} : |\mathbf{x}| = 1\}$ be the unit sphere in $\mathbb{R}^n$, and let $f : S \to \mathbb{R}$ be a continuous function. Assuming the fact that S is connected (see Exercise 2), show that there must be a pair of diametrically opposite points on S at which f assumes the same value. (*Hint:* Consider $g(\mathbf{x}) = f(\mathbf{x}) - f(-\mathbf{x})$.)

10. Suppose S is a connected set in $\mathbb{R}^2$ that contains $(1, 3)$ and $(4, -1)$. Show that S contains at least one point on the line $x = y$. (*Hint:* Consider $f(x, y) = x - y$.)

11. Let $S \subset \mathbb{R}^2$ be given by (1.29).
   a. Show that S is connected. (*Hint:* The curve $y = \sin(\pi/x)$, $x > 0$, is arcwise connected. Use Exercise 8.)
   b. Show that S is not arcwise connected. (Suppose $\mathbf{f} : [0, 1] \to S$ is continuous and satisfies $\mathbf{f}(0) = (2, 0)$ and $\mathbf{f}(1) = (0, 1)$. Show that the x-coordinate of $\mathbf{f}(t)$ must assume all values between 2 and 0 as t ranges from 0 to 1, and conclude that for each positive integer k there exists $t_k \in [0, 1]$ such that $\mathbf{f}(t_k) = (1/2k, 0)$. By passing to a convergent subsequence, you can suppose that $t_0 = \lim_{k \to \infty} t_k$ exists. Show that the y-coordinate of $\mathbf{f}(t)$ must assume all values between -1 and 1 as t ranges from t_k to t_{k+1}, and derive a contradiction.)

1.8   Uniform Continuity

Suppose S is a subset of $\mathbb{R}^n$. We recall that a function $\mathbf{f} : S \to \mathbb{R}^m$ is said to be continuous on S if, for each $\mathbf{x} \in S$, $\mathbf{f}(\mathbf{y})$ can be made as close as we wish to $\mathbf{f}(\mathbf{x})$ by taking $\mathbf{y}$ sufficiently close to $\mathbf{x}$. In general, the meaning of "sufficiently close" will depend on $\mathbf{x}$: If $\mathbf{f}$ is nearly constant near $\mathbf{x}$, we may be able to move quite a distance away from $\mathbf{x}$ without changing the value of $\mathbf{f}$ much, but if $\mathbf{f}$ is rapidly varying near $\mathbf{x}$, we will need to stay close to $\mathbf{x}$ to ensure that the value of $\mathbf{f}$ remains close to $\mathbf{f}(\mathbf{x})$. For some purposes, however, it is important to have some control over the rate at which $\mathbf{f}(\mathbf{y})$ approaches $\mathbf{f}(\mathbf{x})$ as $\mathbf{y}$ approaches $\mathbf{x}$ that is *independent of* $\mathbf{x}$. Functions for which this is possible are called *uniformly continuous*.

   More precisely, a function $\mathbf{f} : S \to \mathbb{R}^m$ is said to be **uniformly continuous** on S if for every $\epsilon > 0$ there is a $\delta > 0$ so that

$$|\mathbf{f}(\mathbf{x}) - \mathbf{f}(\mathbf{y})| < \epsilon \text{ whenever } \mathbf{x}, \mathbf{y} \in S \text{ and } |\mathbf{x} - \mathbf{y}| < \delta.$$

The crucial point is that for simple continuity the number δ may depend on $\mathbf{x}$, but for uniform continuity it does not. This is a rather subtle point, and the reader should not be discouraged if its significance is not immediately clear; some very eminent mathematicians of the past also had trouble with it!

   Some readers may find it enlightening to see these conditions rewritten in a symbolic way that makes them as concise as possible. We employ the logical symbols $\forall$ and $\exists$, which mean "for all" and "there exists," respectively. With this understanding, the condition for $\mathbf{f}$ to be continuous on S is that

(1.31) $\forall \epsilon > 0 \, \forall \mathbf{x} \in S \, \exists \delta > 0 : \forall \mathbf{y} \in S \quad |\mathbf{x} - \mathbf{y}| < \delta \implies |\mathbf{f}(\mathbf{x}) - \mathbf{f}(\mathbf{y})| < \epsilon,$

whereas the condition for $\mathbf{f}$ to be uniformly continuous on S is that

(1.32) $\forall \epsilon > 0 \, \exists \delta > 0 : \forall \mathbf{x}, \mathbf{y} \in S \quad |\mathbf{x} - \mathbf{y}| < \delta \implies |\mathbf{f}(\mathbf{x}) - \mathbf{f}(\mathbf{y})| < \epsilon.$

The difference between (1.31) and (1.32) is that the "$\forall \mathbf{x}$" has been interchanged with the "$\exists \delta$," so that in (1.31) the δ is allowed to depend on $\mathbf{x}$, whereas in (1.32) the same δ must work for every $\mathbf{x}$.

   EXAMPLE 1. The function $f(x) = \sin x$ is uniformly continuous on $\mathbb{R}$. Indeed, since $|f'(x)| = |\cos x| \le 1$ for all x, the mean value theorem (reviewed in §2.1) shows that $|f(x) - f(y)| \le |x - y|$ for all x, y. Thus, we can take $\delta = \epsilon$, independent of x: If $|x - y| < \epsilon$, then $|f(x) - f(y)| < \epsilon$.

   EXAMPLE 2. The function $g(x) = x^2$ is not uniformly continuous on $\mathbb{R}$, essentially because the slope of the graph at $x = a$ increases without bound as $a \to \infty$. To be more precise, let us suppose that $a > 0$ and $h > 0$. Since $g(a+h)-g(a) = 2ah+h^2 > 2ah$, there is no hope to get $|g(a+h)-g(a)| < \epsilon$ unless $h < \epsilon/2a$. Thus, the allowable δ in (1.31) at $x = a$ must be smaller than $\epsilon/2a$, which gets smaller as a gets larger. On the other hand, g is uniformly continuous on every bounded interval, because on such an interval there is a finite upper bound for $|g'|$, and the mean value theorem can be applied as in Example 1.

Example 2 exemplifies the typical situation, in the following sense. On a set that is not bounded or not closed, things can get worse and worse as one goes off to infinity or to the boundary of the set; but on a *compact* set such pathologies cannot occur.

1.33 Theorem. *Suppose $S \subset \mathbb{R}^n$ and $\mathbf{f} : S \to \mathbb{R}^m$ is continuous at every point of S. If S is compact, then $\mathbf{f}$ is uniformly continuous on S.*

Proof. Suppose $\mathbf{f}$ is not uniformly continuous on S; we shall derive a contradiction. The negation of the uniform continuity condition (1.32) is that

$$\exists \epsilon > 0 \; \forall \delta > 0 \; \exists \mathbf{x}, \mathbf{y} \in S : \; |\mathbf{x} - \mathbf{y}| < \delta \text{ and } |\mathbf{f}(\mathbf{x}) - \mathbf{f}(\mathbf{y})| \geq \epsilon.$$

Taking $\delta = 1, \frac{1}{2}, \frac{1}{3}, \ldots$, we see that for each positive integer k there exist $\mathbf{x}_k, \mathbf{y}_k \in S$ such that $|\mathbf{x}_k - \mathbf{y}_k| < k^{-1}$ and $|\mathbf{f}(\mathbf{x}_k) - \mathbf{f}(\mathbf{y}_k)| \geq \epsilon$. By the Bolzano-Weierstrass theorem, by passing to a subsequence we may assume that $\{\mathbf{x}_k\}$ converges, say to $\mathbf{a} \in S$. Since $|\mathbf{x}_k - \mathbf{y}_k| \to 0$, we also have $\mathbf{y}_k \to \mathbf{a}$. But then $\mathbf{f}(\mathbf{x}_k) - \mathbf{f}(\mathbf{y}_k) \to \mathbf{f}(\mathbf{a}) - \mathbf{f}(\mathbf{a}) = \mathbf{0}$, contradicting the assertion that $|\mathbf{f}(\mathbf{x}_k) - \mathbf{f}(\mathbf{y}_k)| \geq \epsilon$. $\square$

It is remarkable that continuity is the only condition that must be imposed on $\mathbf{f}$ in this theorem. In particular, in contrast to what Examples 1 and 2 might suggest, no conditions on the derivatives of $\mathbf{f}$ enter the picture, even their existence! See Exercise 2.

EXERCISES

1. A function $\mathbf{f} : S \to \mathbb{R}^m$ that satisfies

$$|\mathbf{f}(\mathbf{x}) - \mathbf{f}(\mathbf{y})| \leq C|\mathbf{x} - \mathbf{y}|^\lambda \text{ for all } \mathbf{x}, \mathbf{y} \in S,$$

where C and λ are positive constants, is said to be **Hölder continuous on** S (with exponent λ). Show that if $\mathbf{f}$ is Hölder continuous on S, then $\mathbf{f}$ is uniformly continuous on S.

2. Suppose $0 < \lambda < 1$.
   a. Show that $(a+b)^\lambda < a^\lambda + b^\lambda$ for all $a, b > 0$. (*Hint:* Since $\lambda - 1 < 0$, for $t > 0$ we have $(a+t)^{\lambda-1} < t^{\lambda-1}$. Integrate both sides from 0 to b.)
   b. Let $f_\lambda(x) = |x|^\lambda$. Show that f satisfies the condition in Exercise 1, with $S = \mathbb{R}$ and $C = 1$, and hence conclude that f is uniformly continuous on $\mathbb{R}$. (Note that f is unbounded on $\mathbb{R}$ and that the slope of its graph becomes infinite at the origin.)

3. Suppose that $\mathbf{f} : S \to \mathbb{R}^m$ and $\mathbf{g} : S \to \mathbb{R}^m$ are both uniformly continuous on S. Show that $\mathbf{f} + \mathbf{g}$ is uniformly continuous on S.

4. Show that if $\mathbf{f} : S \to \mathbb{R}^m$ is uniformly continuous on S and $\{\mathbf{x}_k\}$ is a Cauchy sequence in S, then $\{\mathbf{f}(\mathbf{x}_k)\}$ is also a Cauchy sequence. On the other hand, give an example of a Cauchy sequence $\{x_k\}$ in $(0, \infty)$ and a continuous function $f : (0, \infty) \to \mathbb{R}$ (of necessity, *not* uniformly continuous) such that $\{f(x_k)\}$ is not Cauchy.

5. Show that if $\mathbf{f} : S \to \mathbb{R}^m$ is uniformly continuous and S is bounded, then $\mathbf{f}(S)$ is bounded.

Chapter 2

DIFFERENTIAL CALCULUS

The main theme of this chapter is the theory and applications of differential calculus for functions of several variables. The reader is expected to be familiar with differential calculus for functions of one variable. However, we offer a review of the one-variable theory that contains a few features that the reader may not have seen before, and the one-variable theory makes another appearance in the section on Taylor's theorem.

2.1   Differentiability in One Variable

We begin with an approach to the notion of derivative that is a bit different from the one usually found in elementary calculus books. This point of view is very useful in more advanced work, and it is the one that leads to the proper notion of differentiability for functions of several variables.

The basic idea is that a function $f : \mathbb{R} \to \mathbb{R}$ is differentiable at $x = a$ if it is *approximately linear* near $x = a$. Geometrically, this means that the graph of f has a tangent line at $x = a$. Analytically, it means that there is a linear function $l(x) = mx + b$ satisfying the following two conditions:

- $l(a) = f(a)$, so that $b = f(a) - ma$ and hence $l(x) = f(a) + m(x - a)$;

- the difference $f(x) - l(x)$ tends to zero *at a faster rate than* $x - a$ as $x \to a$, that is,

$$\frac{f(x) - l(x)}{x - a} \to 0 \text{ as } x \to a.$$

It will be convenient to denote the increment $x - a$ by h, so that

$$f(x) - l(x) = f(a + h) - f(a) - mh.$$

We think of this difference as a function of h and denote it by $E(h)$; thus $E(h)$ is the error when we approximate $f(a + h)$ by the linear function $f(a) + mh$.

We proceed to the formal definition. Suppose f is a real-valued function defined on some open interval in $\mathbb{R}$ containing the point a. We say that f is **differentiable** at a if there is a number m such that

$$(2.1) \qquad f(a + h) = f(a) + mh + E(h), \quad \text{where} \quad \lim_{h \to 0} \frac{E(h)}{h} = 0;$$

in other words, if $f(a+h)$ is the sum of the linear function $f(a) + mh$ and an error term that tends to zero more rapidly than h as $h \to 0$. In this case we have

$$m = \frac{f(a+h) - f(a) - E(h)}{h} = \frac{f(a+h) - f(a)}{h} - \frac{E(h)}{h}.$$

As $h \to 0$ the last term on the right vanishes, so we see that

$$(2.2) \qquad m = \lim_{h \to 0} \frac{f(a+h) - f(a)}{h}.$$

Thus the number m is uniquely determined, and it is the **derivative** of f at a as usually defined in elementary calculus books, denoted by $f'(a)$. Conversely, if the limit m in (2.2) exists, then (2.1) holds with $E(h) = f(a+h) - f(a) - mh$. Thus, our definition of differentiability is equivalent to the usual one; it simply puts more emphasis on the idea of linear approximation.

Observe that if $E(h)/h$ vanishes as $h \to 0$, then so does $E(h)$ itself and hence so does $f(a + h) - f(a)$. That is, *differentiability at a implies continuity at a.*

It is often convenient to express the relation $\lim_{h \to 0} E(h)/h = 0$ by saying that "$E(h)$ is $o(h)$" (pronounced "little oh of h"), meaning that $E(h)$ is of *smaller order of magnitude* than h. Thus the differentiability of f at $x = a$ can be expressed by saying that $f(a + h)$ is the sum of a linear function of h and an error term that is $o(h)$.

The standard rules for differentiation are easily derived from (2.1). We illustrate the ideas by working out the product rule.

The Product Rule: Suppose f and g are differentiable at $x = a$. Then

$$f(a + h) = f(a) + f'(a)h + E_1(h), \qquad g(a + h) = g(a) + g'(a)h + E_2(h),$$

where $E_1(h)$ and $E_2(h)$ are $o(h)$. Multiplying these equations together yields

$$(2.3) \qquad f(a+h)g(a+h) = f(a)g(a) + \big[f'(a)g(a) + f(a)g'(a)\big]h + E_3(h),$$

where

$$E_3(h) = \left[f(a) + f'(a)h + E_1(h)\right]E_2(h) + E_1(h)\left[g(a) + g'(a)h\right].$$

Clearly $E_3(h)$ is $o(h)$ since $E_1(h)$ and $E_2(h)$ are, so (2.3) is of the form (2.1) with f replaced by fg and $m = f'(a)g(a) + f(a)g'(a)$. In other words, fg is differentiable at a and $(fg)'(a) = f'(a)g(a) + f(a)g'(a)$.

The chain rule can also be derived in this way; we shall do so, in a more general setting, in §2.3.

We can also define "one-sided derivatives" of a function f at a point a. To wit, the **left-hand derivative** $f'_-(a)$ and the **right-hand derivative** $f'_+(a)$ are the one-sided limits

$$(2.4) \qquad f'_\pm(a) = \lim_{h \to 0\pm} \frac{f(a+h) - f(a)}{h}.$$

Clearly f is differentiable at a if and only if its left-hand and right-hand derivatives at a exist and are equal. These notions are particularly useful in two situations: (i) in discussing functions whose graphs have "corners" such as $f(x) = |x|$, which has one-sided derivatives at the origin although it is not differentiable there, and (ii) in discussing functions whose domain is a closed interval $[a, b]$, where the one-sided derivatives $f'_+(a)$ and $f'_-(b)$ may be significant.

The Mean Value Theorem. The definition of the derivative involves passing from the "local" information given by the values of $f(x)$ for x near a to the "infinitesimal" information $f'(a)$, which (intuitively speaking) gives the infinitesimal change in f corresponding to an infinitesimal change in x. To reverse the process and pass from "infinitesimal" information to "local" information — that is, to extract information about f from a knowledge of f' — the principal tool is the mean value theorem, one of the most important theoretical results of elementary calculus. The derivation begins with the following result, which is important in its own right.

2.5 Proposition. *Suppose f is defined on an open interval I and $a \in I$. If f has a local maximum or minimum at the point $a \in I$ and f is differentiable at a, then $f'(a) = 0$.*

Proof. Suppose f has a local minimum at a; the argument at a maximum is similar. In the difference quotient $[f(a+h) - f(a)]/h$, the numerator is ≥ 0 for all h near 0 since $f(a+h) \geq f(a)$, so the quotient has the same sign as h. It follows that the one-sided limits as $h \to 0$ from the left and right must be ≤ 0 and ≥ 0, respectively; since they are both equal to $f'(a)$, the only possibility is that $f'(a) = 0$. $\square$

2.6 Lemma (Rolle's Theorem). *Suppose f is continuous on $[a, b]$ and differentiable on (a, b). If $f(a) = f(b)$, there is at least one point $c \in (a, b)$ such that $f'(c) = 0$.*

Proof. By the extreme value theorem (1.23), f assumes a maximum value and a minimum value on $[a, b]$. If the maximum and minimum each occur at an endpoint, then f is constant on $[a, b]$ since the values at the endpoints are equal, so $f'(x) = 0$ for all $x \in (a, b)$. Otherwise, at least one of them occurs at some interior point $c \in (a, b)$, and then $f'(c) = 0$ by Proposition 2.5. □

2.7 Theorem (Mean Value Theorem I). *Suppose f is continuous on $[a, b]$ and differentiable on (a, b). There is at least one point $c \in (a, b)$ such that*

$$f'(c) = \frac{f(b) - f(a)}{b - a}.$$

Proof. The straight line joining $(a, f(a))$ to $(b, f(b))$ is the graph of the function

$$l(x) = f(a) + \frac{f(b) - f(a)}{b - a}(x - a),$$

and the assertion is that there is a point $c \in (a, b)$ where the slope of the graph $y = f(x)$ is the same as the slope of this line, in other words, where the derivative of the *difference* $g(x) = f(x) - l(x)$ is zero. But f and l have the same values at a and b, so $g(a) = g(b) = 0$, and the conclusion then follows by applying Rolle's theorem to g. □

The mean value theorem is nonconstructive; that is, although it asserts the existence of a certain point $c \in (a, b)$, it gives no clue as to how to find that point. Students often find this perplexing at first, but in fact the whole power of the mean value theorem comes from situations where there is no need to know precisely where c is. In many applications, one has information about the behavior of f' on some interval, and one deduces information about f on that same interval. The following theorem comprises the most important of them.

We say that a function f is **increasing** (resp. **strictly increasing**) on an interval I if $f(a) \leq f(b)$ (resp. $f(a) < f(b)$) whenever $a, b \in I$ and $a < b$; similarly for **decreasing** and **strictly decreasing**.

2.8 Theorem. *Suppose f is differentiable on the open interval I.*
a. *If $|f'(x)| \leq C$ for all $x \in I$, then $|f(b) - f(a)| \leq C|b - a|$ for all $a, b \in I$.*
b. *If $f'(x) = 0$ for all $x \in I$, then f is constant on I.*
c. *If $f'(x) \geq 0$ (resp. $f'(x) > 0$, $f'(x) \leq 0$, or $f'(x) < 0$) for all $x \in I$, then f is increasing (resp. strictly increasing, decreasing, or strictly decreasing) on I.*

Proof. Given $a, b \in I$, we have $f(b) - f(a) = f'(c)(b - a)$ for some $c \in I$. In (a) or (b) we know that $|f'(c)| \leq C$ or $f'(c) = 0$, respectively, and we conclude that $|f(b) - f(a)| \leq C|b - a|$ or $f(b) = f(a)$. In (c), if we know that $f'(c) \geq 0$, we conclude that $f(b) - f(a) \geq 0$ for $b > a$, and similarly for the other cases. $\square$

In case the reader feels that we are belaboring the obvious here, we should point out that the mere differentiability of f at a single point a gives less information about the behavior of f near $x = a$ than we would like. For example, if $f'(a) > 0$, it does *not* follow that f is increasing in some neighborhood of a; see Exercises 3 and 4.

The mean value theorem admits the following important generalization, of which we shall present some applications below.

2.9 Theorem (Mean Value Theorem II). *Suppose that f and g are continuous on $[a, b]$ and differentiable on (a, b), and $g'(x) \neq 0$ for all $x \in (a, b)$. Then there exists $c \in (a, b)$ such that*

$$\frac{f'(c)}{g'(c)} = \frac{f(b) - f(a)}{g(b) - g(a)}.$$

Proof. Let

$$h(x) = [f(b) - f(a)][g(x) - g(a)] - [g(b) - g(a)][f(x) - f(a)].$$

Then h is continuous on $[a, b]$ and differentiable on (a, b), and $h(a) = h(b) = 0$. By Rolle's theorem, there is a point $c \in (a, b)$ such that

$$0 = h'(c) = [f(b) - f(a)]g'(c) - [g(b) - g(a)]f'(c).$$

Since g' is never 0 on (a, b), we have $g'(c) \neq 0$ and also $g(b) - g(a) \neq 0$ (by the mean value theorem, since $g(b) - g(a) = g'(\tilde{c})(b - a)$ for some $\tilde{c} \in (a, b)$). Hence we can divide by both these quantities to obtain the desired result. $\square$

L'Hôpital's Rule. Often one is faced with the evaluation of limits of quotients $f(x)/g(x)$ where f and g both tend to zero or infinity. The collection of related results that go under the name of "l'Hôpital's rule" enable one to evaluate such limits in many cases by examining the quotient of the derivatives, $f'(x)/g'(x)$. The cases involving the indeterminate form $0/0$ can be summarized as follows.

2.10 Theorem (L'Hôpital's Rule I). *Suppose f and g are differentiable functions on (a, b) and*

$$\lim_{x \to a+} f(x) = \lim_{x \to a+} g(x) = 0.$$

If g' never vanishes on (a, b) and the limit

$$\lim_{x \to a+} \frac{f'(x)}{g'(x)} = L$$

exists, then g never vanishes on (a, b) and

$$\lim_{x \to a+} \frac{f(x)}{g(x)} = L.$$

The same result holds for

- *the left-hand limit $\lim_{x \to a-}$, if f and g are differentiable on an interval (d, a),*
- *the two-sided limit $\lim_{x \to a}$, if f and g are differentiable on intervals (d, a) and (a, b), and*
- *the limit $\lim_{x \to \infty}$ or $\lim_{x \to -\infty}$, if f and g are differentiable on an interval (b, ∞) or $(-\infty, b)$.*

Proof. If we (re)define $f(a)$ and $g(a)$ to be 0, then f and g are continuous on the interval $[a, x]$ for $x < b$. By Theorem 2.9, for each $x \in (a, b)$ there exists $c \in (a, x)$ (depending on x) such that

$$\frac{f(x)}{g(x)} = \frac{f(x) - f(a)}{g(x) - g(a)} = \frac{f'(c)}{g'(c)}.$$

Since $c \in (a, x)$, c approaches $a+$ as x does, so

$$\lim_{x \to a+} \frac{f(x)}{g(x)} = \lim_{c \to a+} \frac{f'(c)}{g'(c)} = L.$$

The proof for left-hand limits is similar, and the case of two-sided limits is obtained by combining right-hand and left-hand limits. Finally, for the case $a = \pm\infty$, we set $y = 1/x$ and consider the functions $F(y) = f(1/y)$ and $G(y) = g(1/y)$. Since $F'(y) = -f'(1/y)/y^2$ and $G'(y) = -g'(1/y)/y^2$, we have $F'(y)/G'(y) = f'(1/y)/g'(1/y)$, so by the results just proved,

$$\lim_{x \to \pm\infty} \frac{f(x)}{g(x)} = \lim_{y \to 0\pm} \frac{F(y)}{G(y)} = \lim_{y \to 0\pm} \frac{F'(y)}{G'(y)} = \lim_{x \to \pm\infty} \frac{f'(x)}{g'(x)}.$$

$\square$

Under the conditions of Theorem 2.10, it may well happen that $f'(x)$ and $g'(x)$ tend to zero also, so that the limit of $f'(x)/g'(x)$ cannot be evaluated immediately. In this case we can apply Theorem 2.10 again to evaluate the limit by examining $f''(x)/g''(x)$. More generally, if the functions $f, f', \ldots, f^{(k-1)}, g, g', \ldots, g^{(k-1)}$ all tend to zero as x tends to $a+$ or $a-$ or $\pm\infty$, but $f^{(k)}(x)/g^{(k)}(x) \to L$, then $f(x)/g(x) \to L$.

EXAMPLE 1. Let $f(x) = 2x - \sin 2x$, $g(x) = x^2 \sin x$, $a = 0$. Then f, g, and their first two derivatives vanish at $x = a$, but the third derivatives do not, so

$$\lim_{x \to 0} \frac{2x - \sin 2x}{x^2 \sin x} = \lim_{x \to 0} \frac{2 - 2\cos 2x}{2x \sin x + x^2 \cos x} = \lim_{x \to 0} \frac{4 \sin 2x}{(2 - x^2) \sin x + 4x \cos x}$$

$$= \lim_{x \to 0} \frac{8 \cos 2x}{(6 - x^2) \cos x - 6x \sin x} = \frac{4}{3}.$$

The corresponding result for limits of the form ∞/∞ is also true.

2.11 Theorem (L'Hôpital's Rule II). *Theorem 2.10 remains valid when the hypothesis that* $\lim f(x) = \lim g(x) = 0$ *(as* $x \to a+$, $x \to a-$, *etc.) is replaced by the hypothesis that* $\lim |f(x)| = \lim |g(x)| = \infty$.

Proof. We consider the case of left-hand limits as $x \to a-$; the other cases follow as in Theorem 2.10.

Given $\epsilon > 0$, we wish to show that $\left| [f(x)/g(x)] - L \right| < \epsilon$ provided that x is sufficiently close to a on the left. Since $f'(x)/g'(x) \to L$ and $|g(x)| \to \infty$, we can choose $x_0 < a$ so that

$$\left| \frac{f'(x)}{g'(x)} - L \right| < \frac{\epsilon}{2} \text{ and } g(x) \neq 0 \text{ for } x_0 < x < a.$$

Moreover, by Theorem 2.9, if $x_0 < x < a$ we have

$$\frac{f(x) - f(x_0)}{g(x) - g(x_0)} = \frac{f'(c)}{g'(c)} \text{ for some } c \in (x_0, x),$$

and hence, since $x_0 < c < a$,

$$\left| \frac{f(x) - f(x_0)}{g(x) - g(x_0)} - L \right| < \frac{\epsilon}{2} \text{ for } x_0 < x < a.$$

Next, division of top and bottom by $g(x)$ yields

$$\frac{f(x) - f(x_0)}{g(x) - g(x_0)} = \frac{\dfrac{f(x)}{g(x)} - \dfrac{f(x_0)}{g(x)}}{1 - \dfrac{g(x_0)}{g(x)}}.$$

Since $|g(x)| \to \infty$ as $x \to a$, the quotients $f(x_0)/g(x)$ and $g(x_0)/g(x)$ can be made as close to zero as we please by taking x sufficiently close to a. It follows that for x sufficiently close to a we have

$$\left| \frac{f(x) - f(x_0)}{g(x) - g(x_0)} - \frac{f(x)}{g(x)} \right| < \frac{\epsilon}{2},$$

and hence, by the preceding estimate,

$$\left| \frac{f(x)}{g(x)} - L \right| < \epsilon,$$

which is what we needed to show.   □

The following special cases of Theorem 2.11 are of fundamental importance.

2.12 Corollary. *For any $a > 0$ we have*

$$\lim_{x \to +\infty} \frac{x^a}{e^x} = \lim_{x \to +\infty} \frac{\log x}{x^a} = \lim_{x \to 0+} \frac{\log x}{x^{-a}} = 0.$$

That is, the exponential function e^x grows more rapidly than any power of x as $x \to +\infty$, whereas $|\log x|$ grows more slowly than any positive power of x as $x \to +\infty$ and more slowly than any negative power of x as $x \to 0+$.

Proof. For the first limit, let k be the smallest integer that is $\geq a$. A k-fold application of Theorem 2.11 yields

$$\lim_{x \to +\infty} \frac{x^a}{e^x} = \lim_{x \to +\infty} \frac{a(a-1)\cdots(a-k+1)x^{a-k}}{e^x},$$

and the latter limit is zero because $a - k \leq 0$. For the other two limits, a single application of Theorem 2.11 suffices:

$$\lim_{x \to +\infty} \frac{\log x}{x^a} = \lim_{x \to +\infty} \frac{1}{ax^a} = 0, \qquad \lim_{x \to 0+} \frac{\log x}{x^{-a}} = \lim_{x \to 0+} \frac{x^a}{a} = 0.$$

□

By raising the quantities in Corollary 2.12 to a positive power b and replacing a by a/b, we obtain the more general formulas

$$(2.13) \qquad \lim_{x \to +\infty} \frac{x^a}{e^{bx}} = \lim_{x \to +\infty} \frac{(\log x)^b}{x^a} = \lim_{x \to 0+} \frac{|\log x|^b}{x^{-a}} = 0 \qquad (a, b > 0).$$

Vector-Valued Functions. The differential calculus generalizes easily to functions of a real variable with values in $\mathbb{R}^n$ rather than $\mathbb{R}$. If $\mathbf{f} = (f_1, \ldots, f_n)$ is such a function, its derivative at the point a is defined to be

$$\mathbf{f}'(a) = \lim_{h \to 0} \frac{\mathbf{f}(a+h) - \mathbf{f}(a)}{h}.$$

The jth component of the difference quotient on the right is $h^{-1}[f_j(a+h)-f_j(a)]$. It follows that $\mathbf{f}$ is differentiable if and only if each of its component functions f_j is differentiable, and that differentiation is simply performed componentwise:

$$\mathbf{f}'(a) = \left(f_1'(a), \ldots f_n'(a)\right).$$

The usual rules of differentiation generalize easily to this situation. In particular, there are two forms of the product rule: one for the product of a scalar function φ and a vector function $\mathbf{f}$, and one for the dot product of two vector functions $\mathbf{f}$ and $\mathbf{g}$:

$$(\varphi \mathbf{f})' = \varphi' \mathbf{f} + \varphi \mathbf{f}', \qquad (\mathbf{f} \cdot \mathbf{g})' = \mathbf{f}' \cdot \mathbf{g} + \mathbf{f} \cdot \mathbf{g}'.$$

The first of these is just the ordinary product rule applied to each component φf_j of $\varphi \mathbf{f}$, and the second one is almost as easy (Exercise 8). Similarly, when $n = 3$ we have the product rule for cross products:

$$(\mathbf{f} \times \mathbf{g})' = \mathbf{f}' \times \mathbf{g} + \mathbf{f} \times \mathbf{g}'.$$

(The only point that needs attention here is that the factors $\mathbf{f}$ and $\mathbf{g}$ must be in the same order in all three products.)

The most common geometric interpretation of a function $\mathbf{f} : \mathbb{R} \to \mathbb{R}^n$ ($n > 1$) is as the parametric representation of a curve in $\mathbb{R}^n$. That is, the independent variable t is interpreted as time, and $\mathbf{f}(t)$ is the position of a particle moving in $\mathbb{R}^n$ at time t that traces out a curve as t varies. In this setting, the derivative $\mathbf{f}'(t)$ represents the **velocity** of the particle at time t.

Of particular importance are the straight lines in $\mathbb{R}^n$. If $\mathbf{a}, \mathbf{c} \in \mathbb{R}^n$ and $\mathbf{c} \neq \mathbf{0}$, the line through $\mathbf{a}$ in the direction parallel to the vector $\mathbf{c}$ is represented parametrically by $\mathbf{l}(t) = \mathbf{a} + t\mathbf{c}$. In particular, for the line passing through two points $\mathbf{a}$ and $\mathbf{b}$ we have $\mathbf{c} = \mathbf{b} - \mathbf{a}$, and the line is given by $\mathbf{l}(t) = \mathbf{a} + t(\mathbf{b} - \mathbf{a})$; the line segment from $\mathbf{a}$ to $\mathbf{b}$ is obtained by restricting t to the interval $[0, 1]$.

If $\mathbf{f} : \mathbb{R} \to \mathbb{R}^n$ gives a parametric representation of a curve in $\mathbb{R}^n$ and $\mathbf{f}'(a) \neq \mathbf{0}$, the function $\mathbf{l}(t) = \mathbf{f}(a) + t\mathbf{f}'(a)$ gives a parametric representation of the **tangent line** to the curve at the point $\mathbf{f}(a)$. (If $\mathbf{f}'(a) = \mathbf{0}$, the curve may not have a tangent line at $\mathbf{f}(a)$. For example, if $\mathbf{f}(t) = (t^3, |t|^3)$, then $\mathbf{f}'(0) = (0,0)$, but the curve in question is the graph $y = |x|$.) We shall discuss these matters more thoroughly in Chapter 3.

It should be pointed out that *the mean value theorem is not valid for vector-valued functions.* For example, the function $\mathbf{f}(t) = (\cos t, \sin t)$ satisfies $\mathbf{f}(0) = \mathbf{f}(2\pi)$, but $\mathbf{f}'(t) = (-\sin t, \cos t)$, so there is no point t where $\mathbf{f}'(t) = \mathbf{0}$. However, some of the corollaries of the mean value theorem remain valid. In particular, if $|\mathbf{f}'(t)| \leq M$ for all $t \in [a, b]$, then

$$|\mathbf{f}(b) - \mathbf{f}(a)| \leq M|b - a|.$$

We shall prove this for the more general case of functions of several variables in §2.10.

EXERCISES

1. Suppose that f is differentiable on the interval I and that $f'(x) > 0$ for all $x \in I$ except for finitely many points at which $f'(x) = 0$. Show that f is strictly increasing on I.

2. Define the function f by $f(x) = x^2 \sin(1/x)$ if $x \neq 0$ and $f(0) = 0$. Show that f is differentiable at every $x \in \mathbb{R}$, including $x = 0$, but that f' is discontinuous at $x = 0$. (Calculating $f'(x)$ for $x \neq 0$ is easy; to calculate $f'(0)$ you need to go back to the definition of derivative.)

3. Let f be the function in Exercise 2, and let $g(x) = f(x) + \frac{1}{2}x$. Show that $g'(0) > 0$ but that there is no neighborhood of 0 on which g is increasing. (More precisely, every interval containing 0 has subintervals on which g is decreasing.)

4. Define the function h by $h(x) = x^2$ if x is rational, $h(x) = 0$ if x is irrational. Show that h is differentiable at $x = 0$, even though it is discontinuous at every other point.

5. Suppose that f is continuous on $[a, b]$ and differentiable on (a, b), and that the right-hand limit $L = \lim_{x \to a+} f'(x)$ exists. Show that the right-hand derivative $f'_+(a)$ exists and equals L. (*Hint:* Consider the difference quotients defining $f'_+(a)$ and use the mean value theorem.) Of course, the analogous result for left-hand limits at b also holds.

6. Suppose that f is three times differentiable on an interval containing a. Show that

$$\lim_{h \to 0} \frac{f(a + 2h) - 2f(a + h) + f(a)}{h^2} = f''(a),$$

$$\lim_{h \to 0} \frac{f(a + 3h) - 3f(a + 2h) + 3f(a + h) - f(a)}{h^3} = f^{(3)}(a).$$

Can you find the generalization to higher derivatives?

7. Show that for any $a, b \in \mathbb{R}$, $\lim_{x \to 0}(1 + ax)^{b/x} = e^{ab}$. (*Hint:* Take logarithms.)

8. Suppose $\mathbf{f}$ and $\mathbf{g}$ are differentiable functions on $\mathbb{R}$ with values in $\mathbb{R}^n$.
   a. Show that $(\mathbf{f} \cdot \mathbf{g})' = \mathbf{f}' \cdot \mathbf{g} + \mathbf{f} \cdot \mathbf{g}'$.
   b. Suppose also that $n = 3$, and show that $(\mathbf{f} \times \mathbf{g})' = \mathbf{f}' \times \mathbf{g} + \mathbf{f} \times \mathbf{g}'$.

9. Define the function f by $f(x) = e^{-1/x^2}$ if $x \neq 0$, $f(0) = 0$.

a. Show that $\lim_{x \to 0} f(x)/x^n = 0$ for all $n > 0$. (You'll find that a simple-minded application of Theorem 2.10 doesn't work. Try setting $y = 1/x^2$ instead.)

b. Show that f is differentiable at $x = 0$ and that $f'(0) = 0$.

c. Show by induction on k that for $x \neq 0$, $f^{(k)}(x) = P(1/x)e^{-1/x^2}$, where P is a polynomial of degree $3k$.

d. Show by induction on k that $f^{(k)}(0)$ exists and equals 0 for all k. (Use the results of (a) and (c) to compute the derivative of $f^{(k-1)}$ at $x = 0$ directly from the definition, as in (b).)

The upshot is that f possesses derivatives of all orders at every point and that $f^{(k)}(0) = 0$ for all k.

10. Exercise 2 shows that it is possible for f' to exist at every point of an interval I but to have discontinuities. It is an intriguing fact that when f' exists at *every* point of I, it has the intermediate value property whether or not it is continuous. More precisely:

Darboux's Theorem. Suppose f is differentiable on $[a, b]$. If v is any number between $f'(a)$ and $f'(b)$, there is a point $c \in (a, b)$ such that $f'(c) = v$.

Prove Darboux's theorem, as follows: To simplify the notation, consider the case $a = 0$, $b = 1$. Define $h : [0, 2] \to \mathbb{R}$ by setting $h(0) = f'(0)$,

$$h(x) = \frac{f(x) - f(0)}{x} \text{ if } 0 < x \leq 1, \quad h(x) = \frac{f(1) - f(x - 1)}{2 - x} \text{ if } 1 \leq x < 2,$$

and $h(2) = f'(1)$. Show that h is continuous on $[0, 2]$ and apply the intermediate value theorem to it. (This argument has a simple geometric interpretation, which you can find if you think of $h(x)$ as the slope of the chord joining a certain pair of points on the graph of f.)

2.2   Differentiability in Several Variables

The simplest notion of derivative for a function of several variables is that of *partial derivatives,* which are just the derivatives of the function with respect to each of its variables when the others are held fixed. That is, the **partial derivative** of a function $f(x_1, \ldots, x_n)$ with respect to the variable x_j is

$$\lim_{h \to 0} \frac{f(x_1, \ldots, x_j + h, \ldots, x_n) - f(x_1, \ldots, x_j, \ldots, x_n)}{h},$$

provided that the limit exists.

The most common notations for the partial derivative just defined are

$$\frac{\partial f}{\partial x_j}, \qquad f_{x_j}, \qquad f_j, \qquad \partial_{x_j} f, \qquad \partial_j f.$$

The first one is a modification of the Leibniz notation df/dx for ordinary derivatives with the d replaced by the "curly d" ∂. The second one, with the variable of differentiation indicated merely as a subscript on the function, is often used when the first one seems too cumbersome. The third one is a variation on the second one that is used when one does not want to commit oneself to naming the independent variables but wants to speak of "the partial derivative of f with respect to its jth variable." The notations f_{x_j} and f_j have the disadvantage that they may conflict with other uses of subscripts — for example, denoting an ordered list of functions by $f_1, f_2, f_3, \ldots$. It has therefore become increasingly common in advanced mathematics to use the notations $\partial_{x_j} f$ and $\partial_j f$ instead, which are reasonably compact and at the same time quite unambiguous.

EXAMPLE 1. Let $f(x, y, z) = \dfrac{e^{3x} \sin xy}{1 + 5y - 7z}$. Then

$$\partial_x f = \partial_1 f = \frac{\partial f}{\partial x} = \frac{3e^{3x} \sin xy + e^{3x} y \cos xy}{1 + 5y - 7z},$$

$$\partial_y f = \partial_2 f = \frac{\partial f}{\partial y} = \frac{(1 + 5y - 7z)e^{3x} x \cos xy - 5e^{3x} \sin xy}{(1 + 5y - 7z)^2},$$

$$\partial_z f = \partial_3 f = \frac{\partial f}{\partial z} = \frac{7e^{3x} \sin xy}{(1 + 5y - 7z)^2}.$$

The partial derivatives of a function give information about how the value of the function changes when just one of the independent variables changes; that is, they tell how the function varies along the lines parallel to the coordinate axes. Sometimes this is just what is needed, but often we want something more. We may want to know how the function behaves when several of the variables are changed at once; or we may want to consider a new coordinate system, rotated with respect to the old one, and ask how the function varies along the lines parallel to the new axes. Do the partial derivatives provide such information? Without additional conditions on the function, the answer is *no*.

EXAMPLE 2. Let us take another look at the function in Example 1 of §1.3:

$$(2.14) \qquad f(x, y) = \frac{xy}{x^2 + y^2} \text{ for } (x, y) \neq (0, 0), \quad f(0, 0) = 0.$$

We have already observed that f is discontinuous at the origin; it approaches different limits as (x, y) approaches the origin along different straight lines. However, we have $f(x, 0) = 0$ for all x and $f(0, y) = 0$ for all y, so the partial derivatives $f_x(0, 0)$ and $f_y(0, 0)$ both exist and equal zero:

$$f_x(0,0) = \lim \frac{f(h,0) - f(0,0)}{h} = 0 = \lim \frac{f(0,h) - f(0,0)}{h} = f_y(0,0).$$

Clearly $f_x(0, 0)$ and $f_y(0, 0)$ aren't describing the behavior of f near the origin very well: when either x or y is varied while the other is held fixed at 0, f doesn't change at all, but when both are varied at once, f can change quite drastically!

We need to give more thought to what it should mean for a function of several variables to be differentiable. The right idea is provided by the characterization of differentiability in one variable that we developed in the preceding section. Namely, a function $f(\mathbf{x})$ is differentiable at a point $\mathbf{x} = \mathbf{a}$ if there is a linear function $l(\mathbf{x})$ such that $l(\mathbf{a}) = f(\mathbf{a})$ and the difference $f(\mathbf{x}) - l(\mathbf{x})$ tends to zero faster than $\mathbf{x} - \mathbf{a}$ as $\mathbf{x}$ approaches $\mathbf{a}$. Now, the general linear[1] function of n variables has the form

$$l(\mathbf{x}) = b + c_1 x_1 + \cdots + c_n x_n = b + \mathbf{c} \cdot \mathbf{x},$$

and the condition $l(\mathbf{a}) = f(\mathbf{a})$ forces b to be $f(\mathbf{a}) - \mathbf{c} \cdot \mathbf{a}$, so that $l(\mathbf{x}) = f(\mathbf{a}) + \mathbf{c} \cdot (\mathbf{x} - \mathbf{a})$. With this in mind, here is the formal definition.

A function f defined on an open set $S \subset \mathbb{R}^n$ is called **differentiable** at a point $\mathbf{a} \in S$ if there is a vector $\mathbf{c} \in \mathbb{R}^n$ such that

(2.15) $$\lim_{\mathbf{h} \to 0} \frac{f(\mathbf{a} + \mathbf{h}) - f(\mathbf{a}) - \mathbf{c} \cdot \mathbf{h}}{|\mathbf{h}|} = 0.$$

In this case $\mathbf{c}$ (which is uniquely determined by (2.15), as we shall see shortly) is called the **gradient** of f at $\mathbf{a}$ and is denoted by $\nabla f(\mathbf{a})$. Denoting the numerator of the quotient on the left side of (2.15) by $E(\mathbf{h})$, we observe that (2.15) can be rewritten as

(2.16) $$f(\mathbf{a} + \mathbf{h}) = f(\mathbf{a}) + \nabla f(\mathbf{a}) \cdot \mathbf{h} + E(\mathbf{h}), \text{ where } \frac{E(\mathbf{h})}{|\mathbf{h}|} \to 0 \text{ as } \mathbf{h} \to \mathbf{0},$$

which clearly expresses the fact that $f(\mathbf{a} + \mathbf{h})$, as a function of $\mathbf{h}$, is well approximated by the linear function $f(\mathbf{a}) + \nabla f(\mathbf{a}) \cdot \mathbf{h}$ near $\mathbf{h} = \mathbf{0}$.

---

[1]Unfortunately the term "linear" has two common meanings as applied to functions: "first-degree polynomial" and "satisfying $l(a\mathbf{x} + b\mathbf{y}) = al(\mathbf{x}) + bl(\mathbf{y})$." The first meaning — the one used here — allows a constant term; the second does not. See Appendix A, (A.5).

FIGURE 2.1: A tangent plane to a smooth surface.

What does this mean? First, let us establish the geometric intuition. If $n = 2$, the graph of the equation $z = f(\mathbf{x})$ (with $\mathbf{x} = (x, y)$) represents a surface in 3-space, and the graph of the equation $z = f(\mathbf{a}) + \nabla f(\mathbf{a}) \cdot (\mathbf{x} - \mathbf{a})$ ($\mathbf{x}$ is the variable; $\mathbf{a}$ is fixed) represents a plane. These two objects both pass through the point $(\mathbf{a}, f(\mathbf{a}))$, and at nearby points $\mathbf{x} = \mathbf{a} + \mathbf{h}$ we have

$$z_{\text{surface}} - z_{\text{plane}} = f(\mathbf{a} + \mathbf{h}) - f(\mathbf{a}) - \nabla f(\mathbf{a}) \cdot \mathbf{h}.$$

Condition (2.16) says precisely that this difference tends to zero *faster than* $\mathbf{h}$ as $\mathbf{h} \to \mathbf{0}$. Geometrically, this means that the plane $z = f(\mathbf{a}) + \nabla f(\mathbf{a}) \cdot (\mathbf{x} - \mathbf{a})$ is the **tangent plane** to the surface $z = f(\mathbf{x})$ at $\mathbf{x} = \mathbf{a}$, as indicated in Figure 2.1. The same interpretation is valid in any number of variables, with a little stretch of the imagination: The equation $z = f(\mathbf{x})$ represents a "hypersurface" in $\mathbb{R}^{n+1}$ with coordinates $(x_1, \ldots, x_n, z)$, and the equation $z = f(\mathbf{a}) + \nabla f(\mathbf{a}) \cdot (\mathbf{x} - \mathbf{a})$ represents its "tangent hyperplane" at $\mathbf{a}$.

Next, let us establish the connection with partial derivatives and the uniqueness of the vector $\mathbf{c}$ in (2.15). Suppose f is differentiable at $\mathbf{a}$. If we take the increment $\mathbf{h}$ in (2.16) to be of the form $\mathbf{h} = (h, 0, \ldots, 0)$ with $h \in \mathbb{R}$, we have $\mathbf{c} \cdot \mathbf{h} = c_1 h$ and $|\mathbf{h}| = \pm h$ (depending on the sign of h). Thus (2.16) says (after multiplying through by -1 if h is negative) that

$$\lim_{h \to 0} \frac{f(a_1 + h, a_2, \ldots, a_n) - f(a_1, \ldots, a_n)}{h} - c_1 = 0,$$

or in other words, that $c_1 = \partial_1 f(\mathbf{a})$. Likewise, $c_j = \partial_j f(\mathbf{a})$ for $j = 2, \ldots, n$. To summarize:

2.17 Theorem. *If f is differentiable at $\mathbf{a}$, then the partial derivatives $\partial_j f(\mathbf{a})$ all exist, and they are the components of the vector $\nabla f(\mathbf{a})$.*

We also have the following:

2.18 Theorem. *If f is differentiable at $\mathbf{a}$, then f is continuous at $\mathbf{a}$.*

Proof. Multiplying (2.15) through by $|\mathbf{h}|$, we see that $f(\mathbf{a}+\mathbf{h})-f(\mathbf{a})-\nabla f(\mathbf{a})\cdot\mathbf{h} \rightarrow 0$ as $\mathbf{h} \rightarrow \mathbf{0}$. Since $\nabla f(\mathbf{a})\cdot\mathbf{h}$ clearly vanishes as $\mathbf{h}$ does, we have $f(\mathbf{a}+\mathbf{h})-f(\mathbf{a}) \rightarrow 0$ as $\mathbf{h} \rightarrow \mathbf{0}$, which says precisely that f is continuous at $\mathbf{a}$. $\quad\square$

The converses of Theorems 2.17 and 2.18 are false. The continuity of f does not imply the differentiability of f even in dimension $n = 1$ (think of functions like $f(x) = |x|$ whose graphs have corners). When $n > 1$, the mere existence of the partial derivatives of f does not imply the differentiability of f either. The example (2.14) demonstrates this: Its partial derivatives exist, but it is not continuous at the origin, so it cannot be differentiable there.

To restate what we have just shown: For a function f to be differentiable at $\mathbf{a}$ it is necessary for the partial derivatives $\partial_j f(\mathbf{a})$ to exist, but not sufficient. How, then, do we know when a function is differentiable? Fortunately, there is a simple condition, not too much stronger than the existence of the partial derivatives, that guarantees differentiability.

2.19 Theorem. *Let f be a function defined on an open set in $\mathbb{R}^n$ that contains the point $\mathbf{a}$. Suppose that the partial derivatives $\partial_j f$ all exist on some neighborhood of $\mathbf{a}$ and that they are continuous at $\mathbf{a}$. Then f is differentiable at $\mathbf{a}$.*

Proof. Let's consider the case $n = 2$, to keep the notation simple. We wish to show that

$$(2.20) \quad \frac{f(\mathbf{a} + \mathbf{h}) - f(\mathbf{a}) - \mathbf{c} \cdot \mathbf{h}}{|\mathbf{h}|} \rightarrow 0 \text{ as } \mathbf{h} \rightarrow \mathbf{0}, \text{ where } \mathbf{c} = \big(\partial_1 f(\mathbf{a}), \partial_2 f(\mathbf{a})\big).$$

To do this, we shall analyze the increment $f(\mathbf{a} + \mathbf{h}) - f(\mathbf{a})$ by making the change one variable at a time:

$$(2.21) \quad f(\mathbf{a} + \mathbf{h}) - f(\mathbf{a}) = \big[f(a_1 + h_1,\, a_2 + h_2) - f(a_1,\, a_2 + h_2)\big]$$
$$+ \big[f(a_1,\, a_2 + h_2) - f(a_1, a_2)\big].$$

We assume that $\mathbf{h}$ is small enough so that the partial derivatives $\partial_j f(\mathbf{x})$ exist whenever $|\mathbf{x} - \mathbf{a}| \leq |\mathbf{h}|$. In this case, we can use the one-variable mean value theorem to express the differences on the right side of (2.21) in terms of the partial derivatives of f at suitable points. If we set $g(t) = f(t,\, a_2 + h_2)$, we have

$$f(a_1 + h_1,\, a_2 + h_2) - f(a_1,\, a_2 + h_2) = g(a_1 + h_1) - g(a_1)$$
$$= g'(a_1 + c_1)h_1 = \partial_1 f(a_1 + c_1,\, a_2 + h_2)h_1$$

for some number c_1 lying between 0 and h_1. Similarly,

$$f(a_1,\, a_2 + h_2) - f(a_1, a_2) = \partial_2 f(a_1,\, a_2 + c_2)h_2$$

for some c_2 between 0 and h_2. Substituting these results back into (2.21) and then into the left side of (2.20), we obtain

$$\frac{f(\mathbf{a}+\mathbf{h}) - f(\mathbf{a}) - \mathbf{c} \cdot \mathbf{h}}{|\mathbf{h}|} = [\partial_1 f(a_1 + c_1,\, a_2 + h_2) - \partial_1 f(a_1, a_2)]\frac{h_1}{|\mathbf{h}|}$$

$$+ [\partial_2 f(a_1,\, a_2 + c_2) - \partial_2 f(a_1, a_2)]\frac{h_2}{|\mathbf{h}|}.$$

Now let $\mathbf{h} \to \mathbf{0}$. The expressions in brackets tend to 0 because the partial derivatives $\partial_j f$ are continuous at $\mathbf{a}$, and the ratios $h_1/|\mathbf{h}|$ and $h_2/|\mathbf{h}|$ are bounded by 1 in absolute value. Thus (2.20) is valid and f is differentiable at $\mathbf{a}$.

The idea for general n is exactly the same. We write $f(\mathbf{a}+\mathbf{h}) - f(\mathbf{a})$ as the sum of n increments, each of which involves a change in only one variable — for example, the first of them is

$$f(a_1 + h_1,\, a_2 + h_2, \ldots, a_n + h_n) - f(a_1,\, a_2 + h_2, \ldots, a_n + h_n)$$

— and then use the mean value theorem to express each difference in terms of a partial derivative of f and proceed as before. $\square$

A function f whose partial derivatives $\partial_j f$ all exist and are continuous on an open set S is said to be of **class C^1** on S. For short, we shall also say that "f is C^1 on S" or "$f \in C^1(S)$" and refer to "a C^1 function f." Theorems 2.17 and 2.19 then say that

$$C^1 \Longrightarrow \text{differentiable} \Longrightarrow \text{partial derivatives exist.}$$

The reverse implications are false. We already know that existence of partial derivatives does not imply differentiability, and there are differentiable functions whose derivatives are discontinuous. The standard example in one variable is the function in Exercise 2, §2.1, and it is easy to generate higher-dimensional examples from this one.

For most of the elementary functions that we shall work with, the continuity of the partial derivatives is obvious by inspection, so verifying the differentiability of a function is usually no problem. For example, for $(x, y) \neq (0,0)$ the partial derivatives of our old friend (2.14) are

$$\partial_x f(x, y) = \frac{y^3 - x^2 y}{(x^2 + y^2)^2}, \qquad \partial_y f(x, y) = \frac{x^3 - xy^2}{(x^2 + y^2)^2},$$

which are continuous everywhere except at the origin (but *not* at the origin). Thus f is differentiable at every point except the origin.

We conclude this section by examining a few ramifications of the notion of differentiability.

Differentials. Suppose f is differentiable at $\mathbf{a}$, so that

$$f(\mathbf{a} + \mathbf{h}) - f(\mathbf{a}) = \nabla f(\mathbf{a}) \cdot \mathbf{h} + \text{error},$$

where the error term is negligibly small in comparison with $\mathbf{h}$. If we neglect the error term, the resulting approximation to the increment $f(\mathbf{a} + \mathbf{h}) - f(\mathbf{a})$ is called the **differential** of f at $\mathbf{a}$ and is denoted by $df(\mathbf{a}; \mathbf{h})$ or $df_{\mathbf{a}}(\mathbf{h})$:

(2.22) $df(\mathbf{a}; \mathbf{h}) = df_{\mathbf{a}}(\mathbf{h}) = \nabla f(\mathbf{a}) \cdot \mathbf{h} = \partial_1 f(\mathbf{a}) h_1 + \cdots + \partial_n f(\mathbf{a}) h_n.$

If we set $f(\mathbf{x}) = u$ and $\mathbf{h} = d\mathbf{x} = (dx_1, \ldots, dx_n)$, this formula can be written informally as

$$du = \frac{\partial f}{\partial x_1} dx_1 + \frac{\partial f}{\partial x_2} dx_2 + \cdots + \frac{\partial f}{\partial x_n} dx_n.$$

We can think of this in two ways. Intuitively, if we think of $dx_1, \ldots, dx_n$ as infinitesimal increments in the independent variables $x_1, \ldots, x_n$, then du is the corresponding infinitesimal increment in the dependent variable u. Or, if we think of $dx_1, \ldots, dx_n$ as honest, finite increments, du is the corresponding increment in the u value, not on the (hyper)surface $u = f(\mathbf{x})$, but on its tangent (hyper)plane: It is the *linear approximation* to the increment in the function f.

Differentials obey the usual elementary rules of differentiation, such as the sum, product, and quotient rules:

$$d(f + g) = df + dg, \qquad d(fg) = f \, dg + g \, df, \qquad d\left(\frac{f}{g}\right) = \frac{g \, df - f \, dg}{g^2}.$$

This follows from (2.22) and the fact that the partial derivatives obey these rules. We'll see later how differentials interact with the chain rule.

Differentials are handy for approximating small changes in a function. Here's an example:

EXAMPLE 3. A right circular cone has height 5 and base radius 3. (a) About how much does the volume increase if the height is increased to 5.02 and the radius is increased to 3.01? (b) If the height is increased to 5.02, by about how much should the radius be decreased to keep the volume constant?

Solution. The volume of a cone is given by $V = \frac{1}{3}\pi r^2 h$, so $dV = \frac{2}{3}\pi r h \, dr + \frac{1}{3}\pi r^2 \, dh$. (a) If $r = 3$, $h = 5$, $dr = .01$, and $dh = .02$, we have $dV = \frac{2}{3}\pi(3)(5)(.01) + \frac{1}{3}\pi(3^2)(.02) = .16\pi \approx .50$. (b) If $r = 3$, $h = 5$, $dh = .02$, as in (a) we have $dV = 10\pi \, dr + .06\pi$, so $dV = 0$ if $dr = -.006$.

Directional Derivatives. The partial derivatives $\partial_j f$ give information about how $f(\mathbf{x})$ varies as $\mathbf{x}$ moves along lines parallel to the coordinate axes. Sometimes we wish to study the variation of f along oblique lines instead. Thus, given a unit vector $\mathbf{u}$ and a base point $\mathbf{a}$, we consider the line passing through $\mathbf{a}$ in the direction $\mathbf{u}$, which can be represented parametrically by $\mathbf{g}(t) = \mathbf{a} + t\mathbf{u}$. The **directional derivative** of f at $\mathbf{a}$ in the direction $\mathbf{u}$ is defined to be

$$\partial_{\mathbf{u}} f(\mathbf{a}) = \frac{d}{dt} f(\mathbf{a} + t\mathbf{u})\Big|_{t=0} = \lim_{t \to 0} \frac{f(\mathbf{a} + t\mathbf{u}) - f(\mathbf{a})}{t},$$

provided that the limit exists. For example, if $\mathbf{u}$ is the unit vector in the positive jth coordinate direction (that is, $\mathbf{u} = (0, \ldots, 1, \ldots, 0)$ with the 1 in the jth place), then $\partial_{\mathbf{u}} f$ is just the partial derivative $\partial_j f$.

2.23 Theorem. *If f is differentiable at $\mathbf{a}$, then the directional derivatives of f at $\mathbf{a}$ all exist, and they are given by*

$$(2.24) \qquad\qquad\qquad \partial_{\mathbf{u}} f(\mathbf{a}) = \nabla f(\mathbf{a}) \cdot \mathbf{u}.$$

Proof. Differentiability of f means that

$$(2.25) \qquad\qquad \frac{f(\mathbf{a} + \mathbf{h}) - f(\mathbf{a}) - \nabla f(\mathbf{a}) \cdot \mathbf{h}}{|\mathbf{h}|} \to 0 \text{ as } \mathbf{h} \to \mathbf{0}.$$

We take $\mathbf{h} = t\mathbf{u}$. If $t > 0$, then $|\mathbf{h}| = t$ and the expression on the left of (2.25) is

$$\frac{f(\mathbf{a} + t\mathbf{u}) - f(\mathbf{a})}{t} - \nabla f(\mathbf{a}) \cdot \mathbf{u}.$$

If $t < 0$, then $|\mathbf{h}| = -t$ and the expression on the left of (2.25) is

$$-\frac{f(\mathbf{a} + t\mathbf{u}) - f(\mathbf{a})}{t} + \nabla f(\mathbf{a}) \cdot \mathbf{u}.$$

In either case, this quantity tends to 0 as $t \to 0$, which means that $\partial_{\mathbf{u}} f(\mathbf{a})$ exists and equals $\nabla f(\mathbf{a}) \cdot \mathbf{u}$. $\qquad\qquad\qquad\qquad\qquad\qquad\qquad\qquad \square$

It is possible for all the directional derivatives of f to exist even if f is not differentiable, but in that case they cannot be computed from the partial derivatives by the simple formula (2.24); see Exercise 7.

Consideration of directional derivatives leads to a geometric interpretation of the gradient vector $\nabla f(\mathbf{a})$ when this vector is nonzero. Indeed, by (2.24) and Cauchy's inequality, we have $|\partial_{\mathbf{u}} f(\mathbf{a})| \leq |\nabla f(\mathbf{a})|$ for every unit vector $\mathbf{u}$, and the extreme case $\partial_{\mathbf{u}} f(\mathbf{a}) = |\nabla f(\mathbf{a})|$ occurs when $\mathbf{u}$ is the unit vector in the same direction as $\nabla f(\mathbf{a})$. Thus, $\nabla f(\mathbf{a})$ is the vector whose magnitude is the largest directional derivative of f at $\mathbf{a}$, and whose direction is the direction of that derivative. In other words, $\nabla f(\mathbf{a})$ *points in the direction of steepest increase of f at* $\mathbf{a}$*, and its magnitude is the rate of increase of f in that direction.*

EXAMPLE 4. Let $f(x, y) = x^2 + 5xy^2$, $\mathbf{a} = (-2, 1)$. (a) Find the directional derivative of f at $\mathbf{a}$ direction of the vector $\mathbf{v} = (12, 5)$. (b) What is the largest of the directional derivatives of f at $\mathbf{a}$, and in what direction does it occur?

Solution. We have $\nabla f(x, y) = (2x + 5y^2, 10xy)$, so that $\nabla f(-2, 1) = (1, -20)$. The unit vector in the direction of $\mathbf{v}$ is $\mathbf{u} = (\frac{12}{13}, \frac{5}{13})$, so the directional derivative in this direction is $\nabla f(\mathbf{a}) \cdot \mathbf{u} = (1, -20) \cdot (\frac{12}{13}, \frac{5}{13}) = -\frac{88}{13}$. The largest directional derivative at $\mathbf{a}$ is $|\nabla f(\mathbf{a})| = \sqrt{401}$, and it occurs in the direction $\frac{1}{\sqrt{401}}(1, -20)$.

EXERCISES

1. For each of the following functions f, (i) compute ∇f, (ii) find the directional derivative of f at the point $(1, -2)$ in the direction $(\frac{3}{5}, \frac{4}{5})$.
    a. $f(x, y) = x^2 y + \sin \pi x y$.
    b. $f(x, y) = e^{4x - y^2}$.
    c. $f(x, y) = (x + 2y + 4)/(7x + 3y)$.

2. For each of the following functions f, (i) compute the differential df, (ii) use the differential to estimate the difference $f(1.1, 1.2, -0.1) - f(1, 1, 0)$.
    a. $f(x, y, z) = x^2 e^{x-y+3z}$.
    b. $f(x, y, z) = y^3 + \log(x + z^2)$.

3. Let $w = f(x, y, z) = \dfrac{x^2 y^{3/2} z}{z + 1}$. Suppose that, at the outset, $(x, y, z) = (5, 4, 1)$, so that $w = 100$. Use differentials to answer the following questions.
    a. Suppose we change x to 5.03 and y to 3.92. By (about) how much should we change z in order to keep $w = 100$?
    b. Suppose we want to increase the value of w a little bit by changing the value of only one of the independent variables. Which variable should we choose to get the biggest increase in w for the smallest change of the independent variable?

4. Show that $u = f(x, y, z) = xe^{2z} + y^{-1}e^{5z}$ satisfies the differential equation $x\partial_x u + 2y\partial_y u + \partial_z u = 3u$.

5. Show that $u = f(x, y) = xy/(xy - y + 2x)$ satisfies the differential equation $x^2 \partial_x u + y^2 \partial_y u = u^2$.

6. For $j = 1, \ldots, n$, define the function f_j on $\mathbb{R}^n \setminus \{\mathbf{0}\}$ by $f_j(\mathbf{x}) = x_j/|\mathbf{x}|$. Show that $\sum_1^n x_j \, df_j \equiv 0$.

7. Let $f(x, y) = \dfrac{x^2 y}{x^2 + y^2}$ if $(x, y) \neq (0, 0)$ and $f(0, 0) = 0$.

   a. Show that f is continuous at $(0, 0)$. (*Hint:* Since $0 \leq (x \pm y)^2 = x^2 + y^2 \pm 2xy$, we have $|xy| \leq \frac{1}{2}(x^2 + y^2)$ for all x, y.)

   b. Show that the directional derivatives $\partial_{\mathbf{u}} f(0, 0)$ all exist, and compute them. (Work directly with the definition of directional derivative. The best way to write a unit vector in $\mathbb{R}^2$ is as $\mathbf{u} = (\cos \theta, \sin \theta)$.)

   c. Show that f is not differentiable at $(0, 0)$. (*Hint:* If it were, the directional derivatives $\partial_{\mathbf{u}} f(0, 0)$ would be related to the partial derivatives $\partial_x f(0, 0)$ and $\partial_y f(0, 0)$ by (2.24).)

8. Suppose f is a function defined on an open set $S \subset \mathbb{R}^n$. Show that if the partial derivatives $\partial_j f$ exist and are bounded on S, then f is continuous on S. (Exercise 7 provides an example of a function that satisfies these conditions on $S = \mathbb{R}^2$ but is not everywhere differentiable.)

2.3   The Chain Rule

There are several different but closely related versions of the chain rule for functions of several variables. The most basic one concerns the situation where we have a function $f(x_1, \ldots, x_n)$ and the variables $x_1, \ldots, x_n$ are themselves functions of a single real variable t. To be precise, suppose $x_j = g_j(t)$, or $\mathbf{x} = \mathbf{g}(t)$; we then have the composite function $\varphi(t) = f(\mathbf{g}(t))$.

We recall that the derivative $\mathbf{g}'(t)$ is defined componentwise:

$$\mathbf{g}'(t) = \big(g_1'(t), \ldots, g_n'(t)\big).$$

Geometrically speaking, the equation $\mathbf{x} = \mathbf{g}(t)$ represents a parametrized curve in $\mathbb{R}^n$; we may think of a particle moving in $\mathbb{R}^n$ whose position at time t is $\mathbf{g}(t)$. In this case the vector $\mathbf{g}'(t)$ is the velocity of the particle at time t; it is tangent to the curve at $\mathbf{g}(t)$, and its magnitude is the speed at which the particle is traveling along the curve.

2.26 Theorem (Chain Rule I). *Suppose that $\mathbf{g}(t)$ is differentiable at $t = a$, $f(\mathbf{x})$ is differentiable at $\mathbf{x} = \mathbf{b}$, and $\mathbf{b} = \mathbf{g}(a)$. Then the composite function $\varphi(t) = f(\mathbf{g}(t))$ is differentiable at $t = a$, and its derivative is given by*

$$\varphi'(a) = \nabla f(\mathbf{b}) \cdot \mathbf{g}'(a),$$

or, in Leibniz notation, with $w = f(\mathbf{x})$,

$$(2.27) \qquad \frac{dw}{dt} = \frac{\partial w}{\partial x_1} \frac{dx_1}{dt} + \cdots + \frac{\partial w}{\partial x_n} \frac{dx_n}{dt}.$$

Proof. Differentiability of f and $\mathbf{g}$ at the appropriate points means that

$$f(\mathbf{b} + \mathbf{h}) = f(\mathbf{b}) + \nabla f(\mathbf{b}) \cdot \mathbf{h} + E_1(\mathbf{h}), \qquad E_1(\mathbf{h})/|\mathbf{h}| \to 0 \text{ as } \mathbf{h} \to \mathbf{0};$$
$$\mathbf{g}(a + u) = \mathbf{g}(a) + u\mathbf{g}'(a) + \mathbf{E}_2(u), \qquad |\mathbf{E}_2(u)|/u \to 0 \text{ as } u \to 0.$$

In the first equation we take $\mathbf{h} = \mathbf{g}(a + u) - \mathbf{g}(a)$. By the second equation, we also have $\mathbf{h} = u\mathbf{g}'(a) + \mathbf{E}_2(u)$, and we are given that $\mathbf{g}(a) = \mathbf{b}$, so

$$\begin{aligned}
\varphi(a + u) &= f(\mathbf{g}(a + u)) = f(\mathbf{b} + \mathbf{h}) = f(\mathbf{b}) + \nabla f(\mathbf{b}) \cdot \mathbf{h} + E_1(\mathbf{h}) \\
&= f(\mathbf{g}(a)) + \nabla f(\mathbf{b}) \cdot [u\mathbf{g}'(a) + \mathbf{E}_2(u)] + E_1(\mathbf{h}) \\
&= \varphi(a) + u\nabla f(\mathbf{b}) \cdot \mathbf{g}'(a) + E_3(u),
\end{aligned}$$

where

$$E_3(u) = \nabla f(\mathbf{b}) \cdot \mathbf{E}_2(u) + E_1(\mathbf{h}).$$

We claim that the error term $E_3(u)$ satisfies $E_3(u)/u \to 0$ as $u \to 0$. Granted this, we have

$$\frac{\varphi(a + u) - \varphi(a)}{u} = \nabla f(\mathbf{b}) \cdot \mathbf{g}'(a) + \frac{E_3(u)}{u} \to \nabla f(\mathbf{b}) \cdot \mathbf{g}'(a) \text{ as } u \to 0,$$

so that $\varphi'(a) = \nabla f(\mathbf{b}) \cdot \mathbf{g}'(a)$ as claimed.

Showing that $E_3(u)/u \to 0$ is just a matter of sorting out the mess a little. The fact that $|\mathbf{E}_2(u)|/u \to 0$ takes care of the first term in $E_3(u)$, by Cauchy's inequality:

$$\frac{|\nabla f(\mathbf{b}) \cdot \mathbf{E}_2(u)|}{|u|} \leq |\nabla f(\mathbf{b})| \left| \frac{\mathbf{E}_2(u)}{u} \right| \to 0.$$

It also implies that when u is small we have $|\mathbf{E}_2(u)| \leq |u|$ and hence

$$|\mathbf{h}| = |u\mathbf{g}'(a) + \mathbf{E}_2(u)| \leq (|\mathbf{g}'(a)| + 1)|u|.$$

Now the second term in $E_3(u)$, namely $E_1(\mathbf{h})$, becomes negligibly small in comparison to $|\mathbf{h}|$ as $|\mathbf{h}| \to 0$, and the estimate above shows that $|\mathbf{h}|$ in turn is bounded by a constant times $|u|$, so $E_1(\mathbf{h})$ becomes negligibly small in comparison to $|u|$ as $u \to 0$, which means that $E_1(\mathbf{h})/u \to 0$ as desired. $\qquad \square$

EXAMPLE 1. Suppose $w = f(x, y, z)$ is a differentiable function of (x, y, z), and that $x = t^4 - t$, $y = \sin 3t$, and $z = e^{-2t}$. Then w can be regarded as a composite function of t, and we have

$$\frac{dw}{dt} = \frac{d}{dt} f(t^4 - t, \ \sin 3t, \ e^{-2t})$$
$$= (\partial_1 f) \cdot (4t^3 - 1) + (\partial_2 f) \cdot (3 \cos 3t) + (\partial_3 f) \cdot (-2e^{-2t}),$$

where the partial derivatives $\partial_j f$ are all evaluated at $(t^4 - t, \ \sin 3t, \ e^{-2t})$.

Suppose now that the variables $x_1, \ldots, x_n$ are differentiable functions, not of a single real variable t, but of a family of variables $\mathbf{t} = (t_1, \ldots, t_m)$; say, $x_j = g_j(t_1, \ldots, t_m)$, or $\mathbf{x} = \mathbf{g}(\mathbf{t})$. If f is a differentiable function of $\mathbf{x}$, we then have the composite function $\varphi(\mathbf{t}) = f(\mathbf{g}(\mathbf{t}))$. The chain rule, as stated above, can be used to compute the *partial* derivatives of φ with respect to the variables t_k. Indeed, we simply fix all but one of those variables and apply the chain rule to the resulting function of the remaining single variable to obtain

(2.28) $$\frac{\partial \varphi}{\partial t_k}(\mathbf{a}) = \nabla f(\mathbf{b}) \cdot \frac{\partial \mathbf{g}}{\partial t_k}(\mathbf{a}) \qquad (\mathbf{b} = \mathbf{g}(\mathbf{a})),$$

or, setting $w = f(\mathbf{x})$,

$$\frac{\partial w}{\partial t_k} = \frac{\partial w}{\partial x_1} \frac{\partial x_1}{\partial t_k} + \cdots + \frac{\partial w}{\partial x_n} \frac{\partial x_n}{\partial t_k}.$$

To be precise, this calculation shows that *if the partial derivatives $\partial \mathbf{g}/\partial t_k$ exist at $\mathbf{t} = \mathbf{a}$ and if f is differentiable at $\mathbf{x} = \mathbf{b} = \mathbf{g}(\mathbf{a})$, then the partial derivatives $\partial \varphi/\partial t_k$ exist at $\mathbf{t} = \mathbf{a}$ and are given by* (2.28). It also shows that *if $\mathbf{g}$ is of class C^1 near $\mathbf{a}$ and f is of class C^1 near $\mathbf{b} = \mathbf{g}(\mathbf{a})$, then φ is of class C^1, and in particular is differentiable, near $\mathbf{a}$.* Indeed, under these hypotheses, (2.28) shows that the partial derivatives $\partial \varphi/\partial t_k$ are continuous.

It is also natural to ask whether the composite function $f \circ \mathbf{g}$ is differentiable when f and $\mathbf{g}$ are only assumed to be differentiable rather than C^1. The answer is affirmative. When $\mathbf{t}$ is only a single real variable, this result is contained in the chain rule as stated and proved above. The proof for the general case, $\mathbf{t} = (t_1, \ldots, t_m)$, is almost identical except that the notation is a little messier, and we shall not take the trouble to write it out. But we shall give a formal statement of the result:

2.29 Theorem (Chain Rule II). *Suppose that $g_1, \ldots, g_n$ are functions of $\mathbf{t} = (t_1, \ldots, t_m)$ and f is a function of $\mathbf{x} = (x_1, \ldots, x_n)$. Let $\mathbf{b} = \mathbf{g}(\mathbf{a})$ and $\varphi = f \circ \mathbf{g}$. If $g_1, \ldots, g_n$ are differentiable at $\mathbf{a}$ (resp. of class C^1 near $\mathbf{a}$) and f is differentiable* *at* **b** *(resp. of class C^1 near* **b***), then φ is differentiable at* **a** *(resp. of class C^1 near* **a***), and its partial derivatives are given by*

$$(2.30) \qquad \frac{\partial \varphi}{\partial t_k} = \frac{\partial f}{\partial x_1} \frac{\partial x_1}{\partial t_k} + \cdots + \frac{\partial f}{\partial x_n} \frac{\partial x_n}{\partial t_k},$$

where the derivatives $\partial f / \partial x_j$ are evaluated at **b** *and the derivatives $\partial \varphi / \partial t_k$ and $\partial x_j / \partial t_k = \partial g_j / \partial t_k$ are evaluated at* **a***.*

EXAMPLE 2. Suppose that f is a differentiable function of x and y and that $x = s \log(1 + t^2)$ and $y = \cos(s^3 + 5t)$. Then the partial derivatives of the composite function $z = f(s \log(1 + t^2), \cos(s^3 + 5t))$ are given by

$$\frac{\partial z}{\partial s} = \frac{\partial f}{\partial x} \frac{\partial x}{\partial s} + \frac{\partial f}{\partial y} \frac{\partial y}{\partial s} = f_x \cdot \log(1 + t^2) + f_y \cdot (-3s^2) \sin(s^3 + 5t),$$

$$\frac{\partial z}{\partial t} = \frac{\partial f}{\partial x} \frac{\partial x}{\partial t} + \frac{\partial f}{\partial y} \frac{\partial y}{\partial t} = f_x \frac{2st}{1 + t^2} + f_y \cdot (-5) \sin(s^3 + 5t).$$

Here, the partial derivatives of f are to be evaluated at $(s \log(1 + t^2), \cos(s^3 + 5t))$.

The chain rule (2.30) has a neat interpretation in terms of differentials. Let $w = f(\mathbf{x})$. If we regard $x_1, \ldots, x_n$ as independent variables, we have

$$(2.31) \qquad dw = \frac{\partial w}{\partial x_1} dx_1 + \cdots + \frac{\partial w}{\partial x_n} dx_n.$$

On the other hand, if we regard $x_1, \ldots, x_n$ as functions of the variables $t_1, \ldots, t_m$ and w as the composite function $f(\mathbf{x}(\mathbf{t}))$, we have

$$(2.32) \qquad dx_j = \frac{\partial x_j}{\partial t_1} dt_1 + \cdots + \frac{\partial x_j}{\partial t_m} dt_m$$

and

$$(2.33) \qquad dw = \frac{\partial w}{\partial t_1} dt_1 + \cdots + \frac{\partial w}{\partial t_m} dt_m.$$

If we substitute the expressions (2.32) for dx_j into (2.31) and regroup the terms, we obtain

$$dw = \frac{\partial w}{\partial x_1} \left[\frac{\partial x_1}{\partial t_1} dt_1 + \cdots + \frac{\partial x_1}{\partial t_m} dt_m \right] + \cdots + \frac{\partial w}{\partial x_n} \left[\frac{\partial x_n}{\partial t_1} dt_1 + \cdots + \frac{\partial x_n}{\partial t_m} dt_m \right]$$

$$= \left[\frac{\partial w}{\partial x_1} \frac{\partial x_1}{\partial t_1} + \cdots + \frac{\partial w}{\partial x_n} \frac{\partial x_n}{\partial t_1} \right] dt_1 + \cdots + \left[\frac{\partial w}{\partial x_1} \frac{\partial x_1}{\partial t_m} + \cdots + \frac{\partial w}{\partial x_n} \frac{\partial x_n}{\partial t_m} \right] dt_m.$$

The content of the chain rule (2.30) is precisely that this last expression for dw coincides with (2.33). In other words, the differential formalism has the chain rule "built in," just as it does in one variable (where the chain rule $dw/dt = (dw/dx)(dx/dt)$ is just a matter of "canceling the dx's").

The preceding discussion concerns the situation where the variable w depends on a set of variables x_j, and the x_j's depend on a different set of variables t_k. However, in many situations the variables on different "levels" can get mixed up with each other. The typical example is as follows. Consider a physical quantity $w = f(x, y, z, t)$ whose value depends on the position (x, y, z) and the time t (temperature, for example, or air pressure in a region of the atmosphere). Consider also a vehicle moving through space, so that its coordinates (x, y, z) are functions of t. We wish to know how the quantity w varies in time, as measured by an observer on the vehicle; that is, we are interested in the behavior of the composite function

$$w = f\big(x(t), y(t), z(t), t\big).$$

Here t enters not only as a "first-level" variable, as the last argument of f, but also as a "second-level" variable through the t-dependence of x, y, z.

How should this be handled? There is no real problem; the only final independent variable is t, so the chain rule in the form (2.27) can be applied:

$$(2.34) \qquad \frac{dw}{dt} = \frac{\partial w}{\partial x}\frac{dx}{dt} + \frac{\partial w}{\partial y}\frac{dy}{dt} + \frac{\partial w}{\partial z}\frac{dz}{dt} + \frac{\partial w}{\partial t}.$$

In the last term we have omitted the derivative dt/dt, which of course equals 1. (If this makes you nervous, denote the fourth variable in f by u instead of t; then we are considering $w = f(x(t), y(t), z(t), u(t))$ where $u(t) = t$.)

Notice the subtle use of notation: The dw/dt on the left of (2.34) denotes the "total derivative" of w, taking into account all the ways in which w depends on t, whereas the $\partial w/\partial t$ on the right denotes the partial derivative that involves only the explicit dependence of the function f on its fourth variable t. This notation works well enough in this situation, but it becomes inadequate if there is more than one final independent variable.

Suppose, for example, that we are studying a function $w = f(x, y, t, s)$, and that x and y are themselves functions of the independent variables t and s. Then the analogue of (2.34) would be

$$\frac{\partial w}{\partial t} = \frac{\partial w}{\partial x}\frac{\partial x}{\partial t} + \frac{\partial w}{\partial y}\frac{\partial y}{\partial t} + \frac{\partial w}{\partial t},$$

but this is nonsense! The $\partial w/\partial t$'s on the left and on the right denote different things. In such a situation we must use one of the alternative notations for partial

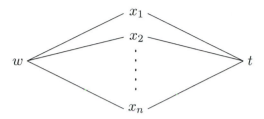

FIGURE 2.2: Diagram of dependence for the basic chain rule.

derivatives that offer more precision, or perhaps add some subscripts to the $\partial w/\partial t$'s to specify their meaning. In this case, if $x = \varphi(t, s)$ and $y = \psi(t, s)$, we could write

(2.35) $$\frac{\partial w}{\partial t} = (\partial_1 f)(\partial_1 \varphi) + (\partial_2 f)(\partial_1 \psi) + \partial_3 f.$$

The mixture of dependent-and-independent-variable notation on the left and functional notation on the right in (2.35) is perhaps inelegant, but it does the job! In general, it is best not to be too doctrinaire about deciding to use one notation for partial derivatives rather than another one; clarity is more important than consistency. We shall be quite free about adopting whichever notation works best in a particular situation, and the exercises aim at encouraging the reader to do likewise.

When the relations among the variables become too complicated for comfort, we can often sort things out by drawing a schematic diagram of the functional relationships. The idea is as follows:

i. Write down the dependent variable on the left of the page, a list of the independent variables on which it ultimately depends on the right, and lists of the intermediate variables in the middle.

ii. Whenever one variable p depends directly on another one q, draw a line joining them; this line represents the partial derivative $\partial p/\partial q$.

iii. To find the derivative of the variable w on the left with respect to one of the variables t on the right, consider all the ways you can go from w to t by following the lines. For each such path, write down the product of partial derivatives corresponding to the lines along the path, then add the results.

The diagram for the basic chain rule (2.27) is shown in Figure 2.2: The path from w to x_j to t gives the term $(\partial w/\partial x_j)(dx_j/dt)$ in (2.27). On the other hand, Figure 2.3 gives the diagram for $w = f(x, y, t, s)$ where x and y depend on t and s: There are three paths from w to t (w to x to t, w to y to t, and w to t directly) that give the three terms on the right of (2.35).

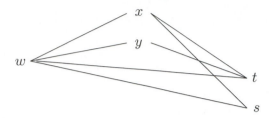

FIGURE 2.3: Diagram of dependence for $w = f(x,y,t,s)$, $x = \varphi(t,s)$, $y = \psi(t,s)$.

Here is another useful corollary of the chain rule. A function f on $\mathbb{R}^n$ is called (positively) **homogeneous** of degree a ($a \in \mathbb{R}$) if $f(t\mathbf{x}) = t^a f(\mathbf{x})$ for all $t > 0$ and $\mathbf{x} \neq \mathbf{0}$.

2.36 Theorem (Euler's Theorem). *If f is homogeneous of degree a, then at any point $\mathbf{x}$ where f is differentiable we have*

$$x_1 \partial_1 f(\mathbf{x}) + x_2 \partial_2 f(\mathbf{x}) + \cdots + x_n \partial_n f(\mathbf{x}) = af(\mathbf{x}).$$

Proof. Consider the function $\varphi(t) = f(t\mathbf{x})$. On the one hand, since $f(t\mathbf{x}) = t^a f(\mathbf{x})$, we have $\varphi'(t) = at^{a-1} f(\mathbf{x}) = at^{-1} f(t\mathbf{x})$. On the other, by the chain rule we have

$$\varphi'(t) = \nabla f(t\mathbf{x}) \cdot \frac{d}{dt}(t\mathbf{x}) = \mathbf{x} \cdot \nabla f(t\mathbf{x}).$$

Setting $t = 1$ and equating the two expressions for $\varphi'(1)$, we obtain the asserted result. $\square$

We conclude this section with an additional geometric insight into the meaning of the gradient of a function. If F is a differentiable function of $(x,y,z) \in \mathbb{R}^3$, the locus of the equation $F(x,y,z) = 0$ is typically a smooth two-dimensional surface in S in $\mathbb{R}^3$. (We shall consider this matter more systematically in Chapter 3.) Suppose that $(x,y,z) = \mathbf{g}(t)$ is a parametric represention of a smooth curve on S. On the one hand, by the chain rule we have $(d/dt)F(\mathbf{g}(t)) = \nabla F(\mathbf{g}(t)) \cdot \mathbf{g}'(t)$. On the other hand, since the curve lies on S, we have $F(\mathbf{g}(t)) = 0$ for all t and hence $(d/dt)F(\mathbf{g}(t)) = 0$. Thus, for any curve on the S, the gradient of F is orthogonal to the tangent vector to the curve at each point on the curve. Since such curves can go in any direction on the surface, we conclude that at any point $\mathbf{a} \in S$, $\nabla F(\mathbf{a})$ is orthogonal to every vector that is tangent to S at $\mathbf{a}$. (Of course, this is interesting only if $\nabla F(\mathbf{a}) \neq \mathbf{0}$.) We summarize:

2.37 Theorem. *Suppose that F is a differentiable function on some open set $U \subset \mathbb{R}^3$, and suppose that the set*

$$S = \big\{(x, y, z) \in U : F(x, y, z) = 0\big\}$$

is a smooth surface. If $\mathbf{a} \in S$ and $\nabla F(\mathbf{a}) \neq \mathbf{0}$, then the vector $\nabla F(\mathbf{a})$ is perpendicular, or normal, to the surface S at $\mathbf{a}$.

2.38 Corollary. *Under the conditions of the theorem, the equation of the tangent plane to S at $\mathbf{a}$ is $\nabla F(\mathbf{a}) \cdot (\mathbf{x} - \mathbf{a}) = 0$.*

This formula for the tangent plane to a surface agrees with the one we gave in §2.2 when the surface is the graph of a function $f(x, y)$. The easy verification is left to the reader (Exercise 5).

A similar result holds if we have two equations $F(x, y, z) = 0$ and $G(x, y, z) = 0$. Each of them (usually) represents a surface, and the intersection of the two surfaces is (usually) a curve. At any point $\mathbf{a}$ on this curve, the vectors $\nabla F(\mathbf{a})$ and $\nabla G(\mathbf{a})$ are both perpendicular to the curve, and if they are linearly independent, they span the normal plane to the curve at $\mathbf{a}$.

These ideas carry over into dimensions other than 3. For $n = 2$, an equation $F(x, y) = 0$ typically represents a curve C, and $\nabla F(a, b)$ is normal to C at each $(a, b) \in C$. For $n > 3$, we simply stretch our imagination to say that $\nabla F(\mathbf{a})$ is normal to the hypersurface defined by $F(\mathbf{x}) = 0$ at $\mathbf{x} = \mathbf{a}$.

EXERCISES

In these exercises, all functions in question are assumed to be differentiable.

1. Find the indicated derivatives of w in terms of the derivatives of f, g, h.
   a. $w = f(x, y, t)$, $x = g(y, t)$, $y = h(t)$. What is dw/dt?
   b. $w = f(x, u, v)$, $u = g(x, y)$, $v = h(x, z)$. What are $\partial_x w$, $\partial_y w$, $\partial_z w$? ($\partial_x w$ refers to the *complete* dependence of w on x, as opposed to $\partial_1 f$.)
   c. $w = f(u)$, $u = g(x, y)$, $y = h(x)$. What is dw/dx?

2. Find $\partial_x w$ and $\partial_y w$ in terms of the partial derivatives $\partial_1 f$, $\partial_2 f$, and $\partial_3 f$.
   a. $w = f(2x - y^2,\ x \sin 3y,\ x^4)$.
   b. $w = f(e^{x-3y},\ \log(x^2 + 1),\ \sqrt{y^4 + 4})$.
   c. $w = \arctan[f(y^2,\ 2x - y,\ -4)]$.

3. Show that the given function u satisfies the given differential equation.
   a. $u = f(3x + 2y)$; $2\partial_x u - 3\partial_y u = 0$.
   b. $u = xy + x f(y/x)$; $x\partial_x u + y\partial_y u - u = xy$.

c. $u = f(xz, yz)$; $x\partial_x u + y\partial_y u = z\partial_z u$.

4. Let $u = f(r)$ and $r = |\mathbf{x}| = (x_1^2 + \cdots + x_n^2)^{1/2}$. Show that $\sum_1^n (\partial u/\partial x_j)^2 = [f'(r)]^2$.

5. Show that the formula for the tangent plane to the surface $z = f(x, y)$ given in §2.2 coincides with the formula for the tangent plane to the surface $F(x, y, z) = 0$ given in this section, when $F(x, y, z) = f(x, y) - z$.

6. Find the tangent plane to the surface in $\mathbb{R}^3$ described by the given equation at the given point $\mathbf{a} \in \mathbb{R}^3$.
   a. $z = x^2 - y^3$, $\mathbf{a} = (2, -1, 5)$.
   b. $x^2 + 2y^2 + 3z^2 = 6$, $\mathbf{a} = (1, 1, -1)$.
   c. $z = \sqrt{x} + \arctan y$, $\mathbf{a} = (9, 0, 3)$.
   d. $xyz^2 - \log(z - 1) = 8$, $\mathbf{a} = (-2, -1, 2)$.

7. Suppose $\varphi(x)$ is defined by a formula in which x occurs in several places. (For example, there are three x's in $\varphi(x) = x^2 e^x/(x + 3)$.) Show that the derivative $\varphi'(x)$ is obtained by differentiating with respect to each of the x's in turn, treating the others as constants, and adding the results. (*Hint:* If x occurs in n places in the formula for φ, let $F(x_1, \ldots, x_n)$ be the function of n variables obtained by replacing each of the x's in the formula by a different variable. How do you express φ in terms of F?) Notice that the rules for differentiating sums and products are special cases of this result, obtained by taking $\varphi(x) = f(x) + g(x)$ or $\varphi(x) = f(x)g(x)$. What is the derivative of $\varphi(x) = f(x)^{g(x)}$?

2.4   The Mean Value Theorem

The mean value theorem for functions of n variables can be stated as follows. We recall that if $\mathbf{a}$ and $\mathbf{b}$ are two points in $\mathbb{R}^n$, the line passing through them can be described parametrically by $\mathbf{g}(t) = \mathbf{a} + t(\mathbf{b} - \mathbf{a})$. In particular, the line segment whose endpoints are $\mathbf{a}$ and $\mathbf{b}$ is the set of points $\mathbf{a} + t(\mathbf{b} - \mathbf{a})$ with $0 \leq t \leq 1$.

2.39 Theorem (Mean Value Theorem III). *Let S be a region in $\mathbb{R}^n$ that contains the points $\mathbf{a}$ and $\mathbf{b}$ as well as the line segment L that joins them. Suppose that f is a function defined on S that is continuous at each point of L and differentiable at each point of L except perhaps the endpoints $\mathbf{a}$ and $\mathbf{b}$. Then there is a point $\mathbf{c}$ on L such that*

$$f(\mathbf{b}) - f(\mathbf{a}) = \nabla f(\mathbf{c}) \cdot (\mathbf{b} - \mathbf{a}).$$

Proof. Let $\mathbf{h} = \mathbf{b} - \mathbf{a}$; then $L = \{\mathbf{a} + t\mathbf{h} : 0 \leq t \leq 1\}$. Define $\varphi(t) = f(\mathbf{a} + t\mathbf{h})$ for $0 \leq t \leq 1$. Since f is continuous on L, φ is continuous on $[0, 1]$. Moreover, by the chain rule, φ is differentiable on $(0,1)$ and

$$\varphi'(t) = \nabla f(\mathbf{a} + t\mathbf{h}) \cdot \frac{d}{dt}(\mathbf{a} + t\mathbf{h}) = \nabla f(\mathbf{a} + t\mathbf{h}) \cdot \mathbf{h} = \nabla f(\mathbf{a} + t\mathbf{h}) \cdot (\mathbf{b} - \mathbf{a}).$$

By the one-variable mean value theorem, there is a point $u \in (0,1)$ such that $\varphi(1) - \varphi(0) = \varphi'(u) \cdot (1 - 0) = \varphi'(u)$. Let $\mathbf{c} = \mathbf{a} + u\mathbf{h}$; then

$$f(\mathbf{b}) - f(\mathbf{a}) = \varphi(1) - \varphi(0) = \varphi'(u) = \nabla f(\mathbf{c}) \cdot (\mathbf{b} - \mathbf{a}).$$

$\square$

To state the principal corollaries of the mean value theorem, we need a definition. A set $S \subset \mathbb{R}^n$ is called **convex** if whenever $\mathbf{a}, \mathbf{b} \in S$, the line segment from $\mathbf{a}$ to $\mathbf{b}$ also lies in S. Clearly every convex set is arcwise connected (line segments are arcs!), but most connected sets are not convex. See Figure 2.4.

EXAMPLE 1. Every ball is convex. Indeed, let $B = \{\mathbf{x} : |\mathbf{x} - \mathbf{c}| < r\}$ be the ball of radius r about $\mathbf{c}$. If $\mathbf{a}, \mathbf{b} \in B$, for $0 \leq t \leq 1$ we have

$$\big|[\mathbf{a} + t(\mathbf{b} - \mathbf{a})] - \mathbf{c}\big| = \big|(1 - t)(\mathbf{a} - \mathbf{c}) + t(\mathbf{b} - \mathbf{c})\big|$$
$$\leq (1 - t)|\mathbf{a} - \mathbf{c}| + t|\mathbf{b} - \mathbf{c}| < (1 - t)r + tr = r,$$

so $\mathbf{a} + t(\mathbf{b} - \mathbf{a}) \in B$. (We have used the fact that t and $1 - t$ are both nonnegative when $0 \leq t \leq 1$.)

2.40 Corollary. *Suppose that f is differentiable on an open convex set S and $|\nabla f(\mathbf{x})| \leq M$ for every $\mathbf{x} \in S$. Then $|f(\mathbf{b}) - f(\mathbf{a})| \leq M|\mathbf{b} - \mathbf{a}|$ for all $\mathbf{a}, \mathbf{b} \in S$.*

Proof. The line segment from $\mathbf{a}$ to $\mathbf{b}$ lies in S, and for some $\mathbf{c}$ on this segment we have $f(\mathbf{b}) - f(\mathbf{a}) = \nabla f(\mathbf{c}) \cdot (\mathbf{b} - \mathbf{a})$. Hence, by Cauchy's inequality, $|f(\mathbf{b}) - f(\mathbf{a})| \leq |\nabla f(\mathbf{c})| \, |\mathbf{b} - \mathbf{a}| \leq M|\mathbf{b} - \mathbf{a}|$. $\square$

2.41 Corollary. *Suppose f is differentiable on an open convex set S and $\nabla f(\mathbf{x}) = 0$ for all $\mathbf{x} \in S$. Then f is constant on S.*

Proof. Pick $\mathbf{a} \in S$ and take $M = 0$ in Corollary 2.40. We conclude that for every $\mathbf{b} \in S$, $|f(\mathbf{b}) - f(\mathbf{a})| = 0$, that is, $f(\mathbf{b}) = f(\mathbf{a})$. $\square$

The hypothesis of convexity is essential in Corollary 2.40. In a situation like that of the set S_2 in Figure 2.4, $|\mathbf{b} - \mathbf{a}|$ is small, but $f(\mathbf{b}) - f(\mathbf{a})$ could be quite large even when $|\nabla f|$ is small in S_2. (Think of a gently sloping spiral ramp.) However, Corollary 2.41 can be generalized substantially.

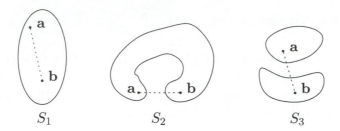

FIGURE 2.4: A convex set (S_1), a set that is connected but not convex (S_2), and a disconnected set (S_3).

2.42 Theorem. *Suppose that f is differentiable on an open connected set S and $\nabla f(\mathbf{x}) = \mathbf{0}$ for all $\mathbf{x} \in S$. Then f is constant on S.*

Proof. Pick $\mathbf{a} \in S$, and define $S_1 = \{\mathbf{x} \in S : f(\mathbf{x}) = f(\mathbf{a})\}$ and $S_2 = \{\mathbf{x} \in S : f(\mathbf{x}) \neq f(\mathbf{a})\}$. We shall show that S_2 must be empty, and hence that f is constant on $S = S_1$, by showing that otherwise (S_1, S_2) would be a disconnection of S.

Clearly S_1 and S_2 are disjoint and their union is S; moreover, $\mathbf{a} \in S_1$. The set S_2 is open (by Theorem 1.13) because the complement of the point $f(\mathbf{a})$ is an open subset of $\mathbb{R}$. The set S_1 is also open, for the following reason. Suppose $\mathbf{x} \in S_1$. Since S is open, there is a ball B centered at $\mathbf{x}$ that is contained in S. Since B is convex, f is constant on B by Corollary 2.41, and hence $B \subset S_1$. That is, every $\mathbf{x} \in S_1$ is an interior point of S_1, so S_1 is open. Since both S_1 and S_2 are open, neither one can intersect the closure of the other one without intersecting the other one itself. But clearly S_1 and S_2 are disjoint, their union is S, and S_1 is nonempty since it contains $\mathbf{a}$. Therefore, (S_1, S_2) is a disconnection of S unless S_2 is empty. $\square$

The hypothesis of connectedness is necessary here. If $S = S' \cup S''$ where S' and S'' are open and disjoint, we obtain a counterexample by taking $f(\mathbf{x}) = 0$ for $\mathbf{x} \in S'$ and $f(\mathbf{x}) = 1$ for $\mathbf{x} \in S''$. (See Figure 2.4. Differentiability of a function f on the set S_3 there affords no control over the relation between the values of f at $\mathbf{a}$ and $\mathbf{b}$.)

EXERCISES

1. State and prove two analogues of Rolle's theorem for functions of several variables, whose hypotheses are, respectively, the following:

   a. f is differentiable on a set containing the line segment from $\mathbf{a}$ to $\mathbf{b}$, and $f(\mathbf{a}) = f(\mathbf{b})$.

   b. f is differentiable on a bounded open set S, continuous on the closure of S, and constant on the boundary of S.

2. Question: If f is differentiable on a connected open set S and $\partial_1 f(\mathbf{x}) = 0$ for all $\mathbf{x} \in S$, must f be independent of x_1 on S (that is, $f(\mathbf{a}) = f(\mathbf{b})$ whenever $\mathbf{a}, \mathbf{b} \in S$ and $a_j = b_j$ for all $j \neq 1$)?

   a. Show that the answer is *yes* when S is convex.

   b. Give a counterexample to show that the answer is *no* in general. (*Hint:* Think of a staircase where you go halfway up on one flight, make a $180°$ turn on a flat landing, then go the rest of the way up on a second flight parallel to the first one.)

2.5 Functional Relations and Implicit Functions: A First Look

Often we are presented with an equation $F(x_1, \ldots, x_n) = 0$ relating a collection of variables $x_1, \ldots, x_n$. (There is no harm in taking the right side to be 0; just move everything over to the left side of the equation.) It may be possible to solve this equation for one of the variables in terms of the remaining ones, say $x_n = g(x_1, \ldots, x_{n-1})$, and we wish to study the resulting function g in terms of the original function F.

To make things clearer, let us change the notation a little, replacing n by $n + 1$ and denoting the last variable x_n by y; thus, the given equation has the form $F(x_1, \ldots, x_n, y) = 0$, and it is supposed to determine y as a function of $\mathbf{x} = (x_1, \ldots, x_n)$.

Let us be clear about what we mean by saying that "it is possible to solve for y." First, we mean that it is possible to solve *in principle,* not necessarily that there is an explicit formula for y. Second, there might be more than one solution, and obtaining y as a function of the x_j's then involves making a definite choice among the solutions; moreover, the domain of this function may be smaller than one would suspect from the original equation.

   EXAMPLE 1.

   a. Consider the equation $x - y - y^5 = 0$. It's easy to solve this for x in terms of y, $x = y + y^5$, but there is no nice algebraic formula for y in terms of x. However, $y + y^5$ is a strictly increasing function of y (its derivative is $1 + 5y^4$, which is positive everywhere), and its values clearly range from $-\infty$ to ∞, so there for each x there is exactly one y satisfying $x = y + y^5$, and we can call it $g(x)$. The object in such a situation is to use the equation $x = y + y^5$ to study the function g.

b. The equation $x^2 + y^2 + z^2 = 1$ can be solved for z as a continuous function of x and y in two ways, $z = \sqrt{1 - x^2 - y^2}$ and $z = -\sqrt{1 - x^2 - y^2}$, both of which are defined only for $x^2 + y^2 \le 1$.

At this stage we are not going to going to worry about these matters, or about the question of when it is possible to solve the equation at all; such questions will be addressed in Chapter 3. Rather, we shall *assume* that there is a differentiable function $g(x_1, \ldots, x_n)$, defined for $x_1, \ldots, x_n$ in some region $S \subset \mathbb{R}^n$, so that the equation $F(x_1, \ldots, x_n, y) = 0$ is satisfied identically when $g(x_1, \ldots, x_n)$ is substituted for y:

$$(2.43) \qquad F\big(x_1, \ldots, x_n, g(x_1, \ldots, x_n)\big) \equiv 0, \qquad (x_1, \ldots, x_n) \in S.$$

In this situation we can use the chain rule to compute the partial derivatives of g in terms of the partial derivatives of F, simply by differentiating the equation (2.43) with respect to the variables x_j:

$$(2.44) \qquad \partial_j F + \partial_{n+1} F \frac{\partial g}{\partial x_j} = 0, \quad \text{so} \quad \frac{\partial g}{\partial x_j} = -\frac{\partial_j F}{\partial_{n+1} F}.$$

EXAMPLE 1 (continued).

a. Differentiation of the equation $x - y - y^5 = 0$ with respect to x yields $1 - (dy/dx) - 5y^4(dy/dx) = 0$, or $(dy/dx) = 1/(1 + 5y^4)$. Of course, this gives dy/dx in terms of y instead of x, and we don't have a formula for y in terms of x, but this is better than nothing!

b. Differentiation of $x^2 + y^2 + z^2 = 1$ with respect to x, with z as as the dependent variable, gives $2x + 2z(\partial z/\partial x) = 0$, or $\partial z/\partial x = -x/z$. It is easily verified that this formula is correct whether we take $z = \sqrt{1 - x^2 - y^2}$ or $z = -\sqrt{1 - x^2 - y^2}$.

In a related situation, we may wish to differentiate a function $\varphi(x_1, \ldots, x_n, y)$ where the variables $x_1, \ldots, x_n, y$ satisfy a relation $F(x_1, \ldots, x_n, y) = 0$. Assuming, as before, that the equation $F(x_1, \ldots, x_n, y) = 0$ can be solved for y, say $y = g(x_1, \ldots, x_n)$, it then becomes a matter of applying the chain rule to the composite function

$$w = \varphi\big(x_1, \ldots, x_n, g(x_1, \ldots, x_n)\big),$$

to obtain

$$\frac{\partial w}{\partial x_j} = \partial_j \varphi + (\partial_{n+1} \varphi)(\partial_j g).$$

The derivatives $\partial_j g$ can then be evaluated by using (2.44).

In such a situation, however, there is a tricky point that must be confronted. Let us explain it in the case of three variables for simplicity. That is, suppose we are given $w = \varphi(x, y, z)$ where x, y, z are constrained to satisfy $F(x, y, z) = 0$, and suppose we can solve the latter equation for any one of the three variables in terms of the other two. If we take x as an independent variable, *the meaning of $\partial w / \partial x$ depends critically on whether we take y or z as the other independent variable.*

EXAMPLE 2. Let $w = x^2 + y^2 + z$, and suppose x, y, z are constrained to satisfy $x + y + z = 0$. If we take x and y as independent variables, then $z = -(x + y)$, so

$$w = x^2 + y^2 - x - y, \qquad \frac{\partial w}{\partial x} = 2x - 1.$$

But if we take x and z as independent variables, then $y = -(x + z)$, and

$$w = x^2 + (x + z)^2 + z = 2x^2 + 2xz + z^2 + z, \qquad \frac{\partial w}{\partial x} = 4x + 2z.$$

Clearly, these two formulas for $\partial w / \partial x$ almost never agree.

The usual way to clarify this situation is to put subscripts on the partial derivatives to indicate which variables are being held fixed:

$$\frac{\partial w}{\partial x}\bigg|_y = \text{derivative of } w \text{ with respect to } x \text{ when } y \text{ is fixed.}$$

Thus, in Example 2,

$$\frac{\partial w}{\partial x}\bigg|_y = 2x - 1, \qquad \frac{\partial w}{\partial x}\bigg|_z = 4x + 2z.$$

The preceding ideas work in much the same way when we are given more than one constraint equation. For example, if we are given two equations $F(x, y, u, v) = 0$ and $G(x, y, u, v) = 0$, we may be able to solve them for the two variables u and v in terms of the other two variables x and y. In this case the partial derivatives of u and v with respect to x, say, can be calculated by differentiating the equations $F = 0$ and $G = 0$, obtaining

$$\partial_x F + \partial_u F \frac{\partial u}{\partial x} + \partial_v F \frac{\partial v}{\partial x} = 0,$$

$$\partial_x G + \partial_u G \frac{\partial u}{\partial x} + \partial_v G \frac{\partial v}{\partial x} = 0,$$

and then solving these (linear!) equations simultaneously for $\partial u/\partial x$ and $\partial v/\partial x$. By Cramer's rule (Appendix A, (A.54)), the result is

$$\frac{\partial u}{\partial x} = -\frac{\det \begin{pmatrix} \partial_x F & \partial_v F \\ \partial_x G & \partial_v G \end{pmatrix}}{\det \begin{pmatrix} \partial_u F & \partial_v F \\ \partial_u G & \partial_v G \end{pmatrix}}, \qquad \frac{\partial v}{\partial x} = -\frac{\det \begin{pmatrix} \partial_u F & \partial_x F \\ \partial_u G & \partial_x G \end{pmatrix}}{\det \begin{pmatrix} \partial_u F & \partial_v F \\ \partial_u G & \partial_v G \end{pmatrix}}.$$

EXAMPLE 3. Suppose the quantities x, y, and z are initially equal to 1, 0, and 2, respectively, and are constrained to satisfy the equations $x^5 + x(y^3 + 1)z - 2yz^5 = 3$ and $yz = \sin(2x + y - z)$. By about how much do y and z change if x is changed to 1.02?

Solution. We need to find dy/dx and dz/dx, which we abbreviate as y' and z'. Differentiating the two equations with respect to x, treating y and z as implicit functions of x, we obtain

$$5x^4 + (y^3 + 1)z + 3xy^2 zy' + x(y^3 + 1)z' - 2z^5 y' - 10yz^4 z' = 0,$$
$$zy' + yz' = \cos(2x + y - z) \cdot (2 + y' - z').$$

We could solve these equations for y' and z' as they stand, but since we are interested in the answer at $(x, y, z) = (1, 0, 2)$, we can simplify matters by substituting in these values right now. The first equation reduces to $7 + z' - 64y' = 0$ and the second one to $2y' = 2 + y' - z'$, or

$$64y' - z' = 7, \quad y' + z' = 2 \qquad \big(\text{when } (x, y, z) = (1, 0, 2)\big).$$

Solving these equations yields $y' = \frac{9}{65}$ and $z' = \frac{121}{65}$, so — returning to the original question — $dy = y'\, dx = \frac{9}{65}(.02) = \frac{9}{3250}$ and $dz = z'\, dx = \frac{121}{65}(.02) = \frac{121}{3250}$.

EXERCISES

1. Compute $\partial z/\partial x$ and $\partial z/\partial y$ when z is determined as a function of y and x by the following equations:
   a. $x + y^2 + z^3 = 3xyz.$
   b. $2x^2 + 3y^2 + z^2 = e^{-z}.$

2. Suppose y and z are determined as functions of x by the equations $z = x^2 - y^2$ and $z = 2x + 4y$. Find dy/dx and dz/dx (a) by solving the equations explicitly for y and z; (b) by implicit differentiation.

3. Compute dy/dt and dz/dt when y and z are determined as functions of t by the equations $y^5 + e^{yz} + zt^2 = 1$ and $y^2 + z^4 = t^2$.

4. If $u = x^2 + 3y^2$ and $y = xz$, there are two possible meanings for $\partial u/\partial x$ depending on whether the independent variables are taken as (x, y) or (x, z). Compute both of them.

5. Let $V = \pi r^2 h$ and $S = 2\pi r(r + h)$ (the volume and surface area of a circular cylinder). Compute

$$\left.\frac{\partial V}{\partial h}\right|_r, \quad \left.\frac{\partial V}{\partial h}\right|_S, \quad \left.\frac{\partial V}{\partial S}\right|_r, \quad \left.\frac{\partial S}{\partial V}\right|_r,$$

where the subscript indicates the variable that is being held fixed.

6. Suppose that $F(x, y, z) = 0$ is an equation that can be solved to yield any of the three variables as a function of the other two. Show that

$$\frac{\partial x}{\partial y}\frac{\partial y}{\partial z}\frac{\partial z}{\partial x} = -1,$$

provided that the symbols are interpreted properly. (Part of the problem is to say what the proper interpretation is.)

7. Suppose that the variables E, T, V, and P are related by a pair of equations, $f(E, T, V, P) = 0$ and $g(E, T, V, P) = 0$, that can be solved for any two of the variables in terms of the other two, and suppose that the differential equation $\partial_V E - T\partial_T P + P = 0$ is satisfied when V and T are taken as the independent variables. Show that $\partial_P E + T\partial_T V + P\partial_P V = 0$ when P and T are taken as the independent variables. (This example comes from thermodynamics, where E, T, V, and P represent energy, temperature, volume, and pressure.)

2.6 Higher-Order Partial Derivatives

If f is a differentiable function on an open set $S \subset \mathbb{R}^n$, its partial derivatives $\partial_j f$ are also functions on S, and they themselves may have partial derivatives. The standard notations for the second-order derivative

$$\frac{\partial}{\partial x_i}\left[\frac{\partial f}{\partial x_j}\right]$$

are

$$\frac{\partial^2 f}{\partial x_i \partial x_j}, \qquad f_{x_j x_i}, \qquad f_{ji}, \qquad \partial_{x_i}\partial_{x_j}f, \qquad \partial_i\partial_j f$$

if $i \neq j$ and

$$\frac{\partial^2 f}{\partial x_j^2}, \qquad f_{x_j x_j}, \qquad f_{jj}, \qquad \partial_{x_j}^2 f, \qquad \partial_j^2 f$$

if $i = j$. The analogues of these notations for higher-order partial derivatives should be pretty clear. However, all of them become quite cumbersome when the order of the derivative is even moderately large. There is a more compact notation for partial derivatives of arbitrary order that we shall introduce below.

A function f is said to be of **class C^k** on an open set U if all of its partial derivatives of order $\leq k$ — that is, all the derivatives $\partial_{i_1} \partial_{i_2} \cdots \partial_{i_l} f$, for all choices of the indices i_j and all $l \leq k$ — exist and are continuous on U. We also say that f is of class C^k on a nonopen set S if it is of class C^k on some open set that includes S. If the partial derivatives of f of *all* orders exist and are continuous on U, f is said to be of **class C^∞** on U.

It is common to refer to the derivatives $\partial_j^2 f$ and $\partial_i \partial_j f$ ($i \neq j$) as **pure** and **mixed** second-order partial derivatives of f, respectively. In this connection, a question that immediately arises is whether the order of differentiation matters. In other words, is $\partial_i \partial_j f$ the same as $\partial_j \partial_i f$? Experimentation with elementary examples suggests that the answer is *yes*.

EXAMPLE 1. If $g(x, y) = x \sin(x^3 + e^{2y})$, we have

$$\partial_x g = \sin(x^3 + e^{2y}) + 3x^3 \cos(x^3 + e^{2y}), \qquad \partial_y g = 2xe^{2y} \cos(x^3 + e^{2y}).$$

Differentiating $\partial_x g$ with respect to y and $\partial_y g$ with respect to x yields

$$\partial_y \partial_x g(x, y) = 2e^{2y} \cos(x^3 + e^{2y}) - 6x^3 e^{2y} \sin(x^3 + e^{2y}) = \partial_x \partial_y g(x, y).$$

However, the following example shows that $\partial_i \partial_j f$ may fail to coincide with $\partial_j \partial_i f$.

EXAMPLE 2. Let

$$f(x, y) = \frac{xy(x^2 - y^2)}{x^2 + y^2} \text{ if } (x, y) \neq (0, 0), \quad f(0, 0) = 0.$$

Since $f(x, 0) = f(0, y) = 0$ for all x, y, we have $\partial_x f(0, 0) = \partial_y f(0, 0) = 0$, and a little calculation shows that for $(x, y) \neq (0, 0)$,

$$\partial_x f(x, y) = \frac{x^4 y + 4x^2 y^3 - y^5}{(x^2 + y^2)^2}, \qquad \partial_y f(x, y) = \frac{x^5 - 4x^3 y^2 - xy^4}{(x^2 + y^2)^2}.$$

In particular, $\partial_x f(0, y) = -y$ and $\partial_y f(x, 0) = x$ for all x, y, so

$$\partial_y \partial_x f(0, 0) = -1 \text{ but } \partial_x \partial_y f(0, 0) = 1.$$

On the other hand, another little calculation shows that

$$\partial_y\partial_x f(x,y) = \partial_x\partial_y f(x,y) = \frac{x^6 + 9x^4y^2 - 9x^3y^4 - y^6}{(x^2+y^2)^3} \quad \text{for } (x,y) \neq (0,0).$$

This last expression has no limit as $(x,y) \to (0,0)$ (approaching $(0,0)$ along different straight lines gives different limits). Thus, we see that $\partial_y\partial_x f$ and $\partial_x\partial_y f$ exist everywhere, are continuous except at the origin, and are equal except at the origin.

Fortunately, the pathological behavior in Example 2 is quite atypical. The following theorem guarantees that the order of differentiation is immaterial in most situations that arise in practice.

2.45 Theorem. *Let f be a function defined in an open set $S \subset \mathbb{R}^n$. Suppose $\mathbf{a} \in S$ and $i,j \in \{1,\dots,n\}$. If the derivatives $\partial_i f$, $\partial_j f$, $\partial_i\partial_j f$, and $\partial_j\partial_i f$ exist in S, and if $\partial_i\partial_j f$ and $\partial_j\partial_i f$ are continuous at $\mathbf{a}$, then $\partial_i\partial_j f(\mathbf{a}) = \partial_j\partial_i f(\mathbf{a})$.*

Proof. Since only the variables x_i and x_j are actually involved here, we may as well assume that $n = 2$ and write $\mathbf{x} = (x,y)$ and $\mathbf{a} = (a,b)$, so that we are studying the derivatives $\partial_x\partial_y f$ and $\partial_y\partial_x f$. These derivatives can be regarded as limits of second-order difference quotients, so we begin by examining the "difference of differences" obtained when x and y are both changed by an amount h:

$$\begin{aligned}
D &= \big[f(a+h, b+h) - f(a+h, b)\big] - \big[f(a, b+h) - f(a,b)\big] \\
&= \big[f(a+h, b+h) - f(a, b+h)\big] - \big[f(a+h, b) - f(a,b)\big].
\end{aligned}$$

That is, if we set

$$\varphi(t) = f(a+h, b+t) - f(a, b+t), \quad \psi(t) = f(a+t, b+h) - f(a+t, b),$$

we have

$$D = \varphi(h) - \varphi(0) = \psi(h) - \psi(0).$$

We apply the (one-variable) mean value theorem twice to the first expression for D, obtaining

$$\begin{aligned}
D = \varphi'(v)h &= \big[\partial_y f(a+h, b+v) - \partial_y f(a, b+v)\big]h \\
&= \partial_x\partial_y f(a+u, b+v)h^2,
\end{aligned}$$

where u and v are some numbers between 0 and h. Likewise, using the second expression for D, we obtain

$$\begin{aligned}
D = \psi'(\tilde{u})h &= \big[\partial_x f(a+\tilde{u}, b+h) - \partial_x f(a+\tilde{u}, b)\big]h \\
&= \partial_y\partial_x f(a+\tilde{u}, b+\tilde{v})h^2,
\end{aligned}$$

where $\widetilde{u}$ and $\widetilde{v}$ are some other numbers between 0 and h. Equating these two expressions and cancelling the h^2, we have

$$\partial_x \partial_y f(a + u, \, b + v) = \partial_y \partial_x f(a + \widetilde{u}, \, b + \widetilde{v}).$$

Now let $h \to 0$. Then $u, v, \widetilde{u}, \widetilde{v} \to 0$ also, so since $\partial_x \partial_y f$ and $\partial_y \partial_x f$ are assumed continuous at (a, b), we obtain $\partial_x \partial_y f(a, b) = \partial_y \partial_x f(a, b)$. $\qquad \square$

2.46 Corollary. *If f is of class C^2 on an open set S, then $\partial_i \partial_j f = \partial_j \partial_i f$ on S, for all i and j.*

Once this is known, an elementary but slightly messy inductive argument shows that the analogous result for higher-order derivatives is also true:

2.47 Theorem. *If f is of class C^k on an open set S, then*

$$\partial_{i_1} \partial_{i_2} \cdots \partial_{i_k} f = \partial_{j_1} \partial_{j_2} \cdots \partial_{j_k} f \text{ on } S$$

whenever the sequence $\{j_1, \ldots, j_k\}$ is a reordering of the sequence $\{i_1, \ldots, i_k\}$.

The fact that the order of differentiation in a mixed partial derivative can occasionally matter is a technicality that is of essentially no importance in applications. In fact, by adopting a more sophisticated viewpoint one can prove a theorem to the effect that, under very general conditions, $\partial_i \partial_j f$ and $\partial_j \partial_i f$ are always equal "almost everywhere," which is enough to allow regarding them as equal for all practical purposes.

The chain rule can be used to compute higher-order partial derivatives of composite functions, but there are some pitfalls to be avoided. To be concrete, suppose that $w = f(x, y)$ and that x and y are functions of s and t. Assume that all the functions in question are at least of class C^2. To begin with, the chain rule for first-order derivatives gives

$$(2.48) \qquad \frac{\partial w}{\partial s} = \frac{\partial w}{\partial x} \frac{\partial x}{\partial s} + \frac{\partial w}{\partial y} \frac{\partial y}{\partial s}.$$

If we want to compute $\partial^2 w / \partial s^2$, we differentiate (2.48) with respect to s, obtaining

$$(2.49) \qquad \frac{\partial^2 w}{\partial s^2} = \frac{\partial}{\partial s} \left[\frac{\partial w}{\partial x} \right] \frac{\partial x}{\partial s} + \frac{\partial w}{\partial x} \frac{\partial^2 x}{\partial s^2} + \frac{\partial}{\partial s} \left[\frac{\partial w}{\partial y} \right] \frac{\partial y}{\partial s} + \frac{\partial w}{\partial y} \frac{\partial^2 y}{\partial s^2}.$$

The first pitfall is to write $\dfrac{\partial}{\partial s} \left[\dfrac{\partial w}{\partial x} \right]$ as a mixed partial derivative $\dfrac{\partial^2 w}{\partial s \partial x}$. This makes no sense because when we write $\partial w / \partial x$ we are thinking of w as a function of x and y, not x and s. Rather, $\partial w/\partial x$ is a function of x and y just like w, and to differentiate it with respect to s we use the chain rule again; and likewise for $\partial w/\partial y$:

$$(2.50) \quad \frac{\partial}{\partial s}\left[\frac{\partial w}{\partial x}\right] = \frac{\partial^2 w}{\partial x^2}\frac{\partial x}{\partial s} + \frac{\partial^2 w}{\partial x \partial y}\frac{\partial y}{\partial s}, \qquad \frac{\partial}{\partial s}\left[\frac{\partial w}{\partial y}\right] = \frac{\partial^2 w}{\partial x \partial y}\frac{\partial x}{\partial s} + \frac{\partial^2 w}{\partial y^2}\frac{\partial y}{\partial s}.$$

Now we plug these results into (2.49) to get the final answer, which thus contains quite a few terms. Pitfall number 2: It's easy to forget some of these terms.

In this situation it's usually advantageous to use the notation f_x and f_y instead of $\partial w/\partial x$ and $\partial w/\partial y$, and likewise for second-order derivatives. This makes (2.48)–(2.50) look a little more manageable:

$$\frac{\partial w}{\partial s} = f_x \frac{\partial x}{\partial s} + f_y \frac{\partial y}{\partial s},$$

$$\frac{\partial^2 w}{\partial s^2} = \frac{\partial f_x}{\partial s}\frac{\partial x}{\partial s} + f_x \frac{\partial^2 x}{\partial s^2} + \frac{\partial f_y}{\partial s}\frac{\partial y}{\partial s} + f_y \frac{\partial^2 y}{\partial s^2},$$

$$\frac{\partial f_x}{\partial s} = f_{xx}\frac{\partial x}{\partial s} + f_{xy}\frac{\partial y}{\partial s}, \qquad \frac{\partial f_y}{\partial s} = f_{xy}\frac{\partial x}{\partial s} + f_{yy}\frac{\partial y}{\partial s}.$$

The final result is then

$$\frac{\partial^2 w}{\partial s^2} = f_{xx}\left[\frac{\partial x}{\partial s}\right]^2 + 2f_{xy}\frac{\partial x}{\partial s}\frac{\partial y}{\partial s} + f_{yy}\left[\frac{\partial y}{\partial s}\right]^2 + f_x \frac{\partial^2 x}{\partial s^2} + f_y \frac{\partial^2 y}{\partial s^2}.$$

Of course, similar results also hold for the other second-order derivatives of w.

EXAMPLE 3. Suppose $u = f(x, y)$, $x = s^2 - t^2$, $y = 2st$. Assuming f is of class C^2, find $\partial^2 u/\partial s \partial t$ in terms of the derivatives of f.

Solution. $\dfrac{\partial u}{\partial t} = f_x \dfrac{\partial x}{\partial t} + f_y \dfrac{\partial y}{\partial t} = -2t f_x + 2s f_y$, so

$$\frac{\partial^2 u}{\partial s \partial t} = -2t[2s f_{xx} + 2t f_{xy}] + 2s[2s f_{xy} + 2t f_{yy}] + 2f_y$$

$$= -4st f_{xx} + 4(s^2 - t^2)f_{xy} + 4st f_{yy} + 2f_y.$$

EXAMPLE 4. Let us see what happens to some derivatives when we change from Cartesian to polar coordinates. Let $u = f(x, y)$, where f is of class C^2, and let $x = r\cos\theta$ and $y = r\sin\theta$. Then

$$\frac{\partial u}{\partial r} = f_x \frac{\partial x}{\partial r} + f_y \frac{\partial y}{\partial r} = (\cos\theta)f_x + (\sin\theta)f_y,$$

$$\frac{\partial u}{\partial \theta} = f_x \frac{\partial x}{\partial \theta} + f_y \frac{\partial y}{\partial \theta} = -(r\sin\theta)f_x + (r\cos\theta)f_y.$$

Proceeding to the second derivatives,

$$\frac{\partial^2 u}{\partial r^2} = (\cos\theta)\frac{\partial f_x}{\partial r} + (\sin\theta)\frac{\partial f_y}{\partial r}$$

$$= (\cos^2\theta)f_{xx} + (2\cos\theta\sin\theta)f_{xy} + (\sin^2\theta)f_{yy},$$

$$\frac{\partial^2 u}{\partial\theta^2} = -(r\cos\theta)f_x - (r\sin\theta)\frac{\partial f_x}{\partial\theta} - (r\sin\theta)f_y + (r\cos\theta)\frac{\partial f_y}{\partial\theta}$$

$$= (r^2\sin^2\theta)f_{xx} - (2r^2\sin\theta\cos\theta)f_{xy} + (r^2\cos^2\theta)f_{yy} - r\frac{\partial u}{\partial r}.$$

The calculation of the mixed derivative $\partial^2 u/\partial r\partial\theta$ is left to the reader (Exercise 2).

Notice, in particular, that by combining the last two equations and using the identity $\sin^2\theta + \cos^2\theta = 1$, we obtain

$$\frac{\partial^2 u}{\partial r^2} + \frac{1}{r}\frac{\partial u}{\partial r} + \frac{1}{r^2}\frac{\partial^2 u}{\partial\theta^2} = f_{xx} + f_{yy}.$$

The expression on the right, the sum of the pure second partial derivatives of f with respect to a Cartesian coordinate system, turns up in many practical and theoretical applications; it is called the **Laplacian** of f. (We shall encounter it again in Chapter 5.) What we have just accomplished is the calculation of the Laplacian in polar coordinates. We state this result formally, with slightly different notation.

2.51. Proposition. *Suppose u is a C^2 function of (x, y) in some open set in $\mathbb{R}^2$. If (x, y) is related to (r, θ) by $x = r\cos\theta$, $y = r\sin\theta$, we have*

$$\frac{\partial^2 u}{\partial x^2} + \frac{\partial^2 u}{\partial y^2} = \frac{\partial^2 u}{\partial r^2} + \frac{1}{r}\frac{\partial u}{\partial r} + \frac{1}{r^2}\frac{\partial^2 u}{\partial\theta^2}.$$

Multi-index Notation. Traditional notations for partial derivatives become rather cumbersome for derivatives of order higher than two, and they make it rather difficult to write Taylor's theorem in an intelligible fashion. However, a better notation, which is now in common usage in the literature of partial differential equations, is available.

A **multi-index** is an n-tuple of nonnegative integers. Multi-indices are generally denoted by the Greek letters α or β:

$$\alpha = (\alpha_1, \alpha_2, \dots, \alpha_n), \quad \beta = (\beta_1, \beta_2, \dots, \beta_n) \qquad (\alpha_j, \beta_j \in \{0, 1, 2, \dots\}).$$

If α is a multi-index, we define

$$|\alpha| = \alpha_1 + \alpha_2 + \cdots + \alpha_n, \qquad \alpha! = \alpha_1!\alpha_2!\cdots\alpha_n!,$$
$$\mathbf{x}^\alpha = x_1^{\alpha_1} x_2^{\alpha_2} \cdots x_n^{\alpha_n} \text{ (where } \mathbf{x} = (x_1, x_2, \ldots, x_n) \in \mathbb{R}^n),$$
$$\partial^\alpha f = \partial_1^{\alpha_1} \partial_2^{\alpha_2} \cdots \partial_n^{\alpha_n} f = \frac{\partial^{|\alpha|} f}{\partial x_1^{\alpha_1} \partial x_2^{\alpha_2} \cdots \partial x_n^{\alpha_n}}$$

The number $|\alpha| = \alpha_1 + \cdots + \alpha_n$ is called the **order** or **degree** of α. Thus, the order of α is the same as the order of $\mathbf{x}^\alpha$ as a monomial or the order of ∂^α as a partial derivative. (The notation $|\alpha| = \alpha_1 + \cdots + \alpha_n$ conflicts with the notation $|\mathbf{x}| = (x_1^2 + \cdots + x_n^2)^{1/2}$ for the norm of an n-tuple of real numbers, but the meaning will be clear from the context.)

If f is a function of class C^k, by Theorem 2.47 the order of differentiation in a kth-order partial derivative of f is immaterial. Thus, the generic kth-order partial derivative of f can be written simply as $\partial^\alpha f$ with $|\alpha| = k$.

EXAMPLE 5. With $n = 3$ and $\mathbf{x} = (x, y, z)$, we have

$$\partial^{(0,3,0)} f = \frac{\partial^3 f}{\partial y^3}, \qquad \mathbf{x}^{(2,1,5)} = x^2 y z^5.$$

As the notation $\mathbf{x}^\alpha$ indicates, multi-indices are handy for writing not only derivatives but also polynomials in several variables. To illustrate their use, we present a generalization of the binomial theorem.

2.52 Theorem (The Multinomial Theorem). *For any* $\mathbf{x} = (x_1, x_2, \ldots x_n) \in \mathbb{R}^n$ *and any positive integer* k,

$$(x_1 + x_2 + \cdots + x_n)^k = \sum_{|\alpha|=k} \frac{k!}{\alpha!} \mathbf{x}^\alpha.$$

Proof. The case $n = 2$ is just the binomial theorem:

$$(x_1 + x_2)^k = \sum_{j=0}^{k} \frac{k!}{j!(k-j)!} x_1^j x_2^{k-j} = \sum_{\alpha_1+\alpha_2=k} \frac{k!}{\alpha_1!\alpha_2!} x_1^{\alpha_1} x_2^{\alpha_2} = \sum_{|\alpha|=k} \frac{k!}{\alpha!} \mathbf{x}^\alpha,$$

where we have set $\alpha_1 = j$, $\alpha_2 = k - j$, and $\alpha = (\alpha_1, \alpha_2)$. The general case follows by induction on n. Suppose the result is true for $n < N$ and $\mathbf{x} = (x_1, \ldots, x_N)$. By using the result for $n = 2$ and then the result for $n = N - 1$, we obtain

$$(x_1 + \cdots + x_N)^k = \left[(x_1 + \cdots + x_{N-1}) + x_N\right]^k$$

$$= \sum_{i+j=k} \frac{k!}{i!j!}(x_1 + \cdots + x_{N-1})^i x_N^j$$

$$= \sum_{i+j=k} \frac{k!}{i!j!} \sum_{|\beta|=i} \frac{i!}{\beta!}\tilde{\mathbf{x}}^\beta x_N^j,$$

where $\beta = (\beta_1, \ldots, \beta_{N-1})$ and $\tilde{\mathbf{x}} = (x_1, \ldots, x_{N-1})$. To conclude, we set $\alpha = (\beta_1, \ldots, \beta_{N-1}, j)$, so that $\beta!j! = \alpha!$ and $\tilde{\mathbf{x}}^\beta x_N^j = \mathbf{x}^\alpha$. Observing that α runs over all multi-indices of order k when β runs over all multi-indices of order $i = k - j$ and j runs from 0 to k, we obtain $\sum_{|\alpha|=k} k! \mathbf{x}^\alpha / \alpha!$. $\square$

EXERCISES

In these exercises, all functions in question are assumed to be of class C^2.

1. Verify by explicit calculation that $\partial_x \partial_y f = \partial_y \partial_x f$:
   a. $f(x, y) = x^2 y + \sin \pi x y$.
   b. $f(x, y) = e^{4x - y^2}$.
   c. $f(x, y) = (x + 2y + 4)/(7x + 3y)$.

2. Calculate $\partial^2 u / \partial r \partial \theta$ if $u = f(x, y)$, $x = r \cos \theta$, $y = r \sin \theta$. (See Example 4.)

3. Compute the indicated derivatives of w in terms of the derivatives of f:
   a. $\partial_x^2 w$ and $\partial_x \partial_y w$, if $w = f(2x - y^2, x \sin 3y, x^4)$.
   b. $\partial_x \partial_y w$ and $\partial_y^2 w$, if $w = f(e^{x-3y}, \log(x^2 + 1), \sqrt{y^4 + 4})$.

4. Show that if $u = F(x + g(y))$, then $u_x u_{xy} = u_y u_{xx}$.

5. Suppose that f is a homogeneous function of degree a on $\mathbb{R}^n$. Show that $\sum_{j,k=1}^n x_j x_k \partial_j \partial_k f = a(a - 1)f$ (cf. Euler's theorem (2.36) and its proof).

6. Suppose $u = f(x, y)$, $x = s^2 - t^2$, $y = 2st$. Show that $\partial_s^2 u + \partial_t^2 u = 4(s^2 + t^2)(\partial_x^2 f + \partial_y^2 f)$ (cf. Example 3).

7. Suppose $u = f(x - ct) + g(x + ct)$, where c is a constant. Show that $\partial_x^2 u = c^{-2} \partial_t^2 u$.

8. For $\mathbf{x} = (x, y, z) \in \mathbb{R}^3 \setminus \{\mathbf{0}\}$ and $t \in \mathbb{R}$, let $F(\mathbf{x}, t) = r^{-1}g(ct - r)$, where c is a constant, g is a C^2 function of one variable, and $r = |\mathbf{x}|$. Show that $\partial_x^2 F + \partial_y^2 F + \partial_z^2 F = c^{-2} \partial_t^2 F$.

9. For $\mathbf{x} \in \mathbb{R}^n \setminus \{\mathbf{0}\}$, let $F(\mathbf{x}) = f(r)$ where f is a C^2 function on $(0, \infty)$ and $r = |\mathbf{x}|$. Show that $\partial_1^2 F + \cdots + \partial_n^2 F = f''(r) + (n-1)r^{-1} f'(r)$.

10. Derive the following version of the product rule for partial derivatives: $\partial^\alpha (fg) = \sum_{\beta+\gamma=\alpha} (\alpha!/\beta!\gamma!) \partial^\beta f \partial^\gamma g$.

11. Prove the following n-dimensional binomial theorem: For all $\mathbf{x}, \mathbf{y} \in \mathbb{R}^n$ we have $(\mathbf{x} + \mathbf{y})^\alpha = \sum_{\beta+\gamma=\alpha} (\alpha!/\beta!\gamma!) \mathbf{x}^\beta \mathbf{y}^\gamma$.

2.7 Taylor's Theorem

In this section we discuss Taylor expansions in their finite form, as polynomial approximations to a function rather than expansions in infinite series. We begin with a review of Taylor's theorem for functions of one real variable.

Taylor's theorem is a higher-order version of the tangent line approximation; it says that a function f of class C^k on an interval I containing the point $x = a$ is the sum of a certain polynomial of degree k and a remainder term that vanishes more rapidly than $|x - a|^k$ as $x \to a$. Specifically, the polynomial $P = P_{a,k}$ of order k such that $P^{(j)}(0) = f^{(j)}(a)$ for $0 \le j \le k$, namely

$$(2.53) \qquad P_{a,k}(h) = \sum_{j=0}^{k} \frac{f^{(j)}(a)}{j!} h^j,$$

is called the kth-order **Taylor polynomial** for f based at a, and the difference

$$(2.54) \qquad R_{a,k}(h) = f(a+h) - P_{a,k}(h) = f(a+h) - \sum_{j=0}^{k} \frac{f^{(j)}(a)}{j!} h^j$$

is called the kth-order **Taylor remainder**. The various versions of Taylor's theorem provide formulas or estimates for $R_{a,k}$ that ensure that the Taylor polynomial $P_{a,k}$ is a good approximation to f near a. The ones most commonly known involve the stronger assumption that f is of class C^{k+1} and yield the stronger conclusion that the remainder vanishes as rapidly as $|x - a|^{k+1}$. We present two of these, as well as one that yields the more general form of the theorem stated above.

The easiest version of Taylor's theorem to derive is the following.

2.55 Theorem (Taylor's Theorem with Integral Remainder, I). *Suppose that f is of class C^{k+1} ($k \ge 0$) on an interval $I \subset \mathbb{R}$, and $a \in I$. Then the remainder $R_{a,k}$ defined by (2.53)–(2.54) is given by*

$$(2.56) \qquad R_{a,k}(h) = \frac{h^{k+1}}{k!} \int_0^1 (1-t)^k f^{(k+1)}(a+th)\, dt.$$

Proof. For $k = 0$ the assertion is just that

$$(2.57) \qquad f(a+h) = f(a) + h \int_0^1 f'(a+th)\, dt,$$

which is easily verified by the substitution $u = a + th$:

$$h \int_0^1 f'(a+th)\, dt = \int_a^{a+h} f'(u)\, du = f(a+h) - f(a).$$

The trick now is to integrate (2.57) by parts, choosing for the antiderivative of the constant function 1 not t but $t - 1$, alias $-(1 - t)$:

$$h \int_0^1 f'(a+th)\, dt = -(1-t)hf'(a+th)\Big|_0^1 + h \int_0^1 (1-t)f''(a+th)h\, dt$$

$$= f'(a)h + h^2 \int_0^1 (1-t)f''(a+th)\, dt.$$

Plugging this into (2.57), we obtain (2.56) in the case $k = 1$. If we integrate by parts again,

$$h^2 \int_0^1 (1-t)f''(a+th)\, dt$$

$$= h^2 \frac{-(1-t)^2}{2} f''(a+th)\Big|_0^1 + h^2 \int_0^1 \frac{(1-t)^2}{2} f'''(a+th)h\, dt$$

$$= \frac{f''(a)}{2}h^2 + \frac{h^3}{2} \int_0^1 (1-t)^2 f'''(a+th)\, dt,$$

we obtain the theorem for $k = 2$. The pattern is now clear: Integrating (2.57) by parts k times yields (2.56). $\qquad \square$

Next we present a modification of Theorem 2.55 that works without assuming that f has any additional derivatives beyond the ones occurring in the Taylor polynomial.

2.58 Theorem (Taylor's Theorem with Integral Remainder, II). *Suppose that f is of class C^k ($k \geq 1$) on an interval $I \subset \mathbb{R}$, and $a \in I$. Then the remainder $R_{a,k}$ defined by (2.53)–(2.54) is given by*

$$(2.59) \qquad R_{a,k}(h) = \frac{h^k}{(k-1)!} \int_0^1 (1-t)^{k-1} \big[f^{(k)}(a+th) - f^{(k)}(a)\big]\, dt.$$

Proof. We begin by using Theorem 2.55, with k replaced by $k - 1$:

$$f(a + h) - \sum_{j=0}^{k-1} \frac{f^{(j)}(a)}{j!} h^j = \frac{h^k}{(k-1)!} \int_0^1 (1-t)^{k-1} f^{(k)}(a + th) \, dt.$$

Subtracting $f^{(k)}(a) h^k / k!$ from both sides gives

$$f(a + h) - \sum_{j=0}^{k} \frac{f^{(j)}(a)}{j!} h^j = \frac{h^k}{(k-1)!} \int_0^1 (1-t)^{k-1} f^{(k)}(a + th) \, dt - f^{(k)}(a) \frac{h^k}{k!}.$$

In view of the fact that

$$\frac{h^k}{k!} = \frac{h^k}{(k-1)!} \int_0^1 (1-t)^{k-1} \, dt,$$

this gives (2.59). $\qquad\square$

The formulas (2.56) and (2.59) are generally used not to obtain the exact value of the remainder but to obtain an estimate for it. The main results are in the following corollaries.

2.60 Corollary. *If f is of class C^k on I, then $R_{a,k}(h)/h^k \to 0$ as $h \to 0$.*

Proof. $f^{(k)}$ is continuous at a, so for any $\epsilon > 0$ there exists $\delta > 0$ such that $|f^{(k)}(y) - f^{(k)}(a)| < \epsilon$ when $|y - a| < \delta$. In particular,

$$\left| f^{(k)}(a + th) - f^{(k)}(a) \right| < \epsilon \text{ for } 0 \le t \le 1 \text{ when } |h| < \delta.$$

Hence, (2.59) gives

$$|R_{a,k}(h)| \le \frac{|h|^k}{(k-1)!} \int_0^1 (1-t)^{k-1} \epsilon \, dt = \frac{\epsilon}{k!} |h|^k \text{ for } |h| < \delta.$$

In other words, $|R_{a,k}(h)/h^k| < \epsilon/k!$ whenever $|h| < \delta$, and hence $R_{a,k}(h)/h^k \to 0$ as $h \to 0$. $\qquad\square$

Thus, if f is of class C^k near $x = a$, we can write $f(x)$ as the sum of a kth-order polynomial (the Taylor polynomial) in $h = x - a$ and a remainder that vanishes at $x = a$ faster than any nonzero term in the polynomial. Notice that for $k = 1$, this is just a restatement of the differentiability of f. If f is actually of class C^{k+1}, we obtain a better estimate from (2.56):

2.61 Corollary. *If f is of class C^{k+1} on I and $|f^{(k+1)}(x)| \leq M$ for $x \in I$, then*

$$|R_{a,k}(h)| \leq \frac{M}{(k+1)!}|h|^{k+1}, \qquad (a+h \in I).$$

Proof. By (2.56),

$$|R_{a,k}(h)| \leq \frac{|h|^{k+1}}{k!} \int_0^1 (1-t)^k M\, dt = \frac{M}{(k+1)!}|h|^{k+1}.$$

$\square$

Finally, we present Lagrange's form of the remainder, which turns Taylor's theorem into a higher-order version of the mean value theorem. Just as we deduced the mean value theorem from Rolle's theorem, we shall obtain Lagrange's formula from the following variant of Rolle's theorem.

2.62 Lemma. *Suppose g is $k+1$ times differentiable on $[a,b]$. If $g(a) = g(b)$ and $g^{(j)}(a) = 0$ for $1 \leq j \leq k$, then there is a point $c \in (a,b)$ such that $g^{(k+1)}(c) = 0$.*

Proof. By Rolle's theorem, there is a point $c_1 \in (a,b)$ such that $g'(c_1) = 0$. Since g' is continuous on $[a, c_1]$ and differentiable on (a, c_1), and $g'(a) = g'(c_1) = 0$, there is a point $c_2 \in (a, c_1)$ such that $g''(c_2) = 0$. Proceeding inductively, we find that for $1 \leq j \leq k+1$ there is a point $c_j \in (a, c_{j-1})$ such that $g^{(j)}(c_j) = 0$, and the final case $j = k+1$ is the desired result. $\square$

2.63 Theorem (Taylor's Theorem with Lagrange's Remainder). *Suppose f is $k+1$ times differentiable on an interval $I \subset \mathbb{R}$, and $a \in I$. For each $h \in \mathbb{R}$ such that $a + h \in I$ there is a point c between 0 and h such that*

$$(2.64) \qquad\qquad R_{a,k}(h) = f^{(k+1)}(a+c)\frac{h^{k+1}}{(k+1)!}.$$

Proof. Let us fix a particular h, and suppose for now that $h > 0$. Let

$$g(t) = R_{a,k}(t) - \frac{R_{a,k}(h)}{h^{k+1}}t^{k+1}$$

$$= f(a+t) - f(a) - f'(a)t - \cdots - \frac{f^{(k)}(a)}{k!}t^k - \frac{R_{a,k}(h)}{h^{k+1}}t^{k+1}.$$

The coefficient of t^{k+1} is chosen to make $g(h) = 0$, and clearly $g(0) = 0$. Similarly, for $j \leq k$ we have

$$g^{(j)}(t) = f^{(j)}(a+t) - f^{(j)}(a) - \cdots - \frac{f^{(k)}(a)}{(k-j)!}t^{k-j}$$

$$- \frac{R_{a,k}(h)}{h^{k+1}}(k+1)\cdots(k+2-j)t^{k+1-j},$$

so $g^{(j)}(0) = 0$. Therefore, by Lemma 2.62, there is a point $c \in (0, h)$ such that

$$0 = g^{(k+1)}(c) = f^{(k+1)}(a + c) - \frac{R_{a,k}(h)}{h^{k+1}}(k+1)!.$$

But this is precisely (2.64). The case $h < 0$ is handled similarly by considering the function $\widetilde{g}(t) = g(-t)$ on the interval $[0, |h|]$. $\square$

Corollary 2.61 is obviously an immediate consequence of (2.64).

Remark. In Theorem 2.55 we assumed that f is of class C^{k+1}, but in Theorem 2.63 we needed only the existence, not the continuity, of $f^{(k+1)}$. Actually, in Theorem 2.55 it is enough to assume that $f^{(k+1)}$ is Riemann integrable.

For the convenience of the reader, we recall a few of the most familiar and useful Taylor expansions, which are easily derived from the definition (2.53). They will be used without comment in the rest of the book.

2.65 Proposition. *The Taylor polynomials of degree k about $a = 0$ of the functions*

$$e^x, \qquad \cos x, \qquad \sin x, \qquad (1 - x)^{-1}$$

are, respectively,

$$\sum_{0 \le j \le k} \frac{x^j}{j!}, \qquad \sum_{0 \le j \le k/2} \frac{(-1)^j x^{2j}}{(2j)!}, \qquad \sum_{0 \le j \le (k-1)/2} \frac{(-1)^j x^{2j+1}}{(2j+1)!}, \qquad \sum_{0 \le j \le k} x^j.$$

Taylor polynomials have many uses. From a practical point of view, they allow one to approximate complicated functions by polynomials that are relatively easy to compute with. On the more theoretical side, it is an important general principle that *the behavior of a function $f(x)$ near $x = a$ is largely determined by the first nonvanishing term, apart from the constant term $f(a)$, in its Taylor expansion.* That is, if $f'(a) \ne 0$, then the tangent line approximation $f(x) \approx f(a) + f'(a)(x - a)$ is a good one. If $f'(a) = 0$ but $f''(a) \ne 0$, the second-order term is decisive, and so forth. This is the basis for the second-derivative test for local extrema: If $f''(a) \ne 0$, then $f(x) \approx f(a) + \frac{1}{2}f''(a)(x - a)^2$, and the expression on the right is a quadratic function with a maximum or minimum at a, depending on the sign of $f''(a)$. (See Exercise 9 and §2.8.) The following example illustrates another application of this principle.

EXAMPLE 1. Use Taylor expansions to evaluate $\displaystyle\lim_{x \to 0} \frac{x^2 - \sin x^2}{x^4 (1 - \cos x)}$.

Solution. We have

$$x^2 - \sin x^2 = x^2 - (x^2 - \tfrac{1}{6}x^6 + \cdots) = \tfrac{1}{6}x^6 + \cdots,$$
$$x^4(1 - \cos x) = x^4\big(1 - (1 - \tfrac{1}{2}x^2 + \cdots)\big) = \tfrac{1}{2}x^6 + \cdots,$$

where the dots denote error terms that vanish faster than x^6 as $x \to 0$. Therefore,

$$\frac{x^2 - \sin x^2}{x^4(1 - \cos x)} = \frac{\tfrac{1}{6}x^6 + \cdots}{\tfrac{1}{2}x^6 + \cdots} = \frac{\tfrac{1}{6} + \cdots}{\tfrac{1}{2} + \cdots},$$

where the dots in the last fraction denote error terms that vanish as $x \to 0$. The limit is therefore $\tfrac{1}{3}$. (To appreciate the efficiency of this calculation, try doing it by l'Hôpital's rule!)

We now generalize these results to functions on $\mathbb{R}^n$. Suppose $f : \mathbb{R}^n \to \mathbb{R}$ is of class C^k on a convex open set S. We can derive a Taylor expansion for $f(\mathbf{x})$ about a point $\mathbf{a} \in S$ by looking at the restriction of f to the line joining $\mathbf{a}$ and $\mathbf{x}$. That is, we set $\mathbf{h} = \mathbf{x} - \mathbf{a}$ and

$$g(t) = f(\mathbf{a} + t(\mathbf{x} - \mathbf{a})) = f(\mathbf{a} + t\mathbf{h}).$$

By the chain rule,

$$g'(t) = \mathbf{h} \cdot \nabla f(\mathbf{a} + t\mathbf{h}),$$

and hence

$$g^{(j)}(t) = (\mathbf{h} \cdot \nabla)^j f(\mathbf{a} + t\mathbf{h}),$$

where the expression on the right denotes the result of applying the operation

$$(2.66) \qquad\qquad \mathbf{h} \cdot \nabla = h_1 \frac{\partial}{\partial x_1} + \cdots + h_n \frac{\partial}{\partial x_n}$$

j times to f. The Taylor formula for g with $a = 0$ and $h = 1$,

$$g(1) = \sum_0^k \frac{g^{(j)}(0)}{j!} 1^j + (\text{remainder}),$$

therefore yields

$$(2.67) \qquad\qquad f(\mathbf{a} + \mathbf{h}) = \sum_0^k \frac{(\mathbf{h} \cdot \nabla)^j f(\mathbf{a})}{j!} + R_{\mathbf{a},k}(\mathbf{h}),$$

where formulas for $R_{\mathbf{a},k}(\mathbf{h})$ can be obtained from the formulas (2.56), (2.59), or (2.64) applied to g.

It is usually preferable, however, to rewrite (2.67) and the accompanying formulas for the remainder so that the partial derivatives of f appear more explicitly. To do this, we apply the multinomial theorem to the expression (2.66) to get

$$(\mathbf{h} \cdot \nabla)^j = \sum_{|\alpha|=j} \frac{j!}{\alpha!} \mathbf{h}^\alpha \partial^\alpha.$$

Substituting this into (2.67) and the remainder formulas, we obtain the following:

2.68 Theorem (Taylor's Theorem in Several Variables). *Suppose* $f : \mathbb{R}^n \to \mathbb{R}$ *is of class* C^k *on an open convex set* S. *If* $\mathbf{a} \in S$ *and* $\mathbf{a} + \mathbf{h} \in S$, *then*

$$(2.69) \qquad f(\mathbf{a} + \mathbf{h}) = \sum_{|\alpha| \le k} \frac{\partial^\alpha f(\mathbf{a})}{\alpha!} \mathbf{h}^\alpha + R_{\mathbf{a},k}(\mathbf{h}),$$

where

$$(2.70) \qquad R_{\mathbf{a},k}(\mathbf{h}) = k \sum_{|\alpha|=k} \frac{\mathbf{h}^\alpha}{\alpha!} \int_0^1 (1-t)^{k-1} \left[\partial^\alpha f(\mathbf{a}+t\mathbf{h}) - \partial^\alpha f(\mathbf{a}) \right] dt.$$

If f *is of class* C^{k+1} *on* S, *we also have*

$$(2.71) \qquad R_{\mathbf{a},k}(\mathbf{h}) = (k+1) \sum_{|\alpha|=k+1} \frac{\mathbf{h}^\alpha}{\alpha!} \int_0^1 (1-t)^k \partial^\alpha f(\mathbf{a}+t\mathbf{h}) \, dt,$$

and

$$(2.72) \qquad R_{\mathbf{a},k}(\mathbf{h}) = \sum_{|\alpha|=k+1} \partial^\alpha f(\mathbf{a}+c\mathbf{h}) \frac{\mathbf{h}^\alpha}{\alpha!} \text{ for some } c \in (0,1).$$

This result bears a pleasing similarity to the single-variable formulas (2.54), (2.56), (2.59), and (2.64) — a triumph for multi-index notation! It may be reassuring, however, to see the formula for the second-order Taylor polynomial written out in the more familiar notation:

(2.73)

$$P_{\mathbf{a},2}(\mathbf{h}) = f(\mathbf{a}) + \sum_{j=1}^n \partial_j f(\mathbf{a}) h_j + \frac{1}{2} \sum_{j,k=1}^n \partial_j \partial_k f(\mathbf{a}) h_j h_k$$

$$(2.74) \quad = f(\mathbf{a}) + \sum_1^n \partial_j f(\mathbf{a}) h_j + \frac{1}{2} \sum_{j=1}^n \partial_j^2 f(\mathbf{a}) h_j^2 + \sum_{1 \le j < k \le n} \partial_j \partial_k f(\mathbf{a}) h_j h_k.$$

The first of these formulas is (2.67) with $k = 2$; the second one is (2.69). (Every multi-index α of order 2 is either of the form $(\dots, 2, \dots)$ or $(\dots, 1, \dots, 1, \dots)$, where the dots denote zero entries, so the sum over $|\alpha| = 2$ in (2.69) breaks up into the last two sums in (2.74).) Notice that the mixed derivatives $\partial_j \partial_k$ ($j \neq k$) occur twice in (2.73) (since $\partial_j \partial_k = \partial_k \partial_j$) but only once in (2.74) (since $j < k$ there); this accounts for the disappearance of the factor of $\frac{1}{2}$ in the last sum in (2.74).

We also have the following analogue of Corollaries 2.60 and 2.61:

2.75. Corollary. *If f is of class C^k on S, then $R_{a,k}(\mathbf{h})/|\mathbf{h}|^k \to 0$ as $\mathbf{h} \to 0$. If f is of class C^{k+1} on S and $|\partial^\alpha f(\mathbf{x})| \leq M$ for $\mathbf{x} \in S$ and $|\alpha| = k+1$, then*

$$|R_{a,k}(\mathbf{h})| \leq \frac{M}{(k+1)!} \|\mathbf{h}\|^{k+1},$$

where

$$\|\mathbf{h}\| = |h_1| + |h_2| + \cdots + |h_n|.$$

Proof. The proof of the first assertion is the same as the proof of Corollary 2.60. As for the second, it follows easily from either (2.71) or (2.72) that

$$|R_{a,k}(\mathbf{h})| \leq M \sum_{|\alpha|=k+1} \frac{|\mathbf{h}^\alpha|}{\alpha!},$$

and this last expression equals $M\|\mathbf{h}\|^{k+1}/(k+1)!$ by the multinomial theorem. $\square$

An essential fact about the Taylor expansion of a function f about a point $\mathbf{a}$ is that it is the *only* way to write f as the sum of a polynomial of degree k and a remainder that vanishes to higher order than $|\mathbf{x} - \mathbf{a}|^k$ as $\mathbf{x} \to \mathbf{a}$. To see this, we need the following lemma.

2.76. Lemma. *If $P(\mathbf{h})$ is a polynomial of degree $\leq k$ that vanishes to order $> k$ as $\mathbf{h} \to 0$ [i.e., $P(\mathbf{h})/|\mathbf{h}|^k \to 0$], then $P \equiv 0$.*

Proof. The hypothesis implies that, for each fixed $\mathbf{h}$, $P(t\mathbf{h})/t^k \to 0$ as $t \to 0$. Write $P = P_0 + P_1 + \cdots + P_k$ where P_j is the sum of the terms of order j in P; thus

$$P(t\mathbf{h}) = P_0 + tP_1(\mathbf{h}) + t^2 P_2(\mathbf{h}) + \cdots + t^k P_k(\mathbf{h}).$$

P_0 is the constant term; since $P(0) = 0$ we must have $P_0 = 0$. Hence, dividing by t,

$$\frac{P(t\mathbf{h})}{t} = P_1(\mathbf{h}) + tP_2(\mathbf{h}) + \cdots + t^{k-1} P_k(\mathbf{h}).$$

Since $P(t\mathbf{h})/t \to 0$, we must have $P_1(\mathbf{h}) = 0$. But then, dividing by t again,

$$\frac{P(t\mathbf{h})}{t^2} = P_2(\mathbf{h}) + \cdots + t^{k-2}P_k(\mathbf{h}),$$

so $P_2(\mathbf{h}) = 0$ since $P(t\mathbf{h})/t^2 \to 0$. Continuing inductively, we conclude that $P_j(\mathbf{h}) = 0$ for all j, so $P \equiv 0$. □

2.77 Theorem. *Suppose f is of class $C^{(k)}$ near $\mathbf{a}$. If $f(\mathbf{a} + \mathbf{h}) = Q(\mathbf{h}) + E(\mathbf{h})$ where Q is a polynomial of degree $\leq k$ and $E(\mathbf{h})/|\mathbf{h}|^k \to 0$ as $\mathbf{h} \to \mathbf{0}$, then Q is the Taylor polynomial $P_{\mathbf{a},k}$.*

Proof. Corollary 2.75 says that $f(\mathbf{a}+\mathbf{h}) = P_{\mathbf{a},k}(\mathbf{h})+R_{\mathbf{a},k}(\mathbf{h})$, where $R_{\mathbf{a},k}(\mathbf{h})/|\mathbf{h}|^k$ tends to zero as $\mathbf{h}$ does. If also $f(\mathbf{a}+\mathbf{h}) = Q(\mathbf{h})+E(\mathbf{h})$, then $Q-P_{\mathbf{a},k} = R_{\mathbf{a},k}-E$, so

$$\frac{Q(\mathbf{h}) - P_{\mathbf{a},k}(\mathbf{h})}{|\mathbf{h}|^k} = \frac{R_{\mathbf{a},k}(\mathbf{h}) - E(\mathbf{h})}{|\mathbf{h}|^k} \to 0.$$

By Lemma 2.76, $Q = P_{\mathbf{a},k}$. □

Theorem 2.77 has the following important practical consequence. If one wants to compute the Taylor expansion of f, it may be very tedious to calculate all the derivatives needed in formula (2.69) directly. But if one can find, by any means whatever, a polynomial Q of degree k such that $[f(\mathbf{a} + \mathbf{h}) - Q(\mathbf{h})]/|\mathbf{h}|^k \to 0$, then Q *must* be the Taylor polynomial. This enables one to generate new Taylor expansions from old ones by operations such as substitution, multiplication, etc.

EXAMPLE 2. Find the 3rd-order Taylor polynomial of $f(x,y) = e^{x^2+y}$ about $(x, y) = (0, 0)$.

Solution. The direct method is to calculate the derivatives f_x, f_y, f_{xx}, f_{xy}, f_{yy}, f_{xxx}, f_{xxy}, f_{xyy}, and f_{yyy}, and then plug the results into (2.69), but only a masochist would do this. Instead, use the familiar expansion for the exponential function (Proposition 2.65), neglecting all terms of order higher than 3:

$$e^{x^2+y} = 1 + (x^2 + y) + \tfrac{1}{2}(x^2 + y)^2 + \tfrac{1}{6}(x^2 + y)^3 + (\text{order} > 3)$$
$$= 1 + x^2 + y + \tfrac{1}{2}(x^4 + 2x^2y + y^2) + \tfrac{1}{6}(x^6 + 3x^4y + 3x^2y^2 + y^3)$$
$$+ (\text{order} > 3)$$
$$= 1 + y + x^2 + \tfrac{1}{2}y^2 + x^2y + \tfrac{1}{6}y^3 + (\text{order} > 3).$$

In the last line we have thrown the terms x^4, x^6, x^4y, and x^2y^2 into the garbage pail, since they are themselves of order > 3. Thus the answer is $1 + y + x^2 +$

$\frac{1}{2}y^2 + x^2y + \frac{1}{6}y^3$. Alternatively,

$$e^{x^2+y} = e^{x^2}e^y = (1 + x^2 + \cdots)(1 + y + \frac{1}{2}y^2 + \frac{1}{6}y^3 + \cdots)$$
$$= 1 + y + x^2 + \frac{1}{2}y^2 + x^2y + \frac{1}{6}y^3 + \cdots$$

where the dots indicate terms of order > 3.

EXERCISES

1. Let $f(x) = x^2(x - \sin x)$ and $g(x) = (e^x - 1)(\cos 2x - 1)^2$.
   a. Compute the Taylor polynomials of order 5 based at $a = 0$ of f and g. (Don't compute any derivatives; use Proposition 2.65 as a starting point.)
   b. Use the result of (a) to find $\lim_{x \to 0} f(x)/g(x)$ without using l'Hôpital's rule.

2. Find the Taylor polynomial $P_{1,3}(h)$ and give a constant C such that $|R_{1,3}(h)| \leq Ch^4$ on the interval $|h| \leq \frac{1}{2}$ for each of the following functions.
   a. $f(x) = \log x$.
   b. $f(x) = \sqrt{x}$.
   c. $f(x) = (x + 3)^{-1}$.

3. Show that $|\sin x - x + \frac{1}{6}x^3| < .08$ for $|x| \leq \frac{1}{2}\pi$. (*Hint:* $x - \frac{1}{6}x^3$ is actually the 4th-order Taylor polynomial of $\sin x$.) How large do you have to take k so that the kth-order Taylor polynomial of $\sin x$ about $a = 0$ approximates $\sin x$ to within .01 for $|x| \leq \frac{1}{2}\pi$?

4. Use a Taylor approximation to e^{-x^2} to compute $\int_0^1 e^{-x^2}\, dx$ to three decimal places, and prove the accuracy of your answer. (*Hint:* It's easier to apply Corollary 2.61 to $f(t) = e^{-t}$ and set $t = x^2$ than to apply Corollary 2.61 to e^{-x^2} directly.)

5. Find the Taylor polynomial of order 4 based at $\mathbf{a} = (0,0)$ for each of the following functions. Don't compute any derivatives; use Proposition 2.65.
   a. $f(x, y) = x\sin(x + y)$.
   b. $e^{xy}\cos(x^2 + y^2)$.
   c. $e^{x-2y}/(1 + x^2 - y)$.

6. Find the 3rd-order Taylor polynomial of $f(x, y) = x + \cos \pi y + x\log y$ based at $\mathbf{a} = (3, 1)$.

7. Find the 3rd-order Taylor polynomial of $f(x, y, z) = x^2y + z$ based at $\mathbf{a} = (1, 2, 1)$. The remainder vanishes identically; why? (You can see this either from the Taylor remainder formula or by algebra.)

8. Suppose f is defined on the open interval I and $a \in I$. The Taylor polynomial $P_{a,k}$ is well defined provided merely that f is of class C^{k-1} on I and $f^{(k)}(a)$ exists. Show that under these hypotheses, the remainder $R_{a,k} = f - P_{a,k}$ still satisfies $\lim_{h \to 0} R_{a,k}(h)/h^k = 0$. (*Hint:* Apply l'Hôpital's rule $k - 1$ times, then recall precisely what it means for $f^{(k)}(a)$ to exist.)

9. Suppose that f is of class C^k on an open interval containing the point a, and that $f'(a) = \cdots = f^{(k-1)}(a) = 0$ but $f^{(k)}(a) \neq 0$. Use Corollary 2.60 to show that (i) if k is even, then f has a local maximum or local minimum at a according as $f^{(k)}(a)$ is negative or positive, and (ii) if k is odd, f has neither a maximum nor a minimum at a.

10. Suppose f is of class C^k on an open convex set $S \subset \mathbb{R}^n$ and its kth-order derivatives, $\partial^\alpha f$ with $|\alpha| = k$, satisfy

$$|\partial^\alpha f(\mathbf{y}) - \partial^\alpha f(\mathbf{x})| \leq C|\mathbf{y} - \mathbf{x}|^\lambda \qquad (\mathbf{x}, \mathbf{y} \in S),$$

where C and λ are positive constants (cf. Exercise 1 in §1.8). Use (2.70) to show that there is another positive constant C' such that

$$|R_{\mathbf{a},k}(\mathbf{h})| \leq C'|\mathbf{h}|^{k+\lambda} \qquad (\mathbf{a} \in S \text{ and } \mathbf{a} + \mathbf{h} \in S).$$

2.8 Critical Points

We know from elementary calculus that in studying a differentiable function f of a real variable, it is particularly important to look at the points where the derivative f' vanishes. The same is true for functions of several variables.

Suppose f is a differentiable function on some open set $S \subset \mathbb{R}^n$. The point $\mathbf{a} \in S$ is called a **critical point** for f if $\nabla f(\mathbf{a}) = \mathbf{0}$. Finding the critical points of f is a matter of solving the n equations $\partial_1 f(\mathbf{x}) = 0, \ldots, \partial_n f(\mathbf{x}) = 0$ simultaneously for the n quantities $x_1, \ldots, x_n$.

We say that f has a **local maximum** (or **local minimum**) at $\mathbf{a}$ if $f(\mathbf{x}) \leq f(\mathbf{a})$ (or $f(\mathbf{x}) \geq f(\mathbf{a})$) for all $\mathbf{x}$ in some neighborhood of $\mathbf{a}$. Just as in the one-variable case, we have:

2.78 Proposition. *If f has a local maximum or minimum at $\mathbf{a}$ and f is differentiable at $\mathbf{a}$, then $\nabla f(\mathbf{a}) = \mathbf{0}$.*

Proof. If f has a local maximum or minimum at $\mathbf{a}$, then for any unit vector $\mathbf{u}$, the function $g(t) = f(\mathbf{a} + t\mathbf{u})$ has a local maximum or minimum at $t = 0$, so $g'(0) = \partial_{\mathbf{u}} f(\mathbf{a}) = 0$. In particular, $\partial_j f(\mathbf{a}) = 0$ for all j, so $\nabla f(\mathbf{a}) = \mathbf{0}$. $\qquad \square$

How can we tell whether a function has a local maximum or minimum (or neither) at a critical point? For functions of one variable we have the second derivative
test: If f is of class C^2, then f has a local minimum at a if $f''(a) > 0$ and a local
maximum if $f''(a) < 0$. (If $f''(a) = 0$, no conclusion can be drawn.) Something
similar happens for functions of n variables, but the situation is a good deal more
complicated. The full story involves a certain amount of linear algebra; the reader
who is content to consider the case of two variables and wishes to skip the linear
algebra may proceed directly to Theorem 2.82.

Suppose f is a real-valued function of class C^2 on some open set $S \subset \mathbb{R}$ and
that f has a critical point at $\mathbf{a}$, i.e., $\nabla f(\mathbf{a}) = 0$. Instead of one second derivative to
examine at $\mathbf{a}$, we have a whole $n \times n$ matrix of them, called the **Hessian** of f at $\mathbf{a}$:

$$(2.79) \qquad H = H(\mathbf{a}) = \begin{pmatrix} \partial_1^2 f(\mathbf{a}) & \partial_1 \partial_2 f(\mathbf{a}) & \cdots & \partial_1 \partial_n f(\mathbf{a}) \\ \partial_2 \partial_1 f(\mathbf{a}) & \partial_2^2 f(\mathbf{a}) & \cdots & \partial_2 \partial_n f(\mathbf{a}) \\ \vdots & \vdots & \ddots & \vdots \\ \partial_n \partial_1 f(\mathbf{a}) & \partial_n \partial_2 f(\mathbf{a}) & \cdots & \partial_n^2 f(\mathbf{a}) \end{pmatrix}.$$

The equality of mixed partials (Theorem 2.45) guarantees that this is a *symmetric*
matrix, that is, $H_{ij} = H_{ji}$.

By (2.73), the second-order Taylor expansion of f about $\mathbf{a}$ is

$$f(\mathbf{a} + \mathbf{k}) = f(\mathbf{a}) + \sum_{j=1}^{n} \partial_j f(\mathbf{a}) k_j + \frac{1}{2} \sum_{i,j=1}^{n} \partial_i \partial_j f(\mathbf{a}) k_i k_j + R_{\mathbf{a},2}(\mathbf{k}).$$

(We use $\mathbf{k}$ rather than $\mathbf{h}$ for the increment in this section to avoid a notational clash
with the Hessian H.) If $\nabla f(\mathbf{a}) = 0$, the first-order sum vanishes, and the second-
order sum is $\frac{1}{2} \sum H_{ij} k_i k_j = \frac{1}{2} H\mathbf{k} \cdot \mathbf{k}$. In short,

$$(2.80) \qquad\qquad f(\mathbf{a} + \mathbf{k}) = f(\mathbf{a}) + \tfrac{1}{2} H\mathbf{k} \cdot \mathbf{k} + R_{\mathbf{a},2}(\mathbf{k}).$$

Now we can begin to see how to analyze the behavior of f about $\mathbf{a}$ in terms of
the matrix H. To start with the simplest situation, suppose it happens that all the
mixed partials $\partial_i \partial_j f$ ($i \neq j$) vanish at $\mathbf{a}$. Denoting $\partial_j^2 f(\mathbf{a})$ by λ_j, we then have

$$f(\mathbf{a} + \mathbf{k}) = f(\mathbf{a}) + \sum_{1}^{n} \lambda_j k_j^2 + R_{\mathbf{a},2}(\mathbf{k}).$$

Let us neglect the remainder term for the moment. If all λ_j are positive, then
$\sum \lambda_j k_j^2 > 0$ for all $\mathbf{k} \neq \mathbf{0}$, so f has a local minimum; likewise, if all λ_j are negative, then f has a local maximum. If some λ_j are positive and some are negative,

then $\sum \lambda_j k_j^2$ will be positive for some values of $\mathbf{k}$ and negative for others, so f will have neither a maximum or a minimum. It's not hard to see that these conclusions remain valid when the remainder term is included; we shall present the details below. Only when some of the λ_j are zero is the outcome unclear; it is precisely in this situation that the remainder term plays a significant role.

This is all very well, but the condition that $\partial_i\partial_j f(\mathbf{a}) = 0$ for $i \neq j$ is obviously very special. *However, it can always be achieved by a suitable rotation of coordinates,* that is, by replacing the standard basis for $\mathbb{R}^n$ with another suitably chosen orthonormal basis. This is the content of the spectral theorem, which says that every symmetric matrix has an orthonormal eigenbasis (see Appendix A, (A.56)–(A.58)). With this result in hand, we arrive at the second-derivative test for functions of several variables.

2.81 Theorem. *Suppose f is of class C^2 at $\mathbf{a}$ and that $\nabla f(\mathbf{a}) = 0$, and let H be the Hessian matrix (2.79). For f to have a local minimum at $\mathbf{a}$, is it necessary for the eigenvalues of H all to be nonnegative and sufficient for them all to be strictly positive. For f to have a local maximum at $\mathbf{a}$, it is necessary for the eigenvalues of H all to be nonpositive and sufficient for them all to be strictly negative.*

Proof. We prove only the first assertion; the argument for the second one is similar. Let $\mathbf{u}_1, \ldots, \mathbf{u}_n$ be an orthonormal eigenbasis for H with eigenvalues $\lambda_1, \ldots, \lambda_n$. Our assertion is then that f has a local minimum if all the eigenvalues are (strictly) positive but not if some eigenvalue is negative.

If all eigenvalues are positive, let l be the smallest of them. Writing $\mathbf{k} = c_1\mathbf{u}_1 + \cdots + c_n\mathbf{u}_n$ as before, we have

$$H\mathbf{k} \cdot \mathbf{k} = \sum \lambda_j c_j^2 \geq l \sum c_j^2 = l|\mathbf{k}|^2.$$

But when $\mathbf{k}$ is near 0, the error term in (2.80) is less than $\frac{1}{4}l|\mathbf{k}|^2$ by Corollary 2.75, so

$$f(\mathbf{a} + \mathbf{k}) - f(\mathbf{a}) \geq \tfrac{1}{2}l|\mathbf{k}|^2 - \tfrac{1}{4}l|\mathbf{k}|^2 > 0.$$

Thus f has a local minimum. On the other hand, if some eigenvalue, say λ_1, is negative, the same argument shows that $f(\mathbf{a} + t\mathbf{u}_1) - f(\mathbf{a}) < 0$ for small $t \neq 0$, so f does not have a local minimum. $\qquad\square$

In short, if all eigenvalues are positive, then f has a local minimum; if all eigenvalues are negative, then f has a local maximum. If there are two eigenvalues of opposite signs, then f is said to have a **saddle point**. At a saddle point, f has neither a maximum nor a minimum; its graph goes up in one direction and down in some other direction. The only cases where we can't be sure what's going on are

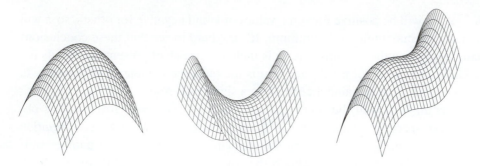

FIGURE 2.5: Left: A local maximum ($z = -x^2 - y^2$). Middle: A saddle point ($z = x^2 - y^2$). Right: A degenerate critical point ($z = x^3 - y^2$).

those where all the eigenvalues of H are nonnegative or nonpositive but at least one of them is zero. When that happens, if $\mathbf{k}$ is an eigenvector with eigenvalue 0 (i.e., $\mathbf{k}$ is in the nullspace of H), the quadratic term in (2.80) vanishes and the remainder term becomes significant; to determine the behavior of f near $\mathbf{a}$ we need to look at the higher-order terms in the Taylor expansion.

Some types of critical points are illustrated in Figure 2.5. A critical point for which zero is an eigenvalue of the Hessian matrix H — or equivalently, for which $\det H = 0$ or H is singular — is called **degenerate**.

In two dimensions it is easy to sort out the various cases:

2.82 Theorem. *Suppose f is of class C^2 on an open set in $\mathbb{R}^2$ containing the point* $\mathbf{a}$, *and suppose $\nabla f(\mathbf{a}) = \mathbf{0}$. Let $\alpha = \partial_1^2 f(\mathbf{a})$, $\beta = \partial_1 \partial_2 f(\mathbf{a})$, $\gamma = \partial_2^2 f(\mathbf{a})$. Then:*

a. If $\alpha\gamma - \beta^2 < 0$, f has a saddle point at $\mathbf{a}$.
b. If $\alpha\gamma - \beta^2 > 0$ and $\alpha > 0$, f has a local minimum at $\mathbf{a}$.
c. If $\alpha\gamma - \beta^2 > 0$ and $\alpha < 0$, f has a local maximum at $\mathbf{a}$.
d. If $\alpha\gamma - \beta^2 = 0$, no conclusion can be drawn.

Proof. The determinant of the Hessian matrix $H = \begin{pmatrix} \alpha & \beta \\ \beta & \gamma \end{pmatrix}$ is $\alpha\gamma - \beta^2$. Since the determinant is the product of the eigenvalues, the two eigenvalues have opposite signs if $\alpha\gamma - \beta^2 < 0$, and they have the same sign if $\alpha\gamma - \beta^2 > 0$. In the latter case, H is positive (or negative) definite when the eigenvalues are positive (or negative), and since $\alpha = H\mathbf{u} \cdot \mathbf{u}$ where $\mathbf{u} = (1, 0)$, these cases occur precisely when $\alpha > 0$ or $\alpha < 0$. The result now follows from Theorem 2.81. $\quad\square$

EXAMPLE 1. Find and classify the critical points of the function $f(x, y) = xy(12 - 3x - 4y)$.

Solution. We have

$$\partial_x f = 12y - 6xy - 4y^2 = y(12 - 6x - 4y),$$
$$\partial_y f = 12x - 3x^2 - 8xy = x(12 - 3x - 8y).$$

Thus, if $\partial_x f = 0$ then $y = 0$ or $12 - 6x - 4y = 0$, and if $\partial_y f = 0$ then $x = 0$ or $12 - 3x - 8y = 0$. So there are four possibilities:

$$y = x = 0, \quad y = 12 - 3x - 8y = 0,$$
$$12 - 6x - 4y = x = 0, \text{ and } 12 - 6x - 4y = 12 - 3x - 8y = 0.$$

Solving these gives the critical points $(0, 0)$, $(4, 0)$, $(0, 3)$, and $(\frac{4}{3}, 1)$. Since $\partial_x^2 f = -6y$, $\partial_y^2 f = -8x$, and $\partial_x \partial_y f = 12 - 6x - 8y$, Theorem 2.82 shows that the first three of these are saddle points and the last is a local maximum.

The geometry of this example is quite simple. The set where $f = 0$ is the union of the three lines $x = 0$, $y = 0$, and $3x + 4y = 12$. These lines separate the plane into regions on which f is alternately positive and negative. The three saddle points are the points where these lines intersect, and the local maximum is the "peak" in the middle of the triangle defined by these lines.

EXAMPLE 2. Find and classify the critical points of the function $f(x, y) = y^3 - 3x^2y$.

Solution. We have $\partial_x f = -6xy$ and $\partial_y f = 3y^2 - 3x^2$. Thus, if $\partial_x f = 0$, then either $x = 0$ or $y = 0$, and the equation $\partial_y f = 0$ then forces $x = y = 0$. So $(0, 0)$ is the only critical point. The reader may readily verify that all the second derivatives of f vanish at $(0, 0)$, so Theorem 2.82 is of no use. But since $f(x, y) = y(y - \sqrt{3}\,x)(y + \sqrt{3}\,x)$, the lines $y = 0$ and $y = \pm\sqrt{3}\,x$ separate the plane into six regions on which f is alternately positive and negative, and these regions all meet at the origin. Thus f has neither a maximum nor a minimum at the origin. This configuration is called a "monkey saddle." (The three regions where $f < 0$ provide places for the two legs and tail of a monkey sitting on the graph of f at the origin.)

EXERCISES

1. Find all the critical points of the following functions. Tell whether each nondegenerate critical point is a local maximum, local minimum, or saddle point. If possible, tell whether the degenerate critical points are local extrema too.

   a. $f(x, y) = x^2 + 3y^4 + 4y^3 - 12y^2$.
   b. $f(x, y) = x^4 - 2x^2 + y^3 - 6y$.
   c. $f(x, y) = (x - 1)(x^2 - y^2)$.
   d. $f(x, y) = x^2 y^2 (2 - x - y)$.
   e. $f(x, y) = (2x^2 + y^2)e^{-x^2 - y^2}$.
   f. $f(x, y) = ax^{-1} + by^{-1} + xy$, $a, b \neq 0$. (The nature of the critical point depends on the signs of a and b.)
   g. $f(x, y, z) = x^3 - 3x - y^3 + 9y + z^2$.
   h. $f(x, y, z) = (3x^2 + 2y^2 + z^2)e^{-x^2 - y^2 - z^2}$.
   i. $f(x, y, z) = xyz(4 - x - y - z)$.

2. What are the conditions on a, b, c for $f(x, y) = ax^2 + bxy + cy^2$ to have a minimum, maximum, or saddle point at the origin?

3. The origin is a degenerate critical point of the functions $f_1(x, y) = x^2 + y^4$, $f_2(x, y) = x^2 - y^4$, and $f_3(x, y) = x^2 + y^3$. Describe the graphs of these three functions near the origin. Is the origin a local extremum for any of them?

4. Let $f(x, y) = (y - x^2)(y - 2x^2)$.
   a. Show that the origin is a degenerate critical point of f.
   b. Show that the restriction of f to any line through the origin (i.e., the function $g(t) = f(at, bt)$ for any $(a, b) \neq (0, 0)$) has a local minimum at the origin, but f does not have a local minimum at the origin. (*Hint:* Consider the regions where $f > 0$ or $f < 0$.)

5. Let H be the Hessian of f. Show that for any unit vector $\mathbf{u}$, $H\mathbf{u} \cdot \mathbf{u}$ is the second directional derivative of f in the direction $\mathbf{u}$.

2.9　Extreme Value Problems

In the previous section we studied the critical points of a differentiable function, which include its local maxima and minima. In this section we consider the problem of finding the absolute maximum or minimum of a differentiable function on a set $S \subset \mathbb{R}^n$, which has a somewhat different flavor.

    The fundamental theoretical fact that underlies this study is the extreme value theorem (1.23), whose statement we now recall: *If S is a compact subset of $\mathbb{R}^n$ and f is a continuous function on S, then f assumes a minimum and a maximum value on S — that is, there are points $\mathbf{a}, \mathbf{b} \in S$ such that $f(\mathbf{a}) \leq f(\mathbf{x}) \leq f(\mathbf{b})$ for all $\mathbf{x} \in S$.* As the examples that we presented in §1.6 show, the conclusion is generally invalid if S fails to be both closed and bounded. Accordingly, we shall assume throughout this section that S is closed, but we shall include some discussion of the situation when S is unbounded. Moreover, to keep the problem within the realm of calculus, we shall assume that S is either (i) the closure of an open set with a smooth or piecewise smooth boundary, or (ii) a smooth submanifold, such as a curve or surface, defined by one or more constraint equations. (These geometric notions will be studied in more detail in Chapter 3.)

Suppose, to begin with, that S is the closure of an open set in $\mathbb{R}^n$, and that we wish to find the absolute maximum or minimum of a differentiable function f on S. We assume that the boundary of S is a smooth submanifold (a curve if $n = 2$, a surface if $n = 3$) that can be described as the level set of a differentiable function G, or that it is the union of a finite number of pieces of this form. (For example, if S is a cube, its boundary is the union of six faces, each of which is a region in a smooth surface, viz., a plane.) If S is bounded, the extreme values are guaranteed to exist, and we can proceed as follows.

i. If an extreme value occurs at an interior point of S, that point must be a critical point of f. So, we find all the critical points of f inside S and compute the values of f at these points.

ii. To find candidates for extreme values on the boundary, we can apply the techniques for solving extremal problems with constraints presented below.

iii. Finally, we pick the smallest and largest of the values of f at the points found in steps (i) and (ii); these will be the minimum and maximum of f on S. There is usually no need to worry about the second derivative test in this situation.

If S is unbounded, the procedure is the same, but we must add an extra argument to show that the desired extremum actually exists. This must be done on a case-by-case basis, as there is no general procedure available; however, here are a couple of simple results that cover many situations in practice and illustrate the sort of reasoning that must be employed.

2.83 Theorem. *Let f be a continuous function on an unbounded closed set $S \subset \mathbb{R}^n$.*

a. *If $f(\mathbf{x}) \to +\infty$ as $|\mathbf{x}| \to \infty$ ($\mathbf{x} \in S$), then f has an absolute minimum but no absolute maximum on S.*

b. *If $f(\mathbf{x}) \to 0$ as $|\mathbf{x}| \to \infty$ ($\mathbf{x} \in S$) and there is a point $\mathbf{x}_0 \in S$ where $f(\mathbf{x}_0) > 0$ (resp. $f(\mathbf{x}_0) < 0$), then f has an absolute maximum (resp. minimum) on S.*

Proof. (a) If $f(\mathbf{x}) \to \infty$ as $|\mathbf{x}| \to \infty$, then clearly f has no maximum. On the other hand, pick a point $\mathbf{x}_0 \in S$ and let $V = \{\mathbf{x} \in S : f(\mathbf{x}) \leq f(\mathbf{x}_0)\}$. Then V is closed (by Theorem 1.13) and bounded (since $f(\mathbf{x}) > f(\mathbf{x}_0)$ when $|\mathbf{x}|$ is large). By the extreme value theorem, f has a minimum on V, say at $\mathbf{a} \in V$. But then $f(\mathbf{a})$ is the absolute minimum of f on V because $f(\mathbf{x}) > f(\mathbf{x}_0) \geq f(\mathbf{a})$ for $\mathbf{x} \in S \setminus V$.

The proof of (b) is similar. If $f(\mathbf{x}_0) > 0$, let $V = \{\mathbf{x} : f(\mathbf{x}) \geq f(\mathbf{x}_0)\}$. Then V is closed (by Theorem 1.13) and bounded (since $f(\mathbf{x}) \to 0$ as $|\mathbf{x}| \to \infty$). By the extreme value theorem, f has a maximum on V, say at $\mathbf{a} \in V$. But then $f(\mathbf{a})$ is the absolute maximum of f on S because $f(\mathbf{x}) < f(\mathbf{x}_0) \leq f(\mathbf{a})$ for $\mathbf{x} \in S \setminus V$.   □

EXAMPLE 1. Find the absolute maximum and minimum values of the function
$$f(x,y) = \frac{x}{x^2 + (y-1)^2 + 4} \text{ on the first quadrant } S = \{(x,y) : x, y \geq 0\}.$$
Solution. Clearly $f(x,y) \geq 0$ for $x, y \geq 0$, and $f(0, y) = 0$, so the minimum is zero, achieved at all points on the y-axis. Moreover, $f(x, y)$ is less than the smaller of x^{-1} and $(y-1)^{-2}$, so it vanishes as $|(x, y)| \to \infty$. Hence, by Theorem 2.83, f has a maximum on S, which must occur either in the interior of S or on the positive x-axis. A short calculation that we leave to the reader shows that the only critical point of f in S is at $(2, 1)$, and $f(2, 1) = \frac{1}{4}$. Also, $f(x, 0) = x/(x^2 + 5)$, and the critical points of this function of one variable are at $x = \pm\sqrt{5}$. Only $x = \sqrt{5}$ is relevant for our purposes, and $f(\sqrt{5}, 0) = \sqrt{5}/10$, which is a bit less than $\frac{1}{4}$. Thus the maximum value of f on S is $\frac{1}{4}$.

Let us turn to the study of extremum problems with constraints. To be precise, we consider the following situation: We wish to minimize or maximize a differentiable function f on the set

$$S = \{\mathbf{x} : G(\mathbf{x}) = 0\},$$

where G is of class C^1 and $\nabla G(\mathbf{x}) \neq \mathbf{0}$ on S. (The latter assumption guarantees that the set S is smooth in the sense that it possesses a tangent (hyper)plane at every point $\mathbf{a} \in S$, namely, the (hyper)plane through $\mathbf{a}$ that is perpendicular to the vector $\nabla G(\mathbf{a})$; see Theorem 2.37 and §§3.3–4.) Most applied max-min problems are of this sort, including the ones one first meets in freshman calculus — for example, "Find the maximum area of a rectangle with a given perimeter P," i.e., maximize xy subject to the constraint $2x + 2y = P$.

There are several methods for attacking such a problem. The most obvious one is to solve the constraint equation $G(\mathbf{x}) = 0$ for one of the variables, either explicitly or implicitly, and thus reduce the problem to finding the critical points of a function of the remaining $n-1$ variables. (Of course, this is what one always does in freshman calculus.) Another possibility is to describe the set S parametrically and thus obtain an $(n-1)$-variable problem with the parameters as independent variables. This is particularly effective when S is a closed curve or surface such as a circle or sphere that cannot be described in its entirety as the graph of a function.

There is yet another method, however, which may derived from the following considerations. Suppose that f, as a function on the set $S = \{\mathbf{x} : G(\mathbf{x}) = 0\}$, has a local extremum at $\mathbf{x} = \mathbf{a}$. If $\mathbf{x} = \mathbf{h}(t)$ is a curve on S passing through $\mathbf{a}$ at $t = 0$, the composite function $\varphi(t) = f(\mathbf{h}(t))$ has a local extremum at $t = 0$, so $\nabla f(\mathbf{a}) \cdot \mathbf{h}'(0) = \varphi'(0) = 0$. Thus, $\nabla f(\mathbf{a})$ is orthogonal to the tangent vector to every curve on S passing through $\mathbf{a}$; in other words, $\nabla f(\mathbf{a})$ is normal to S at $\mathbf{a}$. But we already know that $\nabla G(\mathbf{a})$ is normal to S at $\mathbf{a}$ since S is a level set of G. It follows that ∇f *is proportional to* ∇G at $\mathbf{a}$:

$$\nabla f(\mathbf{a}) = \lambda \nabla G(\mathbf{a}) \text{ for some } \lambda \in \mathbb{R}.$$

This is the key to the method. The n equations $\partial_j f = \lambda \partial_j G$ together with the constraint equation $G = 0$ give $n+1$ equations in the $n+1$ variables $x_1, \ldots, x_n$ and λ, and solving them simultaneously will locate the local extrema of f on S. (It will also produce the appropriate values of λ, which are usually not of much interest, although one may have to find them in the process of solving for the x_j's.) This method is called **Lagrange's method**, and the parameter λ is called the **Lagrange multiplier** for the problem.

The other methods described above involve reducing the original n-variable problem to an $(n-1)$-variable problem, whereas Lagrange's method deals directly with the original n variables. This may be advantageous when the reduction is awkward or when it would involve breaking some symmetry of the original problem. The disadvantage is that, whereas the other methods lead to solving $n-1$ equations in $n-1$ variables, Lagrange's method requires solving $n+1$ equations in $n+1$ variables.

EXAMPLE 2. Let's try out Lagrange's method on the simple problem of maximizing the area of a rectangle with perimeter P. Here $f(x, y) = xy$ and $G(x, y) = 2x + 2y - P$, so the equations $\partial_x f = \lambda \partial_x G$, $\partial_y f = \lambda \partial_y G$, and $G = 0$ become

$$y = 2\lambda, \qquad x = 2\lambda, \qquad 2x + 2y = P.$$

The first two equations give $y = x$; substituting into the third equation shows that $x = y = \frac{1}{4}P$, so the maximum of f is $\frac{1}{16}P^2$. (Note that the only relevant values of x and y are $0 \le x, y \le \frac{1}{2}P$, so we're working on a compact set and the existence of the maximum is not in question. The minimum on this set, namely 0, is achieved when $x = 0$, $y = \frac{1}{2}P$, or vice versa.)

EXAMPLE 3. Find the absolute maximum and minimum of $f(x, y) = x^2 + y^2 + y$ on the disc $x^2 + y^2 \leq 1$.

Solution. We have $f_x = 2x$, $f_y = 2y + 1$. Thus the only critical point is at $(0, -\frac{1}{2})$ (which lies in the disc), at which $f = -\frac{1}{4}$. To see what happens on the boundary, we can use Lagrange's method with $G(x, y) = x^2 + y^2 - 1$. We have to solve

$$2x = 2\lambda x, \qquad 2y + 1 = 2\lambda y, \qquad x^2 + y^2 = 1.$$

The first equation implies that *either* $x = 0$ *or* $\lambda = 1$. The latter alternative is impossible since the equation $2y + 1 = 2y$ has no solutions, so $x = 0$ and then $y = \pm 1$ (since $x^2 + y^2 = 1$). We have $f(0, 1) = 2$, $f(0, -1) = 0$. So the absolute maximum is 2 (at $(0, 1)$) and the absolute minimum is $-\frac{1}{4}$ (at $(0, -\frac{1}{2})$).

We could also analyze f on the boundary by parametrizing the latter as $x = \cos\theta$, $y = \sin\theta$. Then $f(\cos\theta, \sin\theta) = 1 + \cos\theta$, which has a maximum value of 2 at $\theta = 0$ and a minimum value of 0 at $\theta = \pi$.

Similar ideas work when there is more than one constraint equation. Let's consider the case of two equations:

$$S = \big\{ \mathbf{x} : G_1(\mathbf{x}) = G_2(\mathbf{x}) = 0 \big\}.$$

Here G_1 and G_2 are differentiable functions (the subscripts are labels for the functions, not partial derivatives), and we assume that the vectors $\nabla G_1(\mathbf{x})$ and $\nabla G_2(\mathbf{x})$ are linearly independent for $\mathbf{x} \in S$. (Again, this guarantees that S is a "smooth" set, as we shall see in Chapter 3.) To find the extreme values of a differentiable function on S, we have three methods:

- Solve the equations $G_1(\mathbf{x}) = G_2(\mathbf{x}) = 0$ for two of the variables and find the critical points of the resulting function of the remaining $n - 2$ variables.

- Find a parametrization of the set S in terms of parameters $t_1, \ldots, t_{n-2}$, and find the critical points of f as a function of these variables.

- (Lagrange's method) At a local extremum, ∇f must be normal to S and hence must be a linear combination of ∇G_1 and ∇G_2:

$$\nabla f = \lambda \nabla G_1 + \mu \nabla G_2 \text{ for some } \lambda, \mu \in \mathbb{R}.$$

The n equations $\partial_j f = \lambda \partial_j G_1 + \mu \partial_j G_2$ together with the two constraint equations $G_1 = G_2 = 0$ can be solved for the $n + 2$ variables $x_1, \ldots, x_n$, λ, and μ, yielding the points where local extrema can occur.

The generalization to k constraint equations should now be pretty clear.

EXERCISES

1. Find the extreme values of $f(x, y) = 2x^2 + y^2 + 2x$ on the set $\{(x, y) : x^2 + y^2 \le 1\}$.

2. Find the extreme values of $f(x, y) = 3x^2 - 2y^2 + 2y$ on the set $\{(x, y) : x^2 + y^2 \le 1\}$.

3. Find the extreme values of $f(x, y) = x^3 - x + y^2 - 2y$ on the closed triangular region with vertices at $(-1, 0)$, $(1, 0)$, and $(0, 2)$.

4. Find the extreme values of $f(x, y) = 3x^2 - 8xy - 4y^2 + 2x + 16y$ on the set $\{(x, y) : 0 \le x \le 4, \ 0 \le y \le 3\}$.

5. Let $f(x, y) = (A - bx - cy)^2 + x^2 + y^2$, where A, b, c are positive constants. Show that f has an absolute minimum on $\mathbb{R}^2$ and find it.

6. Show that $f(x, y) = (x^2 + 2y^2)e^{-x^2 - y^2}$ has an absolute minimum and maximum on $\mathbb{R}^2$, and find them.

7. Show that $f(x, y) = (x^2 - 2y^2)e^{-x^2 - y^2}$ has an absolute minimum and maximum on $\mathbb{R}^2$, and find them.

8. Let $f(x, y) = xy + 3x^{-1} + 4y^{-1}$. Show that f has a minimum but no maximum on the set $\{(x, y) : x, y > 0\}$, and find the minimum.

9. Find the extreme values of $f(x, y, z) = x^2 + 2y^2 + 3z^2$ on the unit sphere $\{(x, y, z) : x^2 + y^2 + z^2 = 1\}$.

10. Let $(x_1, y_1), \ldots, (x_k, y_k)$ be points in the plane whose x-coordinates are not all equal. The linear function $f(x) = ax + b$ such that the sum of the squares of the vertical distances from the given points to the line $y = ax + b$ (namely, $\sum_1^k (y_j - ax_j - b)^2$) is minimized is called the **linear least-squares fit** to the points (x_j, y_j). Show that it is given by

$$a = \frac{k^{-1} \sum_1^k x_j y_j - \overline{x}\,\overline{y}}{k^{-1} \sum_1^n x_j^2 - \overline{x}^2}, \qquad b = \overline{y} - a\overline{x},$$

where $\overline{x} = k^{-1} \sum_1^k x_j$ and $\overline{y} = k^{-1} \sum_1^k y_j$ are the averages of the x_j's and y_j's.

11. Let x, y, z be positive variables and a, b, c positive constants. Find the minimum of $x + y + z$ subject to the constraint $(a/x) + (b/y) + (c/z) = 1$.

12. Find the minimum possible value of the sum of the three linear dimensions (length, breadth, and width) of a rectangular box whose volume is a given constant V. Is there a maximum possible value?

13. Find the point on the line through $(1, 0, 0)$ and $(0, 1, 0)$ that is closest to the line through $(0, 0, 0)$ and $(1, 1, 1)$. (*Hint:* Minimize the square of the distance.)

14. Find the maximum possible volume of a rectangular solid if the sum of the areas of the bottom and the four vertical sides is a constant A, and find the dimensions of the box that has the maximum volume.

15. The two planes $x + z = 4$ and $3x - y = 6$ intersect in a line L. Use Lagrange's method to find the point on L that is closest to the origin. (*Hint:* Minimize the square of the distance.)

16. Find the maximum value of $(xv - yu)^2$ subject to the constraints $x^2 + y^2 = a^2$ and $u^2 + v^2 = b^2$. Do this (a) by Lagrange's method, (b) by the parametrization $x = a\cos\theta$, $y = a\sin\theta$, $u = b\cos\varphi$, $v = b\sin\varphi$.

17. Let $P_1 = (x_1, y_1)$ and $P_2 = (x_2, y_2)$ be two points in the plane such that $x_1 \neq x_2$ and $y_1 > 0 > y_2$. A particle travels in a straight line from P_1 to a point Q on the x-axis with speed v_1, then in a straight line from Q to P_2 with speed v_2. The point Q is allowed to vary. Use Lagrange's method to show that the total travel time from P_1 to P_2 is minimized when $(\sin\theta_1)/(\sin\theta_2) = v_1/v_2$, where θ_1 (resp. θ_2) is the angle between the line $P_1 Q$ (resp. $Q P_2$) and the vertical line through Q. (*Hint:* Take θ_1, θ_2 as the independent variables.)

18. Let $x_1, x_2, \ldots, x_n$ denote nonnegative numbers. For $c > 0$, maximize the product $x_1 x_2 \cdots x_n$ subject to the constraint $x_1 + x_2 + \cdots + x_n = c$, and hence derive the inequality of geometric and arithmetic means,

$$\left(x_1 x_2 \cdots x_n\right)^{1/n} \leq \frac{x_1 + x_2 + \cdots + x_n}{n} \qquad (x_1, \ldots, x_n \geq 0),$$

where equality holds if and only if the x_j's are all equal.

19. Let A be a symmetric $n \times n$ matrix, and let $f(\mathbf{x}) = (A\mathbf{x}) \cdot \mathbf{x}$ for $\mathbf{x} \in \mathbb{R}^n$. Show that the maximum and minimum of f on the unit sphere $\{\mathbf{x} : |\mathbf{x}| = 1\}$ are the largest and smallest eigenvalues of A.

2.10 Vector-Valued Functions and Their Derivatives

So far our focus has been on real-valued functions on $\mathbb{R}^n$, that is, mappings from $\mathbb{R}^n$ to $\mathbb{R}$. In a number of situations, however, it is useful to consider vector-valued functions, that is, mappings (or maps, for short) from $\mathbb{R}^n$ to $\mathbb{R}^m$ where n and m are any positive integers. We shall denote such functions or mappings by boldface letters such as $\mathbf{f}$:

$$\mathbf{f}(\mathbf{x}) = \left(f_1(\mathbf{x}), f_2(\mathbf{x}), \ldots, f_m(\mathbf{x})\right).$$

Examples of the uses of such mappings include the following:

- Functions from $\mathbb{R}$ to $\mathbb{R}^m$ can be interpreted as parametrized curves in $\mathbb{R}^m$. Similarly, maps from $\mathbb{R}^2$ to $\mathbb{R}^m$ give parametrizations of 2-dimensional surfaces in $\mathbb{R}^m$, and so forth.

- In the situation of the chain rule, where w is a function of $x_1, \ldots, x_n$ and the x_j's are functions of other variables $t_1, \ldots, t_k$, we are dealing with a map $\mathbf{x} = \mathbf{g}(\mathbf{t})$ from $\mathbb{R}^k$ to $\mathbb{R}^n$.

- A map $\mathbf{f} : \mathbb{R}^n \to \mathbb{R}^n$ can represent a vector field, that is, a map that assigns to each point $\mathbf{x}$ a vector quantity $\mathbf{f}(\mathbf{x})$ such as a force or a magnetic field.

- A map $\mathbf{f} : \mathbb{R}^n \to \mathbb{R}^n$ can represent a transformation of a region of space obtained by applying geometric operations such as dilations and rotations. For example, under the transformation $\mathbf{f}(\mathbf{x}) = 2\mathbf{x} + \mathbf{a}$, a region in $\mathbb{R}^n$ is expanded by a factor of 2 and then moved over by the amount $\mathbf{a}$.

- A map $\mathbf{f} : \mathbb{R}^n \to \mathbb{R}^n$ can represent the transformation from one coordinate system to another — for example, the polar coordinate map $\mathbf{f}(r, \theta) = (r\cos\theta, \, r\sin\theta)$.

We shall have more to say about all of these interpretations in Chapter 3.

The simplest mappings from $\mathbb{R}^n$ to $\mathbb{R}^m$ are the *linear*[2] ones, that is, maps $\mathbf{f} : \mathbb{R}^n \to \mathbb{R}^m$ that satisfy

$$\mathbf{f}(a\mathbf{x} + b\mathbf{y}) = a\mathbf{f}(\mathbf{x}) + b\mathbf{f}(\mathbf{y}) \qquad (a, b, \in \mathbb{R}, \, \mathbf{x}, \mathbf{y} \in \mathbb{R}^n).$$

Such a map is represented by an $m \times n$ matrix $A = (A_{jk})$, in such a way that if elements of $\mathbb{R}^n$ and $\mathbb{R}^m$ are represented as column vectors, $\mathbf{f}(\mathbf{x})$ is just the matrix product $A\mathbf{x}$. In other words,

$$f_j(\mathbf{x}) = \sum_{k=1}^{n} A_{jk} x_k.$$

You can see that the study of mappings from $\mathbb{R}^n$ to $\mathbb{R}^m$ is complicated, as the study of the linear ones already constitutes the subject of linear algebra! However, the basic ideas of differential calculus generalize easily from the scalar case. The only bits of linear algebra we need for present purposes are the correspondence between linear maps and matrices, the notion of addition and multiplication of matrices, and the notion of determinant; see Appendix A, (A.3)–(A.15) and (A.24)-(A.33).

---

[2]Here we use the word "linear" in the more restrictive sense; see Appendix A, (A.5).

A mapping $\mathbf{f}$ from an open set $S \subset \mathbb{R}^n$ into $\mathbb{R}^m$ is said to be **differentiable** at $\mathbf{a} \in S$ if there is an $m \times n$ matrix L such that

$$(2.84) \qquad \lim_{\mathbf{h} \to 0} \frac{|\mathbf{f}(\mathbf{a} + \mathbf{h}) - \mathbf{f}(\mathbf{a}) - L\mathbf{h}|}{|\mathbf{h}|} = 0.$$

There can only be one such matrix L (the reason is given in the next paragraph), and it is called the **(Fréchet) derivative** of $\mathbf{f}$ at $\mathbf{a}$. Commonly used notations for it include $D\mathbf{f}(\mathbf{a})$, $D_{\mathbf{a}}\mathbf{f}$, $\mathbf{f}'(\mathbf{a})$, and $d f_{\mathbf{a}}$. We shall denote it by $D\mathbf{f}(\mathbf{a})$. Thus, if $\mathbf{f}$ is differentiable on S, the map $D\mathbf{f}$ that assigns to each $\mathbf{a} \in S$ the derivative $D\mathbf{f}(\mathbf{a})$ is a matrix-valued function on S.

We need to verify that there is at most one matrix L satisfying (2.84). If L' is another such matrix, we have

$$|L\mathbf{h} - L'\mathbf{h}| = \big|[\mathbf{f}(\mathbf{a} + \mathbf{h}) - \mathbf{f}(\mathbf{a}) - L'\mathbf{h}] - [\mathbf{f}(\mathbf{a} + \mathbf{h}) - \mathbf{f}(\mathbf{a}) - L\mathbf{h}]\big|$$
$$\leq |\mathbf{f}(\mathbf{a} + \mathbf{h}) - \mathbf{f}(\mathbf{a}) - L'\mathbf{h}| + |\mathbf{f}(\mathbf{a} + \mathbf{h}) - \mathbf{f}(\mathbf{a}) - L\mathbf{h}|,$$

so that $|L\mathbf{h} - L'\mathbf{h}|/|\mathbf{h}| \to 0$. But if $L' \neq L$, we can pick a unit vector $\mathbf{u}$ with $L\mathbf{u} \neq L'\mathbf{u}$. Setting $\mathbf{h} = s\mathbf{u}$, we have $\mathbf{h} \to 0$ as $s \to 0$, but

$$\frac{|L(s\mathbf{u}) - L'(s\mathbf{u})|}{|s\mathbf{u}|} = \frac{|s(L\mathbf{u} - L'\mathbf{u})|}{|s|} = |L\mathbf{u} - L'\mathbf{u}| \not\to 0.$$

This is a contradiction, so $L' = L$.

In the scalar case $m = 1$ (where $\mathbf{f} = f$), the definition of differentiability above coincides with the old one, and $D\mathbf{f}(\mathbf{a})$ is just $\nabla f(\mathbf{a})$, considered as a *row* vector, i.e., a $1 \times n$ matrix. (If we think of $\nabla f(\mathbf{a})$ as a column vector, then $D\mathbf{f}(\mathbf{a}) = [\nabla f(\mathbf{a})]^*$.) Something similar happens when $m > 1$. Indeed, a vector $\mathbf{v}$ approaches the vector $\mathbf{0}$ precisely when each of its components approaches the number 0, so (2.84) is equivalent to the equations

$$\lim_{\mathbf{h} \to 0} \frac{|f_j(\mathbf{a} + \mathbf{h}) - f_j(\mathbf{a}) - L^j \cdot \mathbf{h}|}{|\mathbf{h}|} = 0 \qquad (j = 1, \dots, m)$$

where L^j is the jth row of the matrix L. But these equations say that the components f_j are differentiable at $\mathbf{x} = \mathbf{a}$ and that $\nabla f_j(\mathbf{a}) = L^j$. In short, we have:

2.85. Proposition. *An $\mathbb{R}^m$-valued function $\mathbf{f}$ is differentiable at $\mathbf{a}$ precisely when each of its components $f_1, \dots, f_m$ is differentiable at $\mathbf{a}$. In this case, $D\mathbf{f}(\mathbf{a})$ is the matrix whose jth row is the row vector $\nabla f_j(\mathbf{a})$. In other words,*

$$D\mathbf{f} = \begin{pmatrix} \partial f_1/\partial x_1 & \cdots & \partial f_1/\partial x_n \\ \vdots & & \vdots \\ \partial f_m/\partial x_1 & \cdots & \partial f_m/\partial x_n \end{pmatrix}.$$

The general form of the chain rule can now be stated very simply:

2.86 Theorem (Chain Rule III). *Suppose* $\mathbf{g} : \mathbb{R}^k \to \mathbb{R}^n$ *is differentiable at* $\mathbf{a} \in \mathbb{R}^k$ *and* $\mathbf{f} : \mathbb{R}^n \to \mathbb{R}^m$ *is differentiable at* $\mathbf{g}(\mathbf{a}) \in \mathbb{R}^n$. *Then* $\mathbf{H} = \mathbf{f} \circ \mathbf{g} : \mathbb{R}^k \to \mathbb{R}^m$ *is differentiable at* $\mathbf{a}$, *and*

$$D\mathbf{H}(\mathbf{a}) = D\mathbf{f}(\mathbf{g}(\mathbf{a}))D\mathbf{g}(\mathbf{a}),$$

where the expression on the right is the product of the matrices $D\mathbf{f}(\mathbf{g}(\mathbf{a}))$ *and* $D\mathbf{g}(\mathbf{a})$.

Proof. Differentiability of $\mathbf{H}$ is equivalent to the differentiability of each of its components $H_i = f_i \circ \mathbf{g}$, and for these we have, by Theorem 2.29,

$$\partial_k H_i = (\partial_1 f_i)(\partial_k g_1) + \cdots + (\partial_n f_i)(\partial_k g_n) = \sum_{j=1}^{n} (\partial_j f_i)(\partial_k g_j).$$

($\partial_k H_i$ and $\partial_k g_j$ are to be evaluated at $\mathbf{a}$, $\partial_j f_i$ at $\mathbf{g}(\mathbf{a})$.) But $\partial_k H_i$ is the ikth entry of the matrix $D\mathbf{H}$, and the sum on the right is the ikth entry of the product matrix $(D\mathbf{f})(D\mathbf{g})$, so we are done. $\qquad\square$

Since the product of two matrices gives the composition of the linear transformations defined by those matrices, the chain rule just says that *the linear approximation of a composition is the composition of the linear approximations.*

As we pointed out at the end of §2.1, the mean value theorem is false for vector-valued functions. That is, for a differentiable $\mathbb{R}^m$-valued function $\mathbf{f}$ with $m > 1$, given two points $\mathbf{a}$ and $\mathbf{b}$ there is usually no $\mathbf{c}$ on the line segment between $\mathbf{a}$ and $\mathbf{b}$ such that $\mathbf{f}(\mathbf{b}) - \mathbf{f}(\mathbf{a}) = [D\mathbf{f}(\mathbf{c})][\mathbf{b} - \mathbf{a}]$. However, the main corollary of the mean value theorem, an estimate on $|\mathbf{f}(\mathbf{a}) - \mathbf{f}(\mathbf{b})|$ in terms of a bound on the derivative of $\mathbf{f}$, is still valid. To state it, we employ the following terminology: The **norm** of a linear mapping $A : \mathbb{R}^n \to \mathbb{R}^m$ is the smallest constant C such that $|A\mathbf{x}| \le C|\mathbf{x}|$ for all $\mathbf{x} \in \mathbb{R}^n$. The norm of A is denoted by $\|A\|$; thus,

$$(2.87) \qquad\qquad |A\mathbf{x}| \le \|A\| \, |\mathbf{x}| \qquad (\mathbf{x} \in \mathbb{R}^n).$$

Equivalently, $\|A\| = \max\{|A\mathbf{x}| : |\mathbf{x}| = 1\}$; see Exercise 9. An estimate for $\|A\|$ in terms of the entries A_{jk} is given in Exercise 10.

2.88 Theorem. *Suppose* $\mathbf{f}$ *is a differentiable* $\mathbb{R}^m$-*valued function on an open convex set* $S \subset \mathbb{R}^n$, *and suppose that* $\|D\mathbf{f}(\mathbf{x})\| \le M$ *for all* $\mathbf{x} \in S$. *Then*

$$|\mathbf{f}(\mathbf{b}) - \mathbf{f}(\mathbf{a})| \le M|\mathbf{b} - \mathbf{a}| \text{ for all } \mathbf{a}, \mathbf{b} \in S.$$

Proof. Given a unit vector $\mathbf{u} \in \mathbb{R}^m$, let us consider the scalar-valued function $f_{\mathbf{u}}(\mathbf{x}) = \mathbf{u} \cdot \mathbf{f}(\mathbf{x})$. Clearly $f_{\mathbf{u}}$ is differentiable on S and $\partial_k f_{\mathbf{u}} = \mathbf{u} \cdot \partial_k \mathbf{f} = \sum_{j=1}^{m} u_j \partial_k f_j$. By the mean value theorem (2.39) applied to $f_{\mathbf{u}}$, then, there is a point $\mathbf{c}$ on the line segment between $\mathbf{a}$ and $\mathbf{b}$ (depending on $\mathbf{u}$) such that

$$\mathbf{u} \cdot [\mathbf{f}(\mathbf{b}) - \mathbf{f}(\mathbf{a})] = f_{\mathbf{u}}(\mathbf{b}) - f_{\mathbf{u}}(\mathbf{a}) = [\nabla f_{\mathbf{u}}(\mathbf{c})] \cdot [\mathbf{b} - \mathbf{a}]$$

$$= \sum_{j,k} u_j \partial_k f_j(\mathbf{c})(b_k - a_k) = \mathbf{u} \cdot [(D\mathbf{f}(\mathbf{c}))(\mathbf{b} - \mathbf{a})].$$

Hence, by Cauchy's inequality, the fact that $|\mathbf{u}| = 1$, and (2.87),

$$\left| \mathbf{u} \cdot [\mathbf{f}(\mathbf{b}) - \mathbf{f}(\mathbf{a})] \right| \leq |\mathbf{u}| \, \|D\mathbf{f}(\mathbf{c})\| \, |\mathbf{b} - \mathbf{a}| \leq M |\mathbf{b} - \mathbf{a}|.$$

The desired result now follows by taking $\mathbf{u}$ to be the unit vector in the direction of $\mathbf{f}(\mathbf{b}) - \mathbf{f}(\mathbf{a})$, so that $\mathbf{u} \cdot [\mathbf{f}(\mathbf{b}) - \mathbf{f}(\mathbf{a})] = |\mathbf{f}(\mathbf{b}) - \mathbf{f}(\mathbf{a})|$. (Of course, if $\mathbf{f}(\mathbf{b}) - \mathbf{f}(\mathbf{a}) = \mathbf{0}$, the result is trivial.) □

In the case $m = n$, the Fréchet derivative $D\mathbf{f}$ of a function $\mathbf{f} : \mathbb{R}^n \to \mathbb{R}^n$ is an $n \times n$ matrix of functions, defined on the set S where $\mathbf{f}$ is differentiable, so we can form its determinant. This determinant, a scalar-valued function on S, is called the **Jacobian** of the mapping $\mathbf{f}$. It is sometimes denoted by $J_{\mathbf{f}}$, or, if $\mathbf{y} = \mathbf{f}(\mathbf{x})$, by $\partial(y_1, \ldots, y_n)/\partial(x_1, \ldots, x_n)$:

$$(2.89) \qquad \det D\mathbf{f} = J_{\mathbf{f}} = \frac{\partial(y_1, \ldots, y_n)}{\partial(x_1, \ldots, x_n)}.$$

(The last notation may look peculiar at first, but it is actually quite handy.) Since the determinant of a product of two matrices is the product of the determinants, the chain rule implies that if $\mathbf{y} = \mathbf{f}(\mathbf{x})$ and $\mathbf{x} = \mathbf{g}(\mathbf{t})$ ($\mathbf{t}, \mathbf{x}, \mathbf{y} \in \mathbb{R}^n$), then

$$J_{\mathbf{f} \circ \mathbf{g}}(\mathbf{t}) = J_{\mathbf{f}}(\mathbf{g}(\mathbf{t})) J_{\mathbf{g}}(\mathbf{t}), \quad \text{or}$$

$$(2.90) \qquad \frac{\partial(y_1, \ldots, y_n)}{\partial(t_1, \ldots, t_n)} = \frac{\partial(y_1, \ldots, y_n)}{\partial(x_1, \ldots, x_n)} \frac{\partial(x_1, \ldots, x_n)}{\partial(t_1, \ldots, t_n)}.$$

If $\mathbf{f} : \mathbb{R}^n \to \mathbb{R}^m$ with $n > m$, we can form a number of different Jacobians by singling out m of the independent variables for attention and treating the others as constants, thereby considering $\mathbf{f}$ as a function from $\mathbb{R}^m$ to $\mathbb{R}^m$. In other words, we can look at the determinants of all the $m \times m$ submatrices of the $m \times n$ matrix $D\mathbf{f}$. The last notation in (2.89) is handy in this situation because it allows us to name the m independent variables that have been singled out. Similarly, if $n < m$, we can consider the determinants of the $n \times n$ submatrices of $D\mathbf{f}$ obtained by singling out n of the components of $\mathbf{f}$.

EXAMPLE 1. Let $(u, v) = \mathbf{f}(x, y, z) = (2x + y^3, xe^{5y-7z})$. Then

$$Df(x, y, z) = \begin{pmatrix} 2 & 3y^2 & 0 \\ e^{5y-7z} & 5xe^{5y-7z} & -7xe^{5y-7z} \end{pmatrix},$$

so

$$\frac{\partial(u, v)}{\partial(x, y)} = (10x - 3y^2)e^{5y-7z}, \qquad \frac{\partial(u, v)}{\partial(y, z)} = -21xy^2 e^{5y-7z},$$

$$\frac{\partial(u, v)}{\partial(x, z)} = -14xe^{5y-7z}.$$

EXERCISES

1. Let $(u, v) = \mathbf{f}(x, y, z) = (xyz^2 - 4y^2, 3xy^2 - yz)$. Compute $Df(x, y, z)$, $\partial(u, v)/\partial(x, y)$, $\partial(u, v)/\partial(y, z)$, and $\partial(u, v)/\partial(x, z)$

2. Let $(u, v, w) = \mathbf{f}(x, y) = (x + 6y, 3xy, x^2 - 3y^2)$. Compute $Df(x, y)$, $\partial(u, v)/\partial(x, y)$, $\partial(v, w)/\partial(x, y)$, and $\partial(u, w)/\partial(x, y)$.

3. Define $\mathbf{f} : \mathbb{R}^2 \rightarrow \mathbb{R}^3$ by $\mathbf{f}(u, v) = (u^2 - 5v, ve^{2u}, 2u - \log(1 + v^2))$.
   a. Compute $Df(u, v)$. What is $Df(0, 0)$?
   b. Suppose $\mathbf{g} : \mathbb{R}^2 \rightarrow \mathbb{R}^2$ is of class C^1, $\mathbf{g}(1, 2) = (0, 0)$, and $Dg(1, 2) = \begin{pmatrix} 1 & 2 \\ 3 & 4 \end{pmatrix}$. Compute $D(\mathbf{f} \circ \mathbf{g})(1, 2)$.

4. Define $\mathbf{f} : \mathbb{R}^3 \rightarrow \mathbb{R}^2$ by $\mathbf{f}(x, y, z) = (2x + (y-1)^2 - \sin z, 3x + 2e^{2y-5z})$.
   a. Compute $Df(x, y, z)$. What are $\mathbf{f}(0, 0, 0)$ and $Df(0, 0, 0)$?
   b. Let $\mathbf{g}$ be as in Exercise 3b. Compute $D(\mathbf{g} \circ \mathbf{f})(0, 0, 0)$.

5. Show that if $\mathbf{f} : \mathbb{R}^n \rightarrow \mathbb{R}^m$ is defined by $\mathbf{f}(\mathbf{x}) = A\mathbf{x} + \mathbf{b}$, where A is an $m \times n$ matrix and $\mathbf{b} \in \mathbb{R}^m$, then $Df(\mathbf{x}) = A$ for all $\mathbf{x}$.

6. Suppose $f : \mathbb{R}^n \rightarrow \mathbb{R}$ is of class C^2; then ∇f is a C^1 mapping from $\mathbb{R}^n$ to itself. Show that $D(\nabla f)$ is the Hessian matrix of f.

7. Suppose $\mathbf{f}$ and $\mathbf{g}$ are differentiable mappings from $\mathbb{R}^n$ to $\mathbb{R}^m$. Show that their dot product, $h(\mathbf{x}) = \mathbf{f}(\mathbf{x}) \cdot \mathbf{g}(\mathbf{x})$, is a differentiable real-valued function on $\mathbb{R}^n$, and that

$$\nabla h(\mathbf{x}) = [Df(\mathbf{x})]^* \mathbf{g}(\mathbf{x}) + [Dg(\mathbf{x})]^* \mathbf{f}(\mathbf{x}),$$

if we think of $\nabla h(\mathbf{x})$, $\mathbf{f}(\mathbf{x})$, and $\mathbf{g}(\mathbf{x})$ as column vectors. (Here A^* denotes the transpose of the matrix A; see Appendix A, (A.15).)

8. Suppose that $w = f(x, y, t, s)$ and x and y are also functions of t and s (the situation depicted in Figure 2.3). The total dependence of w on t and s can be expressed by writing $w = f(\mathbf{g}(t, s))$ where $\mathbf{g}(t, s) = (x(t, s), y(t, s), t, s)$. Show that the chain rule (2.86), applied to the composite function $f \circ \mathbf{g}$, yields the same result as the one obtained in §2.3.

9. Let $A : \mathbb{R}^n \to \mathbb{R}^m$ be a linear map.
  a. Show that the function $\varphi(\mathbf{x}) = |A\mathbf{x}|$ has a maximum value on the set $\{\mathbf{x} : |\mathbf{x}| = 1\}$.
  b. Let M be the maximum in part (a). Show that $|A\mathbf{x}| \leq M|\mathbf{x}|$ for all $\mathbf{x} \in \mathbb{R}^n$, with equality for at least one unit vector $\mathbf{x}$. Deduce that $M = \|A\|$.

10. Let $A : \mathbb{R}^n \to \mathbb{R}^m$ be a linear map.
  a. Show that $\|A\| \leq \sqrt{m} \max_{j=1}^{m} (\sum_{k=1}^{n} |A_{jk}|)$. (*Hint:* Use (1.3).)
  b. Show that this inequality is an equality when the matrix of A is given by $A_{j1} = 1$ and $A_{jk} = 0$ for $k > 1$ ($1 \leq j \leq m$).

Chapter 3

THE IMPLICIT FUNCTION
THEOREM AND ITS APPLICATIONS

In this chapter we take up the general question of the local solvability of systems of equations involving nonlinear differentiable functions. The main result is the implicit function theorem, one of the major theoretical results of advanced calculus. Among other things, it provides the key to answering many questions about relations between analytic properties of functions and geometric properties of the sets they define. We shall present some of its applications to the study of geometric transformations, coordinate systems, and various ways of representing curves, surfaces, and smooth sets of higher dimension.

3.1   The Implicit Function Theorem

In this section we consider the problem of solving an equation $F(x_1, \ldots, x_n) = 0$ for one of the variables x_j as a function of the remaining $n-1$ variables, or more generally of solving a system of k such equations for k of the variables as functions of the remaining $n-k$ variables.

We begin with the case of a single equation, and to develop some feeling for the geometry of the problem we consider the cases $n = 2$ and $n = 3$. For $n = 2$ we are given an equation $F(x, y) = 0$ relating the variables x and y, and we ask when we can solve for y as a function of x or vice versa. Geometrically, the set $S = \{(x, y) : F(x, y) = 0\}$ will usually be some sort of curve, and our question is: When can S be represented as the graph of a function $y = f(x)$ or $x = g(y)$? Likewise, for $n = 3$, the set where $F(x, y, z) = 0$ will usually be a surface, and we ask when this surface can be represented as the graph of a function $z = f(x, y)$, $y = g(x, z)$, or $x = h(y, z)$.

Simple examples show that it is usually impossible to represent the whole set $S = \{\mathbf{x} : F(\mathbf{x}) = 0\}$ as the graph of a function. For example, if $n = 2$ and $F(x, y) = x^2 + y^2 - 1$, the set S is the unit circle. We can represent the upper or lower semicircle as the graph of $f(x) = \pm\sqrt{1 - x^2}$, and the right or left semicircle as the graph of $g(y) = \pm\sqrt{1 - y^2}$, but the whole circle is not a graph. Thus, in order to get reasonable results, we must be content only to represent pieces of S as graphs. More specifically, our object will be to represent a piece of S in the neighborhood of a given point $\mathbf{a} \in S$ as a graph.

Since we want to single out one of the variables as the one to be solved for, we make a little change of notation: We denote the number of variables by $n + 1$ and denote the last variable by y rather than x_{n+1}. We then have the following precise analytical statement of the problem:

> *Given a function $F(\mathbf{x}, y)$ of class C^1 and a point $(\mathbf{a}, b)$ satisfying $F(\mathbf{a}, b) = 0$, when is there*
> *i. a function $f(\mathbf{x})$, defined in some open set in $\mathbb{R}^n$ containing $\mathbf{a}$, and*
> *ii. an open set $U \subset \mathbb{R}^{n+1}$ containing $(\mathbf{a}, b)$,*
> *such that for $(\mathbf{x}, y) \in U$,*

$$F(\mathbf{x}, y) = 0 \quad \Longleftrightarrow \quad y = f(\mathbf{x})?$$

We do *not* try to specify in advance how big the open sets in question will be; that will depend strongly on the nature of the function F.

The key to the answer is to look at the linear case. If

$$L(x_1, \ldots, x_n, y) = \alpha_1 x_1 + \cdots + \alpha_n x_n + \beta y + c,$$

the solution is obvious: The equation $L(\mathbf{x}, y) = 0$ can be solved for y if and only if the coefficient β is nonzero. But *near a given point $(\mathbf{a}, b)$, every differentiable function $F(\mathbf{x}, y)$ is approximately linear;* in fact, if $F(\mathbf{a}, b) = 0$,

$$F(\mathbf{x}, y) = [\partial_1 F(\mathbf{a}, b)](x_1 - a_1) + \cdots + [\partial_n F(\mathbf{a}, b)](x_n - a_n)$$
$$+ [\partial_y F(\mathbf{a}, b)](y - b) + \text{small error}.$$

If the "small error" were not there, the equation $F(\mathbf{x}, y) = 0$ could be solved for y precisely when $\partial_y F(\mathbf{a}, b) \neq 0$. We now show that the condition $\partial_y F(\mathbf{a}, b) \neq 0$ is still the appropriate one when the error term is taken into account.

3.1 Theorem (The Implicit Function Theorem for a Single Equation).   *Let $F(\mathbf{x}, y)$ be a function of class C^1 on some neighborhood of a point $(\mathbf{a}, b) \in \mathbb{R}^{n+1}$. Suppose that $F(\mathbf{a}, b) = 0$ and $\partial_y F(\mathbf{a}, b) \neq 0$. Then there exist positive numbers r_0, r_1 such that the following conclusions are valid.*

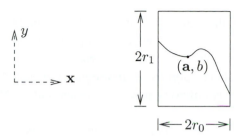

FIGURE 3.1: The geometry of the implicit function theorem. $\partial_y F > 0$ in the box, $F > 0$ on the top side, $F < 0$ on the bottom side, and $F = 0$ on the curve.

a. *For each $\mathbf{x}$ in the ball $|\mathbf{x} - \mathbf{a}| < r_0$ there is a unique y such that $|y - b| < r_1$ and $F(\mathbf{x}, y) = 0$. We denote this y by $f(\mathbf{x})$; in particular, $f(\mathbf{a}) = b$.*
b. *The function f thus defined for $|\mathbf{x} - \mathbf{a}| < r_0$ is of class C^1, and its partial derivatives are given by*

$$(3.2) \qquad \partial_j f(\mathbf{x}) = -\frac{\partial_j F(\mathbf{x}, f(\mathbf{x}))}{\partial_y F(\mathbf{x}, f(\mathbf{x}))}.$$

Notes.
 i. The number r_0 may be very small, and there is no way to estimate its size without further hypotheses on F.
 ii. The formula (3.2) for $\partial_j f$ is, of course, the one obtained via the chain rule by differentiating the equation $F(\mathbf{x}, f(\mathbf{x})) = 0$.

Proof. We first prove (a). We may assume that $\partial_y F(\mathbf{a}, b) > 0$ (by replacing F by $-F$ if necessary). Since $\partial_y F$ is continuous, it remains positive in some neighborhood of $(\mathbf{a}, b)$, say for $|\mathbf{x}-\mathbf{a}| < r_1$ and $|y-b| < r_1$. On this set, $F(\mathbf{x}, y)$ is a strictly increasing function of y for each fixed $\mathbf{x}$. In particular, since $F(\mathbf{a}, b) = 0$ we have $F(\mathbf{a}, b + r_1) > 0$ and $F(\mathbf{a}, b - r_1) < 0$. The continuity of F then implies that for some $r_0 \le r_1$ we have $F(\mathbf{x}, b + r_1) > 0$ and $F(\mathbf{x}, b - r_1) < 0$ for $|\mathbf{x} - \mathbf{a}| < r_0$.

In short, for each $\mathbf{x}$ in the ball $B = \{\mathbf{x} : |\mathbf{x}-\mathbf{a}| < r_0\}$ we have $F(\mathbf{x}, b-r_1) < 0$, $F(\mathbf{x}, b + r_1) > 0$, and $F(\mathbf{x}, y)$ is strictly increasing as a function of y for $|y - b| < r_1$. It follows from the intermediate value theorem that there is a unique y for each $\mathbf{x} \in B$ that satisfies $|y - b| < r_1$ and $F(\mathbf{x}, y) = 0$, which establishes (a). See Figure 3.1.

Next we observe that the function $y = f(\mathbf{x})$ thus defined is continuous at $\mathbf{x} = \mathbf{a}$; in other words, for any $\epsilon > 0$ there is a $\delta > 0$ such that $|f(\mathbf{x}) - f(\mathbf{a})| < \epsilon$

whenever $|\mathbf{x} - \mathbf{a}| < \delta$. Indeed, the argument just given shows that $|f(\mathbf{x}) - f(\mathbf{a})| = |y - b| < r_1$ whenever $|\mathbf{x} - \mathbf{a}| < r_0$, and we could repeat that argument with r_1 replaced by any smaller number ϵ to obtain an appropriate δ in place of r_0.

In fact, this argument can also be applied with $\mathbf{a}$ replaced by any other point $\mathbf{x}_0$ in the ball B to show that f is continuous at $\mathbf{x}_0$. To recapitulate it briefly: Given $\epsilon > 0$, there exists $\delta > 0$ such that if $|\mathbf{x} - \mathbf{x}_0| < \delta$ we have $F(\mathbf{x}, y_0 - \epsilon) < 0$ and $F(\mathbf{x}, y_0 + \epsilon) > 0$, where $y_0 = f(\mathbf{x}_0)$. For each such $\mathbf{x}$ there is a unique y such that $|y - y_0| < \epsilon$ and $F(\mathbf{x}, y) = 0$, and that y is $f(\mathbf{x})$; hence $|f(\mathbf{x}) - f(\mathbf{x}_0)| = |y - y_0| < \epsilon$.

Now that we know that f is continuous on B, we can show that its partial derivatives $\partial_j f$ exist on B and are given by (3.2) — which also shows that they are continuous. Given $\mathbf{x} \in B$ and a (small) real number h, let $y = f(\mathbf{x})$ and

$$k = f(\mathbf{x} + \mathbf{h}) - f(\mathbf{x}), \text{ where}$$
$$\mathbf{h} = (0, \ldots, 0, h, 0, \ldots, 0) \text{ with the } h \text{ in the } j\text{th place.}$$

Then $y + k = f(\mathbf{x} + \mathbf{h})$, so $F(\mathbf{x} + \mathbf{h}, y + k) = F(\mathbf{x}, y) = 0$. Hence, by the mean value theorem,

$$0 = F(\mathbf{x} + \mathbf{h}, y + k) - F(\mathbf{x}, y)$$
$$= h\partial_j F(\mathbf{x} + t\mathbf{h}, y + tk) + k\partial_y F(\mathbf{x} + t\mathbf{h}, y + tk)$$

for some $t \in (0, 1)$. Rearranging this equation gives

$$\frac{f(\mathbf{x} + \mathbf{h}) - f(\mathbf{x})}{h} = \frac{k}{h} = -\frac{\partial_j F(\mathbf{x} + t\mathbf{h}, y + tk)}{\partial_y F(\mathbf{x} + t\mathbf{h}, y + tk)}.$$

Now let $h \to 0$. Since f is continuous we also have $k \to 0$, and then since $\partial_j F$ and $\partial_y F$ are continuous and $\partial_y F \neq 0$, passage to the limit yields (3.2). $\square$

3.3 Corollary. *Let F be a function of class C^1 on $\mathbb{R}^n$, and let $S = \{\mathbf{x} : F(\mathbf{x}) = 0\}$. For every $\mathbf{a} \in S$ such that $\nabla F(\mathbf{a}) \neq \mathbf{0}$ there is a neighborhood N of $\mathbf{a}$ such that $S \cap N$ is the graph of a C^1 function.*

Proof. Since $\nabla F(\mathbf{a}) \neq \mathbf{0}$, we have $\partial_j F(\mathbf{a}) \neq 0$ for some j. The equation $F = 0$ can then be solved to yield x_j as a C^1 function of the remaining variables near the point $\mathbf{a}$. $\square$

EXAMPLE 1. Let $F(x, y) = x - y^2 - 1$, for which $\partial_x F(x, y) = 1$ and $\partial_y F(x, y) = -2y$. First, $\partial_x F$ is never 0, so the implicit function theorem guarantees that the equation $F(x, y) = 0$ can be solved for x locally near any point (a, b) for which $F(a, b) = 0$. Of course, for this particular F it is easy to solve for x explicitly — namely, $x = y^2 + 1$ — and this solution is valid not just locally but globally. Next, $\partial_y F(a, b) = 0$ precisely when $b = 0$, so the implicit function theorem guarantees that the equation $F(x, y) = 0$ can be solved uniquely for y near any point (a, b) such that $F(a, b) = 0$ and $b \neq 0$. In fact, the possible solutions are $y = \sqrt{x - 1}$ and $y = -\sqrt{x - 1}$. For x very close to a only one of these solutions will be very close to b — namely, $\sqrt{x - 1}$ if $b > 0$ and $-\sqrt{x - 1}$ if $b < 0$ — and this solution is the one that figures in the implicit function theorem. Also, these solutions are defined only for $x \geq 1$, so the number r_0 in the statement of the implicit function theorem is $a - 1$. Finally, we have $F(1, 0) = 0$, but the equation $F(x, y) = 0$ cannot be solved uniquely for y as a function of x in any neighborhood of $(1, 0)$: If $x > 1$ there are two solutions, both equally close to 0, and if $x < 1$ there are none.

EXAMPLE 2. For a contrast with Example 1, let $G(x, y) = x - e^{1-x} - y^3$. First, $\partial_x G(a, b) = 1 + e^{1-a} > 1$ for all (a, b), so the implicit function theorem guarantees that the equation $G(x, y) = 0$ can be solved for x locally near any point (a, b) such that $G(a, b) = 0$. It is not hard to see (Exercise 4) that there is a single solution that works globally, but there is no nice formula for this solution in terms of elementary functions. Next, $\partial_y G(a, b) = -3b^2$, so the implicit function theorem guarantees that the equation $G(x, y) = 0$ can be solved for y as a C^1 function of x locally near any point (a, b) such that $G(a, b) = 0$ and $b \neq 0$. In fact, the solution is $y = (x - e^{1-x})^{1/3}$, which is globally uniquely defined but fails to be differentiable at the point where $y = 0$ (i.e., $x = 1$).

We now turn to the more general problem of solving several equations simultaneously for some of of the variables occurring in them. This will require some facts about invertible matrices and determinants, for which we refer to Appendix A, (A.24)–(A.33) and (A.50)–(A.55). To fix the notation, we shall consider k functions $F_1, \ldots, F_k$ of $n + k$ variables $x_1, \ldots, x_n, y_1, \ldots, y_k$, and ask when we can solve the equations

$$F_1(x_1, \ldots, x_n, y_1, \ldots, y_k) = 0,$$
(3.4)
$$\vdots$$
$$F_k(x_1, \ldots, x_n, y_1, \ldots, y_k) = 0$$

for the y's in terms of the x's. We shall use vector notation to abbreviate (3.4) as

(3.5)
$$\mathbf{F}(\mathbf{x}, \mathbf{y}) = \mathbf{0}.$$

We assume that $\mathbf{F}$ is of class C^1 near a point $(\mathbf{a}, \mathbf{b})$ such that $\mathbf{F}(\mathbf{a}, \mathbf{b}) = \mathbf{0}$, and we ask when (3.5) determines $\mathbf{y}$ as a C^1 function of $\mathbf{x}$ in some neighborhood of $(\mathbf{a}, \mathbf{b})$.

Again the key to the problem is to consider the linear case,

$$(3.6) \qquad\qquad A\mathbf{x} + B\mathbf{y} + \mathbf{c} = \mathbf{0},$$

where A is a $k \times n$ matrix, B is a $k \times k$ matrix, and $\mathbf{c} \in \mathbb{R}^k$. Here the criterion for solvability is obvious: The matrix B must be invertible, in which case the solution is $\mathbf{y} = -B^{-1}(A\mathbf{x}+\mathbf{c})$. Now, the linear approximation to the equation (3.5) near the point $(\mathbf{a}, \mathbf{b})$ is an equation of the form (3.6) in which the matrix B is the (partial) Fréchet derivative of $\mathbf{F}$ with respect to the variables $\mathbf{y}$, evaluated at $(\mathbf{a}, \mathbf{b})$:

$$(3.7) \qquad\qquad B_{ij} = \frac{\partial F_i}{\partial y_j}(\mathbf{a}, \mathbf{b}) \qquad (1 \le i, j \le k).$$

Hence, the crucial requirement is that

$$(3.8) \qquad\qquad \text{the matrix } B \text{ defined by (3.7) is invertible.}$$

Invertibility of a matrix can be characterized in a number of different ways, as discussed in Appendix A, (A.52). For example, (3.8) can be expressed more geometrically as the condition that the gradient vectors $\nabla_{\mathbf{y}} F_j = (\partial_{y_1} F_j, \ldots, \partial_{y_k} F_j)$, $1 \le j \le k$, are linearly independent at $(\mathbf{a}, \mathbf{b})$. However, the version of (3.8) that is directly used in the proof of the following theorem, as well as in many of its applications, is that $\det B \ne 0$. We therefore state the theorem in these terms.

3.9 Theorem (The Implicit Function Theorem for a System of Equations).
Let $\mathbf{F}(\mathbf{x}, \mathbf{y})$ be an $\mathbb{R}^k$-valued function of class C^1 on some neighborhood of a point $(\mathbf{a}, \mathbf{b}) \in \mathbb{R}^{n+k}$ and let $B_{ij} = (\partial F_i / \partial y_j)(\mathbf{a}, \mathbf{b})$. Suppose that $\mathbf{F}(\mathbf{a}, \mathbf{b}) = \mathbf{0}$ and $\det B \ne 0$. Then there exist positive numbers r_0, r_1 such that the following conclusions are valid.
  a. *For each $\mathbf{x}$ in the ball $|\mathbf{x} - \mathbf{a}| < r_0$ there is a unique $\mathbf{y}$ such that $|\mathbf{y} - \mathbf{b}| < r_1$ and $\mathbf{F}(\mathbf{x}, \mathbf{y}) = \mathbf{0}$. We denote this $\mathbf{y}$ by $\mathbf{f}(\mathbf{x})$; in particular, $\mathbf{f}(\mathbf{a}) = \mathbf{b}$.*
  b. *The function $\mathbf{f}$ thus defined for $|\mathbf{x}-\mathbf{a}| < r_0$ is of class C^1, and its partial derivatives $\partial_{x_j} \mathbf{f}$ can be computed by differentiating the equations $\mathbf{F}(\mathbf{x}, \mathbf{f}(\mathbf{x})) = \mathbf{0}$ with respect to x_j and solving the resulting linear system of equations for $\partial_{x_j} f_1, \ldots, \partial_{x_j} f_k$.*

Proof. The proof is presented in Appendix B.2 (Theorem B.2). In a nutshell, it proceeds by induction on k. The hypothesis that $\det B \ne 0$ implies that at least one of the $(k-1) \times (k-1)$ submatrices of B is invertible. By inductive hypothesis, one can solve the corresponding system of $k-1$ equations for $k-1$ of the variables

y_j; then, after substituting the results into the remaining equation, one solves that equation for the remaining variable. The main difficulty is in showing that the implicit function theorem can be applied to the last equation. □

EXAMPLE 3. Consider the problem of solving the equations

(3.10) $$x - yu^2 = 0, \qquad xy + uv = 0$$

for u and v as functions of x and y. Setting $F = x - yu^2$ and $G = xy + uv$, we see that

$$\frac{\partial(F, G)}{\partial(u, v)} = \det \begin{pmatrix} -2yu & 0 \\ v & u \end{pmatrix} = -2yu^2,$$

so the implicit function theorem guarantees a local solution near any point (x_0, y_0, u_0, v_0) at which (3.10) holds provided that $-2y_0 u_0^2 \neq 0$, that is, $y_0 \neq 0$ and $u_0 \neq 0$. Notice that under this condition, the first equation in (3.10) implies that $x_0 \neq 0$ and that x_0 and y_0 have the same sign; the second equation then implies that $v_0 \neq 0$ and that u_0 and v_0 have opposite signs.

It is not hard to find the solution explicitly:

$$u = \pm\sqrt{\frac{x}{y}}, \qquad v = \mp\sqrt{xy^3},$$

the signs of u and v being the same as the signs of u_0 and v_0, respectively. This solution is valid for all (x, y) in the same quadrant as (x_0, y_0). The problems that arise if $y_0 = 0$ or $u_0 = 0$ are evident: If $y_0 = 0$, then the formula for u does not even make sense for $y = y_0$; if $u_0 = 0$, then x_0 must also be 0, and the square roots present the same sort of problem as in Example 1.

EXERCISES

1. Investigate the possibility of solving the equation $x^2 - 4x + 2y^2 - yz = 1$ for each of its variables in terms of the other two near the point $(2, -1, 3)$. Do this both by checking the hypotheses of the implicit function theorem and by explicitly computing the solutions.

2. Show that the equation $x^2 + 2xy + 3y^2 = c$ can be solved either for y as a C^1 function of x or for x as a C^1 function of y (but perhaps not both) near any point (a, b) such that $a^2 + 2ab + 3b^2 = c$, provided that $c > 0$. What happens if $c = 0$ or if $c < 0$?

3. Can the equation $(x^2 + y^2 + 2z^2)^{1/2} = \cos z$ be solved uniquely for y in terms of x and z near $(0, 1, 0)$? For z in terms of x and y?

4. Sketch the graph of the equation $x - e^{1-x} - y^3 = 0$ in Example 2. Show graphically that for each x there is a unique y satisfying this equation, and vice versa.

5. Suppose $F(x, y)$ is a C^1 function such that $F(0, 0) = 0$. What conditions on F will guarantee that the equation $F(F(x, y), y) = 0$ can be solved for y as a C^1 function of x near $(0, 0)$?

6. Investigate the possibility of solving the equations $xy + 2yz - 3xz = 0$, $xyz + x - y = 1$ for two of the variables as functions of the third near the point $(x, y, z) = (1, 1, 1)$.

7. Investigate the possibility of solving the equations $u^3 + xv - y = 0$, $v^3 + yu - x = 0$ for any two of the variables as functions of the other two near the point $(x, y, u, v) = (0, 1, 1, -1)$.

8. Investigate the possibility of solving the equations $xy^2 + xzu + yv^2 = 3$ and $u^3yz + 2xv - u^2v^2 = 2$ for u and v as functions of x, y, and z near $x = y = z = u = v = 1$.

9. Can the equations $x^2 + y^2 + z^2 = 6$, $xy + tz = 2$, $xz + ty + e^t = 0$ be solved for x, y, and z as C^1 functions of t near $(x, y, z, t) = (-1, -2, 1, 0)$?

3.2   Curves in the Plane

In this section we examine the relations between various ways of representing smooth curves in the plane. Here we shall take "smooth" to mean that the curve possesses a tangent line at each point and that the tangent line varies continuously with the point of tangency. (Don't worry if this last continuity condition seems a little unclear; we will reformulate it more precisely below.) Thus "smooth" is the geometric equivalent of "C^1."

There are three common ways of representing smooth curves in the plane $\mathbb{R}^2$:

i. as the graph of a function, $y = f(x)$ or $x = f(y)$, where f is of class C^1;

ii. as the locus[1] of an equation $F(x, y) = 0$, where F is of class C^1;

iii. parametrically, as the range of a C^1 function $\mathbf{f} : (a, b) \to \mathbb{R}^2$.

Of these, (i) is the simplest, and it a special case of the other two. Indeed, the curve given by $y = f(x)$ is the locus of the equation $F(x, y) = 0$ where $F(x, y) =$

---

[1]The **locus** of an equation $F(\mathbf{x}) = c$ is the set of all $\mathbf{x}$ that satisfy the equation.

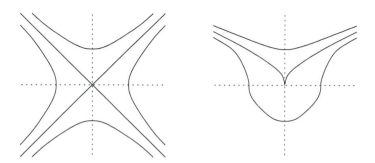

FIGURE 3.2: Left: The sets $x^2 - y^2 = c$ for $c = \pm 1$ (the hyperbolas) and $c = 0$ (the cross). Right: The sets $y^3 = x^2 + c$ for $c = 1$ (top), $c = 0$ (middle), and $c = -1$ (bottom).

$y - f(x)$, and it is also the range of the map $\mathbf{f}(t) = (t, f(t))$. The representations (ii) and (iii) are more flexible, but they are also too general as they stand because the sets represented by them may not be smooth curves. Consider the following examples, in which c denotes an arbitrary real constant:

EXAMPLE 1. Let $F(x, y) = x^2 + y^2 - c$. The set where $F(x, y) = 0$ is a smooth curve (a circle) if $c > 0$, but it is a single point if $c = 0$ and it is the empty set if $c < 0$.

EXAMPLE 2. Let $G(x, y) = x^2 - y^2 - c$. The set where $G(x, y) = 0$ is a hyperbola (the union of two disjoint smooth curves) if $c \neq 0$, but if $c = 0$ it is the union of the two lines $y = x$ and $y = -x$. The latter set looks like a smooth curve in a neighborhood of any of its points *except* the origin, where the two lines cross. See Figure 3.2.

EXAMPLE 3. Let $H(x, y) = y^3 - x^2 - c$. The set where $H(x, y) = 0$ is a smooth curve if $c \neq 0$, but when $c = 0$ it is a curve with a sharp cusp at the origin. The latter set can also be described parametrically by $\mathbf{f}(t) = (t^3, t^2)$. See Figure 3.2.

EXAMPLE 4. The function $\mathbf{g}(t) = (\sin^2 t, \cos^2 t)$ is C^1, but its range is the line segment from $(0, 1)$ to $(1, 0)$. The point $\mathbf{g}(t)$ traverses this line segment from $(0, 1)$ to $(1, 0)$ as t goes from 0 to $\frac{1}{2}\pi$, then traverses it in the reverse direction as t goes from $\frac{1}{2}\pi$ to π, and this pattern is repeated on every interval $[n\pi, (n+1)\pi]$.

In these examples, the functions in question are all of class C^1, but the sets they describe fail to be smooth curves at certain points. However, they share a common feature: The points where smoothness fails — namely, the origin in Examples 1–3 and the points $(0, 1)$ and $(1, 0)$ in Example 4 — are the points where the derivatives of the relevant functions vanish. That is, the origin is the one and only point where the gradients ∇F, ∇G, and ∇H vanish, and it is the image under $\mathbf{f}$ of the one and only point $(t = 0)$ where $\mathbf{f}'$ vanishes. Moreover, $(0, 1)$ and $(1, 0)$ are the images under $\mathbf{g}$ of the points $t = n\pi$ and $t = (n + \frac{1}{2})\pi$ where $\mathbf{g}'(t) = \mathbf{0}$.

This suggests that it might be a good idea to impose the extra conditions that $\nabla F \neq \mathbf{0}$ on the set where $F = 0$ in (ii) and that $\mathbf{f}'(t) \neq \mathbf{0}$ in (iii). And indeed, with the help of the implicit function theorem, it is easy to see that under these extra conditions the representations (i)–(iii) are all *locally* equivalent. That is, if a curve is represented in one of the forms (i)–(iii) and $\mathbf{a}$ is a point on the curve, at least a small piece of the curve including the point $\mathbf{a}$ can also be represented in the other two forms.

We now make this precise. Since (i) is more special than either (ii) or (iii), as we have observed above, it is enough to see that a curve given by (ii) or (iii) can also be represented in the form (i).

3.11 Theorem.

a. Let F be a real-valued function of class C^1 on an open set in $\mathbb{R}^2$, and let $S = \{(x, y) : F(x, y) = 0\}$. If $\mathbf{a} \in S$ and $\nabla F(\mathbf{a}) \neq \mathbf{0}$, there is a neighborhood N of $\mathbf{a}$ in $\mathbb{R}^2$ such that $S \cap N$ is the graph of a C^1 function f (either $y = f(x)$ or $x = f(y)$).

b. Let $\mathbf{f} : (a, b) \to \mathbb{R}^2$ be a function of class C^1. If $\mathbf{f}'(t_0) \neq \mathbf{0}$, there is an open interval I containing t_0 such that the set $\{\mathbf{f}(t) : t \in I\}$ is the graph of a C^1 function f (either $y = f(x)$ or $x = f(y)$).

Proof. Part (a) is a special case of Corollary 3.3. As for (b), let $\mathbf{f} = (\varphi, \psi)$. If $\mathbf{f}'(t_0) \neq \mathbf{0}$, then either $\varphi'(t_0) \neq 0$ or $\psi'(t_0) \neq 0$; let's assume that the former condition holds. Let $F(x, t) = x - \varphi(t)$ and $x_0 = \varphi(t_0)$. Since $\partial_t F(x_0, t_0) = -\varphi'(t_0) \neq 0$, the implicit function theorem guarantees that the equation $x = \varphi(t)$ can be solved for t as a C^1 function of x, say $t = \omega(x)$, in some neighborhood of the point (x_0, t_0). But then $(\varphi(t), \psi(t)) = (x, \psi(\omega(x)))$ for t in some neighborhood I of t_0; that is, the set $\{\mathbf{f}(t) : t \in I\}$ is the graph of the C^1 function $f = \psi \circ \omega$. (If $\psi'(t_0) \neq 0$ instead, one can make the same argument with x and y switched.) $\square$

It should be noted that the conditions of nonvanishing derivatives in Theorem 3.11 are automatically satisfied in the special case where the curve is given in the form (i). That is, if $F(x, y) = y - f(x)$, then $\partial F/\partial y = 1$, so ∇F never vanishes; similarly, if $\mathbf{f}(t) = (t, f(t))$, then $\mathbf{f}'(t) = (1, f'(t)) \neq (0, 0)$.

With this in mind, we may make the following more formal definition of a smooth curve: A set $S \subset \mathbb{R}^2$ is a **smooth curve** if (a) S is connected, and (b) every $\mathbf{a} \in S$ has a neighborhood N such that $S \cap N$ is the graph of a C^1 function f (either $y = f(x)$ or $x = f(y)$). This agrees with the notion of smooth curve indroduced at the beginning of this section: The curve described by $y = f(x)$ has a tangent line at each point $(x_0, f(x_0))$, and that line is given by an equation $y - f(x_0) = f'(x_0)(x - x_0)$ whose coefficients depend continuously on x_0.

It should be emphasized that the conditions $\nabla F \neq \mathbf{0}$ and $\mathbf{f}' \neq \mathbf{0}$ in Theorem 3.11, are *sufficient* for the smoothness of the associated curves but not *necessary*. In other words, the condition $\nabla F(\mathbf{a}) = \mathbf{0}$ or $\mathbf{f}'(t_0) = \mathbf{0}$ allows the possibility of non-smoothness at $\mathbf{a}$ or $\mathbf{f}(t_0)$ but does not guarantee it. For example, suppose $G(x, y)$ is a C^1 function whose gradient does not vanish on the set $S = \{(x, y) : G(x, y) = 0\}$, so that S is a smooth curve, and let $F = G^2$. Then the set where $F = 0$ coincides with S, but $\nabla F = 2G\nabla G \equiv \mathbf{0}$ on S! Similarly, as t ranges over the interval $(-1, 1)$, the functions $\mathbf{f}(t)$ and $\mathbf{g}(t) = \mathbf{f}(t^3)$ describe the same curve, but $\mathbf{g}'(0) = \mathbf{0}$ no matter what $\mathbf{f}$ is.

The following question remains: Suppose S is a subset of $\mathbb{R}^2$ that is described in one of the forms (i)–(iii), and suppose that the regularity condition $\nabla F \neq \mathbf{0}$ on S (in case (ii)) or $\mathbf{f}'(t) \neq \mathbf{0}$ for all $t \in (a, b)$ (in case (iii)) is satisfied. Theorem 3.11 shows that every sufficiently small piece of S is a smooth curve, but is the entire set S a smooth curve? In case (i) the answer is clearly *yes*. However, in cases (ii) and (iii) the answer may be *no*.

The trouble in case (ii) is that S may be disconnected. For example, if $F = GH$, then S is the union of the sets $\{(x, y) : G(x, y) = 0\}$ and $\{(x, y) : H(x, y) = 0\}$, and these sets may well be disjoint and form a disconnection of S. (Also see Exercise 6.)

EXAMPLE 5. Let $F(x, y) = (x^2 + y^2 - 1)(x^2 + y^2 - 2)$. Then the set where $F = 0$ is the union of two disjoint circles centered at the origin. See Figure 3.3.

EXAMPLE 6. Let $F(x, y) = (x^2 + y^2 - 1)(x^2 + y^2 - 2x)$. Then the set S where $F = 0$ is the union of the circles of radius 1 about $(0, 0)$ and $(1, 0)$. These circles intersect at the points $(\frac{1}{2}, \pm\frac{1}{2}\sqrt{3})$, and S is not a smooth curve at these points. The reader may verify that $\nabla F = (0, 0)$ at these points, in accordance with Theorem 3.11. See Figure 3.3 and also Exercise 6.

As for the representation (iii), a set of the form $\{\mathbf{f}(t) : a < t < b\}$ is necessarily connected if $\mathbf{f}$ is continuous (Theorem 1.26). However, the function $\mathbf{f}(t)$ may not be one-to-one, in which case the curve it describes may be traced more than once (as we observed in Example 4) or may cross itself. *These phenomena can happen*

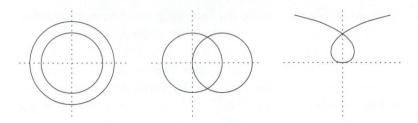

FIGURE 3.3: The sets in Examples 5 (left), 6 (middle), and 8 (right).

even if $\mathbf{f}'(t)$ *never vanishes.* Consequently, the condition $\mathbf{f}'(t) \neq \mathbf{0}$ is not sufficient to guarantee that the set $S = \{\mathbf{f}(t) : t \in (a, b)\}$ is a smooth curve, only that the pieces of it obtained by restricting t to small intervals are smooth curves. In practice, sometimes one simply imposes the extra assumption that $\mathbf{f}$ is one-to-one in order to avoid various pitfalls.

EXAMPLE 7. Let $\mathbf{f}(t) = (\cos t, \sin t)$. Then $\mathbf{f}'(t) = (-\sin t, \cos t)$ is never zero since the sine and cosine functions have no common zeros, but $\mathbf{f}$ is one-to-one on the interval (a, b) only when $b - a \leq 2\pi$. The range $\{\mathbf{f}(t) : t \in \mathbb{R}\}$ of $\mathbf{f}$ is a smooth curve (namely, the unit circle), but in order to obtain a one-to-one correspondence between points on the circle and values of the parameter t, one must restrict t to an interval of the form $[a,\, a + 2\pi)$ or $(a,\, a + 2\pi]$.

EXAMPLE 8. Let $\mathbf{f}(t) = (t^3 - t,\, t^2)$. Then $\mathbf{f}'(t) = (3t^2 - 1,\, 2t)$ never vanishes, but $\mathbf{f}(-1) = \mathbf{f}(1) = (0, 1)$. The curve $\{\mathbf{f}(t) : t \in \mathbb{R}\}$ loops around and crosses itself at $(0, 1)$, so it fails to be a smooth curve at that point. However, $\{\mathbf{f}(t) : t \in I\}$ is a smooth curve as long as I is an interval that does not contain both -1 and 1. See Figure 3.3.

The reader with access to a computer graphics program may find it entertaining to experiment with examples similar to the ones in this section to obtain a better understanding of the relations between analytic and geometric properties of functions and to see the various types of singularities that can arise when the regularity condition $\nabla F \neq \mathbf{0}$ or $\mathbf{f}(t) \neq \mathbf{0}$ is violated.

EXERCISES

1. For each of the following functions $F(x, y)$, determine whether the set $S = \{(x, y) : F(x, y) = 0\}$ is a smooth curve. Draw a sketch of S. Examine the

nature of S near any points where $\nabla F = \mathbf{0}$. Near which points of S is S the graph of a function $y = f(x)$? $x = f(y)$?
  a. $F(x, y) = x^2 + 3y^2 - 3$.
  b. $F(x, y) = x^2 - 3y^2 - 3$.
  c. $F(x, y) = x - \sqrt{3(y^2 + 1)}$.
  d. $F(x, y) = xy(x + y - 1)$.
  e. $F(x, y) = (x^2 + y^2)(y - x^2 - 1)$.
  f. $F(x, y) = (x^2 + y^2)(y - x^2)$.
  g. $F(x, y) = (e^x - 1)^2 + (\sin y - 1)^2$.

2. Let $S_p = \{(x, y) : x^p + y^p = 1\}$, where p is a positive integer.
  a. Show that S_p is a smooth curve for all p.
  b. Draw a sketch of S_p. (The geometry of S_p depends strongly on whether p is even or odd.)
  c. Which portions of S_p can be represented as the graph of a continuous function $y = f(x)$? $x = f(y)$? What if f is required to be C^1? (Again, the cases p even, p odd and > 1, and $p = 1$ are different.)

3. For each of the following functions $\mathbf{f}(t)$, determine whether the set $S = \{\mathbf{f}(t) : t \in \mathbb{R}\}$ is a smooth curve. Draw a sketch of S. Examine the nature of S near any points $\mathbf{f}(t)$ where $\mathbf{f}'(t) = \mathbf{0}$.
  a. $\mathbf{f}(t) = (t^2 - 1, \, t + 1)$.
  b. $\mathbf{f}(t) = (t^2 - 1, \, t^2 + 1)$.
  c. $\mathbf{f}(t) = (t^3 - 1, \, t^3 + 1)$
  d. $\mathbf{f}(t) = (\cos^3 t, \sin^3 t)$.
  e. $\mathbf{f}(t) = (\cos t + \cos 2t, \, \sin t + \sin 2t)$.

4. Let $\varphi(s) = s^2$ if $s \geq 0$, $\varphi(s) = -s^2$ if $s < 0$.
  a. Show that φ is of class C^1, even at $s = 0$.
  b. Let $\mathbf{f}(t) = (\varphi(\cos t), \varphi(\sin t))$. Show that $\{\mathbf{f}(t) : t \in \mathbb{R}\}$ is the square with vertices at $(\pm 1, 0)$ and $(0, \pm 1)$. For which values of t is $\mathbf{f}'(t) = \mathbf{0}$? What are the corresponding points $\mathbf{f}(t)$?

5. Let $\mathbf{f}(t) = \left((t^2 - 1)/(t^2 + 1), \, t(t^2 - 1)/(t^2 + 1)\right)$ and $S = \{\mathbf{f}(t) : t \in \mathbb{R}\}$.
  a. Show that S is the locus of the equation $y^2(1 - x) = x^2(1 + x)$.
  b. Draw a sketch of S. (S is a curve containing a loop; it is called a **strophoid**.) Show that S is asymptotic to the line $x = 1$.
  c. Discuss the nature of the point $(0, 0)$ where S crosses itself in terms of the parametric and nonparametric representations of S in (a).

6. Let F_1 and F_2 be C^1 functions on some open set U in the plane, and let $F_3 = F_1 F_2$. For $j = 1, 2, 3$, let $S_j = \{\mathbf{x} \in U : F_j(\mathbf{x}) = 0\}$.
  a. Show that $S_3 = S_1 \cup S_2$.
  b. Show that if $\mathbf{a} \in S_1 \cap S_2$, then $\nabla F_3(\mathbf{a}) = \mathbf{0}$.

3.3   Surfaces and Curves in Space

In this section we discuss ways of representing smooth surfaces and curves in $\mathbb{R}^3$, with a brief sketch of the situation in higher dimensions.

Surfaces in $\mathbb{R}^3$.   The standard ways of representing surfaces in 3-space are analogous to the standard ways of representing curves in the plane:

i. as the graph of a function, $z = f(x, y)$ (or $y = f(x, z)$ or $x = f(y, z)$), where f is of class C^1;

ii. as the locus of an equation $F(x, y, z) = 0$, where F is of class C^1;

iii. parametrically, as the range of a C^1 function $\mathbf{f} : \mathbb{R}^2 \to \mathbb{R}^3$.

As before, (i) is a special case of (ii) and (iii), with $F(x, y, z) = z - f(x, y)$ and $\mathbf{f}(u, v) = (u, v, f(u, v))$, and as before, some additional conditions need to be imposed in cases (ii) and (iii) in order to guarantee the smoothness of the surface. The condition in case (ii) is exactly the same as for curves, namely, that

$$(3.12) \qquad \nabla F(x, y, z) \neq (0, 0, 0) \text{ whenever } F(x, y, z) = 0.$$

The situation in case (iii) needs to be examined a little more closely.

To be precise, we assume that $\mathbf{f}$ is a C^1 map from some open set $U \subset \mathbb{R}^2$ into $\mathbb{R}^3$, and we consider the set

$$S = \{\mathbf{x} \in \mathbb{R}^3 : \mathbf{x} = \mathbf{f}(\mathbf{u}), \ \mathbf{u} \in U\}.$$

Here $\mathbf{x} = (x, y, z)$ and $\mathbf{u} = (u, v)$; the variables u and v are the parameters used to represent the surface S. We can think of them as giving a coordinate system on S, with the coordinate grid being formed by the images of the lines $v = $ constant and $u = $ constant, that is, the curves given parametrically $\mathbf{x} = \mathbf{f}(u, c)$ and $\mathbf{x} = \mathbf{f}(c, v)$. The picture is as in Figure 3.4.

What is the appropriate nondegeneracy condition on the derivatives of $\mathbf{f}$? A first guess might be that the Fréchet derivative $D\mathbf{f}$ (a 3×2 matrix) should be nonzero, but this is not enough. We can obtain more insight by looking at the case where $\mathbf{f}$ is linear, that is, $\mathbf{f}(u, v) = u\mathbf{a} + v\mathbf{b} + \mathbf{c}$ for some $\mathbf{a}, \mathbf{b}, \mathbf{c} \in \mathbb{R}^3$. Typically the range of such an $\mathbf{f}$ is a plane, but if the vectors $\mathbf{a}$ and $\mathbf{b}$ are linearly dependent — that is, if one is a scalar multiple of the other — it will only be a line (unless $\mathbf{a} = \mathbf{b} = \mathbf{0}$, in which case it is a single point). Now, for a general smooth $\mathbf{f}$, the linear approximation to $\mathbf{f}$ near a point (u_0, v_0) is $\mathbf{f}(u, v) \approx u\mathbf{a} + v\mathbf{b} + \mathbf{c}$ where the

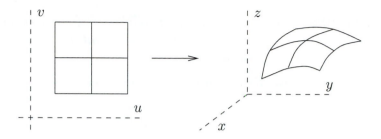

FIGURE 3.4: Parametric representation of a surface.

vectors $\mathbf{a}$, $\mathbf{b}$, and $\mathbf{c}$ are $\partial_u \mathbf{f}$, $\partial_v \mathbf{f}$, and $\mathbf{f}$ evaluated at (u_0, v_0). Hence we are led to the regularity hypothesis:

(3.13)   the vectors $\dfrac{\partial \mathbf{f}}{\partial u}(u, v)$ and $\dfrac{\partial \mathbf{f}}{\partial v}(u, v)$ are linearly independent

at each $(u, v) \in U$,

Since two vectors in $\mathbb{R}^3$ are linearly independent if and only if their cross product is nonzero, (3.13) can be restated as

(3.14) $$\left[\frac{\partial \mathbf{f}}{\partial u} \times \frac{\partial \mathbf{f}}{\partial v} \right](u, v) \neq \mathbf{0} \text{ at each } (u, v) \in U.$$

If S is the graph of a function f and we take the standard parametrization $\mathbf{f}(u, v) = (u, v, f(u, v))$, the condition (3.13) or (3.14) is automatically satisfied, because $\partial_u \mathbf{f} = (1, 0, \partial_u f)$ and $\partial_v \mathbf{f} = (0, 1, \partial_v f)$.

Notice that $\partial_u \mathbf{f}$ and $\partial_v \mathbf{f}$ are the tangent vectors to the "coordinate curves" $\mathbf{x} = \mathbf{f}(u, c)$ and $\mathbf{x} = \mathbf{f}(c, v)$ described above. Thus, the condition (3.13) means that these tangent vectors, at each point of the surface, are nonzero and point in different directions; this implies that the coordinate curves are smooth and intersect nontangentially.

With these things in mind, we arrive at the analogue of Theorem 3.11 for surfaces.

3.15 Theorem.

a. *Let F be a real-valued function of class C^1 on an open set in $\mathbb{R}^3$, and let $S = \{(x, y, z) : F(x, y, z) = 0\}$. If $\mathbf{a} \in S$ and $\nabla F(\mathbf{a}) \neq \mathbf{0}$, there is a neighborhood N of $\mathbf{a}$ in $\mathbb{R}^3$ such that $S \cap N$ is the graph of a C^1 function f (either $z = f(x, y)$, $y = f(x, z)$, or $x = f(y, z)$).*

b. *Let $\mathbf{f}$ be a C^1 mapping from an open set in $\mathbb{R}^2$ into $\mathbb{R}^3$. If $[\partial_u\mathbf{f} \times \partial_v\mathbf{f}](u_0, v_0) \neq 0$, there is a neighborhood N of (u_0, v_0) in $\mathbb{R}^2$ such that the set $\{\mathbf{f}(u, v) : (u, v) \in N\}$ is the graph of a C^1 function.*

Proof. Part (a) is a special case of Corollary 3.3. As for (b), let $\mathbf{f} = (\varphi, \psi, \chi)$. The components of the cross product $\partial_u\mathbf{f} \times \partial_v\mathbf{f}$ are just the Jacobians $\partial(\varphi, \psi)/\partial(u, v)$, $\partial(\varphi, \chi)/\partial(u, v)$, and $\partial(\psi, \chi)/\partial(u, v)$. Under the hypothesis of (b), at least one of them — let us say $\partial(\varphi, \psi)/\partial(u, v)$ — is nonzero at (u_0, v_0). The implicit function theorem then guarantees that the pair of equations $x = \varphi(u, v)$, $y = \psi(u, v)$ can be solved to yield u and v as C^1 functions of x and y near $u = u_0$, $v = v_0$, $x = \varphi(u_0, v_0)$, $y = \psi(u_0, v_0)$. Substituting these functions for u and v in the equation $z = \chi(u, v)$ then yields z as a C^1 function of x and y whose graph is the range of $\mathbf{f}$. $\square$

Thus the representations (i)–(iii) for surfaces are locally equivalent in the presence of the regularity conditions (3.12) and (3.13); a **smooth surface** is a connected subset of $\mathbb{R}^2$ that can be locally described in any of these three forms. The potential global problems with the representations (ii) and (iii) are the same as for plane curves; namely, the set where a C^1 function F vanishes may be disconnected, and a map $\mathbf{f}$ that is locally one-to-one need not be globally one-to-one.

EXAMPLE 1.   Let $\mathbf{f}(u, v) = \big((u + v)\cos(u - v), (u + v)\sin(u - v), u + v\big)$. The set $S = \mathbf{f}(\mathbb{R}^2)$ is a right circular cone with vertex at the origin; it is described nonparametrically by the equation $x^2 + y^2 = z^2$. The set S is a smooth surface except at the origin, which accords with the fact that the gradient of $F(x, y, z) = x^2 + y^2 - z^2$ vanishes at the origin and nowhere else. Correspondingly, the vectors

$$\partial_u\mathbf{f}$$
$$= \big(\cos(u - v) - (u + v)\sin(u - v),\ \sin(u - v) + (u + v)\cos(u - v),\ 1\big)$$

and

$$\partial_v\mathbf{f}$$
$$= \big(\cos(u - v) + (u + v)\sin(u - v),\ \sin(u - v) - (u + v)\cos(u - v),\ 1\big)$$

are linearly independent except when $u + v = 0$, in which case they coincide. The map $\mathbf{f}$ is locally one-to-one except along the line $u + v = 0$, and this entire line is mapped to the origin. (The reader will recognize that $u + v$ and $u - v$ are really the r and θ of cylindrical coordinates in $\mathbb{R}^3$. We have chosen to disguise them a little in order to display a situation where $\partial_u\mathbf{f}$ and $\partial_v\mathbf{f}$ are both nonzero but are linearly dependent where the singularities occur.)

EXAMPLE 2. The unit sphere $S = \{(x, y, z) : x^2 + y^2 + z^2 = 1\}$ can be parametrized by spherical coordinates,

$$\mathbf{f}(\theta, \varphi) = (\cos \theta \sin \varphi, \, \sin \theta \sin \varphi, \, \cos \varphi).$$

Here θ is the longitude and φ is the co-latitude, i.e., the latitude as measured from the north pole rather than the equator. The longitude θ is only well defined up to multiples of 2π, but the co-latitude is usually restricted to the interval $[0, \pi]$. The sphere is a smooth surface, but the map $\mathbf{f}$ does not provide a "good" parametrization of the whole sphere because it is not locally one-to-one when $\sin \varphi = 0$. (That is, the longitude is completely undetermined at the north and south poles.) This degeneracy is also reflected in the tangent vectors

$$\partial_\theta \mathbf{f} = (-\sin \theta \sin \varphi, \, \cos \theta \sin \varphi, \, 0),$$
$$\partial_\varphi \mathbf{f} = (\cos \theta \cos \varphi, \, \sin \theta \cos \varphi, \, -\sin \varphi);$$

they are linearly independent when $\sin \varphi \neq 0$, but $\partial_\theta \mathbf{f} = \mathbf{0}$ when $\sin \varphi = 0$. However, if we restrict θ and φ to the rectangle $-\pi < \theta < \pi, 0 < \varphi < \pi$, we obtain a good parametrization of the sphere with the "international date line" removed.

Finally, a few words about finding the tangent plane to a smooth surface S at a point $\mathbf{a} \in S$. In general, the tangent plane is given by the equation $\mathbf{n} \cdot (\mathbf{x} - \mathbf{a}) = 0$, where $\mathbf{n}$ is a (nonzero) normal vector to S at $\mathbf{a}$. We have already observed in Theorem 2.37 that when S is given by an equation $F = 0$, then the vector $\nabla F(\mathbf{a})$ is normal to S at $\mathbf{a}$. On the other hand, when S is given parametrically as the range of a map $\mathbf{f}(u, v)$, the vectors $\partial_u \mathbf{f}(\mathbf{a})$ and $\partial_v \mathbf{f}(\mathbf{a})$ are tangent to certain curves in S and hence to S itself at $\mathbf{a}$; we therefore obtain a normal at $\mathbf{a}$ by taking their cross product. In both cases, the conditions on F or $\mathbf{f}$ that guarantee the smoothness of S also guarantee that these normal vectors are nonzero.

Curves in $\mathbb{R}^3$. Curves in $\mathbb{R}^3$ are generally described either parametrically or as the intersection of two surfaces. The situation where two of the coordinates are given as functions of the third one can be considered as a special case of either of these. Thus, once again we have three kinds of representation for curves:

i. as a graph, $y = f(x)$ and $z = g(x)$ (or similar expressions with the coordinates permuted), where f and g are C^1 functions;

ii. as the locus of two equations $F(x, y, z) = G(x, y, z) = 0$, where F and G are C^1 functions;

iii. parametrically, as the range of a C^1 function $\mathbf{f} : \mathbb{R} \to \mathbb{R}^3$.

The form (ii) describes the curve as the intersection of the two surfaces $F = 0$ and $G = 0$, and (i) is a special case of (ii) (with $F(x, y, z) = y - f(x)$ and $G(x, y, z) = z - g(x)$) and of (iii) (with $\mathbf{f}(t) = (t, f(t), g(t))$).

By now the reader should be able to guess what the appropriate regularity condition for cases (ii) and (iii) is. In (iii) it is simply that $\mathbf{f}'(t) \neq \mathbf{0}$, and in (ii) it is that

$$\nabla F(\mathbf{x}) \text{ and } \nabla G(\mathbf{x}) \text{ are linearly independent}$$
$$\text{at every } \mathbf{x} \text{ at which } F(\mathbf{x}) = G(\mathbf{x}) = 0.$$

(Geometrically, this means that the surfaces $F = 0$ and $G = 0$ are nowhere tangent to each other.) With these conditions we have an analogue of Theorems 3.11 and 3.15. Rather than give another precise statement and proof, we sketch the ideas and leave the details to the reader (Exercise 7).

First, if ∇F and ∇G are linearly independent, then at least one of the Jacobians $\partial(F, G)/\partial(x, y)$, $\partial(F, G)/\partial(x, z)$, and $\partial(F, G)/\partial(y, z)$ must be nonzero; let us say the last one. Then the implicit function theorem guarantees that the equations $F = G = 0$ can be solved for y and z as functions of x. Second, if $\mathbf{f}'(t) \neq \mathbf{0}$, then one of the components of $\mathbf{f}'(t)$ must be nonzero; let us say the first one. Then the equation $x = f_1(t)$ can be solved for t in terms of x, and then the equations $y = f_2(t)$ and $z = f_3(t)$ yield y and z as functions of x. In either case we end up with the representation (i).

Let us say a little more about what can go wrong in case (ii) when ∇F and ∇G are linearly dependent. The potential problems are clearly displayed in the following situation: Let $F(x, y, z) = z - \varphi(x, y)$, where φ is a C^1 function, and let $G(x, y, z) = z$. Then the sets where $F = 0$ and $G = 0$ are smooth surfaces; the former is the graph of φ, whereas the latter is the xy-plane. The intersection of these two surfaces is the curve in the xy-plane described by the equation $\varphi(x, y) = 0$. Now, this curve can have all sorts of singularities if there are points on it where $\nabla \varphi = (0, 0)$, as we have discussed in §3.2. But since $\nabla F = (-\partial_x \varphi, -\partial_y \varphi, 1)$ and $\nabla G = (0, 0, 1)$, the points where $\nabla \varphi = (0, 0)$ are precisely the points where ∇F and ∇G are linearly dependent.

If a curve S is represented parametrically by a function $\mathbf{f}(t)$, the derivative $\mathbf{f}'(t)$ furnishes a tangent vector to S at the point $\mathbf{f}(t)$. On the other hand, if S is given by a pair of equations $F = G = 0$ and $\mathbf{a} \in S$, the vectors $\nabla F(\mathbf{a})$ and $\nabla G(\mathbf{a})$ are both normal to S at $\mathbf{a}$ and hence span the normal plane to S at $\mathbf{a}$. One can therefore obtain a tangent vector to S at $\mathbf{a}$ by taking their cross product.

Higher Dimensions. The pattern for representations of curves and surfaces that we have established in this section and the preceding one should be pretty clear by now, and it generalizes readily to higher dimensions. We sketch the main points briefly and leave it to the ambitious reader to work out the details.

The general name for a "smooth k-dimensional object" is **manifold**; thus, a curve is a 1-dimensional manifold and a surface is a 2-dimensional manifold. Here we consider the question of representing k-dimensional manifolds in $\mathbb{R}^n$, for any positive integers k and n with $n > k$. The two general forms, corresponding to (ii) and (iii) above for curves and surfaces, are as follows.

The Nonparametric Form: A k-dimensional manifold S in $\mathbb{R}^n$ can be described as the set of simultaneous solutions of $n - k$ equations. That is, given C^1 functions $F_1, \ldots, F_{n-k}$ defined on some open set $U \subset \mathbb{R}^n$, or equivalently a C^1 mapping $\mathbf{F} = (F_1, \ldots, F_{n-k})$ from U into $\mathbb{R}^{n-k}$, we can consider the set

$$(3.16) \qquad\qquad S = \big\{\mathbf{x} : \mathbf{F}(\mathbf{x}) = \mathbf{0}\big\}.$$

The regularity condition that guarantees that S is a smooth k-dimensional manifold is that

$$\nabla F_1(\mathbf{x}), \ldots, \nabla F_{n-k}(\mathbf{x}) \text{ are linearly independent at each } \mathbf{x} \in S,$$

or, equivalently,

$$\text{the matrix } D\mathbf{F}(\mathbf{x}) \text{ has rank } n - k \text{ at every } \mathbf{x} \in S.$$

This condition implies that, for each $\mathbf{x}_0 \in S$, some $(n - k) \times (n - k)$ submatrix of $D\mathbf{F}(\mathbf{x}_0)$ is nonsingular, and the implicit function theorem then implies that the equations $\mathbf{F}(\mathbf{x}) = \mathbf{0}$ can be solved near $\mathbf{x}_0$ for $n-k$ of the variables as C^1 functions of the remaining k variables. This leads to the more special representation of (small pieces of) S by the equations analogous to (i) for curves and surfaces, namely, $\mathbf{x}'' = \mathbf{g}(\mathbf{x}')$, where $\mathbf{x}''$ represents an ordered $(n - k)$-tuple of coordinates and $\mathbf{x}'$ is the ordered k-tuple of remaining coordinates.

The Parametric Form: Given a C^1 map $\mathbf{f}$ from some open set $V \subset \mathbb{R}^k$ into $\mathbb{R}^n$, we can consider the set

$$(3.17) \qquad\qquad S = \big\{\mathbf{f}(\mathbf{u}) : \mathbf{u} \in V\big\}.$$

The regularity condition that guarantees that S is a smooth k-dimensional manifold is that

$$\partial_{u_1}\mathbf{f}(\mathbf{u}), \ldots, \partial_{u_k}\mathbf{f}(\mathbf{u}) \text{ are linearly independent at each } \mathbf{u} \in V,$$

or equivalently,

$$\text{the matrix } Df(\mathbf{u}) \text{ has rank } k \text{ at each } \mathbf{u} \in V.$$

This condition implies that, for each $\mathbf{u}_0 \in V$, some $k \times k$ submatrix of $Df(\mathbf{u}_0)$ is invertible, say the one formed from the rows $i_1, \ldots, i_k$. The implicit function theorem then implies that the equations $x_{i_j} = f_{i_j}(u_1, \ldots, u_k)$ $(1 \leq j \leq k)$ can be solved near $\mathbf{u}_0$ to yield the u_j's as C^1 functions of $\mathbf{x}' = (x_{i_1}, \ldots, x_{i_k})$. Substituting these functions for the u_j's in the remaining equations $x_l = f_l(u_1, \ldots, u_k)$ again yields a representation of (small pieces of) S analogous to (i) for curves and surfaces.

It is perhaps worth pointing out what these two representations boil down to in the linear case. That is, suppose S is a k-dimensional vector subspace of $\mathbb{R}^n$; then S can be represented in the forms (3.16) or (3.17) where the functions $\mathbf{F}$ and $\mathbf{f}$ are linear and hence are given by matrices. (3.16) is the representation of S as the nullspace of an $(n - k) \times n$ matrix, and (3.17) is the representation of S as the column space of an $n \times k$ matrix; in both cases the regularity condition is that the rank of the matrix in question is as large as possible.

EXERCISES

1. For each of the following maps $\mathbf{f} : \mathbb{R}^2 \to \mathbb{R}^3$, describe the surface $S = \mathbf{f}(\mathbb{R}^2)$ and find a description of S as the locus of an equation $F(x, y, z) = 0$. Find the points where $\partial_u \mathbf{f}$ and $\partial_v \mathbf{f}$ are linearly dependent, and describe the singularities of S (if any) at these points.
    a. $\mathbf{f}(u, v) = (2u + v, \ u - v, \ 3u)$.
    b. $\mathbf{f}(u, v) = (au \cos v, \ bu \sin v, \ u)$ $(a, b > 0)$.
    c. $\mathbf{f}(u, v) = (\cos u \cosh v, \ \sin u \cosh v, \ \sinh v)$.
    d. $\mathbf{f}(u, v) = (u \cos v, \ u \sin v, \ u^2)$.

2. Find an equation for the tangent plane to the following parametrized surfaces at the point $(1, -2, 1)$. (The first step is to find the values of the parameters u, v that yield this point.)
    a. $x = e^{u-v}$, $y = u - 3v$, $z = \frac{1}{2}(u^2 + v^2)$.
    b. $x = 1/(u + v)$, $y = -(u + e^v)$, $z = u^3$.

3. Find a parametrization for each of the following surfaces (perhaps involving an angular variable that is defined only up to multiples of 2π).
    a. The surface obtained by revolving the curve $z = f(x)$ $(a < x < b)$ in the xz-plane around the z-axis, where $a > 0$.

b. The surface obtained by revolving the curve $z = f(x)$ $(a < x < b)$ in the xz-plane around the x-axis, where $f(x) > 0$.

c. The lower sheet of the hyperboloid $z^2 - 2x^2 - y^2 = 1$.

d. The cylinder $x^2 + z^2 = 9$.

4. Find a parametric description of the following lines:

   a. The intersection of the planes $x - 2y + z = 3$ and $2x - y - z = -1$.

   b. The intersection of the planes $x + 2y = 3$ and $y - 3z = 2$.

5. Let S be the circle formed by intersecting the plane $x + z = 1$ with the sphere $x^2 + y^2 + z^2 = 1$.

   a. Find a parametrization of S.

   b. Find parametric equations for the tangent line to S at the point $(\frac{1}{2}, -\frac{1}{\sqrt{2}}, \frac{1}{2})$.

6. Let S be the intersection of the cone $z^2 = x^2 + y^2$ and the plane $z = ax + 1$, where $a \in \mathbb{R}$.

   a. Show that S is a circle if $a = 0$, an ellipse if $|a| < 1$, a parabola if $|a| = 1$, and a hyperbola if $|a| > 1$.

   b. Find a parametrization for S in the first two cases and for the part of S lying above the xy-plane in the third case.

7. Give a precise statement and proof of the analogue of Theorem 3.11 for curves in $\mathbb{R}^3$.

3.4 Transformations and Coordinate Systems

In this section we study smooth mappings from $\mathbb{R}^n$ to itself in more detail, with emphasis on geometric intuition for the cases $n = 2$ and $n = 3$.

Suppose $\mathbf{f} : \mathbb{R}^n \to \mathbb{R}^n$ is a map of class C^1. We can regard $\mathbf{f}$ as a *transformation* of $\mathbb{R}^n$, that is, an operation that moves the points in $\mathbb{R}^n$ around in some definite fashion. When $n > 1$, such transformations are usually best pictured with "before and after" sketches. That is, if $\mathbf{x} = \mathbf{f}(\mathbf{u})$, we think of $\mathbf{u}$ and $\mathbf{x}$ as living in two separate copies of $\mathbb{R}^n$. We draw a sketch of $\mathbf{u}$-space with some geometric figures in it, such as a grid of coordinate lines, then draw a sketch of $\mathbf{x}$-space with the images of those figures under the transformation $\mathbf{f}$.

EXAMPLE 1. Define $\mathbf{f} : \mathbb{R}^2 \to \mathbb{R}^2$ by $\mathbf{f}(u, v) = \frac{1}{2}(\sqrt{3}u - v, u + \sqrt{3}v)$. The map $\mathbf{f}$ represents a counterclockwise rotation through the angle $\frac{1}{6}\pi$ about the origin (since $\frac{1}{2}\sqrt{3} = \cos\frac{1}{6}\pi$ and $\frac{1}{2} = \sin\frac{1}{6}\pi$). See Figure 3.5.

EXAMPLE 2. Define $\mathbf{f} : \mathbb{R}^2 \to \mathbb{R}^2$ by $\mathbf{f}(u, v) = (2u, v)$. $\mathbf{f}$ simply stretches out the u coordinate by a factor of 2. See Figure 3.6.

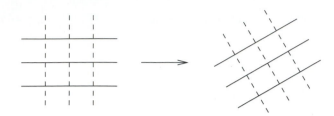

FIGURE 3.5: The transformation $\mathbf{f}(u, v) = \frac{1}{2}(\sqrt{3}u - v,\ u + \sqrt{3}v)$.

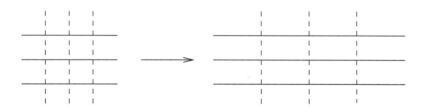

FIGURE 3.6: The transformation $\mathbf{f}(u, v) = (2u, v)$.

EXAMPLE 3. Define $\mathbf{f} : \mathbb{R}^2 \to \mathbb{R}^2$ by $\mathbf{f}(u, v) = (u^2 - v^2, 2uv)$. Unlike the previous two examples, this $\mathbf{f}$ is not one-to-one; it maps (u, v) and $(-u, -v)$ to the same point. (It's not hard to check that this is the only duplication of values: If $\mathbf{f}(u, v) = \mathbf{f}(z, w)$ then $(z, w) = \pm(u, v)$.) In order to draw an intelligible picture, we restrict attention to the region $u > 0$. We also denote $\mathbf{f}(u, v)$ by (x, y), so the "before" and "after" pictures are the uv-plane and the xy-plane. The image of the vertical line $u = c$ under $\mathbf{f}$ is given by $x = c^2 - v^2$, $y = 2cv$. Elimination of v yields $x = c^2 - y^2/4c^2$, the equation of a parabola that opens out to the left. On the other hand, the image of the horizontal line $v = c$ is given by $x = u^2 - c^2$, $y = 2cu$, which yields $x = y^2/4c^2 - c^2$. Since we are assuming $u > 0$, we have $y > 0$ or $y < 0$ depending on whether $c > 0$ or $c < 0$; in either case this curve is half of a parabola opening to the right. See Figure 3.7: The v-axis is mapped to the negative x-axis (both $(0, v)$ and $(0, -v)$ being mapped to $(-v^2, 0)$), as indicated by the dotted lines, and the right half of the uv-plane is bent to the left to fill up the rest of the xy-plane.

We can also draw the reverse picture. The horizontal line $y = c$ in the xy-plane corresponds to the curve $2uv = c$ in the uv-plane, which is a hyperbola whose asymptotes are the coordinate axes. The vertical line $x = c$ corresponds to the curve $u^2 - v^2 = c$, which is a hyperbola whose asymptotes are the

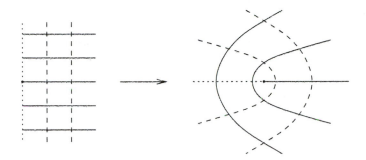

FIGURE 3.7: The transformation $(x, y) = (u^2 - v^2, 2uv)$, showing the image in the xy-plane of the coordinate grid in the half-plane $u > 0$.

FIGURE 3.8: The transformation $(x, y) = (u^2 - v^2, 2uv)$, showing the curves in the half-plane $u > 0$ that map to the coordinate grid in the xy-plane.

lines $v = \pm u$ when $c \neq 0$ and the union of these two lines when $c = 0$. See Figure 3.8.

We can think of mappings from $\mathbb{R}^3$ to itself pictorially in the same way, though the pictures are harder to draw. Figure 3.9 shows what happens to a cube under the transformation $\mathbf{f}(u, v, w) = (-2u, v, \frac{1}{2}w)$.

Another common interpretation of a map $\mathbf{f} : \mathbb{R}^n \to \mathbb{R}^n$ is as a coordinate system on $\mathbb{R}^n$. For example, we usually think of $\mathbf{f}(r, \theta) = (r\cos\theta, r\sin\theta)$ as representing polar coordinates in the plane. In the preceding discussion we thought in terms of moving the points in $\mathbb{R}^n$ around without changing the labeling system (namely, Cartesian coordinates); here we are thinking of leaving the points alone but giving them different labels (polar rather than Cartesian coordinates.) It's just a matter of point of view; the same transformation $\mathbf{f}$ can be interpreted either way. For example, the systems of parabolas and hyperbolas in Figures 3.7 and 3.8 can

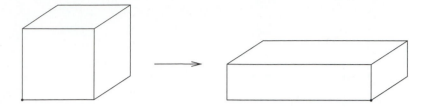

FIGURE 3.9: The transformation $\mathbf{f}(u, v, w) = (-2u, v, \frac{1}{2}w)$. (The u and w axes are horizontal and vertical, respectively.)

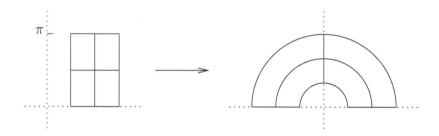

FIGURE 3.10: The polar coordinate transformation $(x, y) = (r\cos\theta, r\sin\theta)$.

be interpreted as the grids for curvilinear coordinate systems in the appropriate parts of the plane, and the map $\mathbf{f}(r, \theta) = (r\cos\theta, r\sin\theta)$ can be interpreted as a transformation. Figure 3.10 shows a representative piece of it.

Not all mappings $\mathbf{f} : \mathbb{R}^n \to \mathbb{R}^n$ can be used as coordinate systems, however. A "good" coordinate system should have the property that there is a one-to-one correspondence between points and their coordinates; that is, each set of coordinates should specify a unique point in $\mathbb{R}^n$, and two different sets of coordinates should specify different points. Polar coordinates, for example, do not satisfy this condition — (r, θ) and (r, φ) are polar coordinates of the same point whenever $\theta - \varphi$ is an integer multiple of 2π, or whenever $r = 0$ — and this fact always has the potential to cause problems when polar coordinates are used. However, if we restrict r and θ to satisfy $r > 0$ and $-\pi < \theta < \pi$, we do get a "good" coordinate system, not on the whole plane, but on the plane with the negative real axis removed. Likewise, the map $(u, v) = (x^2 - y^2, 2xy)$ in Example 3, restricted to the half-plane $x > 0$, gives a "good" coordinate system on the uv-plane with the negative u axis removed.

In short, our attention is directed to transformations $\mathbf{f}$ of class C^1 that map an open set $U \subset \mathbb{R}^n$ in a one-to-one fashion onto another open set $V \subset \mathbb{R}^n$. There is one further requirement that is natural to impose, namely, that the inverse mapping $\mathbf{f}^{-1} : V \to U$ should also be of class C^1, so that the correspondence is smooth in both directions. Hence, the question arises: Given a C^1 transformation $\mathbf{f} : U \to V$, when does $\mathbf{f}$ possess a C^1 inverse $\mathbf{f}^{-1} : V \to U$? That is, when can the equation $\mathbf{f}(\mathbf{x}) = \mathbf{y}$ be solved uniquely for $\mathbf{x}$ as a C^1 function of $\mathbf{y}$?

This question is clearly closely related to the ones that led to the implicit function theorem, and indeed, if we restrict attention to the solvability of the equation $\mathbf{f}(\mathbf{x}) = \mathbf{y}$ in a small neighborhood of a point, its answer becomes a special case of that theorem. As we did before, we can guess what the answer should be by looking at the linear approximation. If $\mathbf{f}(\mathbf{a}) = \mathbf{b}$, the linear approximation to the equation $\mathbf{f}(\mathbf{x}) = \mathbf{y}$ at the point $(\mathbf{a}, \mathbf{b})$ is $T(\mathbf{x} - \mathbf{a}) = \mathbf{y} - \mathbf{b}$ where the matrix T is the Fréchet derivative $D\mathbf{f}(\mathbf{a})$, and the latter equation can be solved for $\mathbf{x}$ precisely when T is invertible, that is, when the Jacobian $\det D\mathbf{f}(\mathbf{a})$ is nonzero. We are therefore led to the following theorem.

3.18 Theorem (The Inverse Mapping Theorem). *Let U and V be open sets in $\mathbb{R}^n$, $\mathbf{a} \in U$, and $\mathbf{b} = \mathbf{f}(\mathbf{a})$. Suppose that $\mathbf{f} : U \to V$ is a mapping of class C^1 and the Fréchet derivative $D\mathbf{f}(\mathbf{a})$ is invertible (that is, the Jacobian $\det D\mathbf{f}(\mathbf{a})$ is nonzero). Then there exist neighborhoods $M \subset U$ and $N \subset V$ of $\mathbf{a}$ and $\mathbf{b}$, respectively, so that $\mathbf{f}$ is a one-to-one map from M onto N, and the inverse map $\mathbf{f}^{-1}$ from N to M is also of class C^1. Moreover, if $\mathbf{y} = \mathbf{f}(\mathbf{x}) \in N$, $D(\mathbf{f}^{-1})(\mathbf{y}) = [D\mathbf{f}(\mathbf{x})]^{-1}$.*

Proof. The existence of the inverse map is equivalent to the unique solvability of the equation $\mathbf{F}(\mathbf{x}, \mathbf{y}) = \mathbf{0}$ for $\mathbf{x}$, where $\mathbf{F}(\mathbf{x}, \mathbf{y}) = \mathbf{f}(\mathbf{x}) - \mathbf{y}$. Since the derivative of $\mathbf{F}$ as a function of $\mathbf{x}$ is just $D\mathbf{f}(\mathbf{x})$, the implicit function theorem (3.9) guarantees that this unique solvability will hold for $(\mathbf{x}, \mathbf{y})$ near $(\mathbf{a}, \mathbf{b})$ provided that $D\mathbf{f}(\mathbf{a})$ is invertible. (In referring to the statement of the implicit function theorem, however, note that the roles of the variables $\mathbf{x}$ and $\mathbf{y}$ have been reversed here.) Moreover, since $\mathbf{f}^{-1}(\mathbf{f}(\mathbf{x})) = \mathbf{x}$ for $\mathbf{x} \in M$, the chain rule gives $D(\mathbf{f}^{-1})(\mathbf{f}(\mathbf{x})) \cdot D\mathbf{f}(\mathbf{x}) = I$ where I is the $n \times n$ identity matrix; in other words, $D(\mathbf{f}^{-1})(\mathbf{y}) = [D\mathbf{f}(\mathbf{x})]^{-1}$ where $\mathbf{y} = \mathbf{f}(\mathbf{x})$. $\qquad\square$

It is to be emphasized that the inverse mapping theorem is local in nature; the *global* invertibility of $\mathbf{f}$ is a more delicate matter. To be more precise, consider the following question: *Suppose $\mathbf{f} : U \to V$ is of class C^1 and that $D\mathbf{f}(\mathbf{x})$ is invertible for every $\mathbf{x} \in U$. Is $\mathbf{f}$ one-to-one on U?*

When $n = 1$, the answer is *yes*, provided that U is an interval. Here we are considering a C^1 function $f(x)$ such that $f'(x) \neq 0$ on the interval $U = (a, b)$. Since f' is continuous, we must have either $f'(x) > 0$ for all $x \in (a, b)$, so that f is strictly increasing, or $f'(x) < 0$ for all $x \in (a, b)$, so that f is strictly decreasing. In either case, f is one-to-one.

When $n > 1$, however, the answer is *no*. The simplest counterexample is our old friend the polar coordinate map, $\mathbf{f}(r, \theta) = (r \cos \theta, r \sin \theta)$, on the set $U = \{(r, \theta) : r > 0\}$. We have

$$Df(r, \theta) = \begin{pmatrix} \cos \theta & -r \sin \theta \\ \sin \theta & r \cos \theta \end{pmatrix}, \qquad \det Df(r, \theta) = r(\cos^2 \theta + \sin^2 \theta) = r,$$

so $\det Df \neq 0$ on U, but $\mathbf{f}$ is not one-to-one since $\mathbf{f}(r, \theta + 2k\pi) = \mathbf{f}(r, \theta)$. It is, however, *locally* one-to-one, in that it is one-to-one if one restricts θ to any interval of length less than 2π. (Notice also that the Jacobian of the polar coordinate map vanishes when $r = 0$. This accords with the fact that the polar coordinate map is not even locally invertible there; the angular coordinate is completely undetermined at the origin.)

The question of global invertibility is a delicate one. Consider the following situation: Let $\mathbf{f} : \mathbb{R}^n \to \mathbb{R}^n$ be a map whose component functions are all polynomials, and suppose that the Jacobian $\det Df$ is identically equal to 1. Is $\mathbf{f}$ globally invertible? The answer is so far unknown; this is a famous unsolved problem.

We should also point out that the invertibility of $Df(\mathbf{a})$ is not necessary for the existence of an inverse map, although it is necessary for the differentiability of that inverse. (Example: Let $f(x) = x^3$. Then f has the global inverse $f^{-1}(y) = y^{1/3}$, but $f(0) = f'(0) = 0$ and f^{-1} is not differentiable at 0.)

EXERCISES

1. For each of the following transformations $(u, v) = \mathbf{f}(x, y)$, (i) compute the Jacobian $\det Df$, (ii) draw a sketch of the images of some of the lines $x = $ constant and $y = $ constant in the uv-plane, (iii) find formulas for the local inverses of $\mathbf{f}$ when they exist.
    a. $u = e^x \cos y$, $v = e^x \sin y$.
    b. $u = x^2$, $v = y/x$.
    c. $u = x^2 + 2xy + y^2$, $v = 2x + 2y$.

2. Let $(u, v) = \mathbf{f}(x, y) = (x - 2y, 2x - y)$.
    a. Compute the inverse transformation $(x, y) = \mathbf{f}^{-1}(u, v)$.
    b. Find the image in the uv-plane of the triangle bounded by the lines $y = x$, $y = -x$, and $y = 1 - 2x$.
    c. Find the region in the xy-plane that is mapped to the triangle with vertices $(0, 0)$, $(-1, 2)$, and $(2, 1)$ in the uv-plane.

3. Let $u = \sin x \cosh y$, $v = \cos x \sinh y$.

   a. Show that the images of the lines $x = $ constant (resp. $y = $ constant) in the uv-plane are hyperbolas (resp. ellipses).

   b. Show that $\partial(u, v)/\partial(x, y) = \cos^2 x + \sinh^2 y$. At what points (x, y) does this Jacobian vanish? Show that the corrsponding points in the uv-plane are $(\pm 1, 0)$.

   c. (optional) Show that the ellipses and hyperbolas in (a) all have foci at $(\pm 1, 0)$.

4. Let $(u, v) = \mathbf{f}(x, y) = (x - y, xy)$.

   a. Sketch some of the curves $x - y = $ constant and $xy = $ constant in the xy-plane. Which regions in the xy-plane map onto the rectangle in the uv-plane given by $0 \le u \le 1$, $1 \le v \le 4$? There are two of them; draw a picture of them.

   b. Compute the derivative $D\mathbf{f}$ and the Jacobian $J = \det D\mathbf{f}$.

   c. The Jacobian J vanishes at (a, b) precisely when the gradients $\nabla u(a, b)$ and $\nabla v(a, b)$ are linearly dependent, i.e., when the level sets of u and v passing through a and b are tangent to each other. (If this doesn't seem obvious at first, think about it!) Use your sketch of the level sets in (a) to show pictorially that this assertion is correct.

   d. Notice that $\mathbf{f}(2, -3) = (5, -6)$. Compute explicitly the local inverse $\mathbf{g}$ of $\mathbf{f}$ such that $\mathbf{g}(5, -6) = (2, -3)$ and also compute its derivative $D\mathbf{g}$.

   e. Show by explicit calculation that the matrices $D\mathbf{f}(2, -3)$ and $D\mathbf{g}(5, -6)$ are inverses of each other.

5. Find a one-to-one C^1 mapping $\mathbf{f}$ from the first quadrant of the xy-plane to the first quadrant of the uv-plane such that the region where $x^2 \le y \le 2x^2$ and $1 \le xy \le 3$ is mapped to a rectangle. Compute the Jacobian $\det D\mathbf{f}$ and the inverse mapping $\mathbf{f}^{-1}$. (*Hint:* Map all the regions where $ax^2 \le y \le bx^2$ and $c \le xy \le d$ to rectangles.)

6. Let $\mathbf{f} : \mathbb{R}^3 \to \mathbb{R}^3$ be the spherical coordinate map,

$$(x, y, z) = \mathbf{f}(r, \varphi, \theta) = (r \sin \varphi \cos \theta, \; r \sin \varphi \sin \theta, \; r \cos \varphi).$$

   Thus r is the distance to the origin, φ is the co-latitude (the angle from the positive z-axis), and θ is the longitude.

   a. Describe the surfaces in xyz-space that are the images of the planes $r = $ constant, $\varphi = $ constant, and $\theta = $ constant.

   b. Compute the derivative $D\mathbf{f}$ and show that $\det D\mathbf{f}(r, \varphi, \theta) = r^2 \sin \varphi$.

   c. What is the condition on the point $(r_0, \varphi_0, \theta_0)$ for $\mathbf{f}$ to be locally invertible about this point? What is the corresponding condition on $(x_0, y_0, z_0) = \mathbf{f}(r_0, \varphi_0, \theta_0)$?

7. We have obtained the inverse mapping theorem as a corollary of the implicit function theorem. It is also possible to prove the inverse mapping theorem directly and then obtain the implicit function theorem as a corollary of it. Do this last step; that is, assume the inverse mapping theorem and deduce the implicit function theorem from it. (*Hint:* Let $\mathbf{F}(\mathbf{x}, \mathbf{y})$ be as in Theorem 3.9. Apply the inverse mapping theorem to the transformation $\mathbf{G} : \mathbb{R}^{n+k} \to \mathbb{R}^{n+k}$ defined by $\mathbf{G}(\mathbf{x}, \mathbf{y}) = (\mathbf{x}, \mathbf{F}(\mathbf{x}, \mathbf{y}))$.)

3.5 Functional Dependence

In the implicit function theorem and its applications discussed in the preceding sections, we have drawn consequences from the nonvanishing of various Jacobians. In this section we consider the opposite situation, in which a Jacobian vanishes identically.

For motivation, let us first consider the linear case. Let A be an $n \times n$ matrix, and define $\mathbf{F} : \mathbb{R}^n \to \mathbb{R}^n$ by $\mathbf{F}(\mathbf{x}) = A\mathbf{x}$ (where $\mathbf{x}$ is considered as a column vector). If A is nonsingular, $\mathbf{F}$ is a one-to-one map from $\mathbb{R}^n$ onto itself, whose inverse is $\mathbf{F}^{-1}(\mathbf{y}) = A^{-1}\mathbf{y}$. However, if $\det A = 0$, the range of T (namely, the column space of A) is a proper linear subspace of $\mathbb{R}^n$, and the components $(f_1, \ldots, f_n)$ of $\mathbf{F}$ satisfy at least one nontrivial linear relation. More precisely, if the rank of A is k, where $k < n$, then the range of $\mathbf{F}$ is a k-dimensional subspace of $\mathbb{R}^n$, and the components of $\mathbf{F}$ satisfy $n - k$ independent linear relations (namely, the relations satisfied by the rows of A).

EXAMPLE 1. Let $\mathbf{F} = (f_1, f_2, f_3)$ be given by the matrix

$$A = \begin{pmatrix} 1 & 2 & -1 \\ 1 & -3 & 4 \\ 2 & -1 & 3 \end{pmatrix},$$

that is,

$$f_1(x, y, z) = x + 2y - z,$$
$$f_2(x, y, z) = x - 3y + 4z,$$
$$f_3(x, y, z) = 2x - y + 3z.$$

It is easily verified that $\det A = 0$, that the first two rows of A are independent, and that the third row is the sum of the first two. This last relation means that the functions f_1, f_2, f_3 satisfy the linear relation $f_3 = f_1 + f_2$. Equivalently, the range of $\mathbf{F}$ is the plane defined by the equation $y_3 = y_1 + y_2$.

EXAMPLE 2. Let $\mathbf{F} = (f_1, f_2, f_3)$ be given by the matrix

$$A = \begin{pmatrix} 1 & 2 & -1 \\ 2 & 4 & -2 \\ -3 & -6 & 3 \end{pmatrix},$$

that is,

$$f_1(x, y, z) = x + 2y - z,$$
$$f_2(x, y, z) = 2x + 4y - 2z,$$
$$f_3(x, y, z) = -x - 2y + z.$$

Here the rank of A is 1, and the functions f_j satisfy the relations $f_2 = 2f_1$, $f_3 = -3f_1$. The range of $\mathbf{F}$ is the line passing through the origin and the point $(1, 2, -3)$.

More generally, one can consider linear maps $\mathbf{F} : \mathbb{R}^m \to \mathbb{R}^n$ defined by $n \times m$ matrices A. The range of such a map is a linear subspace of $\mathbb{R}^n$ whose dimension is the rank of A. It must happen when $n > m$, and may happen when $n \leq m$, that this subspace is a *proper* subspace of $\mathbb{R}^n$, in which case the components of $\mathbf{F}$ satisfy nontrivial linear relations.

Now let us return to the study of more general functions. The appropriate analogue of "linear dependence" for nonlinear functions is "functional dependence," which means that the functions in question satisfy a nontrivial *functional* relation, in other words, that one of them must be expressible as a function of the others. We shall formulate this idea precisely in a way that is appropriate for C^1 functions, although the notion of functional dependence does not really depend on any differentiability conditions.

Suppose $f_1, \ldots, f_n$ are C^1 real-valued functions on an open set $U \subset \mathbb{R}^m$. We say that $f_1, \ldots, f_n$ are **functionally dependent** on U if there is a C^1 function $\Phi : \mathbb{R}^n \to \mathbb{R}$ such that

$$(3.19) \quad \Phi\big(f_1(\mathbf{x}), \ldots, f_n(\mathbf{x})\big) = 0 \text{ and } \nabla\Phi\big(f_1(\mathbf{x}), \ldots, f_n(\mathbf{x})\big) \neq \mathbf{0} \text{ for } \mathbf{x} \in U.$$

The nonvanishing of $\nabla\Phi$ guarantees, via the implicit function theorem, that the equation $\Phi = 0$ can be solved locally for one of the variables in terms of the others; in other words, one of the functions f_j can be expressed in terms of the remaining ones.

Geometrically, (3.19) means that the range of the map $\mathbf{f} = (f_1, \ldots, f_n)$ is contained in the hypersurface $\{\mathbf{y} : \Phi(\mathbf{y}) = 0\}$ in $\mathbb{R}^n$, so that it is at most $(n-1)$-dimensional. (It might be even smaller, of course; the functions f_j might satisfy other relations in addition to the equation $\Phi(\mathbf{f}(\mathbf{x})) = 0$.)

EXAMPLE 3. The functions

$$f_1(x, y, z) = x + y + z,$$
$$f_2(x, y, z) = xy + xz + yz,$$
$$f_3(x, y, z) = x^2 + y^2 + z^2$$

are functionally dependent on $\mathbb{R}^3$, for $f_3 = f_1^2 - 2f_2$.

EXAMPLE 4. The functions $f_1(x, y) = 3x + 1$, $f_2(x, y) = x^2 - y$ are not functionally dependent on any open set in $\mathbb{R}^2$. Indeed, the transformation $\mathbf{f} = (f_1, f_2)$ is a one-to-one map from $\mathbb{R}^2$ onto itself whose inverse $\mathbf{g} = (g_1, g_2)$ is given by $g_1(u, v) = \frac{1}{3}(u - 1)$, $g_2(u, v) = \frac{1}{9}(u - 1)^2 - v$; hence the values of $\mathbf{f}$ are not subject to any restrictions.

It should be noted that the question of functional dependence is interesting only when the number of functions does not exceed the number of independent variables; when it does, functional dependence is almost automatic. For example, if f and g are *any* two C^1 functions of one variable, then f and g are functionally dependent on any interval I on which either $f' \neq 0$ or $g' \neq 0$. Indeed, if $f' \neq 0$ on I, then f is one-to-one on I and so has an inverse; then $\Phi(f(x), g(x)) = 0$ on I where $\Phi(u, v) = g(f^{-1}(u)) - v$.

The main results of this section concern the close relation between the functional dependence of a family of functions and the linear dependence of their linear approximations. To begin with, we consider the case where the number of functions equals the number of independent variables.

3.20 Theorem. *Suppose* $\mathbf{f} = (f_1, \ldots, f_n)$ *is a* C^1 *map on some open set* $U \subset \mathbb{R}^n$. *If* $f_1, \ldots, f_n$ *are functionally dependent on* U, *then the Jacobian* $\det D\mathbf{f}$ *vanishes identically on* U.

Proof. Functional dependence of the f_j's means that there is a C^1 function Φ such that $\Phi(\mathbf{f}(\mathbf{x})) = 0$ and $\nabla\Phi(\mathbf{f}(\mathbf{x})) \neq \mathbf{0}$ for $\mathbf{x} \in U$. Differentiation of the equation $\Phi(\mathbf{f}(\mathbf{x})) = 0$ with respect to the variables $x_1, \ldots, x_n$ via the chain rule yields

$$(\partial_1\Phi)(\partial_1 f_1) + (\partial_2\Phi)(\partial_1 f_2) + \cdots + (\partial_n\Phi)(\partial_1 f_n) = 0,$$
$$\vdots$$
$$(\partial_1\Phi)(\partial_n f_1) + (\partial_2\Phi)(\partial_n f_2) + \cdots + (\partial_n\Phi)(\partial_n f_n) = 0,$$

where the derivatives of Φ are evaluated at $\mathbf{f}(\mathbf{x})$ and the derivatives of the f_j's are evaluated at $\mathbf{x}$. Thus, at each $\mathbf{x} \in U$, the system of equations

$$(\partial_1 f_1)y_1 + (\partial_1 f_2)y_2 + \cdots + (\partial_1 f_n)y_n = 0,$$
$$\vdots$$
$$(\partial_n f_1)y_1 + (\partial_n f_2)y_2 + \cdots + (\partial_n f_n)y_n = 0,$$

has a nonzero solution, namely $\mathbf{y} = \nabla\Phi(\mathbf{f}(\mathbf{x}))$. Therefore, its coefficient matrix $(\partial_j f_k(\mathbf{x}))$, which is nothing but the transpose of $D\mathbf{f}(\mathbf{x})$, must be singular, and hence $\det D\mathbf{f}(\mathbf{x}) = 0$. $\qquad\square$

More interesting is the fact that the converse of this theorem is also true: The vanishing of the Jacobian $\det D\mathbf{f}$ implies the functional dependence of the f_j's. We now present a version of this result with an additional hypothesis (the constancy of the rank of $D\mathbf{f}$) that yields a sharper conclusion. We formulate it so that it also covers the case when the number of functions differs from the number of independent variables.

3.21 Theorem. *Let $\mathbf{f} = (f_1, \ldots, f_n)$ be a C^1 map from a connected open set $U \subset \mathbb{R}^m$ into $\mathbb{R}^n$. Suppose that the matrix $D\mathbf{f}(\mathbf{x})$ has rank k at every $\mathbf{x} \in U$, where $k < n$. Then every $\mathbf{x}_0 \in U$ has a neighborhood N such that $f_1, \ldots, f_n$ are functionally dependent on N and $\mathbf{f}(N)$ is a smooth k-dimensional submanifold of $\mathbb{R}^n$.*

(The restriction to a small neighborhood N is necessary because the set $\mathbf{f}(U)$ can cross itself, as in Example 8 in §3.2.)

Since $D\mathbf{f}(\mathbf{x})$ is an $n \times m$ matrix, its rank k always satisfies $k \leq m$ and $k \leq n$. When $k = m$, the situation described here is simply the representation in parametric form of an m-dimensional submanifold of $\mathbb{R}^n$, as discussed in §§3.2–3, and the conclusion of the theorem is that such a submanifold can also be described as the locus of a system of equations. In other words, the case $k = m$ boils down to Theorems 3.11b and 3.15b and their generalizations to higher dimensions. The case where more needs to be said is the one where $k < m$.

Rather than proving this theorem in complete generality, we shall restrict attention to the case where $m = n = 3$ and k is 1 or 2. The ideas in the general case are the same; only the details are more cumbersome. (See also Exercise 2.) Let us restate the theorem for the special case:

3.22 Theorem. *Let $\mathbf{f} = (f, g, h)$ be a C^1 map from a connected open set $U \subset \mathbb{R}^3$ into $\mathbb{R}^3$. Suppose that the matrix $D\mathbf{f}(\mathbf{x})$ has rank k at every $\mathbf{x} \in U$, where $k = 1$ or 2. Then every $\mathbf{x}_0 \in U$ has a neighborhood N such that the functions f, g, h are functionally dependent on N and $\mathbf{f}(N)$ is a smooth curve (if $k = 1$) or a smooth surface (if $k = 2$).*

Proof. Let $\mathbf{x} = (x, y, z)$, $u = f(\mathbf{x})$, $v = g(\mathbf{x})$, and $w = h(\mathbf{x})$, and fix $\mathbf{x}_0 = (x_0, y_0, z_0) \in U$.

First suppose $k = 1$. Since the matrix $D\mathbf{f}(\mathbf{x}_0)$ has rank 1, it has at least one nonzero entry; by relabeling the functions and variables, we may assume that the

$(1, 1)$ entry is nonzero, that is, $\partial_x f(\mathbf{x}_0) \neq 0$. By the implicit function theorem, then, the equation $u = f(x, y, z)$ can be solved near $\mathbf{x} = \mathbf{x}_0$, $u = u_0 = f(\mathbf{x}_0)$, to yield x as a function of y, z, and u. Then v and w turn into functions of y, z, and u also. Implicit differentiation of the equations $u = f(x, y, z)$ and $v = g(x, y, z)$ with respect to y (taking y, z, and u as the independent variables) yields

$$0 = (\partial_x f)(\partial_y x) + (\partial_y f),$$
$$\partial_y v = (\partial_x g)(\partial_y x) + (\partial_y g).$$

Solving the first equation for $\partial_y x$ and substituting the result into the second equation then yields

$$\partial_y v = \partial_x g \frac{-\partial_y f}{\partial_x f} + \partial_y g = \frac{1}{\partial_x f} \cdot \frac{\partial(f, g)}{\partial(x, y)}.$$

But since $D\mathbf{f}$ has rank 1, all of its 2×2 submatrices are singular; therefore, $\partial(f, g)/\partial(x, y) \equiv 0$ and hence $\partial_y v \equiv 0$. Restricting to a convex neighborhood of (y_0, z_0, u_0), we conclude that v is independent of y. For exactly the same reason, v is independent of z, and w is independent of y and z. That is, v and w are functions of u alone, say $v = \varphi(u)$ and $w = \psi(u)$. This shows that f, g, h are functionally dependent — $g(\mathbf{x}) = \varphi(f(\mathbf{x}))$ and $h(\mathbf{x}) = \psi(f(\mathbf{x}))$ — and that the image of a neighborhood of $\mathbf{x}_0$ under $\mathbf{f}$ is the locus of the equations $v = \varphi(u)$, $w = \psi(u)$, which is a smooth curve.

Now let us turn to the case $k = 2$. Here some 2×2 submatrix of $D\mathbf{f}(\mathbf{x}_0)$ is nonsingular; by relabeling the functions and variables, we can assume that it is the one in the upper left corner, so that $\partial(f, g)/\partial(x, y)$ is nonzero at $\mathbf{x}_0$. By the implicit function theorem, the equations $u = f(x, y, z)$ and $v = g(x, y, z)$ can be solved near $\mathbf{x} = \mathbf{x}_0$, $u = u_0 = f(\mathbf{x}_0)$, $v = v_0 = g(\mathbf{x}_0)$, to yield x and y as functions of u, v, and z. Taking u, v, and z as the independent variables, then, we differentiate the equations $u = f(x, y, z)$, $v = g(x, y, z)$, and $w = h(x, y, z)$ implicitly with respect to z to obtain

$$0 = (\partial_x f)(\partial_z x) + (\partial_y f)(\partial_z y) + (\partial_z f),$$
$$0 = (\partial_x g)(\partial_z x) + (\partial_y g)(\partial_z y) + (\partial_z g),$$
$$\partial_z w = (\partial_x h)(\partial_z x) + (\partial_y h)(\partial_z y) + (\partial_z h),$$

or

$$(\partial_x f)(\partial_z x) + (\partial_y f)(\partial_z y) \qquad\qquad = -\partial_z f,$$
$$(\partial_x g)(\partial_z x) + (\partial_y g)(\partial_z y) \qquad\qquad = -\partial_z g,$$
$$(\partial_x h)(\partial_z x) + (\partial_y h)(\partial_z y) - (\partial_z w) = -\partial_z h.$$

These equations may be solved simultaneously for $\partial_z x$, $\partial_z y$, and $\partial_z w$. By Cramer's rule (Appendix A, (A.54)),

$$
\partial_z w = \det \begin{pmatrix} \partial_x f & \partial_y f & -\partial_z f \\ \partial_x g & \partial_y g & -\partial_z g \\ \partial_x h & \partial_y h & -\partial_z h \end{pmatrix} \bigg/ \det \begin{pmatrix} \partial_x f & \partial_y f & 0 \\ \partial_x g & \partial_y g & 0 \\ \partial_x h & \partial_y h & -1 \end{pmatrix}
$$

$$
= \frac{\partial(f, g, h)}{\partial(x, y, z)} \bigg/ \frac{\partial(f, g)}{\partial(x, y)}.
$$

The denominator is nonzero by assumption, but the numerator is zero because $D\mathbf{f}$ has rank 2. Hence w is independent of z; that is, w depends only on u and v, say $w = \varphi(u, v)$. This shows that f, g, h are functionally dependent — $h(\mathbf{x}) = \varphi(f(\mathbf{x}), g(\mathbf{x}))$ — and that the image of a neighborhood of $\mathbf{x}_0$ under $\mathbf{f}$ is the locus of the equation $w = \varphi(u, v)$, which is a smooth surface. $\qquad\square$

We conclude with a few words about the assumption that the rank of $D\mathbf{f}$ is constant. Suppose that $A(\mathbf{x})$ is a matrix whose entries depend continuously on $\mathbf{x} \in U$ (U an open subset of $\mathbb{R}^m$), and the rank of $A(\mathbf{x}_0)$ is k. Since a set of linearly independent vectors remains linearly independent if the vectors are perturbed slightly, the rank of $A(\mathbf{x})$ is at least k when $\mathbf{x}$ is sufficiently close to $\mathbf{x}_0$. In other words, for each k the set $\{\mathbf{x} \in U : \operatorname{rank}(A(\mathbf{x})) \geq k\}$ is open. In particular, if k_0 is the maximum rank of $A(\mathbf{x})$ as $\mathbf{x}$ ranges over U, then $\{\mathbf{x} \in U : \operatorname{rank}(A(\mathbf{x})) = k_0\}$ is open.

Now, in this chapter we have been concerned with C^1 maps $\mathbf{f} : U \to \mathbb{R}^n$ (U an open subset of $\mathbb{R}^m$) and the matrix in question is the derivative $D\mathbf{f}(\mathbf{x})$. If k_0 is the maximum rank of this matrix as $\mathbf{x}$ ranges over U, the set $V = \{\mathbf{x} \in U : \operatorname{rank}(D\mathbf{f}(\mathbf{x})) = k_0\}$ is open, and the theorems of this chapter can be applied on V. (The implicit function and inverse mapping theorems deal with the case when k_0 is as large as possible, namely, $k_0 = \min(m, n)$; the theorems of this section provide information for smaller values of k.) The typical situation is that V is dense in U, that is, the set $U \setminus V$ has no interior points. Thus, the structure of the mapping $\mathbf{f}$ near "most" points of U (the ones in V) is fairly simple to understand, but at the remaining points, various kinds of singularities can occur. The study of such singularities is a substantial and rather intricate branch of mathematical analysis.

EXERCISES

1. For each of the following maps $\mathbf{f} = (f, g, h)$, determine whether f, g, h are functionally dependent on some open set $U \subset \mathbb{R}^3$ by examining the Jacobian

$\partial(f, g, h)/\partial(x, y, z)$. If they are, determine the rank of $D\mathbf{f}$ on U and find functional relations (one relation if rank($D\mathbf{f}$) = 2, two relations if rank($D\mathbf{f}$) = 1) satisfied by f, g, h.

   a. $f(x, y, z) = x + y - z$, $g(x, y, z) = x - y + z$, $h(x, y, z) = x^2 + y^2 + z^2 - 2yz$.

   b. $f(x, y, z) = x^2 + y^2 + z^2$, $g(x, y, z) = x + y + z$, $h(x, y, z) = y - z$.

   c. $f(x, y, z) = y^{1/2} \sin x$, $g(x, y, z) = y \cos^2 x - y$, $h(x, y, z) = z - 3$.

   d. $f(x, y, z) = xy + z$, $g(x, y, z) = x^2y^2 + 2xyz + z^2$, $h(x, y, z) = 2 - xy - z$.

   e. $f(x, y, z) = \log x - \log y + z$, $g(x, y, z) = \log x - \log y - z$, $h(x, y, z) = (x^2 + 2y^2)/xy$.

   f. $f(x, y, z) = x - y + z$, $g(x, y, z) = x^2 - y^2$, $h(x, y, z) = x + z$.

2. Write out the statement and give a precise proof for the following special cases of Theorem 3.21, along the lines of Theorem 3.22.

   a. $m = n = 2$, $k = 1$.

   b. $m = 2$, $n = 3$, $k = 1$.

Chapter 4

INTEGRAL CALCULUS

In this chapter we study the integration of functions of one and several real variables. As we assume that the reader is already familiar with the standard techniques of integration for functions of one variable, our discussion of integration on the line is limited to theoretical issues. On the other hand, some of these issues arise also in higher dimensions, and we shall sometimes invoke the careful treatment of the one-variable case as an excuse for being somewhat sketchy in developing the theory for several variables.

In elementary calculus, the term "integral" can refer either to the antiderivative of a function f or to a limit of sums of the form $\sum f(x_j)\Delta x_j$; one speaks of indefinite or definite integrals. At the more advanced level, and in particular in this book, "integral" almost always carries the latter meaning. The notion of integration as a sophisticated form of summation is one of the truly fundamental ideas of mathematical analysis, and it arises in many contexts where the connection with differentiation is tenuous or nonexistent.

4.1   Integration on the Line

Recall that for a nonnegative function f, the basic geometric interpretation of the integral $\int_a^b f(x)\,dx$ is as the area of the region between the graph of f and the x-axis over the interval $[a, b]$. The idea for computing this area is to subdivide the interval $[a, b]$ into small subintervals $[x_0, x_1], [x_1, x_2], \ldots, [x_{J-1}, x_J]$, with $x_0 = a$ and $x_J = b$, and to approximate the region under the graph of f by a union of rectangles based on the intervals $[x_{j-1}, x_j]$. If we choose the height h_j of the jth rectangle to be smaller (resp. larger) than all the values of f on the interval $[x_{j-1}, x_j]$, the corresponding sum $\sum_1^k h_j(x_j - x_{j-1})$ will be a lower (resp. upper) bound for the area under the graph of f. If all goes well, these lower and upper approximations will approach each other as we subdivide the interval $[a, b]$ into smaller and smaller pieces, and their common limit will be the integral of f.

Let us make this more precise, introducing some useful definitions along the way. A **partition** P of the interval $[a, b]$ is a subdivision of $[a, b]$ into nonoverlapping subintervals, specified by giving the subdivision points $x_1, \ldots, x_{J-1}$ along with the endpoints $x_0 = a$ and $x_J = b$. In symbols, we shall write

$$P = \{x_0, x_1, \ldots, x_J\}, \text{ with } a = x_0 < x_1 < \cdots < x_J = b.$$

If P and P' are partitions of $[a, b]$, we say that P' is a **refinement** of P if P' is obtained from P by adding in more subdivision points, that is, if $P \subset P'$.

Observe that if P and Q are any two partitions of $[a, b]$, they can be combined into a single partition $P \cup Q$ whose subdivision points are those of P together with those of Q; $P \cup Q$ is a refinement of both P and Q.

Now let f be a bounded real-valued function on $[a, b]$. (We make no continuity assumptions on f at this point.) Given a partition $P = \{x_0, \ldots, x_J\}$ of $[a, b]$, for $1 \le j \le J$ we set
(4.1)
$$m_j = \inf\{f(x) : x_{j-1} \le x \le x_j\}, \qquad M_j = \sup\{f(x) : x_{j-1} \le x \le x_j\}.$$

(If f is continuous, m_j and M_j are just the minimum and maximum values of f on $[x_{j-1}, x_j]$, which exist by the extreme value theorem.) We then define the **lower Riemann sum** $s_P f$ and the **upper Riemann sum** $S_P f$ corresponding to the partition P by

$$(4.2) \qquad s_P f = \sum_1^J m_j(x_j - x_{j-1}), \qquad S_P f = \sum_1^J M_j(x_j - x_{j-1}).$$

See Figure 4.1, where the lower and upper Riemann sums are the sums of the areas of the rectangles, an area being counted as negative if the rectangle is below the x-axis.

If m and M are the infimum and supremum of the values of f over the whole interval $[a, b]$, we clearly have $m_j \ge m$ and $M_j \le M$ for all j, and hence

$$s_P f \ge m \sum_1^J (x_j - x_{j-1}) = m(b - a),$$

$$S_P f \le M \sum_1^J (x_j - x_{j-1}) = M(b - a).$$

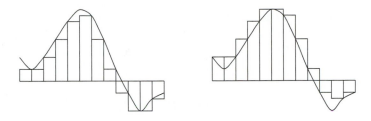

FIGURE 4.1: Lower and upper Riemann sums.

The same argument shows that if one of the subintervals $[x_{j-1}, x_j]$ is subdivided further, the lower sum $s_P f$ becomes larger while the upper sum $S_P f$ becomes smaller. In short:

4.3 Lemma. *If P' is a refinement of P, then $s_{P'} f \geq s_P f$ and $S_{P'} f \leq S_P f$.*

An immediate consequence of this is that any lower Riemann sum for f is less than any upper Riemann sum for f:

4.4 Lemma. *If P and Q are any partitions of $[a, b]$, then $s_P f \leq S_Q f$.*

Proof. Consider the common refinement $P \cup Q$. By Lemma 4.3,

$$s_P f \leq s_{P \cup Q} f \leq S_{P \cup Q} f \leq S_Q f.$$

$\square$

Next, we define the **lower** and **upper integrals** of f on $[a, b]$ by

$$\underline{I}_a^b(f) = \sup_P s_P f, \qquad \overline{I}_a^b(f) = \inf_P S_P f,$$

the supremum and infimum being taken over all partitions P of $[a, b]$. By Lemma 4.4, we have $\underline{I}_a^b(f) \leq \overline{I}_a^b(f)$. If the upper and lower integrals coincide, f is called **Riemann integrable** on $[a, b]$, and the common value of the upper and lower integrals is the **Riemann integral** $\int_a^b f(x)\, dx$. We shall generally omit the eponym "Riemann," as the Riemann integral is the only one we shall use in this book, but it is significant not only for historical reasons but in order to distinguish the Riemann integral from the more sophisticated Lebesgue integral.

At first sight it would seem difficult to determine whether a function f is integrable and to evaluate its integral, as the definitions involve all possible partitions of $[a, b]$. The following lemma is the key to making these calculations more manageable.

4.5 Lemma. *If f is a bounded function on $[a, b]$, the following conditions are equivalent:*

   a. f is integrable on $[a, b]$.

   b. For every $\epsilon > 0$ there is a partition P of $[a, b]$ such that $S_P f - s_P f < \epsilon$.

Proof. If $S_P f - s_P f < \epsilon$ for some partition P, then $\overline{I}_a^b f - \underline{I}_a^b f < \epsilon$, and since ϵ is arbitrary, it follows that $\overline{I}_a^b f = \underline{I}_a^b f$, i.e., f is integrable. Conversely, if f is a bounded function and ϵ is positive, we can find partitions Q and Q' of $[a, b]$ such that $S_Q f < \overline{I}_a^b f + \frac{1}{2}\epsilon$ and $s_{Q'} f > \underline{I}_a^b f - \frac{1}{2}\epsilon$. Thus, if f is integrable, we have $S_Q f - s_{Q'} f < \epsilon$. Let $P = Q \cup Q'$; then by Lemma 4.3, $s_Q' f \leq s_P f \leq S_P f < S_Q f$, so $S_P f - s_P f < s_Q f - s_{Q'} f < \epsilon$. $\square$

The condition (b) in Lemma 4.5 not only gives a workable criterion for integrability but also gives us some leverage for computing the integral of an integrable function f. Indeed, for any partition P we have

$$s_P f \leq \int_a^b f(x)\, dx \leq S_P f,$$

so if $S_P f - s_P f < \epsilon$, $S_P f$ and $s_P f$ are both within ϵ of $\int_a^b f(x)\, dx$. The latter quantity is therefore the limit of the sums $S_P f$ or $s_P f$ as P runs through any sequence of partitions such that $S_P f - s_P f \to 0$.

We next present the fundamental additivity properties of the integral, which are are easy but not quite trivial consequences of the definitions:

4.6 Theorem.

   a. Suppose $a < b < c$. If f is integrable on $[a, b]$ and on $[b, c]$, then f is integrable on $[a, c]$, and

$$(4.7) \qquad \int_a^c f(x)\, dx = \int_a^b f(x)\, dx + \int_b^c f(x)\, dx.$$

   b. If f and g are integrable on $[a, b]$, then so is $f + g$, and

$$(4.8) \qquad \int_a^b [f(x) + g(x)]\, dx = \int_a^b f(x)\, dx + \int_a^b g(x)\, dx.$$

Proof. (a) Given $\epsilon > 0$, let P and Q be partitions of $[a, b]$ and $[b, c]$, respectively, such that $S_P f - s_P f < \frac{1}{2}\epsilon$ and $S_Q f - s_Q f < \frac{1}{2}\epsilon$. Then $P \cup Q$ is a partition of $[a, c]$ and

$$S_{P \cup Q} f = S_P f + S_Q f, \qquad s_{P \cup Q} f = s_P f + s_Q f.$$

It follows that $S_{P \cup Q} f - s_{P \cup Q} f < \epsilon$, so that f is integrable on $[a, c]$ by Lemma 4.5. Moreover, $\int_a^c f(x)\,dx$ is within ϵ of $S_{P \cup Q} f$, and $\int_a^b f(x)\,dx$, and $\int_b^c f(x)\,dx$ are within $\frac{1}{2}\epsilon$ of $S_P f$ and $S_Q f$, respectively, so $\int_a^c f(x)\,dx$ is within 2ϵ of $\int_a^b f(x)\,dx +$ $\int_b^c f(x)\,dx$. Since ϵ is arbitrary, (4.7) follows.

(b) Given $\epsilon > 0$, choose partitions P and Q of $[a, b]$ such that $S_P f - s_P f < \frac{1}{2}\epsilon$ and $S_Q g - s_Q g < \frac{1}{2}\epsilon$, and let $R = P \cup Q$ be the common refinement of P and Q. Then by Lemma 4.3 we have $S_R f - s_R f \leq S_P f - s_P f$ and $S_R g - s_R g \leq S_Q g - s_Q g$. Moreover, the maximum of the sum of two functions is at most the sum of the maxima, and the minimum of the sum is at least the sum of the minima, so

$$S_R(f + g) \leq S_R f + S_R g, \qquad s_R(f + g) \geq s_R f + s_R g.$$

Hence,

$$S_R(f + g) \leq S_R f + S_R g \leq s_R f + \tfrac{1}{2}\epsilon + s_R g + \tfrac{1}{2}\epsilon \leq s_R(f + g) + \epsilon.$$

In other words, $S_R(f + g) - s_R(f + g) < \epsilon$, so that $f + g$ is integrable by Lemma 4.5. The formula (4.8) then follows in much the same way as (4.7). $\qquad\square$

Remark. We make the usual convention that

$$\int_b^a f(x)\,dx = -\int_a^b f(x)\,dx;$$

then (4.7) holds no matter how the points a, b, c are ordered.

The following theorem lists some more standard properties of integrals. They are all quite easy to derive from the definitions with the help of Lemma 4.5, and we leave their proofs as Exercises 2–5.

4.9 Theorem. *Suppose f is integrable on $[a, b]$.*

a. *If $c \in \mathbb{R}$, then cf is integrable on $[a, b]$, and $\int_a^b cf(x)\,dx = c \int_a^b f(x)\,dx$.*

b. *If $[c, d] \subset [a, b]$, then f is integrable on $[c, d]$.*

c. *If g is integrable on $[a, b]$ and $f(x) \leq g(x)$ for $x \in [a, b]$, then $\int_a^b f(x)\,dx \leq \int_a^b g(x)\,dx$.*

d. *$|f|$ is integrable on $[a, b]$, and $\left| \int_a^b f(x)\,dx \right| \leq \int_a^b |f(x)|\,dx$.*

We now derive some useful criteria for integrability. The first one has a very simple proof, and in conjunction with Theorem 4.6a it establishes the integrability of most of the functions that arise in elementary calculus. (Such functions have only a finite number of local maxima and minima on any bounded interval $[a, b]$, so one can break $[a, b]$ up into finitely many subintervals on which the function in question is monotone, apply Theorem 4.10 on each subinterval, and then add the results by Theorem 4.6a.)

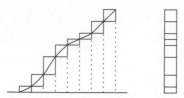

FIGURE 4.2: An increasing function and a partition with equal subintervals. The difference between the upper and lower Riemann sums is
the sum of the areas of the solid rectangles, which is easily found by
stacking them.

4.10 Theorem. *If f is bounded and monotone on* $[a, b]$, *then f is integrable on* $[a, b]$.

Proof. Suppose f is increasing on $[a, b]$; the proof is similar if f is decreasing. Consider the partition P_k of $[a, b]$ into k equal subintervals of length $(b - a)/k$. Since f is increasing, the quantities m_j and M_j in (4.1) are given by

$$m_j = f(x_{j-1}), \qquad M_j = f(x_j),$$

and hence the lower and upper Riemann sums are

$$s_{P_k} f = \frac{b - a}{k} \sum_0^{k-1} f(x_j), \qquad S_{P_k} f = \frac{b - a}{k} \sum_1^{k} f(x_j),$$

and their difference is

$$S_{P_k} f - s_{P_k} f = \frac{b - a}{k} [f(x_k) - f(x_0)] = \frac{(b - a)[f(b) - f(a)]}{k}.$$

This can be made as small as we please by taking k sufficiently large, so f is integrable by Lemma 4.5. (The geometry of this calculation is shown in Figure 4.2.)   □

The next criterion for integrability is the one that is most commonly stated in calculus books. Its proof, however, is frequently omitted because it relies on the notion of uniform continuity that we studied in §1.8.

4.11 Theorem. *If f is continuous on* $[a, b]$, *then f is integrable on* $[a, b]$.

Proof. First, f is bounded on $[a, b]$ by Theorem 1.23, so the upper and lower Riemann sums for any partition exist. By Theorem 1.33, f is uniformly continuous on $[a, b]$; thus, given $\epsilon > 0$, we can find $\delta > 0$ so that $|f(x) - f(y)| < \epsilon/(b - a)$ whenever $x, y \in [a, b]$ and $|x - y| < \delta$. Let P be any partition of $[a, b]$ whose subintervals $[x_{j-1}, x_j]$ all have length less than δ. Then $|f(x) - f(y)| < \epsilon/(b - a)$ whenever x and y both lie in the same subinterval, and in particular the maximum and minimum values of f on that subinterval differ by less than $\epsilon/(b - a)$. But this means that

$$S_P f - s_P f = \sum_{1}^{J} (M_j - m_j)(x_j - x_{j-1})$$

$$< \frac{\epsilon}{b - a} \sum_{1}^{J} (x_j - x_{j-1}) = \frac{\epsilon}{b - a}(b - a) = \epsilon.$$

By Lemma 4.5, then, f is integrable. $\qquad \square$

Theorem 4.11 can be extended to functions that have some discontinuities, as long as the set of discontinuities is "small." The following result suffices for most practical purposes.

4.12 Theorem. *If f is bounded on $[a, b]$ and continuous at all except finitely many points in $[a, b]$, then f is integrable on $[a, b]$.*

Proof. Let $y_1, \ldots, y_L$ be the points in $[a, b]$ where f is discontinuous, and let m and M be the infimum and supremum of $\{f(x) : a \le x \le b\}$, the set of values of f on $[a, b]$. Given $\delta > 0$, let

$$I_l = [a, b] \cap [y_l - \delta, \, y_l + \delta],$$

and let

$$U = \bigcup_{1}^{L} I_l, \qquad V = [a, b] \setminus U^{\text{int}}.$$

Thus U is a union of small intervals that contain the discontinuities of f, and V is the remainder of $[a, b]$. Each interval I_m has length at most 2δ, and there are L of these intervals, so the total length of the set U is at most $2L\delta$. On the other hand, V is a finite union of closed intervals, on each of which f is continuous.

Let P be any partition of $[a, b]$ that includes the endpoints of the intervals I_m among its subdivision points. Then we can write

$$S_P f = S_P^U f + S_P^V f, \qquad s_P f = s_P^U f + s_P^V f,$$

where $S_P^U f$ (resp. $S_P^V f$) is the sum of the terms $M_j(x_j - x_{x-1})$ in $S_P f$ for which
the interval $[x_{j-1}, x_j]$ is contained in U (resp. V), and likewise for $s_P^U f$ and $s_P^V f$.

Now, let $\epsilon > 0$ be given. Since f is continuous on each of the closed intervals
that constitute V, Theorem 4.11 shows that we can make

$$S_P^V f - s_P^V f < \tfrac{1}{2}\epsilon$$

by choosing the partition P sufficiently fine. On the other hand,

$$S_P^U f - s_P^U f = \sum_{[x_{j-1}, x_j] \subset U} (M_j - m_j)(x_j - x_{j-1})$$

$$\leq (M - m)(\text{length of } U) \leq (M - m)2L\delta,$$

and we can make this less than $\tfrac{1}{2}\epsilon$ by taking $\delta < \epsilon/2L(M - m)$. In short, for a
suitably chosen P we have $S_P f - s_P f < \epsilon$, so f is integrable by Lemma 4.5. $\square$

The preceding argument actually proves more than is stated in Theorem 4.12.
It is not necessary that the set of discontinuities of f be finite, only that it can be
covered by finitely many intervals $I_1, \ldots, I_L$ whose total length is as small as we
please. Certain infinite sets, such as convergent sequences, also have this property
(Exercise 6). We make it into a formal definition: A set $Z \subset \mathbb{R}$ is said to have **zero
content** if for any $\epsilon > 0$ there is a finite collection of intervals $I_1, \ldots, I_L$ such that
(i) $Z \subset \bigcup_1^L I_l$, and (ii) the sum of the lengths of the I_l's is less than ϵ. The proof of
Theorem 4.12 now yields the following result:

4.13 Theorem. *If f is bounded on $[a, b]$ and the set of points in $[a, b]$ at which f is
discontinuous has zero content, then f is integrable on $[a, b]$.*

Theorem 4.13 is only a technical refinement of Theorem 4.12, and the reader
should not attach undue importance to it.[1] We mention it because its analogue in
higher dimensions does play a significant role in the theory, as we shall see. We
also remark that neither of Theorems 4.10 and 4.13 includes the other; the set of
discontinuities of a monotone function need not have zero content, and there are
continuous functions that are not monotone on any interval.

If f is an integrable function on $[a, b]$, the value of $\int_a^b f(x)\, dx$ is somewhat
insensitive to the values of f at individual points, in the following sense:

4.14 Proposition. *Suppose f and g are integrable on $[a, b]$ and $f(x) = g(x)$ for
all except finitely many points $x \in [a, b]$. Then $\int_a^b f(x)\, dx = \int_a^b g(x)\, dx$.*

---

[1] It does, however, point the way toward a necessary and sufficient condition for a function to be
integrable, which we shall describe at the end of §4.8.

Proof. First suppose g is identically zero. That is, we are assuming that $f(x) = 0$ for all $x \in [a, b]$ except for finitely many points $y_1, \ldots, y_L$. Let P_k be the partition of $[a, b]$ into k equal subintervals, and take k large enough so that the points y_l all lie in different subintervals. Then

$$S_{P_k} f = \frac{b - a}{k} \sum_1^L \max\left(f(y_l), 0\right), \qquad s_{P_k} f = \frac{b - a}{k} \sum_1^L \min\left(f(y_l), 0\right).$$

Both these quantities tend to zero as $k \to \infty$, and hence $\int_a^b f(x)\, dx = 0$.

The general case follows by applying this argument to the difference $f - g$. $\square$

The main use of Proposition 4.14 is in the context of functions with finitely many discontinuities, as in Theorem 4.12. For such a function f there is often no "right" way to define f at the points where it is discontinuous. Proposition 4.14 assures us that this problem is of no consequence as far as integration is concerned; we may define f at these points however we like, or indeed leave f undefined there, without any effect on $\int_a^b f(x)\, dx$.

Next, we present a general version of the fundamental theorem of calculus. Its two parts say in effect that differentiating an integral or integrating a derivative leads back to the original function.

4.15 Theorem (The Fundamental Theorem of Calculus).

a. *Let f be an integrable function on $[a, b]$. For $x \in [a, b]$, let $F(x) = \int_a^x f(t)\, dt$ (which is well defined by Theorem 4.9b). Then F is continuous on $[a, b]$; moreover, $F'(x)$ exists and equals $f(x)$ at every x at which f is continuous.*

b. *Let F be a continuous function on $[a, b]$ that is differentiable except perhaps at finitely many points in $[a, b]$, and let f be a function on $[a, b]$ that agrees with F' at all points where the latter is defined. If f is integrable on $[a, b]$, then $\int_a^b f(t)\, dt = F(b) - F(a)$.*

Proof. (a) If $x, y \in [a, b]$, by (4.7) we have

$$F(y) - F(x) = \int_x^y f(t)\, dt.$$

Let $C = \sup\{|f(t)| : t \in [a, b]\}$; then by Theorem 4.9d,

$$|F(y) - F(x)| \le \int_x^y |f(t)|\, dt \le C \int_x^y dt = C|y - x|,$$

which implies that F is continuous. Next, suppose that f is continuous at x; thus, given $\epsilon > 0$, there is a $\delta > 0$ so that $|f(t) - f(x)| < \epsilon$ whenever $|t - x| < \delta$. Since

$$f(x) = f(x) \frac{1}{y - x} \int_x^y dt = \frac{1}{y - x} \int_x^y f(x)\, dt,$$

we have

$$\frac{F(y) - F(x)}{y - x} - f(x) = \frac{1}{y - x} \int_x^y [f(t) - f(x)]\, dt.$$

Hence, if $|y - x| < \delta$, we have $|f(t) - f(x)| < \epsilon$ for all t between y and x, so

$$\left| \frac{F(y) - F(x)}{y - x} - f(x) \right| \le \frac{1}{|y - x|} \left| \int_x^y \epsilon\, dt \right| = \epsilon.$$

It follows that $\lim_{y \to x}[F(y) - F(x)]/(y - x) = f(x)$, as claimed.

(b) Let $P = \{x_0, \ldots, x_J\}$ be a partition of $[a, b]$; by adding in extra points, we may assume that all the points where F is not differentiable are among the subdivision points x_j. Then, for each j, F is continuous on the interval $[x_{j-1}, x_j]$ and differentiable on its interior, so by the mean value theorem,

$$F(x_j) - F(x_{j-1}) = F'(t_j)(x_j - x_{j-1}) = f(t_j)(x_j - x_{j-1})$$

for some point $t_j \in (x_{j-1}, x_j)$. Adding up these equalities yields

$$F(b) - F(a) = F(x_J) - F(x_0) = \sum_1^J f(t_j)(x_j - x_{j-1}),$$

which implies that

$$s_P f \le F(b) - F(a) \le S_P f.$$

Since f is integrable, we can make $s_P f$ and $S_P f$ as close to $\int_a^b f(x)\, dx$ as we like by choosing P suitably, and the desired result follows immediately. $\qquad \square$

We have developed the notion of the integral of a function f in terms of the upper and lower Riemann sums $S_P f$ and $s_P f$. More generally, if $P = \{x_0, \ldots, x_J\}$ is a partition of $[a, b]$ and t_j is any point in the interval $[x_{j-1}, x_j]$ $(1 \le j \le J)$, the quantity

$$\sum_1^J f(t_j)(x_j - x_{j-1})$$

is called a **Riemann sum** for f associated to the partition P. Clearly, if m_j and M_j are as in (4.1) we have $m_j \le f(t_j) \le M_j$, so that

$$s_P f \le \sum_1^J f(t_j)(x_j - x_{j-1}) \le S_P f.$$

Thus, if f is integrable and we choose the partition P so that $s_P f$ and $S_P f$ are good approximations to $\int_a^b f(x)\, dx$, all the Riemann sums corresponding to P will also be good approximations to $\int_a^b f(x)\, dx$.

One last question should be addressed: Given an integrable function f on $[a, b]$, for which partitions P do the sums $s_P f$ and $S_P f$ furnish a good approximation to $\int_a^b f(x)\, dx$? It might seem that the answer might depend strongly on the nature of the function f, but in fact, any partition whose subintervals are sufficiently small will do the job. More precisely:

4.16 Proposition. *Suppose f is integrable on $[a, b]$. Given $\epsilon > 0$, there exists $\delta > 0$ such that if $P = \{x_0, \ldots, x_J\}$ is any partition of $[a, b]$ satisfying*

$$\max_{1 \le j \le J} (x_j - x_{j-1}) < \delta,$$

the sums $s_P f$ and $S_P f$ differ from $\int_a^b f(x)\, dx$ by at most ϵ.

Proof. The proof is presented in Appendix B.3 (Theorem B.7). $\qquad\square$

Proposition 4.16 shows, in particular, that one can always compute $\int_a^b f(x)\, dx$ as the limit as $k \to \infty$ of $s_{P_k} f$ or $S_{P_k} f$, where P_k is the partition of $[a, b]$ into k equal subintervals.

One final remark: The definite integral, which is defined as a limit of Riemann sums, may be considered on the intuitive level as a sum of infinitely many infinitesimal terms. This notion, which is probably quite obvious to the alert reader, is often not stated explicitly in mathematics texts because of its lack of rigorous meaning. But the fact is that in many situations — and we shall encounter several of them later on — the interpretation of the integral as a sum of infinitesimals is the clearest way to understand what is going on.

EXERCISES

1. Let $f(x) = 1$ if x is rational, $f(x) = 0$ if x is irrational. Show that f is not integrable on any interval.

2. Prove Theorem 4.9a. (*Hint:* Show that $s_P(cf) = c s_P f$ and $S_P(cf) = c S_P f$ if $c \ge 0$, and $s_P(cf) = c S_P f$ and $S_P(cf) = c s_P f$ if $c < 0$.)

3. Prove Theorem 4.9b. (*Hint:* Consider partitions of $[a, b]$ for which c and d are among the subdivision points.)

4. Prove Theorem 4.9c.

5. Prove Theorem 4.9d. (*Hint:* To prove that $|f|$ is integrable, show that $S_P|f| - s_P|f| \le S_P f - s_P f$. For the inequality $|\int f| \le \int |f|$, observe that $\pm f \le |f|$ and use Theorem 4.9c.)

6. Let $\{x_k\}$ be a convergent sequence in $\mathbb{R}$. Show that the set $\{x_1, x_2, \ldots\}$ has zero content.

7. Let f be an integrable function on $[a, b]$. Suppose that $f(x) \geq 0$ for all x and there is at least one point $x_0 \in [a, b]$ at which f is continuous and strictly positive. Show that $\int_a^b f(x)\, dx > 0$.

8. Let f be an integrable function on $[a, b]$. Prove the following formulas directly from the definitions:

   a. For any $c > 0$, $\int_a^b f(x)\, dx = c \int_{a/c}^{b/c} f(cx)\, dx$.

   b. $\int_a^b f(x)\, dx = \int_{-b}^{-a} f(-x)\, dx$.

   c. For any $c \in \mathbb{R}$, $\int_a^b f(x)\, dx = \int_{a-c}^{b-c} f(x + c)\, dx$.

9. Suppose g and h are continuous functions on $[a, b]$, and f is a continuous function on $\mathbb{R}^2$. Show that for any $\epsilon > 0$ there is a $\delta > 0$ such that if $P = \{x_0, \ldots, x_J\}$ is any partition of $[a, b]$ satisfying $\max_{1 \leq j \leq J}(x_j - x_{j-1}) < \delta$, then

$$\left| \int_a^b f(g(x), h(x))\, dx - \sum_{j=1}^J f(g(x_j'), h(x_j''))(x_j - x_{j-1}) \right| < \epsilon$$

for any choice of x_j', x_j'' in the interval $[x_{j-1}, x_j]$. (The point is that x_j' and x_j'' need not be equal, so the sum in this inequality may not be a genuine Riemann sum for the integral.)

4.2   Integration in Higher Dimensions

In this section we develop the theory of multiple integrals. The basic ideas are much the same as for single integrals; the most serious complication comes from the greater variety of regions over which integration is to be performed. To minimize the complexity of the notation, we first develop the two-dimensional case and then sketch the extension to higher dimensions.

Here and in what follows we shall employ the following notation. If S and T are sets, their **Cartesian product** $S \times T$ is the set of all ordered pairs (s, t) with $s \in S$ and $t \in T$. For example, the plane is the Cartesian product of the line with itself: $\mathbb{R}^2 = \mathbb{R} \times \mathbb{R}$. This idea extends in the obvious way to products of n sets, with ordered n-tuples replacing ordered pairs; for example, $\mathbb{R}^3 = \mathbb{R} \times \mathbb{R} \times \mathbb{R}$. We can also think of $\mathbb{R}^3$ as $\mathbb{R}^2 \times \mathbb{R}$ or as $\mathbb{R} \times \mathbb{R}^2$.

Double Integrals. We begin by defining the double integral of a function over a rectangular region in the plane. In this chapter, by a **rectangle** we shall mean a set of the form

$$R = [a, b] \times [c, d] = \{(x, y) \in \mathbb{R}^2 : x \in [a, b],\ y \in [c, d]\}.$$

(Thus, a "rectangle" in this sense is always closed, and its sides are always parallel to the coordinate axes.) A **partition** of R is a subdivision of R into subrectangles obtained by partitioning both sides of R. Thus, a partition P is specified by its subdivision points,

$$P = \{x_0, \ldots, x_J; y_0, \ldots, y_K\}, \qquad \begin{cases} a = x_0 < \cdots < x_J = b, \\ c = y_0 < \cdots < y_K = d, \end{cases}$$

and it yields a decomposition of R into the subrectangles

$$R_{jk} = [x_{j-1}, x_j] \times [y_{k-1}, y_k]$$

with area

$$\Delta A_{jk} = (x_j - x_{j-1})(y_k - y_{k-1}).$$

Now let f be a bounded function on the rectangle R. Given a partition P as above, we set

$$m_{jk} = \inf\{f(x, y) : (x, y) \in R_{jk}\}, \qquad M_{jk} = \sup\{f(x, y) : (x, y) \in R_{jk}\},$$

and define the **lower** and **upper Riemann sums** of f corresponding to P by

$$s_P f = \sum_{j=1}^{J} \sum_{k=1}^{K} m_{jk} \Delta A_{jk}, \qquad S_P f = \sum_{j=1}^{J} \sum_{k=1}^{K} M_{jk} \Delta A_{jk}.$$

The **lower** and **upper integrals** of f on R are

$$\underline{I}_R(f) = \sup_P s_P f, \qquad \overline{I}_R(f) = \inf_P S_P f,$$

the supremum and infimum being taken over all partitions P of R. If the lower and upper integrals coincide, f is called (**Riemann**) **integrable** on R, and the common value of the upper and lower integrals is called the (**Riemann**) **integral** of f over R and is denoted by

$$\iint_R f\, dA \quad \text{or} \quad \iint_R f(x, y)\, dx\, dy.$$

These notions are entirely analogous to their one-dimensional counterparts. The reader should refer back to §4.1 for a more detailed discussion, which can

easily be adapted to the present situation. However, we have not yet built a satisfactory definition of two-dimensional integrals, because we often wish to integrate
functions over regions other than rectangles. The solution to this problem is simple,
in principle: To integrate a function f over a bounded region $S \subset \mathbb{R}^2$, we draw a
large rectangle R that contains S, (re)define f to be zero outside of S, and integrate
the resulting function over R.

To express this neatly, it is convenient to introduce another definition. If S is a
subset of $\mathbb{R}^2$ (or $\mathbb{R}^n$, or indeed any set), the **characteristic function** or **indicator
function** of S is the function χ_S defined by

$$\chi_S(\mathbf{x}) = \begin{cases} 1 & \text{if } \mathbf{x} \in S, \\ 0 & \text{otherwise.} \end{cases}$$

Now, suppose S is a bounded subset of $\mathbb{R}^2$ and f is a bounded function on $\mathbb{R}^2$.
Let R be a rectangle that contains S. We say that f is **integrable on** S if $f\chi_S$ is
integrable on R, in which case we define the integral of f over S by

$$\iint_S f\, dA = \iint_R f\chi_S\, dA.$$

It is easily verified that this definition does not depend on the choice of the en-
veloping rectangle R, since the integrand $f\chi_S$ vanishes outside of S. (It also does
not depend on the values of f outside of S. We could just as well assume that
f is only defined on S or on some set containing S, with the understanding that
$(f\chi_S)(\mathbf{x}) = 0$ for $\mathbf{x} \notin S$.)

The properties of integrals in two dimensions are very similar to those in one;
the following theorem provides a list of the most basic ones. The proof is essentially
identical to that of Theorems 4.6 and 4.9; we leave the details to the interested
reader.

4.17 Theorem.

a. *If f_1 and f_2 are integrable on the bounded set S and $c_1, c_2 \in \mathbb{R}$, then $c_1 f_1 + c_2 f_2$ is integrable on S, and*

$$\iint_S [c_1 f_1 + c_2 f_2]\, dA = c_1 \iint_S f_1\, dA + c_2 \iint_S f_2\, dA.$$

b. *Let S_1 and S_2 be bounded sets with no points in common, and let f be a bounded function. If f is integrable on S_1 and on S_2, then f is integrable on $S_1 \cup S_2$, in which case*

$$\iint_{S_1 \cup S_2} f\, dA = \iint_{S_1} f\, dA + \iint_{S_2} f\, dA.$$

c. If f and g are integrable on S and $f(\mathbf{x}) \leq g(\mathbf{x})$ for $\mathbf{x} \in S$, then $\iint_S f \, dA \leq \iint_S g \, dA$.

d. If f is integrable on S, then so is $|f|$, and $\left| \iint_S f \, dA \right| \leq \iint_S |f| \, dA$.

At this point we need to say more about the conditions under which a function is integrable. In the one-variable situation, we can get along quite well by restricting attention to continuous functions, but that is not the case here: Even if the function f is continuous, the function χ_S that enters into the definition of $\iint_S f \, dA$ is not. The starting point is the analogue of Theorem 4.13. The notion of "zero content" transfers readily to sets in the plane; namely, a set $Z \subset \mathbb{R}^2$ is said to have **zero content** if for any $\epsilon > 0$ there is a finite collection of rectangles $R_1, \ldots, R_M$ such that (i) $Z \subset \bigcup_1^M R_m$, and (ii) the sum of the areas of the R_m's is less than ϵ. We then have:

4.18 Theorem. *Suppose f is a bounded function on the rectangle R. If the set of points in R at which f is discontinuous has zero content, then f is integrable on R.*

Proof. The proof is essentially identical to that of Theorem 4.13. That is, one first shows that f is integrable if f is continuous on all of R by the argument that proves Theorem 4.11, then encompasses the general case by the argument that proves Theorem 4.12. Details are left to the reader. $\square$

The notion of "zero content" is considerably more interesting in the plane than on the line, as the sets having this property include not only finite sets but things such as smooth curves (that is, curves parametrized by C^1 functions $\mathbf{f} : [a, b] \to \mathbb{R}^2$). The following proposition summarizes the results we will need; see also Exercise 2.

4.19 Proposition.
a. If $Z \subset \mathbb{R}^2$ has zero content and $U \subset Z$, then U has zero content.
b. If $Z_1, \ldots, Z_k$ have zero content, then so does $\bigcup_1^k Z_j$.
c. If $\mathbf{f} : (a_0, b_0) \to \mathbb{R}^2$ is of class C^1, then $\mathbf{f}([a, b])$ has zero content whenever $a_0 < a < b < b_0$.

Proof. Parts (a) and (b) are easy, and their proofs are left as an exercise. As for (c), let $P_k = \{t_0, \ldots, t_k\}$ be the partition of $[a, b]$ into k equal subintervals of length $\delta = (b - a)/k$, and let C be an upper bound for $\{|\mathbf{f}'(t)| : t \in [a, b]\}$. By the mean value theorem applied to the two components $x(t), y(t)$ of $\mathbf{f}(t)$, we have $|x(t) - x(t_j)| \leq C\delta$ and $|y(t) - y(t_j)| \leq C\delta$ for $t \in [t_{j-1}, t_j]$. In other words, $\mathbf{f}([t_{j-1}, t_j])$ is contained in the square of side length $2C\delta$ centered at $\mathbf{f}(t_j)$. Hence, $\mathbf{f}([a, b])$ is contained in the union of these squares, and the sum of their areas is $k(2C\delta)^2 = 4C^2(b - a)^2/k$. This can be made as small as we please by taking k sufficiently large, so $\mathbf{f}([a, b])$ has zero content. $\square$

To apply Theorem 4.18 to the integrand $f\chi_S$ in the definition of $\iint_S f\,dA$, we need to know about the discontinuities of χ_S. The following lemma provides the answer.

4.20 Lemma. *The function χ_S is discontinuous at* $\mathbf{x}$ *if and only if* $\mathbf{x}$ *is in the boundary of* S.

Proof. If $\mathbf{x}$ is in the interior of S, then χ_S is identically 1 on some ball containing $\mathbf{x}$, so it is continuous at $\mathbf{x}$. Likewise, if $\mathbf{x}$ is in the interior of the complement S^c, then f is identically 0 near $\mathbf{x}$ and hence continuous at $\mathbf{x}$. But if $\mathbf{x}$ is in the boundary of S, then there are points arbitrarily close to $\mathbf{x}$ where $\chi_S = 1$ and other such points where $\chi_S = 0$, so χ_S is discontinuous at $\mathbf{x}$.   □

In view of Theorem 4.18 and Lemma 4.20, to have a good notion of integration over a set S, we should require the boundary of S to have zero content. We make this condition into a formal definition: A set $S \subset \mathbb{R}^2$ is **Jordan measurable** if it is bounded and its boundary has zero content. (We shall comment further on this nomenclature below.) We shall generally say "**measurable**" rather than "Jordan measurable," but we advise the reader that in more advanced works the term "measurable" refers to the more general concept of *Lebesgue measurability* (see §4.8).

By Proposition 4.19, any bounded set whose boundary is a finite union of pieces of smooth curves is measurable; these are the sets that we almost always encounter in practice. The following theorem gives a convenient criterion for integrability.

4.21 Theorem. *Let S be a measurable subset of $\mathbb{R}^2$. Suppose $f : \mathbb{R}^2 \to \mathbb{R}$ is bounded and the set of points in S at which f is discontinuous has zero content. Then f is integrable on S.*

Proof. The only points where $f\chi_S$ can be discontinuous are those points in the closure of S where either f or χ_S is discontinuous. By Lemma 4.20 and Proposition 4.19b, the set of such points has zero content. By Theorem 4.18, $f\chi_S$ is integrable on any rectangle R containing S, and hence f is integrable on S.   □

To complete the picture, we need the following generalization of Proposition 4.14, which shows that sets of zero content are negligible for the purposes of integration.

4.22 Proposition. *Suppose $Z \subset \mathbb{R}^2$ has zero content. If $f : \mathbb{R}^2 \to \mathbb{R}$ is bounded, then f is integrable on Z and $\int_Z f\,dA = 0$.*

Proof. Given $\epsilon > 0$, there is a finite collection of rectangles $R_1, \ldots, R_m$ such that $Z \subset \bigcup_1^M R_m$ and the sum of the areas of the R_m's is less than ϵ. By subdividing these rectangles if necessary, we can assume that they have disjoint[2] interiors and form part of a grid obtained by partitioning some large rectangle R. Denoting this partition by P, the area of R_j by $|R_j|$, and $\sup_{\mathbf{x}} |f(\mathbf{x})|$ by C, we have

$$-C\epsilon < -C \sum_1^M |R_j| \leq s_P(f\chi_Z) \leq S_P(f\chi_Z) \leq C \sum_1^M |R_j| < C\epsilon.$$

Since ϵ is arbitrary, the desired conclusion follows directly from the definition of the integral. $\square$

4.23 Corollary.
  a. *Suppose that f is integrable on the set $S \subset \mathbb{R}^2$. If $g(\mathbf{x}) = f(\mathbf{x})$ except for $\mathbf{x}$ in a set of zero content, then g is integrable on S and $\int_S g \, dA = \int_S f \, dA$.*
  b. *Suppose that f is integrable on S and on T, and $S \cap T$ has zero content. Then f is integrable on $S \cup T$, and $\int_{S \cup T} f \, dA = \int_S f \, dA + \int_T f \, dA$.*

Proof. For (a), apply Proposition 4.22 to the function $f - g$. For (b), we are assuming that $f\chi_S$ and $f\chi_T$ are integrable; moreover, by Proposition 4.22, $f\chi_{S \cap T}$ is integrable and its integral is zero. But $f\chi_{S \cup T} = f\chi_S + f\chi_T - f\chi_{S \cap T}$, so the result follows. $\square$

Area. The problem of determining the area of regions in the plane goes back to antiquity. The first effective general method of attacking this problem was provided by the integral calculus in one variable, which yields the area of a region under a graph, or of a region between two graphs. It therefore produces a theory of area for regions that can be broken up into finitely many subregions bounded by graphs of (nice) functions. However, the two-variable theory of integration contains, as a special case, a theory of area (due to the French mathematician Jordan) that encompasses more complicated sorts of regions too. Namely, if S is any Jordan measurable set in the plane, its **area** is the integral over S of the constant function $f(\mathbf{x}) \equiv 1$:

$$(\text{area})(S) = \iint_S 1 \, dA = \iint \chi_S \, dA,$$

the latter integral being taken over any rectangle that contains S.

Let us pause to see just what this means. Given any bounded set $S \subset \mathbb{R}^2$, to compute $\iint_S \chi_S \, dA$ we enclose S in a large rectangle R and consider a partition P

---

[2]A collection $\{S_j\}$ of sets is **disjoint** if $S_j \cap S_k = \varnothing$ for $j \neq k$.

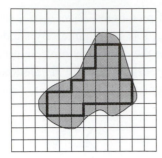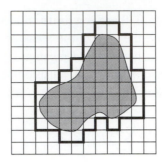

FIGURE 4.3: Approximations to the inner and outer areas of a region.

of R, which produces a grid of small rectangles that cover S. The lower sum for this partition is simply the sums of the areas of the small rectangles that are contained in S, whereas the upper sum is the sum of the areas of the small rectangles that intersect S. Taking the supremum of the lower sums and the infimum of the upper sums yields quantities that may be called the **inner area** and **outer area** of S:

$$\underline{A}(S) = \underline{I}_R(\chi_S), \qquad \overline{A}(S) = \overline{I}_R(\chi_S).$$

When these two quantities coincide, that is, when the characteristic function χ_S is integrable, their common value is the **area** of S. See Figure 4.3.

When do we have $\underline{A}(S) = \overline{A}(S)$? It is not hard to see (Exercises 3–5) that for any bounded set S,

- S and its interior S^{int} have the same inner area;

- S and its closure $\overline{S}$ have the same outer area;

- the inner area of S^{int} plus the outer area of the boundary ∂S equals the outer area of the closure $\overline{S}$.

It follows that *the inner and outer areas of S coincide precisely when the outer area of the boundary ∂S is zero.* But a moment's thought shows that this is nothing but the condition that ∂S should have zero content. In short, *the inner and outer area of S coincide precisely when S is measurable.* This is the explanation for the name "measurable": The measurable sets are the ones that have a well-defined area.

Although the class of Jordan measurable sets is much more extensive than the class of sets whose area can be computed by one-variable calculus, it is not as big as we would ideally wish. It does not include all bounded open sets or all compact sets, for example. Moreover, it does not behave well with respect to passage to limits: The union of a sequence of measurable sets, all contained in a common rectangle, need not be measurable.

A simple example of the latter phenomenon can be obtained by considering the sets S_k of all points in the unit square whose x-coordinate is an integer multiple of 2^{-k}. Each S_k is the union of a finite collection of line segments, so it is measurable and its area is zero. However, the union $\bigcup_1^\infty S_k$ is the set of all points in the unit square whose x-coordinate has a terminating base-2 decimal expansion. This set is dense in the unit square but has no interior, from which it is easy to see that its inner area is 0 but its outer area is 1 (Exercises 3 and 4). By "fattening up" the sets S_k (replacing the line segments in them by thin rectangles), we can also obtain examples of open sets and closed sets that are not measurable (Exercise 6).

The defects of the Jordan theory of area carry over more generally to the theory of integration we are discussing, and for more advanced work one needs the more sophisticated Lebesgue theory of measure and integration, of which we present a brief sketch in §4.8. It is largely for this reason that we are being somewhat cavalier about presenting all the theoretical details in this chapter; there seems to be little virtue in expending an enormous amount of effort on a theory that must be upgraded when one proceeds to a more advanced level.

Higher Dimensions. The theory of n-dimensional integrals is almost identical to the theory of double integrals; the only reason we have not considered an arbitrary n from the beginning is that the notation is simpler, and the geometric intuition is clearer, when $n = 2$. We have merely to replace rectangles by n-dimensional rectangular boxes, that is, regions in $\mathbb{R}^n$ of the form

$$R = [a_1, b_1] \times \cdots \times [a_n, b_n] = \{\mathbf{x} : a_1 \leq x_1 \leq b_1, \ldots, a_n \leq x_n \leq b_n\}.$$

The n-dimensional volume of such a box is the product of the lengths of its sides, $\prod_{j=1}^n (b_j - a_j)$. (Here $\prod$ is the product sign, analogous to $\sum$ for sums.) A partition of such a box is specified by partitioning each of its "sides" $[a_1, b_1], \ldots, [a_n, b_n]$.

The notion of "zero content" generalizes to n dimensions in the obvious way: A bounded set $Z \subset \mathbb{R}^n$ has **zero content** if for any $\epsilon > 0$ there are rectangular boxes $R_1, \ldots, R_K$ whose total volume is less than ϵ, such that $Z \subset \bigcup_1^K R_j$. The analogue of Proposition 4.19c is that smooth submanifolds of dimension $k < n$ in $\mathbb{R}^n$ (given parametrically by C^1 maps $\mathbf{f} : \mathbb{R}^k \to \mathbb{R}^n$) have zero content.

With these modifications, the definition of integrability and Theorems 4.17, 4.18, and 4.21 work just as in the 2-dimensional case. The element of area dA becomes an element of n-dimensional volume, which may be denoted by dV^n, $d^n\mathbf{x}$, or $dx_1 \cdots dx_n$: thus, the notation for n-dimensional integrals over a region

$S \subset \mathbb{R}^n$ is

$$\int \cdots \int_S f \, dV^n = \int \cdots \int_S f(\mathbf{x}) \, d^n\mathbf{x} = \int \cdots \int_S f(x_1, \ldots, x_n) \, dx_1 \cdots dx_n,$$

where $\int \cdots \int$ is shorthand for a row of n integral signs. When $n = 3$, we usually write dV instead of dV^3, the V denoting ordinary 3-dimensional volume.

We conclude with a useful fact about integrals in any number of dimensions.

4.24 Theorem (The Mean Value Theorem for Integrals). *Let S be a compact, connected, measurable susbset of $\mathbb{R}^n$, and let f and g be continuous functions on S with $g \geq 0$. Then there is a point $\mathbf{a} \in S$ such that*

$$\int \cdots \int_S f(\mathbf{x})g(\mathbf{x}) \, d^n\mathbf{x} = f(\mathbf{a}) \int \cdots \int_S g(\mathbf{x}) \, d^n\mathbf{x}.$$

Proof. Let m and M be the minimum and maximum values of f on S, which exist since S is compact. Since $g \geq 0$, we have $mg \leq fg \leq Mg$ on S, and hence

$$m \int \cdots \int_S g(\mathbf{x}) \, d^n\mathbf{x} \leq \int \cdots \int_S f(\mathbf{x})g(\mathbf{x}) \, d^n\delta\mathbf{x} \leq M \int \cdots \int_S g(\mathbf{x}) \, d^n\mathbf{x}.$$

Thus the quotient $(\int \cdots \int fg)/(\int \cdots \int g)$ lies between m and M, so by the intermediate value theorem, it is equal to $f(\mathbf{a})$ for some $\mathbf{a} \in S$. $\qquad\square$

The special case $g \equiv 1$ is of particular interest:

4.25 Corollary. *Let S be a compact, connected, measurable subset of $\mathbb{R}^n$, and let f be a continuous function on S. Then there is a point $\mathbf{a} \in S$ such that*

$$\int \cdots \int_S f(\mathbf{x}) \, d^n\mathbf{x} = f(\mathbf{a})|S|,$$

where $|S|$ denotes the n-dimensional volume of S.

The ratio of $\int \cdots \int_S f(\mathbf{x})d^n\mathbf{x}$ to the n-dimensional volume of S is, by definition, the **average** or **mean value** of f on S. Corollary 4.25 says that when f is continuous and S is compact and connected, there is some point in S at which the actual value of f is the average value.

EXERCISES

1. Prove Proposition 4.19(a,b).

2. Let $f : [a, b] \to \mathbb{R}$ be an integrable function.
   a. Show that the graph of f in $\mathbb{R}^2$ has zero content. (*Hint:* Given a partition P of $[a, b]$, interpret $S_P f - s_P f$ as a sum of areas of rectangles that cover the graph of f.)
   b. Suppose $f \geq 0$ and let $S = \{(x, y) : x \in [a, b], \ 0 \leq y \leq f(x)\}$. Show that S is measurable and that its area (as defined in this section) equals $\int_a^b f(x)\, dx$.

3. Let S be a bounded set in $\mathbb{R}^2$. Show that S and S^{int} have the same inner area. (*Hint:* For any rectangle contained in S, there are slightly smaller rectangles contained in S^{int}.)

4. Let S be a bounded set in $\mathbb{R}^2$. Show that S and $\overline{S}$ have the same outer area. (*Hint:* For any rectangle that does not intersect S, there are slightly smaller rectangles that do not intersect $\overline{S}$.)

5. Let S be a bounded set in $\mathbb{R}^2$. Show that the inner area of S plus the outer area of ∂S equals the outer area of S. (Use Exercises 3 and 4.)

6. Let S be the subset of the x-axis consisting of the union of the open interval of length $\frac{1}{4}$ centered at $\frac{1}{2}$, the open intervals of length $\frac{1}{16}$ centered at $\frac{1}{4}$ and $\frac{3}{4}$, the open intervals of length $\frac{1}{64}$ centered at $\frac{1}{8}$, $\frac{3}{8}$, $\frac{5}{8}$, and $\frac{7}{8}$, and so forth. Let $U = S \times (0, 1)$ be the union of the open rectangles of height 1 based on these intervals. Thus U is the union of one rectangle of area $\frac{1}{4}$, two rectangles of area $\frac{1}{16}$, four rectangles of area $\frac{1}{64}$, ..., some of which overlap.
   a. Show that U is an open subset of the unit square $R = [0, 1] \times [0, 1]$.
   b. Show that the inner area of U is less than $\frac{1}{2}$.
   c. Show that U is dense in R and hence that its outer area is 1. (Use Exercise 4.)
   d. Let $V = R \setminus U$. Show that V is a closed set whose inner area is 0 and whose outer area is bigger than $\frac{1}{2}$.

7. (*The Second Mean Value Theorem for Integrals*) Suppose f is continuous on $[a, b]$ and φ is of class C^1 and increasing on $[a, b]$. Show that there is a point $c \in [a, b]$ such that

$$\int_a^b f(x)\varphi(x)\, dx = \varphi(a) \int_a^c f(x)\, dx + \varphi(b) \int_c^b f(x)\, dx.$$

(*Hint:* First suppose $\varphi(b) = 0$. Set $F(x) = \int_a^x f(t)\, dt$, integrate by parts to show that $\int_a^b f(x)\varphi(x)\, dx = -\int_a^b F(x)\varphi'(x)\, dx$, and apply Theorem 4.24

to the latter integral. To remove the condition $\varphi(b) = 0$, show that if the conclusion is true for f and φ, it is true for f and $\varphi + C$ for any constant C.)

4.3 Multiple Integrals and Iterated Integrals

The next issue to be addressed is the evaluation of n-dimensional integrals. The usual procedure is to reduce them to one-dimensional integrals.

Again we focus on the case $n = 2$, and we begin by considering the integral of a function f over a rectangle R. Given a partition $P = \{x_0, \ldots, x_J; y_0, \ldots, y_K\}$ of R, we pick points $\tilde{x}_j \in [x_{j-1}, x_j]$ and $\tilde{y}_k \in [y_{k-1}, y_k]$ $(1 \leq j \leq J, 1 \leq k \leq K)$ and form the Riemann sum

$$\sum_{j=1}^{J} \sum_{k=1}^{K} f(\tilde{x}_j, \tilde{y}_k) \, \Delta x_j \, \Delta y_k \qquad (\Delta x_j = x_j - x_{j-1}, \ \Delta y_k = y_k - y_{k-1}).$$

If f is integrable on R, this double sum approximates the integral $\iint_R f(x,y) \, dx \, dy$. On the other hand, for each fixed y, the sum $\sum_{j=1}^{J} f(\tilde{x}_j, y) \, \Delta x_j$ is a Riemann sum for the single integral $g(y) = \int_a^b f(x,y) \, dx$, and then the sum $\sum_{k=1}^{K} g(\tilde{y}_k) \, \Delta y_k$ is a Riemann sum for the integral $\int_c^d g(y) \, dy$. Thus, in an approximate sense,

$$\iint_R f(x,y) \, dx \, dy \approx \sum_{j=1}^{J} \sum_{k=1}^{K} f(x_j, y_k) \, \Delta x_j \, \Delta y_k$$

$$\approx \sum_{k=1}^{K} \int_a^b f(x, y_k) \, dx \, \Delta y_k \approx \int_c^d \left[\int_a^b f(x,y) \, dx \right] dy.$$

In short, if there are no unexpected pitfalls we should have

$$\iint_R f \, dA = \int_c^d \left[\int_a^b f(x,y) \, dx \right] dy.$$

We could also play the same game with x and y switched, obtaining

$$\iint_R f \, dA = \int_a^b \left[\int_c^d f(x,y) \, dy \right] dx.$$

If f is continuous on the rectangle R, it is not hard to make this argument rigorous by using the uniform continuity of f. However, we need to allow discontinuous functions in order to encompass integrals over more general regions, and here there is one potential pitfall: The integrability of f on R need not imply the integrability of $f(x, y_0)$, as a function of x for fixed y_0, on $[a, b]$. The line segment $\{(x, y) : a \leq x \leq b, \, y = y_0\}$ is a set of zero content, after all, so f could be discontinuous at every point on it, and its behavior as a function of x could be quite wild. This problem is actually not too serious, and we shall sweep it under the rug by making the assumption — quite harmless in practice — that it does not occur. The resulting theorem is as follows. It is sometimes referred to as **Fubini's theorem**, although that name belongs more properly to the generalization of the theorem to Lebesgue integrals.

4.26 Theorem. *Let $R = \{(x, y) : a \leq x \leq b, \, c \leq y \leq d\}$, and let f be an integrable function on R. Suppose that, for each $y \in [c, d]$, the function f_y defined by $f_y(x) = f(x, y)$ is integrable on $[a, b]$, and the function $g(y) = \int_a^b f(x, y) \, dx$ is integrable on $[c, d]$. Then*

$$(4.27) \qquad \iint_R f \, dA = \int_c^d \left[\int_a^b f(x, y) \, dx \right] dy.$$

Likewise, if $f^x(y) = f(x, y)$ is integrable on $[c, d]$ for each $x \in [a, b]$, and $h(x) = \int_c^d f(x, y) \, dy$ is integrable on $[a, b]$, then

$$(4.28) \qquad \iint_R f \, dA = \int_a^b \left[\int_c^d f(x, y) \, dy \right] dx.$$

Proof. The proof is presented in Appendix B.4 (Theorem B.9). The issue that must be addressed is the permissibility of *first* letting the x-subdivisions get finer and finer, and *then* doing the same for the y-subdivisions, or vice versa, as opposed to requiring both subdivisions to become finer *at the same time.* □

The integrals on the right side of (4.27) and (4.28) are called **iterated integrals**. It is customary to omit the brackets in these integrals and to write, for example,

$$\int_c^d \int_a^b f(x, y) \, dx \, dy,$$

with the understanding that the integration is to be done "from the inside out." That is, the innermost integral $\int_a^b$ corresponds to the innermost differential dx, and the integral with respect to the corresponding variable x is to be performed first. Some people find it clearer to write the differentials dx and dy next to the integral signs to which they pertain, thus:

$$\int_c^d dy \int_a^b dx \, f(x, y).$$

$$\text{FIGURE 4.4: } \iint \cdots dx\,dy \text{ versus } \iint \cdots dy\,dx.$$

If our region of integration is not the whole rectangle R but a subset S, the integration effectively stops at the boundary of S, and the limits of integration should be adjusted accordingly. For example, if S is bounded above and below by the graphs of two functions,

$$(4.29) \qquad S = \{(x,y) : a \le x \le b, \ \varphi(x) \le y \le \psi(x)\},$$

we have

$$(4.30) \qquad \iint_S f\,dA = \int_a^b \int_{\varphi(x)}^{\psi(x)} f(x,y)\,dy\,dx.$$

Here it is essential to integrate first in y, then in x, since the limits $\varphi(x)$ and $\psi(x)$ furnish part of the x-dependence of the integrand for the outer integral $\int_a^b \cdots dx$.

It is important to observe that if S is a region of the form (4.29) where φ and ψ are of class C^1, and f is continuous on S, the hypotheses in Theorem 4.26 that allow integration first in y and then in x are automatically satisfied, so that (4.30) is valid. Indeed, the integrability of $f\chi_S$ on any rectangle $R \supset S$ follows from Proposition 4.19c and Theorem 4.21, and the integrability of the function $(f\chi_S)(x,y)$ as a function of y for fixed x is obvious since it is continuous except at the two points $y = \varphi(x)$ and $y = \psi(x)$.

On the other hand, if S is bounded on the left and right by graphs of functions of y, we obtain a formula similar to (4.30) with the roles of x and y reversed. In general, most of the regions S that arise in practice can be decomposed into a finite number of pieces $S_1, \ldots, S_K$, each of which is of the form (4.29) or of the analogous form with x and y switched. By using the additivity property (Theorem 4.17b), we can reduce the computation of $\iint_S f\,dA$ to the calculation of iterated integrals on these subregions.

Figure 4.4 may be helpful in interpreting iterated integrals. The sketch on the left symbolizes $\iint \cdots dx\,dy$, in which we integrate first over the horizontal lines that run from the left side to the right side of the region, then integrate over the y-interval that comprises the y-coordinates of all these lines. Similarly, the sketch on the right symbolizes $\iint \cdots dy\,dx$.

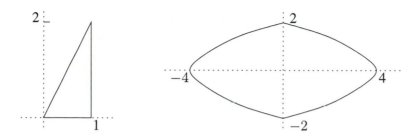

FIGURE 4.5: The regions of integration in Example 1 (left) and Example 2 (right).

EXAMPLE 1. Find the volume of the region in $\mathbb{R}^3$ above the triangle T in the xy-plane with vertices $(0,0)$, $(1,0)$, and $(1,2)$ and below the surface $z = xy + y^2$. (See Figure 4.5.)

Solution. The volume in question is $\iint_T (xy + y^2)\, dA$, which can be expressed as an iterated integral in two ways:

$$\int_0^2 \int_{y/2}^1 (xy + y^2)\, dx\, dy \quad \text{or} \quad \int_0^1 \int_0^{2x} (xy + y^2)\, dy\, dx.$$

For the sake of illustration, we perform both calculations:

$$\int_0^2 \int_{y/2}^1 (xy + y^2)\, dx\, dy = \int_0^2 [\tfrac{1}{2}x^2 y + xy^2]_{y/2}^1\, dy = \int_0^2 (\tfrac{1}{2}y + y^2 - \tfrac{5}{8}y^3)\, dy,$$

$$\int_0^1 \int_0^{2x} (xy + y^2)\, dy\, dx = \int_0^1 [\tfrac{1}{2}xy^2 + \tfrac{1}{3}y^3]_0^{2x}\, dx = \int_0^1 \tfrac{14}{3}x^3\, dx.$$

Both single integrals on the right evaluate to $\tfrac{7}{6}$.

EXAMPLE 2. Let S be the region between the parabolas $x = 4 - y^2$ and $x = y^2 - 4$. (See Figure 4.5.) A double integral $\iint_S f(x, y)\, dA$ can be reduced to iterated integrals in two ways. Integrating first in x is more straightforward:

$$\int_{-2}^2 \int_{y^2-4}^{4-y^2} f(x, y)\, dx\, dy.$$

To integrate first in y, we must break up R into its left and right halves:

$$\int_{-4}^0 \int_{-\sqrt{4+x}}^{\sqrt{4+x}} f(x, y)\, dy\, dx + \int_0^4 \int_{-\sqrt{4-x}}^{\sqrt{4-x}} f(x, y)\, dy\, dx.$$

The ideas in higher dimensions are entirely similar. The analogue of Theorem 4.26 is that an integral over an n-dimensional rectangular solid with sides $[a_1, b_1], \ldots, [a_n, b_n]$ can be evaluated as an n-fold iterated integral,

$$\int \cdots \int_R f \, dV = \int_{a_n}^{b_n} \cdots \int_{a_1}^{b_1} f(x_1, \ldots, x_n) \, dx_1 \cdots dx_n,$$

provided that the indicated integrals exist. The meaning of the iterated integral on the right is that the integration is to be performed first with respect to x_1 and last with respect to x_n. However, the same formula remains valid with the n integrations performed in *any* order. The only thing that needs some care is that the integral signs $\int_{a_j}^{b_j}$ must be matched up with the differentials dx_j in the right order so as to get the right limits of integration, and the convention is the same as in the case $n = 2$: The integrations are to be performed in order from innermost to outermost.

When the region of integration is something other than a rectangular solid, setting up the right limits of integration can be rather complicated. A typical situation in 3 dimensions is as follows: The region of integration S is the region in between two graphs,

$$S = \big\{ (x, y, z) : (x, y) \in U, \; \varphi(x, y) \leq z \leq \psi(x, y) \big\},$$

based on some region U in the xy-plane. The region U in turn is the region between two graphs,

$$U = \big\{ (x, y) : a \leq x \leq b, \; \sigma(x) \leq y \leq \tau(x) \big\},$$

based on an interval $[a, b] \subset \mathbb{R}$. We then have

$$\iiint_S f \, dV = \int_a^b \int_{\sigma(x)}^{\tau(x)} \int_{\varphi(x,y)}^{\psi(x,y)} f(x, y, z) \, dz \, dy \, dx.$$

The rule to remember is that limits of integration in an iterated integral can depend on the remaining "outer" variables whose integration is yet to be performed, but not on the "inner" variables that have already been integrated out. The final answer should be a number, not a function of some of the variables!

EXAMPLE 3. Find the mass of the tetrahedron T formed by the three coordinate planes and the plane $x + y + 2z = 2$ (see Figure 4.6) if the mass density is given by $\rho(x, y, z) = e^{-z}$.

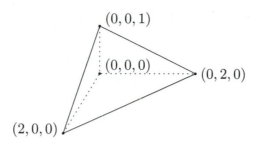

FIGURE 4.6: The tetrahedron in Example 3.

Solution. There are six ways to write the triple integral $\iiint_T e^{-z}\, dV$ as an iterated integral, although only three of them are essentially different, namely,

$$\int_0^2 \int_0^{2-x} \int_0^{1-(x+y)/2} e^{-z}\, dz\, dy\, dx, \qquad \int_0^1 \int_0^{2-2z} \int_0^{2-y-2z} e^{-z}\, dx\, dy\, dz,$$

$$\int_0^2 \int_0^{1-(y/2)} \int_0^{2-y-2z} e^{-z}\, dx\, dz\, dy.$$

(The remaining three can be obtained from these simply by interchanging x and y, since T and the density function are invariant under this interchange.) Using the first of these, we obtain

$$\int_0^2 \int_0^{2-x} \left(1 - e^{(x+y)/2 - 1}\right) dy\, dx = \int_0^2 \left[y - 2e^{(x+y)/2 - 1} \right]_0^{2-x} dx$$

$$= \int_0^2 \left(2e^{(x/2)-1} - x\right) dx = \left[4e^{(x/2)-1} - \tfrac{1}{2}x^2 \right]_0^2 = 2 - 4e^{-1}.$$

The reader may verify that the other two iterated integrals give the same answer.

In the preceding discussion, iterated integrals appeared as a tool for evaluating n-dimensional integrals. However, they also arise in a number of other contexts in advanced analysis where a quantity is defined by performing two or more integrations in succession. In this context, the significance of Theorem 4.26 is that *under suitable hypotheses on the integrand f, the order of integration in an iterated integral can be reversed:*

(4.31)
$$\int_a^b \int_c^d f(x, y)\, dy\, dx = \int_c^d \int_a^b f(x, y)\, dx\, dy.$$

More precisely, (4.31) is valid if f satisfies the conditions in Theorem 4.26 for both (4.27) and (4.28) to hold. (See Exercise 13 for an example to demonstrate the significance of these conditions.) The importance of this result can hardly be overestimated; it is an extremely powerful tool for evaluating quantities defined by integrals. We shall see a number of examples in later chapters.

EXAMPLE 4. Evaluate $\int_0^2 \int_{y/2}^1 y e^{-x^3} \, dx \, dy$.

Solution. The integral cannot be evaluated by elementary methods as it stands, since e^{-x^3} has no elementary antiderivative. However, it can be interpreted as $\iint_T y e^{-x^3} \, dA$ where T is the triangle with vertices $(0,0)$, $(1,0)$, and $(1,2)$ as in Example 1. Writing this double integral as an iterated integral in the other order leads to an easy calculation:

$$\int_0^1 \int_0^{2x} y e^{-x^3} \, dy \, dx = \int_0^1 \tfrac{1}{2} y^2 \big|_0^{2x} e^{-x^3} \, dx = \int_0^1 2x^2 e^{-x^3} \, dx$$

$$= -\tfrac{2}{3} e^{-x^3} \big|_0^1 = \tfrac{2}{3}(1 - e^{-1}).$$

Applications. Double and triple integrals can be used to calculate physical and geometric quantities in much the same way as single integrals. Here are a few standard examples:

- If $f(x,y) \geq 0$, the integral $\iint_S f \, dA$ can be interpreted as the volume of the region in $\mathbb{R}^3$ between the graph of f and the xy-plane that lies over the base region S.

- Suppose that a quantity of some substance (which might be mass, electric charge, a particular chemical compound, etc.) is distributed throughout a region $U \subset \mathbb{R}^3$. It is frequently useful to think of the distribution of the substance as being described by a **density** function ρ; the meaning of this, in practical terms, is that the amount of substance in a set $S \subset U$ is $\iiint_S \rho(\mathbf{x}) \, d^3\mathbf{x}$. This idea works also in other dimensions, for example, to describe distributions of a substance on a planar surface or a line.

  (The reader may wish for a more careful discussion of the meaning of ρ. Informally, $\rho(\mathbf{x})$ represents the ratio of the amount of substance in an infinitesimal cube centered at $\mathbf{x}$ to the volume of that cube. To make this rigorous, one should interpret $\rho(\mathbf{x})$ as the limit of the ratio of the amount of substance in a finite cube centered at $\mathbf{x}$ to the volume of that cube as the side length of the cube tends to zero. One can then prove, under suitable hypotheses, that the amount of substance in any region S is $\iiint_S \rho(\mathbf{x}) d^3\mathbf{x}$. But a complete analysis of these matters is beyond the scope of this book.)

- Suppose that a massive object with mass density $\rho(\mathbf{x})$ occupies the region $S \subset \mathbb{R}^3$, so that its total mass is $m = \iiint_S \rho(\mathbf{x}) d^3\mathbf{x}$. The **center of gravity** of the object is the point $\overline{\mathbf{x}}$ whose coordinates are $\overline{x}_j = m^{-1} \iiint_S x_j \rho(\mathbf{x}) d^3\mathbf{x}$. In the special case where $\rho \equiv 1$, $\overline{\mathbf{x}}$ is the **centroid** of the region S, which is the point whose coordinates are the average values of the coordinate functions on S. The center of mass, in general, can be interpreted similarly as the point whose coordinates are the weighted averages of the coordinate functions on S where the weighting is given by the density ρ.

- Again suppose that a massive object with mass density $\rho(\mathbf{x})$ occupies the region $S \subset \mathbb{R}^3$, and let L be a line in $\mathbb{R}^3$. The **moment of inertia** of the body about the line L, a quantity that is useful in analyzing rotational motion about L, is $\iiint_S d(\mathbf{x})^2 \rho(\mathbf{x}) d^3\mathbf{x}$, where $d(\mathbf{x})$ is the distance from $\mathbf{x}$ to L. (For example, if L is the z-axis, then $d(x, y, z)^2 = x^2 + y^2$.)

EXERCISES

1. Evaluate the following double integrals.
   a. $\iint_S (x + 3y^3) \, dA$, $S = $ the upper half ($y \geq 0$) of the unit disc $x^2 + y^2 \leq 1$.
   b. $\iint_S (x^2 - \sqrt{y}) \, dA$, $S = $ the region between the parabola $x = y^2$ and the line $x = 2y$.

2. Find the volume of the region above the triangle in the xy-plane with vertices $(0, 0)$, $(1, 0)$, and $(0, 1)$, and below the surface $z = 6xy(1 - x - y)$.

3. For the following regions $S \subset \mathbb{R}^2$, express the double integral $\iint_S f \, dA$ in terms of iterated integrals in two different ways.
   a. $S = $ the region in the left half plane between the curve $y = x^3$ and the line $y = 4x$.
   b. $S = $ the triangle with vertices $(0, 0)$, $(2, 2)$, and $(3, 1)$.
   c. $S = $ the region between the parabolas $y = x^2$ and $y = 6 - 4x - x^2$.

4. Express each of the following iterated integrals as a double integral and as an iterated integral in the opposite order. (That is, find the region of integration for the double integral and the limits of integration for the other iterated integral.)
   a. $\int_0^1 \int_{x^2}^{x^{1/3}} f(x, y) \, dy \, dx$.
   b. $\int_0^1 \int_{-y}^{2y} f(x, y) \, dx \, dy$.
   c. $\int_1^2 \int_0^{\log x} f(x, y) \, dy \, dx$.

5. Evaluate the following iterated integrals. (You may need to reverse the order of integration.)

   a. $\int_1^3 \int_1^y y e^{2x} \, dx \, dy$.

   b. $\int_0^1 \int_{\sqrt{x}}^1 \cos(y^3 + 1) \, dy \, dx$.

   c. $\int_1^2 \int_{1/x}^1 y e^{xy} \, dy \, dx$.

6. Fill in the blanks: $\int_0^1 \int_{2x^2}^{x+1} f(y) \, dy \, dx = \int_0^1 [\quad] \, dy + \int_1^2 [\quad] dy$. The expressions you obtain for the [ ]'s should not contain integral signs.

7. Given a continuous function $g : \mathbb{R} \to \mathbb{R}$, let $h(x) = \int_0^x \int_0^y g(t) \, dt \, dy$. That is, h is obtained by integrating g twice, starting the integration at 0. Show that h can be expressed as a single integral, namely, $h(x) = \int_0^x (x - t) g(t) \, dt$. (Note that x can be treated as a constant here; y and t are the variables of integration.)

8. Let $S \subset \mathbb{R}^3$ be the region between the paraboloid $z = x^2 + y^2$ and the plane $z = 1$. Express the triple integral $\iiint_S f \, dV$ as an iterated integral with the order of integration (a) z, y, x; (b) y, z, x; (c) x, y, z. (That is, find the appropriate limits of integration in each case.)

9. Express the iterated integral $\int_0^1 \int_0^{1-y^2} \int_0^y f(x, y, z) \, dz \, dx \, dy$
   a. as a triple integral (i.e., describe the region of integration);
   b. as an iterated integral in the order z, y, x;
   c. as an iterated integral in the order y, z, x.

10. Find the centroid of the tetrahedron bounded by the coordinate planes and the plane $(x/a) + (y/b) + (z/c) = 1$.

11. An object with mass density $\rho(x, y, z) = yz$ occupies the cube $\{(x, y, z) : 0 \le x, y, z \le 2\}$. Find its mass and center of mass.

12. A body with charge density $\rho(x, y, z) = 2z$ occupies the region bounded below by the parabolic cylinder $z = x^2 - 3$, above by the plane $z = x - 1$, and on the sides by the planes $y = 0$ and $y = 2$. Find its net charge (total positive charge minus total negative charge).

13. Let $f(x, y) = y^{-2}$ if $0 < x < y < 1$, $f(x, y) = -x^{-2}$ if $0 < y < x < 1$, and $f(x, y) = 0$ otherwise, and let S be the unit square $[0, 1] \times [0, 1]$.
    a. Show that f is not integrable on S, but that $f(x, y)$ is integrable on $[0, 1]$ as a function of x for each fixed y and as a function of y for each fixed x.
    b. Show by explicit calculation that the iterated integrals $\int_0^1 \int_0^1 f(x, y) \, dx \, dy$ and $\int_0^1 \int_0^1 f(x, y) \, dy \, dx$ both exist and are unequal.

4.4 Change of Variables for Multiple Integrals

To motivate the ideas in this section, we recall the change-of-variable formula for single definite integrals: If g is a one-to-one function of class C^1 on the interval $[a, b]$, then for any continuous function f,

$$(4.32) \qquad \int_a^b f(g(u))g'(u)\, du = \int_{g(a)}^{g(b)} f(x)\, dx.$$

The proof is a simple matter of combining the chain rule and the fundamental theorem of calculus. Indeed, if F is an antiderivative of f, the right side of (4.32) is $F(g(b)) - F(g(a))$, which in turn equals $\int_a^b (F \circ g)'(u)\, du$, and the latter integrand is $f(g(u))g'(u)$. (Formula (4.32) is actually valid when f is merely integrable, but we shall not worry about this refinement here.)

There is one slightly tricky point here, which we point out now because it will be significant later. If g is an increasing function, (4.32) is fine as it stands, but if g is decreasing, the endpoints on the integral on the right are in the "wrong" order, and we might prefer to put them back in the "right" order by introducing a minus sign: $\int_{g(a)}^{g(b)} = -\int_{g(b)}^{g(a)}$. Since g is increasing or decreasing according as g' is positive or negative, we could rewrite (4.32) as

$$(4.33) \qquad \int_{[a,b]} f(g(u))|g'(u)|\, du = \int_{g([a,b])} f(x)\, dx.$$

Here $g([a, b])$ is the interval to which $[a, b]$ is mapped under g, and for any interval I the symbol $\int_I$ means the integral from the *left* endpoint of I to the *right* endpoint. The replacement of g' by $|g'|$ compensates for the extra minus sign that comes from adjusting the order of the endpoints when g is decreasing.

In practice it is often more convenient to have all the g's on one side of the equation. If we set $I = g([a, b])$, we have $[a, b] = g^{-1}(I)$, and (4.33) becomes

$$(4.34) \qquad \int_I f(x)\, dx = \int_{g^{-1}(I)} f(g(u))|g'(u)|\, du.$$

Our object is to find the analogous formula for multiple integrals. It is natural to use (4.34) rather than (4.32) as a starting point, since for multiple integrals the issue of left-to-right or right-to-left disappears and we just speak of integrals over a region, like the integrals over intervals that appear in (4.34). More precisely, suppose $\mathbf{G}$ is a one-to-one transformation from a region R in $\mathbb{R}^n$ to another region $S = \mathbf{G}(R)$ in $\mathbb{R}^n$; then $R = \mathbf{G}^{-1}(S)$, and the formula we are seeking should look

FIGURE 4.7: The element of area in polar coordinates.

something like this:

$$(4.35) \qquad \int \cdots \int_S f(\mathbf{x})\, d^n\mathbf{x} = \int \cdots \int_{\mathbf{G}^{-1}(S)} f(\mathbf{G}(\mathbf{u}))\, [????]\, d^n\mathbf{u}.$$

The missing ingredient is the quantity that will play the role of $|g'(u)|$ in the formula (4.34).

Now, the $g'(u)$ in (4.32) or (4.34) is the factor that relates the differentials du and dx under the transformation $x = g(u)$. In n variables, the n-fold differential $d^n\mathbf{x} = dx_1 \cdots dx_n$ represents the "element of volume," that is, the volume of an infinitesimal piece of n-space. So the question is: How does the volume of a tiny piece of n-space change when one applies the transformation $\mathbf{G}$?

To get a feeling for what is going on, let us look at the polar coordinate map

$$(x, y) = \mathbf{G}(r, \theta) = (r \cos \theta, \, r \sin \theta).$$

A small rectangle in the $r\theta$-plane with lower left corner at (r, θ) and sides of length dr and $d\theta$ is mapped to a small region in the xy-plane bounded by two line segments of length dr and two circular arcs of length $r\, d\theta$ and $(r + dr)\, d\theta$. When dr and $d\theta$ are very small, this is essentially a rectangle with sides dr and $r\, d\theta$, so its area is $r\, dr\, d\theta$. In short, a small bit of the $r\theta$-plane with area $dr\, d\theta$ is mapped to a small bit of the xy-plane with area $r\, dr\, d\theta$; see Figure 4.7. Hence, in this case the missing factor in (4.35) is simply r, and (4.36) becomes

$$(4.36) \qquad \iint_S f(x, y)\, dx\, dy = \iint_R f(r \cos \theta, \, r \sin \theta) r\, dr\, d\theta.$$

Here S is a region in the xy-plane and $R = \mathbf{G}^{-1}(S)$ is the corresponding region in the $r\theta$-plane. Our argument here has been very informal, but this result is correct, and it gives the formula for computing double integrals in polar coordinates.

The case of a *linear* mapping of the plane is also easy to analyze. Given a matrix $A = \begin{pmatrix} a & b \\ c & d \end{pmatrix}$ with $\det A = ad - bc \neq 0$, let $\mathbf{x} = \mathbf{G}(\mathbf{u}) = A\mathbf{u}$, that is,

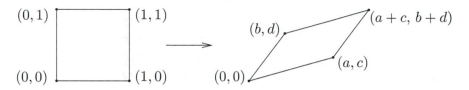

FIGURE 4.8: The linear map $(x, y) = (au + bv, \; cu + dv)$.

$(x, y) = \mathbf{G}(u, v) = (au + bv, \; cu + dv)$. The transformation $\mathbf{G}$ takes the unit vectors $(1, 0)$ and $(0, 1)$ to the vectors (a, c) and (b, d), so it maps the standard coordinate grid to a grid of parallelograms with sides parallel to these vectors. In particular, it maps the square $[0, 1] \times [0, 1]$ to the parallelogram with vertices at $(0, 0)$, (a, c), (b, d), and $(a + b, \; c + d)$, as indicated in Figure 4.8. The area of that parallelogram is $|ad - bc|$, that is, $|\det A|$. (To see this, think of the plane as sitting in $\mathbb{R}^3$ and recall the geometric interpretation of the cross product: The area of the parallelogram is

$$|(a\mathbf{i} + c\mathbf{j}) \times (b\mathbf{i} + d\mathbf{j})| = |(ad - bc)\mathbf{k}| = |ad - bc|.)$$

Since the map $\mathbf{G}$ is linear, it commutes with translations and dilations, so if R is *any* square in the uv-plane, its image under $\mathbf{G}$ is a parallelogram in the xy-plane whose area is $|\det A|$ times that of R. It follows that the missing factor in (4.35) should be simply $|\det A|$, so that for linear maps of the plane, (4.35) becomes

$$\iint_S f(x, y) \, dx \, dy = |ad - bc| \iint_{\mathbf{G}^{-1}(S)} f(au + bv, \; cu + dv) \, du \, dv.$$

The situation is similar for linear mappings of 3-space. Namely, let $\mathbf{x} = \mathbf{G}(\mathbf{u}) = A\mathbf{u}$ where A is an invertible 3×3 matrix. If $\mathbf{i}$, $\mathbf{j}$, and $\mathbf{k}$ are the standard basis vectors for $\mathbb{R}^3$, we have $A\mathbf{i} = \mathbf{a}$, $A\mathbf{j} = \mathbf{b}$, and $A\mathbf{k} = \mathbf{c}$ where $\mathbf{a}$, $\mathbf{b}$, $\mathbf{c}$ are the columns of A, so A maps the unit cube to the parallelepiped generated by these vectors. To find the volume of that parallelepiped, think of the $\mathbf{bc}$-plane as its base. Then the area of the base is $|\mathbf{b} \times \mathbf{c}|$, and the height is the length of the projection of $\mathbf{a}$ onto a line perpendicular to the $\mathbf{bc}$-plane, namely, the line generated by $\mathbf{b} \times \mathbf{c}$. But this length is $|\mathbf{a}| \, |\cos \theta|$ where θ is the angle between $\mathbf{a}$ and $\mathbf{b} \times \mathbf{c}$ (we need the absolute value because θ might be obtuse). Hence,

$$\text{Volume} = |\mathbf{b} \times \mathbf{c}| \, |\mathbf{a}| \, |\cos \theta| = |\mathbf{a} \cdot (\mathbf{b} \times \mathbf{c})|,$$

which is nothing but $|\det A|$ (Exercise 8 in §1.1). As before, we conclude that for the linear map $\mathbf{G}(\mathbf{u}) = A\mathbf{u}$ of $\mathbb{R}^3$, the missing factor in (4.35) should be $|\det A|$.

It is now reasonable to conjecture that the same result should hold for linear mappings of $\mathbb{R}^n$ for any n. We proceed to show that this is correct.

4.37 Theorem. *Let A be an invertible $n \times n$ matrix, and let $\mathbf{G}(\mathbf{u}) = A\mathbf{u}$ be the corresponding linear transformation of $\mathbb{R}^n$. Suppose S is a measurable region in $\mathbb{R}^n$ and f is an integrable function on S. Then $\mathbf{G}^{-1}(S) = \{A^{-1}\mathbf{x} : \mathbf{x} \in S\}$ is measurable and $f \circ \mathbf{G}$ is integrable on $\mathbf{G}^{-1}(S)$, and*

$$(4.38) \qquad \int \cdots \int_S f(\mathbf{x}) \, d^n\mathbf{x} = |\det A| \int \cdots \int_{\mathbf{G}^{-1}(S)} f(A\mathbf{u}) \, d^n\mathbf{u}.$$

Proof. The proof of the measurability of $\mathbf{G}^{-1}(S)$ and the integrability of $f \circ \mathbf{G}$, which is not profound but rather tedious, is given in Appendix B.5 (Corollaries B.16 and B.17). (Actually, what is proved in Appendix B.5 is that if f is continuous except on a set of zero content, a slightly stronger condition than integrability, then the same is true of $f \circ \mathbf{G}$.) Here we concentrate on proving (4.38). The proof naturally requires some linear algebra, in particular, the facts about elementary row operations and determinants in (A.17)–(A.18), (A.28), and (A.30) of Appendix A.

Step 1: Let us agree to (re)define $f(\mathbf{x})$ to be 0 for $\mathbf{x} \notin S$. Then $f(A\mathbf{u}) = 0$ for $\mathbf{u} \notin \mathbf{G}^{-1}(S)$, and we can replace the regions S and $\mathbf{G}^{-1}(S)$ in (4.38) by $\mathbb{R}^n$. This makes the integrals in (4.38) look improper, but they really are not, since the integrands vanish outside bounded sets. The point is that now we don't have to worry about what the limits of integration in each variable are; we can take them to be $\pm\infty$.

Step 2: We prove the theorem when $\mathbf{G}$ is an "elementary transformation," that is, the transformation given by performing a single elementary row operation on the column vector $\mathbf{u}$. There are three kinds of elementary transformations, corresponding to the three types of row operations (see (A17)–(A18)):

1. Multiply the kth component by a nonzero number c, leaving all the other components alone:

$$\mathbf{G}_1(u_1, \ldots, u_k, \ldots, u_n) = (u_1, \ldots, cu_k, \ldots, u_n).$$

2. Add a multiple of the jth component to the kth component, leaving all the other components alone:

$$\mathbf{G}_2(u_1, \ldots, u_k, \ldots, u_n) = (u_1, \ldots, u_k + cu_j, \ldots, u_n).$$

3. Interchange the jth and kth components:

$$\mathbf{G}_3(u_1, \ldots, u_j, \ldots, u_k, \ldots, u_n) = (u_1, \ldots, u_k, \ldots, u_j, \ldots, u_n).$$

The corresponding matrices A_1, A_2, A_3 are obtained by performing the same row operations on the identity matrix. Since $\det I = 1$, the rules that tell how row operations affect determinants (see (A.30)) give

$$(4.39) \qquad \det A_1 = c, \qquad \det A_2 = 1, \qquad \det A_3 = -1.$$

It is easy to verify that (4.38) holds for these three types of transformations. The first two involve a change in only the kth variable, so we can integrate first with respect to that variable and use (4.34) (or, rather, the simple special cases of (4.34) discussed in Exercise 8 of §4.1). Thus, for $\mathbf{G}_1$ we set $x_k = cu_k$ and obtain

$$\int_{-\infty}^{\infty} f(\ldots, x_k, \ldots)\, dx_k = \int_{-c^{-1}\infty}^{c^{-1}\infty} f(\ldots, cu_k, \ldots)\, c\, du_k$$

$$= |c| \int_{-\infty}^{\infty} f(\ldots, cu_k, \ldots)\, du_k.$$

(The endpoints have to be switched if $c < 0$, which accounts for replacing c by $|c|$, as in the discussion preceding (4.34).) Likewise, for $\mathbf{G}_2$ we set $x_k = u_k + cu_j$ and obtain

$$\int_{-\infty}^{\infty} f(\ldots, x_k, \ldots)\, dx_k = \int_{-\infty}^{\infty} f(\ldots, u_k + cu_j, \ldots)\, du_k.$$

(u_j is a *constant* as far as this calculation is concerned.) Now an integration with respect to the remaining variables (for which $x_i = u_i$) yields

$$\int \cdots \int f(\mathbf{x})\, d^n\mathbf{x} = |c| \int \cdots \int f(\mathbf{G}_1(\mathbf{u}))\, d^n\mathbf{u} = \int \cdots \int f(\mathbf{G}_2(\mathbf{u}))\, d^n\mathbf{u}.$$

In view of (4.39), this establishes (4.38) for $\mathbf{G}_1$ and $\mathbf{G}_2$. As for $\mathbf{G}_3$, we have

$$\int_{-\infty}^{\infty}\int_{-\infty}^{\infty} f(\ldots, u_j, \ldots, u_k, \ldots)\, du_j\, du_k$$

$$= \int_{-\infty}^{\infty}\int_{-\infty}^{\infty} f(\ldots, u_k, \ldots, u_j, \ldots)\, du_j\, du_k,$$

simply because the variables u_j and u_k are dummy variables here. That is, we are integrating f with respect to its jth and kth variables, and it doesn't matter what we call them. Now an integration with respect to the remaining variables, together with (4.39), gives (4.38) for $\mathbf{G}_3$.

Step 3: We next verify that if (4.38) is valid for the linear maps $\mathbf{G}(\mathbf{u}) = A\mathbf{u}$ and $\mathbf{H}(\mathbf{u}) = B\mathbf{u}$, then it is also valid for the composition $(\mathbf{G} \circ \mathbf{H})(\mathbf{u}) = AB\mathbf{u}$.

Indeed, if we set $\mathbf{v} = B\mathbf{u}$ and $\mathbf{x} = A\mathbf{v}$, we have

$$\int \cdots \int_S f(\mathbf{x})\, d^n\mathbf{x} = |\det A| \int \cdots \int_{\mathbf{G}^{-1}(S)} f(A\mathbf{v})\, d^n\mathbf{v}$$

$$= |\det A|\,|\det B| \int \cdots \int_{\mathbf{H}^{-1}(\mathbf{G}^{-1}(S))} f(AB\mathbf{u})\, d^n\mathbf{u}.$$

But $(\det A)(\det B) = \det(AB)$ and $\mathbf{H}^{-1}(\mathbf{G}^{-1}(S)) = (\mathbf{G} \circ \mathbf{H})^{-1}(S)$, so the integral on the right equals

$$|\det(AB)| \int \cdots \int_{(\mathbf{G}\circ\mathbf{H})^{-1}(S)} f(AB\mathbf{u})\, d^n\mathbf{u},$$

as claimed.

The Final Step: From Step 3, it follows easily by induction that if (4.38) is valid for $\mathbf{G}_1, \ldots, \mathbf{G}_k$, then it is also valid for the composition $\mathbf{G}_1 \circ \cdots \circ \mathbf{G}_k$. Thus, in view of Step 2, to complete the proof we have merely to observe that *every* invertible linear transformation of $\mathbb{R}^n$ is a composition of elementary transformations. This is equivalent to the fact that every invertible matrix A can be row-reduced to the identity matrix; see (A.52) (in particular, the equivalence of (a) and (i)) and (A.53) in Appendix A. $\qquad\square$

There is one more simple class of transformations for which the change-of-variable formula is easily established, namely the *translations*. These are the mappings of the form $\mathbf{G}(\mathbf{u}) = \mathbf{u} + \mathbf{b}$ where $\mathbf{b}$ is a fixed vector. Indeed, we just make the substitution $x_j = u_j + b_j$, $dx_j = du_j$ in each variable separately to conclude that

$$\int \cdots \int_S f(\mathbf{x})\, d^n\mathbf{x} = \int \cdots \int_{S-\mathbf{b}} f(\mathbf{u} + \mathbf{b})\, d^n\mathbf{u}.$$

Combining this with Theorem 4.37, we see that if $\mathbf{G}(\mathbf{u}) = A\mathbf{u} + \mathbf{b}$, then

(4.40) $$\int \cdots \int_S f(\mathbf{x})\, d^n\mathbf{x} = |\det A| \int \cdots \int_{\mathbf{G}^{-1}(S)} f(A\mathbf{u} + \mathbf{b})\, d^n\mathbf{u}.$$

In particular, by taking $f \equiv 1$, we see that the n-dimensional volume of S is $|\det A|$ times the n-dimensional volume of $\mathbf{G}^{-1}(S)$.

It is now easy to guess what the change-of-variable formula for a general invertible C^1 transformation must be. Indeed, suppose that U and V are open sets in $\mathbb{R}^n$, $\mathbf{G} : U \to V$ is a one-to-one transformation of class C^1 whose derivative $Df(\mathbf{u})$ is invertible for all $\mathbf{u} \in U$, and f is a continuous function on V. To relate the integral of f over a measurable set $S \subset V$ to an integral of $f \circ \mathbf{G}$ over $T = \mathbf{G}^{-1}(S)$, we think of the former as a sum of infinitesimal terms $f(\mathbf{x})\,d^n\mathbf{x}$, each of which is the value of f at a point $\mathbf{x}$ multiplied by the volume $d^n\mathbf{x}$ of an infinitesimal region dS located at $\mathbf{x}$. Under the transformation $\mathbf{x} = \mathbf{G}(\mathbf{u})$, $f(\mathbf{x})$ becomes $f(\mathbf{G}(\mathbf{u}))$, and the region dS is the image under $\mathbf{G}$ of another infinitesimal region dT whose volume is $d^n\mathbf{u}$. But on the infinitesimal level, the differentiable map $\mathbf{G}$ is the same as its linearization:

$$\mathbf{G}(\mathbf{u} + d\mathbf{u}) = \mathbf{x} + D\mathbf{G}(\mathbf{u}) \cdot d\mathbf{u}.$$

Therefore, by (4.40), the elements of volume $d^n\mathbf{x}$ and $d^n\mathbf{u}$ are related by the formula $d^n\mathbf{x} = |\det D\mathbf{G}(\mathbf{u})|\,d^n\mathbf{u}$. Putting this all together, we arrive at the main theorem.

4.41 Theorem. *Given open sets U and V in $\mathbb{R}^n$, let $\mathbf{G} : U \to V$ be a one-to-one transformation of class C^1 whose derivative $D\mathbf{G}(\mathbf{u})$ is invertible for all $\mathbf{u} \in U$. Suppose that $T \subset U$ and $S \subset V$ are measurable sets such that $\mathbf{G}(T) = S$. If f is an integrable function on S, then $f \circ \mathbf{G}$ is integrable on T, and*

$$(4.42) \qquad \int \cdots \int_S f(\mathbf{x})\,d^n\mathbf{x} = \int \cdots \int_T f(\mathbf{G}(\mathbf{u}))|\det D\mathbf{G}(\mathbf{u})|\,d^n\mathbf{u}.$$

Proof. We present a proof in Appendix B.5 (Theorem B.24), under the slightly stronger hypothesis that f is continuous except on a set with zero content. The key idea is explained in the preceding paragraph, but turning it into a solid proof is a surprisingly laborious task. An interesting and quite different approach to the problem can be found in Lax [14], [15]. It shifts the hard work to a different part of the argument; in particular, it uses the notion of partition of unity developed in Appendix B.7. $\qquad\square$

Notice that the results derived earlier in this section are indeed special cases of Theorem 4.41. If $\mathbf{G}$ is a linear map, $\mathbf{G}(\mathbf{u}) = A\mathbf{u}$, then $D\mathbf{G}(\mathbf{u}) = A$ for all $\mathbf{u}$, so $|\det D\mathbf{G}(\mathbf{u})| = |\det A|$ is a constant that can be brought outside the integral sign. And if $\mathbf{G}$ is the polar coordinate map, $\mathbf{G}(r, \theta) = (r\cos\theta, r\sin\theta)$, then $\det D\mathbf{G}(r, \theta) = r$, so we recover (4.36).

Let us record the corresponding results for the standard "polar" coordinate systems in $\mathbb{R}^3$, shown in Figure 4.9. **Cylindrical coordinates** are just polar coordinates in the xy-plane with the z-coordinate added in,

$$\mathbf{G}_{\text{cyl}}(r, \theta, z) = (r\cos\theta, r\sin\theta, z).$$

It is easily verified that $\det D\mathbf{G}_{\text{cyl}}(r, \theta, z) = r$ again, so the formula for integration in cylindrical coordinates is

$$(4.43) \qquad \iiint_S f(x, y, z)\,dx\,dy\,dz = \iiint_{\mathbf{G}_{\text{cyl}}^{-1}(S)} f(r\cos\theta, r\sin\theta, z)\,r\,dr\,d\theta\,dz.$$

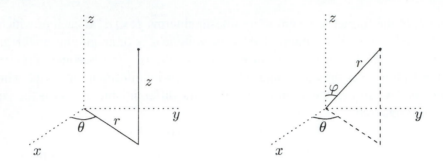

FIGURE 4.9: Cylindrical coordinates (left) and spherical coordinates (right).

Spherical coordinates are given by

$$\mathbf{G}_{\text{sph}}(r, \varphi, \theta) = (r \sin \varphi \cos \theta, \; r \sin \varphi \sin \theta, \; r \cos \varphi).$$

Here r is the distance from the origin, θ is the longitude, and φ is the co-latitude (the angle from the positive z-axis). The reader may check that $\det D\mathbf{G}_{\text{sph}}(r, \varphi, \theta) = r^2 \sin \varphi$ (Exercise 6c, §3.4), so the formula for integration in spherical coordinates is

$$(4.44) \quad \iiint_S f(x, y, z) \, dx \, dy \, dz$$

$$= \iiint_{\mathbf{G}_{\text{sph}}^{-1}(S)} f(r \sin \varphi \cos \theta, \; r \sin \varphi \sin \theta, \; r \cos \varphi) \, r^2 \sin \varphi \, dr \, d\varphi \, d\theta.$$

We conclude with some examples.

EXAMPLE 1. Find the volume and the centroid of the region S above the surface $z = x^2 + y^2$ and below the plane $z = 4$. (See Figure 4.10.)

Solution. Because of the circular symmetry, it is most convenient to use polar coordinates. The projection of S onto the xy-plane is the disc of radius 2 about the origin, so the volume of S is

$$V = \int_0^2 \int_0^{2\pi} (4 - r^2) r \, d\theta \, dr = 2\pi \left[2r^2 - \tfrac{1}{4} r^4 \right]_0^2 = 8\pi.$$

By symmetry, the centroid lies on the z-axis, and its z-coordinate is

$$\bar{z} = \frac{1}{V} \iiint_S z \, dV = \frac{1}{8\pi} \int_0^2 \int_{r^2}^4 \int_0^{2\pi} rz \, d\theta \, dz \, dr = \frac{1}{4} \int_0^2 r \left[\tfrac{1}{2} z^2 \right]_{r^2}^4 dr$$

$$= \frac{1}{4} \left[4r^2 - \tfrac{1}{12} r^6 \right]_0^2 = \frac{8}{3}.$$

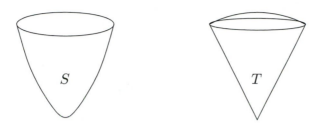

FIGURE 4.10: The regions in Example 1 (left) and Example 2 (right).

EXAMPLE 2. Find the volume of the "ice cream cone" T bounded below by the cone $z = 2\sqrt{x^2 + y^2}$ and above by the sphere $x^2 + y^2 + z^2 = 1$. (See Figure 4.10.)

Solution. In spherical coordinates (r, φ, θ), the equation of the cone is $\tan \varphi = \frac{1}{2}$ and the equation of the sphere is $r = 1$. Hence the volume is

$$\int_0^1 \int_0^{\tan^{-1}(1/2)} \int_0^{2\pi} r^2 \sin \varphi \, d\theta \, d\varphi \, dr = (2\pi)\left[-\cos\varphi \right]_0^{\tan^{-1}(1/2)} \left[\tfrac{1}{3} r^3 \right]_0^1$$

$$= \frac{2\pi}{3}\left(1 - \frac{2}{\sqrt{5}} \right).$$

This can also be done in cylindrical coordinates (r, θ, z) (note that the meaning of r has changed here), in which the equation of the cone is $z = 2r$ and the equation of the sphere is $r^2 + z^2 = 1$. The projection of T onto the xy-plane is the disc $r \leq 1/\sqrt{5}$, so the volume is

$$\int_0^{1/\sqrt{5}} \int_{2r}^{\sqrt{1-r^2}} \int_0^{2\pi} r \, d\theta \, dz \, dr = 2\pi \int_0^{1/\sqrt{5}} (r\sqrt{1-r^2} - 2r^2) \, dr$$

$$= \frac{2\pi}{3}\left[-(1 - r^2)^{3/2} - 2r^3 \right]_0^{1/\sqrt{5}},$$

which yields the same answer as before.

EXAMPLE 3. Let P be the parallelogram bounded by the lines $x - y = 0$, $x + 2y = 0$, $x - y = 1$, and $x + 2y = 6$. (See Figure 4.11.) Compute $\iint_P xy \, dA$.

Solution. The equations of the bounding lines suggest the linear transformation $u = x - y$, $v = x + 2y$, which maps P to the rectangle $0 \leq u \leq 1$, $0 \leq v \leq 6$. In the notation of Theorem 4.37, P plays the role of S and this

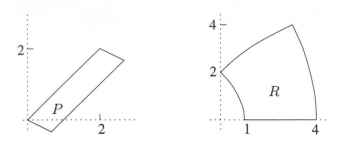

FIGURE 4.11: The regions in Example 3 (left) and Example 4 (right).

transformation is $\mathbf{G}^{-1}$; its inverse $\mathbf{G}$ is easily computed to be $x = \frac{1}{3}(2u + v)$, $y = \frac{1}{3}(v - u)$, whose determinant is $\frac{1}{3}$. Thus, by Theorem 4.37,

$$\iint_P xy \, dA = \frac{1}{3} \int_0^1 \int_0^6 \left(\frac{2u + v}{3}\right) \left(\frac{v - u}{3}\right) dv \, du,$$

which is easily computed to be $\frac{77}{27}$.

Alternatively, one can readily calculate that the vertices of P are $(0, 0)$, $(\frac{8}{3}, \frac{5}{3})$, $(\frac{2}{3}, -\frac{1}{3})$, and $(2, 2)$. It follows that P is the image of the unit square $0 \leq s, t \leq 1$ under the transformation

$$\begin{pmatrix} x \\ y \end{pmatrix} = \begin{pmatrix} \frac{2}{3} & 2 \\ -\frac{1}{3} & 2 \end{pmatrix} \begin{pmatrix} s \\ t \end{pmatrix},$$

where the columns of the 2×2 matrix are the vectors from the origin to the two adjacent vertices. Taking this transformation as $\mathbf{G}$ in Theorem 4.37 yields

$$\iint_P xy \, dA = 2 \int_0^1 \int_0^1 (\tfrac{2}{3}s + 2t)(-\tfrac{1}{3}s + 2t) \, dt \, ds.$$

This integral is essentially the same as the preceding one; the variables (s, t) and (u, v) are related by $u = s$, $v = 6t$.

EXAMPLE 4. Let R be the region in the first quadrant of the xy-plane bounded by the x-axis and the parabolas $x = 1 - \frac{1}{4}y^2$, $x = \frac{1}{4}y^2 - 1$, and $x = 4 - \frac{1}{16}y^2$. (See Figure 4.11.) What is $\iint_S xy \, dx \, dy$?

Solution. Refer back to Example 3 in §3.4: The region R is the image of the rectangle $\{(u, v) : 1 \leq u \leq 2, \ 0 \leq v \leq 1\}$ under the map $\mathbf{G}(u, v) = (u^2 - v^2, 2uv)$. We have $D\mathbf{G}(u, v) = \begin{pmatrix} 2u & -2v \\ 2v & 2u \end{pmatrix}$ and hence $\det D\mathbf{G}(u, v) =$

$4(u^2 + v^2)$. Thus, the substitutions $x = u^2 - v^2$, $y = 2uv$ give

$$\iint_S xy \, dx \, dy = \int_0^1 \int_1^2 (u^2 - v^2)(2uv)4(u^2 + v^2) \, du \, dv$$

$$= \int_0^1 \int_1^2 8(u^5 v - uv^5) \, du \, dv = \int_0^1 (\tfrac{4}{3}u^6 v - 4u^2 v^5)\big|_{u=1}^2 \, dv$$

$$= \int_0^1 (84v - 12v^5) \, dv = (42v^2 - 2v^6)\big|_0^1 = 40.$$

EXERCISES

1. Find the area of the region inside the cardioid $r = 1 + \cos\theta$ (polar coordinates).

2. Find the centroid of the half-cone $\sqrt{x^2 + y^2} \le z \le 1$, $x \ge 0$.

3. Find the volume of the region inside both the sphere $x^2 + y^2 + z^2 = 4$ and the cylinder $x^2 + y^2 = 1$.

4. Find the volume of the region above the xy-plane, below the cone $z = 2 - \sqrt{x^2 + y^2}$, and inside the cylinder $(x-1)^2 + y^2 = 1$.

5. Find the mass of a right circular cylinder of base radius R and height h if the mass density is c times the distance from the bottom of the cylinder.

6. Find the volume of the portion of the sphere $x^2 + y^2 + z^2 = 4$ lying above the plane $z = 1$.

7. Find the mass of a ball of radius R if the mass density is c times the distance from the boundary of the ball.

8. Find the centroid of the portion of the ball $x^2 + y^2 + z^2 \le 1$ lying in the first octant ($x, y, z \ge 0$).

9. Find the centroid of the parallelogram bounded by the lines $x - 3y = 0$, $2x + y = 0$, $x - 3y = 10$, and $2x + y = 15$.

10. Calculate $\iint_S (x+y)^4 (x-y)^{-5} \, dA$ where S is the square $-1 \le x + y \le 1$, $1 \le x - y \le 3$.

11. Find the volume of the ellipsoid $(x + 2y)^2 + (x - 2y + z)^2 + 3z^2 = 1$.

12. Let S be the region in the first quadrant bounded by the curves $xy = 1$, $xy = 4$, and the lines $y = x$, $y = 4x$. Find the area and the centroid of S by using the transformation $u = xy$, $v = y/x$.

13. Let S be the region in the first quadrant bounded by the curves $xy = 1$, $xy = 3$, $x^2 - y^2 = 1$, and $x^2 - y^2 = 4$. Compute $\iint_S (x^2 + y^2) \, dA$. (*Hint:* Let $G(x, y) = (xy, x^2 - y^2)$. What is $|\det DG|$?)

14. Use the transformation $x = u - uv$, $y = uv$ to evaluate $\iint_S (x+y)^{-1} \, dA$ where S is the region in the first quadrant between the lines $x + y = 1$ and $x + y = 4$.

15. Use "double polar coordinates" $x = r\cos\theta$, $y = r\sin\theta$, $z = s\cos\varphi$, $w = s\sin\varphi$ in $\mathbb{R}^4$ to compute the 4-dimensional volume of the ball $x^2 + y^2 + z^2 + w^2 = R^2$.

4.5   Functions Defined by Integrals

Suppose $f(\mathbf{x}, \mathbf{y})$ is a function of $\mathbf{x} \in \mathbb{R}^m$ and $\mathbf{y} \in \mathbb{R}^n$. If $f(\mathbf{x}, \mathbf{y})$ is integrable over the set $S \subset \mathbb{R}^n$ as a function of $\mathbf{y}$ for each fixed $\mathbf{x}$, we can form a new function of $\mathbf{x}$ by integrating out $\mathbf{y}$:

$$(4.45) \qquad\qquad F(\mathbf{x}) = \int \cdots \int_S f(\mathbf{x}, \mathbf{y}) \, d^n\mathbf{y}.$$

The question then arises as to how properties of f such as continuity and differentiability relate to the corresponding properties of F.

Perhaps the most basic question of this sort is the following. Suppose that

$$\lim_{\mathbf{x}\to\mathbf{a}} f(\mathbf{x}, \mathbf{y}) = g(\mathbf{y}) \qquad (\mathbf{y} \in S);$$

is it true that

$$\lim_{\mathbf{x}\to\mathbf{a}} F(\mathbf{x}) = \int \cdots \int_S g(\mathbf{y}) \, d^n\mathbf{y}?$$

In other words, can one interchange the operations of integrating with respect to $\mathbf{y}$ and taking a limit with respect to $\mathbf{x}$? Is the limit of the integral equal to the integral of the limit? In general, the answer is *no*.

EXAMPLE 1. let

$$f(x, y) = \frac{x^2 y}{(x^2 + y^2)^2} \quad ((x, y) \neq (0, 0)), \qquad f(0, 0) = 0.$$

Evidently $\lim_{x\to 0} f(x, y) = 0$ for each y (although for different reasons when $y = 0$ or when $y \neq 0$). However, $\lim_{x\to 0} \int_0^1 f(x, y) \, dy \neq 0$; in fact,

$$\int_0^1 \frac{x^2 y}{(x^2 + y^2)^2} \, dy = -\frac{x^2}{2(x^2 + y^2)}\Big|_0^1 = \frac{1}{2(1 + x^2)},$$

which tends to $\frac{1}{2}$ as $x \to 0$.

Notice, however, that the f in Example 1 is discontinuous, and indeed unbounded, at the origin; for instance, $f(x, x) = 1/4x \to \infty$ as $x \to 0$. ($f(x_0, y)$ is bounded as a function of y for each fixed x_0, but its maximum value tends to infinity as $x_0 \to 0$.) If we impose some stronger conditions on f, we can obtain an affirmative result. The following theorem is not the last word in the subject (see Corollary 4.53), but it suffices for many purposes.

4.46 Theorem. *Suppose S and T are compact subsets of $\mathbb{R}^n$ and $\mathbb{R}^m$, respectively, and S is measurable. If $f(\mathbf{x}, \mathbf{y})$ is continuous on the set $T \times S = \{(\mathbf{x}, \mathbf{y}) : \mathbf{x} \in T, \mathbf{y} \in S\}$, then the function F defined by (4.45) is continuous on T.*

Proof. Given $\epsilon > 0$, we wish to find $\delta > 0$ so that $|F(\mathbf{x}) - F(\mathbf{x}')| < \epsilon$ whenever $|\mathbf{x} - \mathbf{x}'| < \delta$. Let $|S|$ denote the n-dimensional volume of S. Since $T \times S$ is compact, f is uniformly continuous on it by Theorem 1.33, so there is a $\delta > 0$ so that $|f(\mathbf{x}, \mathbf{y}) - f(\mathbf{x}', \mathbf{y})| < \epsilon/|S|$ whenever $\mathbf{y} \in S$, $\mathbf{x}, \mathbf{x}' \in T$, and $|\mathbf{x} - \mathbf{x}'| < \delta$. But then

$$|F(\mathbf{x}) - F(\mathbf{x}')| \le \int \cdots \int_S |f(\mathbf{x}, \mathbf{y}) - f(\mathbf{x}', \mathbf{y})| \, d^n \mathbf{y} < \int \cdots \int_S \frac{\epsilon}{|S|} \, d^n \mathbf{y} = \epsilon,$$

and we are done. $\qquad \square$

Remark. In the statement of Theorem 4.46 we could assume that T is open rather than compact. Indeed, every point $\mathbf{x}$ in an open set T is the center of a closed ball B that is contained in T. Since B is compact, the preceding argument shows that F is continuous on B, and hence F is continuous at every $\mathbf{x} \in T$.

A related question concerns differentiability. Suppose that f is differentiable as a function of $\mathbf{x}$ for each $\mathbf{y} \in S$; is it true that F is differentiable in $\mathbf{x}$ and that its partial derivatives $\partial_{y_j} F$ are the integrals of the derivatives $\partial_{y_j} f$? In other words, is the integral of the derivative equal to the derivative of the integral? This is another question about the interchange of limits and integrals. Indeed, it is always true that the finite difference $F(\mathbf{x} + \mathbf{h}) - F(\mathbf{x})$ is the integral of $f(\mathbf{x} + \mathbf{h}, \mathbf{y}) - f(\mathbf{x}, \mathbf{y})$, simply because integration is a linear operation, and the question is what happens in the limit as $\mathbf{h} \to \mathbf{0}$. As in Example 1, things can go wrong; see Exercise 1. Our main positive result is as follows.

4.47 Theorem. *Suppose $S \subset \mathbb{R}^n$ is compact and measurable, and $T \subset \mathbb{R}^m$ is open. If f is a continuous function on $T \times S$ that is of class C^1 as a function of $\mathbf{x} \in T$ for each $\mathbf{y} \in S$, then the function F defined by (4.45) is of class C^1 on T, and*

(4.48)
$$\frac{\partial F}{\partial x_j}(\mathbf{x}) = \int \cdots \int_S \frac{\partial f}{\partial x_j}(\mathbf{x}, \mathbf{y}) \, d^n \mathbf{y} \qquad (\mathbf{x} \in T).$$

Proof. Given a point $\mathbf{x}_0 \in T$, choose $r > 0$ small enough so that $\mathbf{x} \in T$ whenever $|\mathbf{x} - \mathbf{x}_0| \le 2r$. We shall show that F is of class C^1 on $B(r, \mathbf{x}_0)$ and prove (4.48) for $\mathbf{x} \in B(r, \mathbf{x}_0)$; since $\mathbf{x}_0$ is an arbitrary point in T, this will establish the theorem. For the purpose of computing $\partial_{x_j} F$, the other variables x_k ($k \ne j$) play no role, so we may assume that $m = 1$. In fact, in order to simplify the notation a bit, we shall also assume that $n = 1$; the proof for general n is exactly the same. Accordingly, we write x and y instead of $\mathbf{x}$ and $\mathbf{y}$ henceforth.

For $0 < |h| \le r$ and $|x - x_0| \le r$, we consider the difference quotient

$$\frac{F(x + h) - F(x)}{h} = \int_S \frac{f(x+h, y) - f(x, y)}{h} \, dy.$$

By the mean value theorem, we have $f(x + h, y) - f(x, y) = h \partial_x f(x + th, y)$, where t is some number between 0 and 1 depending on x, h, and y. Hence,

(4.49)
$$\frac{F(x + h) - F(x)}{h} - \int_S \partial_x f(x, y) \, dy = \int_S \left[\partial_x f(x + th, y) - \partial_x f(x, y) \right] dy.$$

The argument now proceeds as in the proof of Theorem 4.46. Since $\partial_x f$ is continuous on the compact set $\overline{B(r, x_0)} \times S$, it is uniformly continuous there by Theorem 1.33. Thus, given $\epsilon > 0$, we can find $\delta > 0$ so that the integrand on the right of (4.49) is less than $\epsilon/|S|$ for all $y \in S$, $x \in B(r, x_0)$, and $t \in (0, 1)$, whenever $|h| < \delta$. It follows that

$$\left| \frac{F(x + h) - F(x)}{h} - \int_S \partial_x f(x, y) \, dy \right| < \int_S \frac{\epsilon}{|S|} \, dy = \epsilon \text{ for } |h| < \delta,$$

and hence that

$$\lim_{h \to 0} \frac{F(x + h) - F(x)}{h} - \int_S \partial_x f(x, y) \, dy = 0,$$

as claimed. $\qquad \square$

EXAMPLE 2. Let $F(x) = \int_0^\pi y^{-1} e^{xy} \sin y \, dy$. This integral cannot be evaluated in elementary terms; however, we have $F'(x) = \int_0^\pi e^{xy} \sin y \, dy$, which can be evaluated by two integrations by parts. The result is that $F'(x) = (e^{\pi x} + 1)/(x^2 + 1)$.

Situations often occur in which the variable $\mathbf{x}$ occurs in the limits of integration as well as the integrand. For simplicity we consider the case where x and y are scalar variables:

(4.50)
$$F(x) = \int_a^{\varphi(x)} f(x, y) \, dy.$$

We suppose that f is continuous in x and y and of class C^1 in x for each y, and that φ is of class C^1. If f does not depend on x, the derivative of F can be computed by the fundamental theorem of calculus together with the chain rule:

$$\frac{d}{dx} \int_a^{\varphi(x)} f(y)\, dy = f(\varphi(x))\varphi'(x).$$

For the more general case (4.50), we can differentiate F by combining this result with Theorem 4.47 according to the recipe in Exercise 7 of §2.3: Differentiate with respect to each x in (4.50) in turn while treating the others as constants, and add the results. The upshot is that

(4.51) $$F'(x) = f(x, \varphi(x))\varphi'(x) + \int_a^{\varphi(x)} \frac{\partial f}{\partial x}(x, y)\, dy.$$

EXAMPLE 3. Given a continuous function g on $\mathbb{R}$, let

$$h(x) = \int_0^x (x - y)g(y)\, dy.$$

Then

$$h'(x) = (x - x)g(x) + \int_0^x g(y)\, dy = \int_0^x g(y)\, dy,$$

and hence $h''(x) = g(x)$. (Cf. Exercise 7 in §4.3, where this result is approached from a different angle.)

The hypotheses of Theorems 4.46 and 4.47 can be weakened considerably, but only at the cost of a more intricate proof. More sophisticated theories of integration (see §4.8) furnish a powerful theorem, the so-called dominated convergence theorem, that generally provides the sharpest results in these situations. The full statement of this theorem requires more background than we have available here, but its restriction to the context of Riemann integrable functions is the following result, in which the crucial condition is the existence of the uniform bound C.

4.52 Theorem (The Bounded Convergence Theorem). *Let S be a measurable subset of $\mathbb{R}^n$ and $\{f_j\}$ a sequence of integrable functions on S. Suppose that $f_j(\mathbf{y}) \to f(\mathbf{y})$ for each $\mathbf{y} \in S$, where f is an integrable function on S, and that there is a constant C such that $|f_j(\mathbf{y})| \le C$ for all j and all $\mathbf{y} \in S$. Then*

$$\lim_{j \to \infty} \int \cdots \int_S f_j(\mathbf{y})\, d^n\mathbf{y} = \int \cdots \int_S f(\mathbf{y})\, d^n\mathbf{y}.$$

An elementary (but not simple) proof for the case where S is an interval in $\mathbb{R}$ can be found in Lewin [17]. The full dominated convergence theorem can be found in Bear [3, p. 68], DePree and Swartz [5, p. 194], Jones [9, p. 133], and Rudin [18, p. 321].

Theorem 4.52 implies the following improvements on Theorems 4.46 and 4.47.

4.53. Corollary. *Let S be a measurable subset of $\mathbb{R}^n$ and T a subset of $\mathbb{R}^m$. Suppose $f(\mathbf{x}, \mathbf{y})$ is a function on $T \times S$ that is integrable as a function of $\mathbf{y} \in S$ for each $\mathbf{x} \in T$, and let F be defined by (4.45).*

  a. *If $f(\mathbf{x}, \mathbf{y})$ is continuous as a function of $\mathbf{x} \in T$ for each $\mathbf{y} \in S$, and there is a constant C such that $|f(\mathbf{x}, \mathbf{y})| \leq C$ for all $\mathbf{x} \in T$ and $\mathbf{y} \in S$, then F is continuous on T.*

  b. *Suppose T is open. If $f(\mathbf{x}, \mathbf{y})$ is of class C^1 as a function of $\mathbf{x} \in T$ for each $\mathbf{y} \in S$, and there is a constant C such that $|\nabla_\mathbf{x} f(\mathbf{x}, \mathbf{y})| \leq C$ for all $\mathbf{x} \in T$ and $\mathbf{y} \in S$, then F is of class C^1 on T and (4.48) holds.*

Proof. To prove part (a), by Theorem 1.15 it is enough to show that $F(\mathbf{x}_j) \to F(\mathbf{x})$ whenever $\{\mathbf{x}_j\}$ is a sequence in S converging to $\mathbf{x} \in S$. This follows by applying the bounded convergence theorem to the sequence of functions $f_j(\mathbf{y}) = f(\mathbf{x}_j, \mathbf{y})$. Similarly, part (b) is proved by applying the bounded convergence theorem to the sequence of difference quotients with increments $\mathbf{h}_j$, where $\{\mathbf{h}_j\}$ is a sequence tending to zero along one of the coordinate axes. The uniform bound on these quotients is obtained by applying the mean value theorem as in the proof of Theorem 4.47; details are left as Exercise 8. $\square$

EXERCISES

1. Let $f(x, y) = x^3 y^{-2} e^{-x^2/y}$ if $y > 0$, $f(x, y) = 0$ if $y \leq 0$.
   a. Show that $f(x, y)$ is of class C^1 as a function of x for each fixed y and as a function of y for each fixed x, but that f is unbounded in any neighborhood of the origin. (For the smoothness in y, cf. Exercise 9 in §2.1.)
   b. Let $F(x) = \int_0^1 f(x, y)\, dy$. Show that $F(x) = xe^{-x^2}$ and hence that $F'(0) = 1$, but that $\int_0^1 \partial_x f(0, y)\, dy = 0$.

2. Compute $F'(x)$ for the functions $F(x)$ defined for $x > 0$ by the following formulas. Your answers should not contain integral signs.
   a. $F(x) = \int_0^1 \log(1 + xe^y)\, dy$.
   b. $F(x) = \int_1^{x^2} y^{-1} \cos(xy^2)\, dy$.
   c. $F(x) = \int_1^{3x} y^{-1} e^{xy}\, dy$.

3. Given a continuous function g on $\mathbb{R}$, let $h(x) = \int_0^x (x-y)e^{x-y}g(y)\,dy$. Show that $h'' - 2h' + h = g$.

4. Given a continuous function g on $\mathbb{R}$, let $h(x) = \frac{1}{2}\int_0^x \sin 2(x-y)g(y)\,dy$. Show that $h'' + 4h = g$.

5. Given $F(x) = \int_{\psi(x)}^{\varphi(x)} f(x,y)\,dy$, find $F'(x)$, assuming suitable smoothness conditions on ψ, φ, and f.

6. (*How to compress n antidifferentiations into one*) Let f be a continuous function on $\mathbb{R}$. For $n \geq 1$, let

$$f^{[n]}(x) = \frac{1}{(n-1)!}\int_0^x (x-y)^{n-1}f(y)\,dy.$$

Show that $\left(f^{[n]}\right)' = f^{[n-1]}$ for $n > 1$ and conclude that $f^{[n]}$ is an nth-order antiderivative of f.

7. Let f be any continuous function on $[0,1]$. For $x \in \mathbb{R}$ and $t > 0$, let

$$u(x,t) = t^{-1/2}\int_0^1 e^{-(x-y)^2/4t}f(y)\,dy, \qquad v(x,t) = t\int_0^1 \frac{f(y)}{(x-y)^2 + t^2}\,dy.$$

    a. Show that $\partial_t u = \partial_x^2 u$.
    b. Show that $\partial_x^2 v + \partial_t^2 v = 0$.

8. Complete the deduction of Corollary 4.53b from the bounded convergence theorem.

4.6 Improper Integrals

In this section we return to integration in one variable. The Riemann theory of integration pertains to bounded functions on finite intervals, but there are many situations where one needs to integrate functions over infinite intervals (i.e., half-lines or the whole line) or functions that are unbounded near some point in the interval of integration. Such integrals are called **improper**, and they are defined in terms of limits of ordinary integrals. To do a really good job with improper integrals, one should adopt the more powerful Lebesgue theory of integration, sketched in §4.8. (Even then, additional limiting procedures are needed to handle integrals such as the one in Example 3 below.) Here we content ourselves with a short discussion of useful results about simple types of improper integrals.

The two most basic types of improper integrals are as follows:

I. $\int_a^\infty f(x)\,dx$, where f is integrable over every finite subinterval $[a,b]$.

II. $\int_a^b f(x)\,dx$, where f is integrable over $[c, b]$ for every $c > a$ but is unbounded near $x = a$.

We study these two types in turn and then consider integrals of more complicated sorts that can be obtained by combining them.

Improper Integrals of Type I. *In this subsection, all functions in question are assumed to be defined on $[a, \infty)$ and integrable on $[a, b]$ for every $b > a$.*
  The definition of the improper integral is

$$\int_a^\infty f(x)\,dx = \lim_{b \to \infty} \int_a^b f(x)\,dx.$$

More precisely, the integral $\int_a^\infty f(x)\,dx$ is said to **converge** if the limit on the right exists, in which case its value is defined to be that limit; otherwise the integral is said to **diverge**, and it is not assigned a numerical value. (However, we may say that $\int_a^\infty f(x)\,dx = \infty$ if $\int_a^b f(x)\,dx$ grows without bound as $b \to \infty$.)

EXAMPLE 1.
  a. $\int_0^\infty e^{-x}\,dx = \lim_{b\to\infty}\left[-e^{-x}\right]_0^b = 1$, since $\lim_{b\to\infty} e^{-b} = 0$.
  b. $\int_0^\infty \cos x\,dx$ diverges, since $\lim_{b\to\infty} \sin b$ does not exist.

Our main concern here is not with the evaluation of $\int_a^\infty f(x)\,dx$ but with the more basic question of whether or not it converges. At the outset, we make one simple but useful remark: If $c > a$, the convergence of $\int_a^\infty f(x)\,dx$ is equivalent to the convergence of $\int_c^\infty f(x)\,dx$, the difference between the two being the ordinary integral $\int_a^c f(x)\,dx$. Thus, the convergence of $\int_a^\infty f(x)\,dx$ depends only on the behavior of $f(x)$ as $x \to \infty$, not on its behavior on a finite interval $[a, c]$.
  We first consider the situation when $f \geq 0$. In this case, the integral $\int_a^b f(x)\,dx$ increases along with the upper endpoint b, so we can exploit the following variant of the monotone sequence theorem.

4.54 Lemma. *If φ is a bounded increasing function on $[a, \infty)$, then $\lim_{x\to\infty} \varphi(x)$ exists and equals $\sup\{\varphi(x) : x \geq a\}$.*

Proof. The proof is left to the reader (Exercise 7); it is essentially identical to the proof of the monotone sequence theorem (1.16).   □

By applying Lemma 4.54 to the function $\varphi(x) = \int_a^x f(t)\,dt$, we see that the integral $\int_a^\infty f(x)\,dx$ converges if and only if $\int_a^b f(x)\,dx$ remains bounded as $b \to \infty$. This immediately leads to the basic comparison test for convergence.

4.55 Theorem. *Suppose that* $0 \le f(x) \le g(x)$ *for all sufficiently large* x. *If* $\int_a^\infty g(x)\, dx$ *converges, so does* $\int_a^\infty f(x)\, dx$. *If* $\int_a^\infty f(x)\, dx$ *diverges, so does* $\int_a^\infty g(x)\, dx$.

Proof. By the remarks following the definition of convergence, we may assume that $0 \le f(x) \le g(x)$ for all $x \ge a$. If $\int_a^\infty g(x)\, dx$ converges, it provides an upper bound for $\varphi(b) = \int_a^b f(x)\, dx$ as $b \to \infty$:

$$\int_a^b f(x)\, dx \le \int_a^b g(x)\, dx \le \int_a^\infty g(x)\, dx.$$

The convergence of $\int_a^\infty f(x)\, dx$ then follows from Lemma 4.54. The second assertion is equivalent to the first one. ☐

The following variant of Theorem 4.55 is sometimes easier to apply:

4.56 Corollary. *Suppose* $f > 0$, $g > 0$, *and* $f(x)/g(x) \to l$ *as* $x \to \infty$. *If* $0 < l < \infty$, *then* $\int_a^\infty f(x)\, dx$ *and* $\int_a^\infty g(x)\, dx$ *are both convergent or both divergent. If* $l = 0$, *the convergence of* $\int_a^\infty g(x)\, dx$ *implies the convergence of* $\int_a^\infty f(x)\, dx$. *If* $l = \infty$, *the divergence of* $\int_a^\infty g(x)\, dx$ *implies the divergence of* $\int_a^\infty f(x)\, dx$.

Proof. If $0 < l < \infty$, the fact that $f(x)/g(x) \to l$ yields the estimates $f(x) \le 2lg(x)$ and $f(x) \ge \frac{1}{2}lg(x)$ for sufficiently large x, so the first assertion follows by comparing f to a multiple of g. If $l = 0$ (resp. $l = \infty$), we have $f(x) \le g(x)$ (resp. $g(x) \ge f(x)$) for sufficiently large x, whence the other assertions follow. ☐

The functions most often used for comparison in Theorem 4.55 and Corollary 4.56 are the power functions x^{-p}. Taking $a = 1$ for convenience, for $p \ne 1$ we have

$$\int_1^b \frac{dx}{x^p} = \frac{b^{1-p} - 1}{1 - p} \to \begin{cases} \infty & \text{if } p < 1, \\ (p-1)^{-1} & \text{if } p > 1, \end{cases}$$

and $\int_1^b x^{-1} dx = \log b \to \infty$. In short, $\int_1^\infty x^{-p}\, dx$ converges if and only if $p > 1$. Combining this fact with Theorem 4.55, we obtain the following handy rule:

4.57 Corollary. *If* $0 \le f(x) \le Cx^{-p}$ *for all sufficiently large* x, *where* $p > 1$, *then* $\int_a^\infty f(x)\, dx$ *converges. If* $f(x) \ge cx^{-1}$ $(c > 0)$ *for all sufficiently large* x, *then* $\int_a^\infty f(x)\, dx$ *diverges.*

EXAMPLE 2. The integral $\int_0^\infty [(2x + 14)/(x^3 + 1)]\, dx$ converges, because

$$\frac{2x + 14}{x^3 + 1} \le \frac{4x}{x^3} = \frac{4}{x^2} \text{ for } x \ge 7.$$

Alternatively, we could observe that

$$\frac{2x + 14}{x^3 + 1} \Big/ \frac{1}{x^2} \to 2 \text{ as } x \to \infty$$

and use Corollary 4.56 with $g(x) = x^{-2}$ to establish the convergence of the integral over, say, $[1, \infty)$. (The integral over $[0, 1]$ is proper.) Note that we are not comparing $\int_0^\infty [(2x + 14)/(x^3 + 1)] \, dx$ to $\int_0^\infty x^{-2} \, dx$, which presents an additional difficulty because x^{-2} is unbounded at $x = 0$; the comparison of $(2x + 14)/(x^3 + 1)$ with x^{-2} is significant only for large x.

It should be noted that the power functions x^{-p} do not quite tell the whole story. There are functions whose rate of decay at infinity is faster than x^{-1} but slower than x^{-p} for $p > 1$, and their integrals may be either convergent or divergent; see Exercises 4 and 5.

Next we remove the assumption that f is nonnegative, and with a view toward future applications, we shall allow f to be complex-valued. The question of convergence can often be reduced to the case where $f \geq 0$ via the following result.

4.58 Theorem. *If $\int_a^\infty |f(x)| \, dx$ converges, then $\int_a^\infty f(x) \, dx$ converges.*

Proof. First suppose f is real-valued. Let $f^+(x) = \max[f(x), 0]$ and $f^-(x) = \max[-f(x), 0]$. Then we have $0 \leq f^+(x) \leq |f(x)|$ and $0 \leq f^-(x) \leq |f(x)|$, so $\int_a^\infty f^+(x) \, dx$ and $\int_a^\infty f^-(x) \, dx$ converge by Theorem 4.55. But $f = f^+ - f^-$, so $\int_a^\infty f(x) \, dx$ converges also.

If f is complex-valued, we have $|\operatorname{Re} f(x)| \leq |f(x)|$ and $|\operatorname{Im} f(x)| \leq |f(x)|$, so the convergence of $\int_a^\infty |f(x)| \, dx$ implies the convergence of $\int_a^\infty |\operatorname{Re} f(x)| \, dx$ and $\int_a^\infty |\operatorname{Im} f(x)| \, dx$ and hence (by the preceding argument) the convergence of the real and imaginary parts of $\int_a^\infty f(x) \, dx$. $\qquad\square$

The integral $\int_a^\infty f(x) \, dx$ is called **absolutely convergent** if $\int_a^\infty |f(x)| \, dx$ converges. Theorem 4.55 and its corollaries can be used to test for absolute convergence, by applying them to $|f|$. It is possible, however, for $\int_a^\infty f(x) \, dx$ to converge even when $\int_a^\infty |f(x)| \, dx$ diverges because of cancellation effects between positive and negative values. Here is an important example.

EXAMPLE 3. The integral $\displaystyle\int_1^\infty \frac{\sin x}{x} \, dx$ is not absolutely convergent (Exercise 8), but it is convergent. To see this, integrate by parts:

$$\int_1^b \frac{\sin x}{x} \, dx = \frac{-\cos x}{x} \Big|_1^b - \int_1^b \frac{\cos x}{x^2} \, dx.$$

Now, $\int_1^\infty |x^{-2} \cos x| \, dx$ converges by Corollary 4.57 since $|x^{-2} \cos x| \le x^{-2}$, so the integral on the right approaches a finite limit as $b \to \infty$; moreover, since $|b^{-1} \cos b| \le b^{-1} \to 0$, so does the other term. Hence $\lim_{b\to\infty} \int_1^b x^{-1} \sin x \, dx$ exists, as claimed.

Improper Integrals of Type II. *In this subsection, all functions in question are assumed to be defined on $(a, b]$ and integrable on $[c, b]$ for every $c > a$.*

The definition of the improper integral in this situation is

$$\int_a^b f(x) \, dx = \lim_{c>a,\ c\to a} \int_c^b f(x) \, dx.$$

That is, $\int_a^b f(x) \, dx$ **converges** if the limit on the right exists, and **diverges** otherwise. The obvious analogues of the results in the preceding subsection are valid in this situation with essentially the same proofs; one has merely to replace conditions like "$x \to \infty$" or "for sufficiently large x" by "$x \to a$" or "for x sufficiently close to a." For instance, here is the basic comparison test:

4.59 Theorem. *Suppose that $0 \le f(x) \le g(x)$ for all x sufficiently close to a. If $\int_a^b g(x) \, dx$ converges, so does $\int_a^b f(x) \, dx$. If $\int_a^b f(x) \, dx$ diverges, so does $\int_a^b g(x) \, dx$.*

The functions most often used for comparison in this situation are the power functions $(x - a)^{-p}$, but now the condition for convergence is $p < 1$ rather than $p > 1$. Indeed, for $p \ne 1$,

$$\int_c^b (x - a)^{-p} \, dx = \frac{(x - a)^{1-p}}{1 - p} \Big|_c^b \to \begin{cases} (1 - p)^{-1}(b - a)^{1-p} & \text{if } p < 1, \\ \infty & \text{if } p > 1, \end{cases}$$

and $\int_c^b (x - a)^{-1} \, dx = \log(x - a)|_c^b \to \infty$. Hence the analogue of Corollary 4.57 is as follows:

4.60 Corollary. *If $0 \le f(x) \le C(x - a)^{-p}$ for x near a, where $p < 1$, then $\int_a^b f(x) \, dx$ converges. If $f(x) > c(x - a)^{-1}$ $(c > 0)$ for x near a, then $\int_a^b f(x) \, dx$ diverges.*

EXAMPLE 4. $\int_0^1 x^{-2} \sin 3x \, dx$ diverges. Indeed, $x^{-1} \sin 3x \to 3$ as $x \to 0$, so $x^{-2} \sin 3x > 2x^{-1}$ for x near 0.

Theorem 4.58 also remains valid in this situation; that is, absolute convergence implies convergence.

EXAMPLE 5. $\int_0^1 x^{-1/2} \sin(x^{-1}) \, dx$ is absolutely convergent, because $|x^{-1/2} \sin(x^{-1})| \le x^{-1/2}$.

Other Types of Improper Integrals. Various other kinds of improper integrals can be built up out of those of types I and II.

First, obviously one can consider the "mirror images" of types I and II; that is, integrals of the form $\int_{-\infty}^{b} f(x)\,dx$ where f is integrable on $[a, b]$ for all $a < b$, or integrals of the form $\int_{a}^{b} f(x)\,dx$ where f is integrable on $[a, c]$ for all $c < b$ but is unbounded near $x = b$. The ideas are exactly the same; only minor notational changes are needed. (In the latter situation, the comparison functions for the analogue of Corollary 4.60 are the power functions $|x - b|^{-p} = (b - x)^{-p}$.)

Second, one can consider improper integrals $\int_{a}^{b} f(x)\,dx$ where a difficulty occurs at both endpoints of the interval of integration, either because the endpoint is at infinity or because the integrand is unbounded there. The trick here is to pick an intermediate point $c \in (a, b)$ and write $\int_{a}^{b} = \int_{a}^{c} + \int_{c}^{b}$, thus reducing the integral to a sum of two integrals that are each of type I or II; the original integral is said to be convergent if and only if each of the two subintegrals is convergent. For example, if f is integrable over every finite interval $[a, b]$, we define

$$\int_{-\infty}^{\infty} f(x)\,dx = \int_{-\infty}^{0} f(x)\,dx + \int_{0}^{\infty} f(x)\,dx$$

$$= \lim_{a \to -\infty} \int_{a}^{0} f(x)\,dx + \lim_{b \to \infty} \int_{0}^{b} f(x)\,dx.$$

The integral on the left converges only when both of the limits on the right exist independently of one another; there is no relation between the variables a and b. The same ideas apply to $\int_{a}^{\infty} f(x)\,dx$ when f is unbounded at a or to $\int_{a}^{b} f(x)\,dx$ when f is unbounded at both a and b.

EXAMPLE 6. $\int_{-\infty}^{\infty} dx/(1 + x^2)$ converges; the integrals over $(-\infty, 0]$ and $[0, \infty)$ are both convergent by comparison to x^{-2}. In fact,

$$\int_{-\infty}^{\infty} \frac{dx}{1 + x^2} = \lim_{a \to -\infty,\ b \to +\infty} \arctan x \Big|_{a}^{b} = \frac{\pi}{2} - \left(-\frac{\pi}{2}\right) = \pi.$$

EXAMPLE 7. $\int_{0}^{\infty} x^{-p}\,dx$ is divergent for every p. Indeed, if $p < 1$, $\int_{0}^{1} x^{-p}\,dx$ converges but $\int_{1}^{\infty} x^{-p}\,dx$ diverges, whereas the reverse is true if $p > 1$. If $p = 1$, these integrals both diverge.

EXAMPLE 8. Consider $\int_{0}^{\infty} f(x)\,dx$ where $f(x) = 1/(x^{1/2} + x^{3/2})$. Since $0 < f(x) < x^{-1/2}$, $\int_{0}^{1} f(x)\,dx$ converges by Corollary 4.60. Since $0 < f(x) < x^{-3/2}$, $\int_{1}^{\infty} f(x)\,dx$ converges by Corollary 4.57. Hence $\int_{0}^{\infty} f(x)\,dx$ converges.

Finally, one can consider improper integrals $\int_a^b f(x)\,dx$ where f is unbounded near one or more interior points of $[a, b]$. Again the trick is to break up $[a, b]$ into subintervals such that the singularities of f occur only at endpoints of the subintervals and consider the integrals of f over the subintervals separately.

EXAMPLE 9. Let $f(x) = (x^3 - 8x^2)^{-1/3}$, and let us consider $\int_0^9 f(x)\,dx$ and $\int_0^\infty f(x)\,dx$. The singularities of f occur at $x = 0$ and $x = 8$, so for the first integral we write

$$\int_0^9 = \int_0^c + \int_c^8 + \int_8^9 \qquad (0 < c < 8).$$

We have $|f(x)| = x^{-2/3}|x - 8|^{-1/3}$, which is approximately $\frac{1}{2}x^{-2/3}$ for x near 0 and approximately $\frac{1}{4}|x - 8|^{-1/3}$ for x near 8. Hence all three subintegrals are absolutely convergent by Corollary 4.60, and the original integral $\int_0^9$ converges. On the other hand, $f(x)$ is positive for $x > 8$ and $f(x)/x^{-1} = (1 - 8x^{-1})^{-1/3} \to 1$ as $x \to \infty$, so $\int_9^\infty f(x)\,dx$ diverges by Corollary 4.56. It follows that $\int_0^\infty f(x)\,dx$ diverges too.

The definition of the improper integral $\int_a^b f(x)\,dx$ given above when f has a singularity in the interior of $[a, b]$ is a little too restrictive for some purposes. Consider, for example, $\int_{-1}^1 x^{-1}dx$. According to our definition, this integral is to be considered as the limit of

$$(4.61) \qquad \int_{-1}^{-\delta} \frac{dx}{x} + \int_\epsilon^1 \frac{dx}{x} = \log \delta - \log \epsilon = \log\left(\frac{\delta}{\epsilon}\right)$$

as δ and ϵ decrease to 0, and this limit does not exist: When δ and ϵ are extremely small, their ratio can be arbitrarily large or arbitrarily small. However, since x^{-1} is an odd function, it seems natural to interpret the value of the integral as 0; the negative infinity of $\int_{-1}^0 x^{-1}\,dx$ should exactly cancel the positive infinity of $\int_0^1 x^{-1}dx$. We can achieve this result by modifying (4.61) so as to preserve the symmetry of the situation, namely, by taking $\delta = \epsilon$, so that $\log(\delta/\epsilon) = 0$.

These considerations lead to the following definition. Suppose $a < c < b$, and supppose f is integrable on $[a, c - \epsilon]$ and on $[c + \epsilon, b]$ for all $\epsilon > 0$. The (**Cauchy**) **principal value** of the integral $\int_a^b f(x)\,dx$ is

$$P.V. \int_a^b f(x)\,dx = \lim_{\epsilon \to 0}\left[\int_a^{c-\epsilon} f(x)\,dx + \int_{c+\epsilon}^b f(x)\,dx\right],$$

provided that the limit exists. Of course, if $\int_a^b f(x)\,dx$ converges, its Cauchy principal value is its ordinary value.

The following proposition describes a typical situation in which principal values occur.

4.62 Proposition. *Suppose $a < 0 < b$. If φ is continuous on $[a, b]$ and differentiable at 0, then $P.V. \int_a^b x^{-1}\varphi(x)\, dx$ exists.*

Proof. First we check the case $\varphi \equiv 1$ by explicit calculation:

$$P.V. \int_a^b \frac{dx}{x} = \lim_{\epsilon \to 0}\left[\int_a^{-e} \frac{dx}{x} + \int_\epsilon^b \frac{dx}{x}\right] = \log|x|\,\Big|_{-a}^{-\epsilon} + \log x\,\Big|_\epsilon^b = \log\left(\frac{b}{|a|}\right).$$

For the general case, we write $\varphi(x) = \varphi(0) + [\varphi(x) - \varphi(0)]$, obtaining

$$P.V. \int_a^b \frac{\varphi(x)}{x}\, dx = \varphi(0)\, P.V. \int_a^b \frac{dx}{x} + \int_a^b \frac{\varphi(x) - \varphi(0)}{x}\, dx.$$

We have just seen that the first quantity on the right exists, and the second one is a proper integral: The integrand is actually continuous on $[a, b]$ if we define its value at $x = 0$ to be $\varphi'(0)$. $\qquad\square$

The notion of principal value is also occasionally applied to integrals of the form $\int_{-\infty}^{\infty} f(x)\, dx$ in which f is integrable over any finite interval:

$$P.V. \int_{-\infty}^{\infty} f(x)\, dx = \lim_{R \to \infty} \int_{-R}^{R} f(x)\, dx.$$

For example, the integral $\int_{-\infty}^{\infty} x(1 + x^2)^{-1}\, dx$ is divergent because the integrand is asymptotically equal to x^{-1} as $x \to \pm\infty$, but its principal value is zero because the integrand is odd.

EXERCISES

1. Determine whether the following improper integrals of type I converge.

   a. $\displaystyle\int_1^\infty \frac{dx}{x\sqrt{x + 3}}$.

   b. $\displaystyle\int_3^\infty \frac{x^2 - 3x - 1}{x(x^2 + 2)}\, dx$.

   c. $\displaystyle\int_0^\infty x^2 e^{-x^2}\, dx$.

   d. $\displaystyle\int_3^\infty \frac{\sin 4x}{x^2 - x - 2}\, dx$.

e. $\displaystyle\int_1^\infty \tan\frac{1}{x}\,dx.$

2. Determine whether the following improper integrals of type II converge.

a. $\displaystyle\int_0^1 \frac{x}{\sqrt{1-x^2}}\,dx.$

b. $\displaystyle\int_{\pi/2}^\pi \cot x\,dx.$

c. $\displaystyle\int_0^1 \frac{\sqrt{1-x}}{x^2-4x+3}\,dx.$

d. $\displaystyle\int_0^1 \frac{dx}{x^{1/2}(x^2+x)^{1/3}}.$

e. $\displaystyle\int_0^1 \frac{1-\cos x}{\sin^3 2x}\,dx.$

3. Determine whether the following improper integrals converge. In each case it will be necessary to break up the integral into a sum of integrals of types I and/or II.

a. $\displaystyle\int_0^\infty x^{-3/4}e^{-x}\,dx.$

b. $\displaystyle\int_0^1 x^{-1/3}(1-x)^{-2}\,dx.$

c. $\displaystyle\int_0^\infty \frac{\sqrt{x}}{e^x-1}\,dx.$

d. $\displaystyle\int_0^\infty \frac{dx}{x(x-1)^{1/3}}.$

e. $\displaystyle\int_0^\infty x^{-1/5}\sin\frac{1}{x}\,dx.$

f. $\displaystyle\int_{-\infty}^\infty \frac{e^x}{e^x+x^2}\,dx.$

4. For $p > 0$, let $f_p(x) = x^{-1}(\log x)^{-p}$.
   a. Given $p > 0$ and $\epsilon > 0$, show that $x^{-1-\epsilon} < f_p(x) < x^{-1}$ for sufficiently large x.
   b. For which p does $\int_2^\infty f_p(x)\,dx$ converge?

5. Let f_p be as in Exercise 4 and $g_p(x) = (x\log x)^{-1}(\log\log x)^{-p}$.
   a. Given $p > 0$ and $\epsilon > 0$, show that $f_{1+\epsilon}(x) < g_p(x) < f_1(x)$ for sufficiently large x.
   b. For which p does $\int_3^\infty g_p(x)\,dx$ converge?

6. Let $f(x) = 1$ on the intervals $[1, 1\frac{1}{2}]$, $[2, 2\frac{1}{4}]$, $[3, 3\frac{1}{8}]$, ..., and $f(x) = 0$ elsewhere.

   a. Show that $\int_0^\infty f(x)\,dx$ converges (and is equal to 1) although $f(x) \not\to 0$ as $x \to \infty$.

   b. Modify f to make an example of a function g such that $\int_0^\infty g(x)\,dx$ converges although $g(x)$ does not remain bounded as $x \to \infty$.

7. Prove Lemma 4.54.

8. Prove that $\int_1^\infty x^{-1}|\sin x|\,dx$ diverges. (*Hint:* Show that there is a constant $c > 0$ such that $\int_{n\pi}^{(n+1)\pi} x^{-1}|\sin x|\,dx > c\int_{n\pi}^{(n+1)\pi} x^{-1}\,dx$ for all $n \geq 1$.)

9. (*Dirichlet's Test for Convergence*) Let f be continuous and let g be C^1 on $[a, \infty)$. Suppose that (i) the function $F(x) = \int_a^x f(t)\,dt$ remains bounded as $x \to \infty$; (ii) $g'(x) \leq 0$ on $[a, \infty)$ and $\lim_{x\to\infty} g(x) = 0$. Show that $\int_a^\infty f(x)g(x)\,dx$ converges. (*Hint:* Example 3 is the case $f(x) = \sin x$, $g(x) = x^{-1}$. Generalize the argument given there.)

10. Evaluate $P.V. \int_{-1}^1 dx/x(x+2)$.

11. Suppose φ is of class C^3 on $[-1, 1]$. Show that $P.V. \int_{-1}^1 x^{-3}\varphi(x)\,dx$ exists if and only if $\varphi'(0) = 0$. (*Hint:* Consider the second-order Taylor expansion of φ.)

4.7   Improper Multiple Integrals

The problem of defining improper integrals in dimensions $n > 1$ is trickier than in dimension 1. Suppose, for example, that f is a continuous function on $\mathbb{R}^2$ and we wish to define $\iint_{\mathbb{R}^2} f\,dA$. The obvious idea is to set

$$\iint_{\mathbb{R}^2} f\,dA = \lim_{r\to\infty} \iint_{S_r} f\,dA,$$

where the S_r's are a family of measurable sets that fill out $\mathbb{R}^2$ as $r \to \infty$. For instance, we could take S_r to be the disc of radius r about the origin, or the square of side length r centered at the origin, or the rectangle of side lengths r and r^2 centered at the origin, or the disc of radius r centered at $(15, -37)$, and so on. The difficulty is evident: There is a bewildering array of possibilities, with no rationale for choosing one over another *and no guarantee that different families S_r will yield the same limit.*

    Evidently there is some work to be done, and we shall not give all the details here. The outcome, in a nutshell, is that everything goes well when the integrand is nonnegative or when the integral is absolutely convergent, but not otherwise.

We begin by considering the situation where a *nonnegative* function f is to be integrated over a set $S \subset \mathbb{R}^n$. We suppose that f is not integrable on S according to the definitions in §4.2, either because S is unbounded or because f is unbounded on S. Instead, we assume the following:

S is the union of an increasing sequence of sets $U_1, U_2, \ldots$,

(4.63)
$$S = \bigcup_1^\infty U_j \qquad (U_1 \subset U_2 \subset U_3 \subset \cdots),$$

where each U_j is measurable and f is integrable on each U_j.

EXAMPLE 1. If $S = \mathbb{R}^n$ and f is continuous on $\mathbb{R}^n$, we can take U_j to be the ball of radius j about the origin. As noted above, there are many other possibilities.

EXAMPLE 2. Suppose f is continuous on $\mathbb{R}^n \setminus \{\mathbf{0}\}$ but $f(\mathbf{x}) \to \infty$ as $\mathbf{x} \to \mathbf{0}$, and S is the ball $\{\mathbf{x} : |\mathbf{x}| \leq 1\}$. Then we can take U_j to be the spherical shell $\{\mathbf{x} : 1/j \leq |\mathbf{x}| \leq 1\}$. (Strictly speaking, the union of the U_j's is $S \setminus \{\mathbf{0}\}$, but this is immaterial: Omission of a single point, or any set of zero content, from a domain has no effect on integration over that domain.)

With S, f, and U_j as in (4.63), the integrals $\int \cdots \int_{U_j} f \, dV^n$ exist for all j, and they increase along with j since the sets U_j do. It therefore follows from the monotone sequence theorem that the limit

$$\lim_{j \to \infty} \int \cdots \int_{U_j} f \, dV^n$$

always exists, provided that we allow $+\infty$ as a value, and this limit is an obvious candidate for the value of the improper integral $\int \cdots \int_S f \, dV^n$.

Here is the crucial point: Suppose that $\{\widetilde{U}_j\}$ is *another* sequence of sets satisfying the conditions of (4.63). *Then the two limits*

$$\lim_{j \to \infty} \int \cdots \int_{U_j} f \, dV^n \quad and \quad \lim_{j \to \infty} \int \cdots \int_{\widetilde{U}_j} f \, dV^n$$

are equal. Therefore, it makes sense to define to define the integral of f over S by

(4.64)
$$\int \cdots \int_S f \, dV^n = \lim_{j \to \infty} \int \cdots \int_{U_j} f \, dV^n,$$

where $\{U_j\}$ is *any* sequence of sets satisfying the conditions of (4.63). It is understood that the value of the integral may be $+\infty$, in which case we say that the integral diverges.

The proof that the limit in (4.64) is independent of the choice of $\{U_j\}$, in full generality, requires the Lebesgue theory of integration. We shall give a proof under some additional restrictions on S and the U_j's, usually easy to satisfy in practice, in Appendix B.6 (Theorem B.25).

It is also true that improper multiple integrals of nonnegative functions can be evaluated as iterated improper integrals under suitable conditions on S and f so that the latter integrals exist. For example,

$$\iint_{\mathbb{R}^2} f \, dA = \int_{-\infty}^{\infty} \int_{-\infty}^{\infty} f(x,y) \, dx \, dy = \int_{-\infty}^{\infty} \int_{-\infty}^{\infty} f(x,y) \, dy \, dx,$$

and if $S = \{(x,y) : 0 \le x \le y\}$,

$$\iint_S f \, dA = \int_0^\infty \int_0^y f(x,y) \, dx \, dy = \int_0^\infty \int_x^\infty f(x,y) \, dy \, dx.$$

We shall not attempt to state a general theorem to cover all the various cases (much less give a precise proof), but we assure the reader that as long as the integrand is nonnegative, there is almost *never* any difficulty.

The analogue of the comparison test, Theorem 4.55, is valid for multiple improper integrals, with the same proof. Again the basic comparison functions are powers of $|\mathbf{x}|$, but the critical exponent depends on the dimension.

4.65 Proposition. *For $p > 0$, define f_p on $\mathbb{R}^n \setminus \{\mathbf{0}\}$ by $f_p(\mathbf{x}) = |\mathbf{x}|^{-p}$. The integral of f_p over a ball $\{\mathbf{x} : |\mathbf{x}| < a\}$ is finite if and only if $p < n$; the integral of f_p over the complement of a ball, $\{\mathbf{x} : |\mathbf{x}| > a\}$, is finite if and only if $p > n$.*

Proof. We present the proof when $n = 2$. The only singularity of f is at the origin, so we may use the annuli $\{\mathbf{x} : \epsilon < |\mathbf{x}| < a\}$ and $\{\mathbf{x} : a < |\mathbf{x}| < b\}$ as approximating regions. In polar coordinates, the integrals then become

$$\int_\epsilon^a \int_0^{2\pi} r^{-p} r \, d\theta \, dr, \qquad \int_a^b \int_0^{2\pi} r^{-p} r \, d\theta \, dr.$$

As $\epsilon \to 0$ and $b \to \infty$ we obtain $2\pi \int_0^a r^{1-p} \, dr$ and $2\pi \int_a^\infty r^{1-p} \, dr$, which are convergent when $p < 2$ and $p > 2$, respectively.

The proof for general n is similar, using spherical coordinates and their analogues in higher dimensions. The reader is invited to work out the case $n = 3$ in Exercise 1. $\square$

As another example of improper double integrals, we now perform a classic calculation that leads to one of the most important formulas in mathematics.

Let us consider the integral

$$\iint_{\mathbb{R}^2} e^{-x^2 - y^2}\, dA.$$

On the one hand, we can take the approximating regions U_j to be discs centered at the origin and switch to polar coordinates:

$$\iint_{\mathbb{R}^2} e^{-x^2-y^2}\, dA = \lim_{R \to \infty} \int_0^R \int_0^{2\pi} e^{-r^2} r\, d\theta\, dr = \int_0^\infty \int_0^{2\pi} e^{-r^2} r\, d\theta\, dr$$

$$= 2\pi \left[-\tfrac{1}{2} e^{-r^2} \right]_0^\infty = \pi.$$

On the other hand, we can take the approximating regions to be squares centered at the origin and stick to Cartesian coordinates:

$$\iint_{\mathbb{R}^2} e^{-x^2-y^2}\, dA = \lim_{R \to \infty} \int_{-R}^R \int_{-R}^R e^{-x^2} e^{-y^2}\, dx\, dy$$

$$= \left(\int_{-\infty}^\infty e^{-x^2}\, dx \right) \left(\int_{-\infty}^\infty e^{-y^2}\, dy \right).$$

The two integrals in parentheses are equal, of course; the name of the variable of integration is irrelevant. We have shown that

$$\left(\int_{-\infty}^\infty e^{-x^2}\, dx \right)^2 = \pi.$$

Since $e^{-x^2} > 0$, we can take the positive square root of both sides to obtain the magic formula:

4.66 Proposition. $\displaystyle \int_{-\infty}^\infty e^{-x^2}\, dx = \sqrt{\pi}.$

The function e^{-x^2} turns up in many contexts. In particular, it is essentially the "bell curve" or "normal distribution" of probability and statistics, but in that setting one must rescale it so that the total area under its graph is 1; Proposition 4.66 provides the appropriate scaling factor. Proposition 4.66 is remarkable not only because it is inaccessible by elementary calculus (the antiderivative of e^{-x^2} is not an elementary function) but because it presents the number π in a starring role that has nothing to do with circles.

Now, what about functions that are not nonnegative? Let us suppose that S, f, and $\{U_j\}$ are as in (4.63), but f is merely assumed to be real-valued. The essential point is that the preceding theory can be applied to $|f|$, so that it makes sense to say that $\int\cdots\int_S |f|\,dV^n$ converges. If this condition holds, the argument used to prove Theorem 4.58 shows that $\lim_{j\to\infty}\int\cdots\int_{U_j} f\,dV^n$ exists and that

$$\lim_{j\to\infty} \int\cdots\int_{U_j} f\,dV^n = \int\cdots\int_S f^+\,dV^n - \int\cdots\int_S f^-\,dV^n,$$

where $f^+(\mathbf{x}) = \max[f(\mathbf{x}),0]$ and $f^-(\mathbf{x}) = \max[-f(\mathbf{x}),0]$. The integrals on the right converge by comparison to the integral of $|f|$, and they are independent of the choice of $\{U_j\}$; hence, so is the limit on the left. In short, if $\int\cdots\int_S |f|\,dV^n$ converges, we may define the improper integral of f over S by formula (4.64); the limit in question exists and is independent of the choice of approximating sequence $\{U_j\}$.

The same result holds if f is complex-valued; we simply consider its real and imaginary parts separately.

In dimensions $n > 1$, however, there is no general theory of improper integrals that are convergent but not absolutely convergent. Such integrals, when they arise, must be defined by specific limiting procedures that are adapted to the situation at hand.

EXERCISES

1. Prove Proposition 4.65 for the case $n = 3$.

2. Determine whether the following improper integrals converge, and evaluate the ones that do.

   a. $\displaystyle\iiint_{\mathbb{R}^3} \frac{dV}{1 + x^2 + y^2 + z^2}.$

   b. $\displaystyle\iint_{x,y>0} \frac{dA}{(1 + x^2 + y^2)^2}.$

   c. $\displaystyle\iiint_{x^2+y^2+z^2<1} \frac{z^2}{(x^2 + y^2 + z^2)^{3/2}}\,dV.$

   d. $\displaystyle\iint_{x>0} x e^{-x^2-y^2}\,dA.$

   e. $\displaystyle\iint_{x^2+y^2<1} \frac{x^2}{(x^2 + y^2)^2}\,dA.$

3. The electrostatic potential generated by a distribution of electric charge in $\mathbb{R}^3$ with density ρ is defined to be

$$\varphi(\mathbf{x}) = \iiint_{\mathbb{R}^3} \frac{\rho(\mathbf{x} - \mathbf{y})}{4\pi|\mathbf{y}|} \, d^3\mathbf{y}.$$

Show that this integral is absolutely convergent if ρ is continuous and vanishes outside a bounded set.

4. Let $f(x,y) = (x^2 - y^2)(x^2 + y^2)^{-2}$, and let S be the unit square $[0,1] \times [0,1]$.

   a. Show that $\iint_S |f| \, dA = \infty$.

   b. Show by explicit calculation that the iterated integrals $\int_0^1 \int_0^1 f(x,y) \, dx \, dy$ and $\int_0^1 \int_0^1 f(x,y) \, dy \, dx$ both exist and are unequal.

4.8 Lebesgue Measure and the Lebesgue Integral

In several places in this book we allude to the fact that in advanced analysis, the Riemann theory of integration that we have developed here is replaced by the more sophisticated theory due to Lebesgue. Detailed accounts of the Lebesgue integral can be found in Bear [3], Jones [9], and Rudin [18]. Here we shall content ourselves with a brief informal description of how it works. (*Note:* There are several ways to develop the Lebesgue theory of integration; in some treatments, the characterization of Lebesgue measure and the Lebesgue integral that we give here are theorems rather than definitions.) In a few places we need the notion of the sum of an infinite series, for which the reader is referred to §6.1.

The starting point is a refined concept of n-dimensional measure, independent of any theory of integration. To keep things on a concrete level, let us explain this concept for the case $n = 2$.

In the Jordan theory of area, described in §4.2, we find the area of a set $S \subset \mathbb{R}^2$ by approximating S from the inside and the outside by unions of rectangles. For the Lebesgue notion of area, we use a two-step approximation process: We first approximate S from the inside by compact sets and from the outside by open sets, then approximate the compact sets from the outside and the open sets from the inside by unions of rectangles. More precisely, let us agree to call a set that is the union of a finite collection of rectangles with disjoint interiors a *tiled set*. The Lebesgue measure $m(S)$ of a set $S \subset \mathbb{R}^2$ is then defined as follows:

- If $T = \bigcup_{k=1}^K R_k$ is a tiled set, where the R_k's are rectangles with disjoint interiors, the Lebesgue measure $m(T)$ is the sum of the areas of the R_k's.

- The Lebesgue measure of a compact set K is

$$m(K) = \inf\{m(T) : T \text{ is a tiled set and } T \supset K\}.$$

- The Lebesgue measure of an open set U is

$$m(U) = \sup\{m(T) : T \text{ is a tiled set and } T \subset U\}.$$

- A set $S \subset \mathbb{R}^2$ is said to be **Lebesgue measurable** if the quantities

$$\sup\{m(K) : K \text{ is compact and } K \subset S\}$$
$$\text{and}$$
$$\inf\{m(U) : U \text{ is open and } U \supset S\}$$

are equal, in which case their common value is the **Lebesgue measure** $m(S)$.

Note that there is no assumption that the sets in question are bounded (although compact sets are bounded by definition); the Lebesgue theory applies equally well to bounded and unbounded sets.

The notion of n-dimensional Lebesgue measure for sets in $\mathbb{R}^n$ is entirely similar; only the terminology needs to be modified a little. Every set that one will ever meet in "real life" — in particular, every open set, every closed set, every intersection of countably many open sets, every union of countably many closed sets, and so on — is Lebesgue measurable.[3] Lebesgue measure has the following fundamental additivity property: *If $\{S_j\}$ is a finite or infinite sequence of disjoint Lebesgue measurable sets, then $\bigcup S_j$ is Lebesgue measurable and $m(\bigcup S_j) = \sum m(S_j)$.* In the Jordan theory, this additivity is guaranteed to hold only for *finitely* many sets; the extension to infinitely many sets is the crucial property that allows the Lebesgue theory to handle various limiting processes more smoothly.

It is not hard to show that every open set $U \subset \mathbb{R}^n$ is the union of a finite or countably infinite family of rectangular boxes R_j (intervals when $n = 1$) with disjoint interiors, and the Lebesgue measure of U is just the sum of the n-dimensional volumes of the boxes. (In general these boxes are not part of a fixed grid of boxes; if there are infinitely many of them, the diameter of R_j generally tends to zero as $j \to \infty$.) It follows that *a set $S \subset \mathbb{R}^n$ has Lebesgue measure zero if and only if for every $\epsilon > 0$, S is contained in the union of a finite or countable family of boxes, the sum of whose volumes is less than ϵ.* The only difference between this and the condition that S have zero content is the fact that here we allow an *infinite* family

---

[3]For those who know some set theory: More precisely, one cannot construct Lebesgue nonmeasurable sets without invoking the axiom of choice.

of boxes, but as with additivity, this difference is significant. In particular, every countable set has Lebesgue measure zero (if $S = \{x_1, x_2, \ldots\}$, let R_j be a box centered at x_j with volume $2^{-j}\epsilon$), whereas many countable sets — the set of points with rational coordinates, for example — are not Jordan measurable.

With the notion of Lebesgue measure in hand, we turn to the Lebesgue integral. First we specify the class of functions to which the theory applies. A function $f : \mathbb{R}^n \to \mathbb{R}$ is called **Lebesgue measurable** if, for every interval $I \subset \mathbb{R}$, the set $\{x \in \mathbb{R}^n : f(x) \in I\}$ is Lebesgue measurable. Again, every function that one will ever meet in "real life" is Lebesgue measurable. In particular, every continuous function is Lebesgue measurable, and if f is Riemann integrable on the set S, then $f\chi_S$ is Lebesgue measurable. Moreover, if $\{f_j\}$ is a sequence of Lebesgue measurable functions such that $f_j(x) \to f(x)$ for every x, then the limit f is Lebesgue measurable. (This last statement is quite false if "Lebesgue measurable" is replaced by "Riemann integrable"!)

Suppose that f is Lebesgue measurable and nonnegative. Rather than partitioning the domain of f, we partition the set $[0, \infty)$ in which f takes its values into small intervals $[0, 2^{-n}), [2^{-n}, 2 \cdot 2^{-n}), [2 \cdot 2^{-n}, 3 \cdot 2^{-n})$, and so on, and form the sum

$$S_n f = \sum_{j=0}^{\infty} \frac{j}{2^n} m\left(\left\{x : \frac{j}{2^n} \leq f(x) < \frac{j+1}{2^n}\right\}\right).$$

(The Lebesgue measurability of f is needed so that the terms in this sum are well defined. One or more of them may be infinite, in which case the value of the sum is $+\infty$.) The sums $S_n f$ increase with n because the associated partitions of $[0, \infty)$ become finer and finer, so they have a limit (possibly $+\infty$), which is defined to be the Lebesgue integral of f (over $\mathbb{R}^n$), denoted by $\int f \, dm$. More generally, we define the Lebesgue integral of f over any Lebesgue measurable set $S \subset \mathbb{R}^n$, denoted by $\int_S f \, dm$, to be $\int (f\chi_S) \, dm$. Note that neither the function f nor the set S needs to be bounded; for nonnegative integrands there are no "improper" integrals in the Lebesgue theory.

Now we drop the assumption that $f \geq 0$. If f is any Lebesgue measurable function, we write it as the difference of the two nonnegative functions

$$f^+(x) = \max[f(x), 0] \quad \text{and} \quad f^-(x) = \max[-f(x), 0]$$

and define the Lebesgue integral $\int f \, dm$ to be $\int f^+ \, dm - \int f^- \, dm$. The integral $\int f \, dm$ is not defined in the case where $\int f^+ \, dm$ and $\int f^- \, dm$ are both infinite, although in some instances one can define it as an "improper" integral by limiting procedures such as those in §4.6. (Example 3 in §4.6 illustrates this phenomenon.)

The Lebesgue integral is an extension of the Riemann integral. That is, if the (proper) Riemann integral $\int_S f \, dV^n$ exists, then so does the Lebesgue integral

$\int_S f \, dm$, and the two are equal; but the class of Lebesgue integrable functions is much bigger than the class of Riemann integrable functions. We conclude with two additional remarks about the relation between the Lebesgue and Riemann integrals.

- The notion of Lebesgue measure provides a definitive answer to the question of which functions are Riemann integrable. Namely, *a function $f : \mathbb{R}^n \to \mathbb{R}$ is Riemann integrable on the bounded set S if and only if f is bounded on S and the set of points at which $f\chi_S$ is discontinuous has Lebesgue measure zero.* (Cf. Theorems 4.13 and 4.18 and the discussion of zero content versus zero measure above.)

- There is a way of giving the Riemann theory of integration an extra twist to obtain an integral, called the *Henstock-Kurzweil integral, generalized Riemann integral*, or *gauge integral*, that is equivalent to the Lebesgue integral for nonnegative functions but also gives a well-defined result for some functions f for which $\int f^+ \, dm$ and $\int f^- \, dm$ are both infinite. See Bartle [2] for a brief introduction and DePree and Swartz [5] for a complete treatment. The virtue of this theory is that it yields a powerful theory of integration within the same conceptual framework as the familiar Riemann integral without the necessity of developing a theory of measure first. The compensating virtue of the Lebesgue theory is that it generalizes readily to yield useful notions of measure and integration in many important situations other than the classical integral on Euclidean space.

Chapter 5

LINE AND SURFACE INTEGRALS; VECTOR ANALYSIS

The themes of this chapter are (1) integrals over curves and surfaces and (2) differential operations on vector fields, which combine to yield (3) a group of theorems relating integrals over curves, surfaces, and regions in space that are among the most powerful and useful results of advanced calculus.

At the outset, let us explain the term "vector field" in more detail. Let $\mathbf{F}$ be an $\mathbb{R}^n$-valued function defined on some subset of $\mathbb{R}^n$. We have encountered such things in previous chapters, where we generally thought of them as representing transformations from one region of $\mathbb{R}^n$ to another or coordinate systems on regions of $\mathbb{R}^n$. In this chapter, however, we think of such an $\mathbf{F}$ as a function that assigns to each point $\mathbf{x}$ in its domain a vector $\mathbf{F}(\mathbf{x})$, represented pictorially as an arrow based at $\mathbf{x}$, and we therefore call it a **vector field**. Two simple vector fields are sketched in Figure 5.1. The primary physical motivation is the idea of a force field. For example, $\mathbf{F}$ could represent a gravitational field, $\mathbf{F}(\mathbf{x})$ being the gravitational force felt by a unit mass located at $\mathbf{x}$, or an electric field, $\mathbf{F}(\mathbf{x})$ being the electrostatic force felt by a unit charge located at $\mathbf{x}$. There are many other physical interpretations; for example, in a moving fluid like a stream of water, $\mathbf{F}(\mathbf{x})$ could represent the velocity of the fluid at position $\mathbf{x}$. (In all these examples, $\mathbf{F}(\mathbf{x})$ may also depend on other parameters such as the time t.)

One other general comment: The notion of differentiability, or being of class C^k, is defined for functions on *open* sets, because to compute the derivative of a function at a point it is necessary to know the values of the function at neighboring points. However, we shall frequently be dealing with functions and vector fields on *closed* sets. *When we say that a function or vector field is of class C^k on a closed*

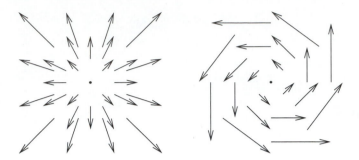

FIGURE 5.1: The vector fields $\mathbf{F}(x, y) = (x, y)$ (left) and $\mathbf{F}(x, y) = (-y, x)$ (right).

set $S \subset \mathbb{R}^n$, we always mean that it is of class C^k on some open set containing S.

5.1 Arc Length and Line Integrals

In this section we discuss integrals over curves, traditionally called "line integrals," which are generalizations of ordinary (one-dimensional) integrals over intervals on the real line. As one would expect, they are based on the idea of cutting up the curve into many tiny pieces, forming appropriate Riemann sums, and passing to the limit. However, there are two species of line integrals, appropriate for integrating real-valued or vector-valued functions, depending on how one adapts the differential dx appearing in $\int_a^b f(x)\,dx$ to the more general situation. Our discussion here will be on the informal, intuitive level where we think of dx as being an infinitesimal increment in the variable x.

Differentials on Curves; Arc Length. Suppose C is a smooth curve in $\mathbb{R}^n$. We consider two nearby points $\mathbf{x}$ and $\mathbf{x} + d\mathbf{x}$ on the curve; here

$$(5.1) \qquad\qquad d\mathbf{x} = (dx_1, \ldots, dx_n)$$

is the *vector difference* between the two points, and we imagine it as being infinitely small. We may, however, be more interested in the *distance* between the two points, traditionally denoted by ds, which is

$$(5.2) \qquad\qquad ds = |d\mathbf{x}| = \sqrt{dx_1^2 + \cdots + dx_n^2}.$$

To give these differentials a precise meaning that can be used for calculations, the best procedure is to parametrize the curve. Thus, we assume that C is given by parametric equations $\mathbf{x} = \mathbf{g}(t)$, $a \le t \le b$, where $\mathbf{g}$ is of class C^1 and $\mathbf{g}'(t) \ne \mathbf{0}$. Then the neighboring points $\mathbf{x}$ and $\mathbf{x} + d\mathbf{x}$ are given by $\mathbf{g}(t)$ and $\mathbf{g}(t + dt)$, so

$$(5.3) \qquad d\mathbf{x} = \mathbf{g}(t + dt) - \mathbf{g}(t) = \mathbf{g}'(t) \, dt = \left(\frac{dx_1}{dt}, \dots, \frac{dx_n}{dt} \right) dt.$$

(The difference between the increment of $\mathbf{g}$ and its linear approximation disappears in the infinitesimal limit.) Moreover,

$$(5.4) \qquad |d\mathbf{x}| = |\mathbf{g}'(t)| \, dt = \sqrt{\left(\frac{dx_1}{dt} \right)^2 + \dots + \left(\frac{dx_n}{dt} \right)^2} \, dt,$$

which is just what one gets by formally multiplying and dividing the expression on the right of (5.2) by dt.

What happens if we sum up all the infinitesimal increments $d\mathbf{x}$ or ds — that is, if we integrate the differentials $d\mathbf{x}$ or $ds = |d\mathbf{x}|$ over the curve? Integration of the vector increments $d\mathbf{x}$ just gives the total vector increment, that is, the vector difference between the initial and final points on the curve:

$$(5.5) \qquad \int_C d\mathbf{x} = \int_a^b \mathbf{g}'(t) \, dt = \mathbf{g}(b) - \mathbf{g}(a).$$

This is nothing but the fundamental theorem of calculus applied to the components of $\mathbf{g}$; it is simple but not very exciting. On the other hand, ds is the straight-line distance between two infinitesimally close points $\mathbf{x}$ and $\mathbf{x} + d\mathbf{x}$ on the curve, and since smooth curves are indistinguishable from their linear approximations on the infinitesimal level, ds is the *arc length* of the bit of curve between $d\mathbf{x}$ and $\mathbf{x} + d\mathbf{x}$. Adding these up gives the total arc length of the curve:

$$(5.6) \qquad \text{Arc length} = \int_C ds = \int_a^b |\mathbf{g}'(t)| \, dt.$$

Our derivation of (5.6) in terms of infinitesimals was meant as motivation rather than as a rigorous proof of anything. Henceforth, we shall take (5.6) as a *definition* of **arc length** for a smooth curve. (There is another, perhaps better, definition that does not require the curve to be C^1; we shall discuss it at the end of this section.) There is, however, one crucial issue that must be addressed: The arc length of a curve C is an intrinsic property of the geometric object C and should not depend on the particular parametrization we use. To see that this is the case, suppose we choose a new parameter u related to t by $t = \varphi(u)$, where φ is a one-to-one smooth

FIGURE 5.2: Two oriented curves.

mapping from the interval $[c, d]$ to the interval $[a, b]$. Then the curve C described by $\mathbf{x} = \mathbf{g}(t)$ is also described by $\mathbf{x} = (\mathbf{g} \circ \varphi)(u)$, $c \le u \le d$, so we should have

$$\text{Arc length} = \int_c^d |(\mathbf{g} \circ \varphi)'(u)| \, du = \int_c^d |\mathbf{g}'(\varphi(u))| \, |\varphi'(u)| \, du,$$

where for the second equality we have used the chain rule. This does indeed agree with (5.6), by formula (4.34).

The same independence of parametrization holds for the related integral (5.5), with one subtle but important difference. The integral $\int_a^b \mathbf{g}'(t) \, dt$ gives the vector difference between the two endpoints of the curve, which is clearly independent of the parametrization *except* insofar as the parametrization determines which is the initial point and which is the final point. If we choose a new parameter u as above so that t is a *decreasing* function of u (thus $a = \varphi(d)$ and $b = \varphi(c)$), then the initial and final points get switched, and so their difference is multiplied by -1.

The issue here is that a parametrization $\mathbf{x} = \mathbf{g}(t)$ determines an **orientation** for the curve C, that is, a determination of which direction along the curve is "forward" and which direction is "backward," the "forward" direction being the direction in which the point $\mathbf{g}(t)$ moves as t increases. The orientation of a curve can be conveniently indicated in a picture by drawing one or more arrowheads along the curve that point in the "forward" direction, as indicated in Figure 5.2. The substance of the preceding paragraph is then that *the integral (5.5) depends on the parametrization only insofar as the parametrization determines a choice of orientation.* In contrast, the arc length of a curve is independent even of the orientation.

The notion of arc length extends in an obvious way to **piecewise smooth** curves, obtained by joining finitely many smooth curves together end-to-end but allowing corners or cusps at the joining points; we simply compute the lengths of the smooth pieces and add them up. We can express this more precisely in terms of parametrizations, as follows: The function $\mathbf{g} : [a, b] \to \mathbb{R}^n$ is called **piecewise smooth** if (i) it is continuous, and (ii) its derivative exists and is continuous except perhaps at finitely many points t_j, at which the one-sided limits $\lim_{t \to t_j \pm} \mathbf{g}'(t)$ exist. (*Note.* In Chapter 8 we shall use the term "piecewise smooth" in a slightly different sense.) In this case $|\mathbf{g}'(t)|$ is an integrable function on $[a, b]$ by Theorem 4.12 (the fact that it may be undefined at a few points is immaterial), and its integral gives the arc length. *The same generalization also applies to the line integrals discussed below.*

Remarks.

i. The parametrization $\mathbf{x} = \mathbf{g}(t)$ may be considered as representing the curve C as the path traced out by a moving particle whose position at time t is $\mathbf{g}(t)$. The derivative $\mathbf{g}'(t)$ is then the velocity of the particle, and its norm $|\mathbf{g}'(t)|$ is the speed of the particle. Integrating the velocity, $\int_a^b \mathbf{g}'(t)\, dt$, gives the net difference in the initial and final positions of the particle, whereas integrating the speed, $\int_a^b |\mathbf{g}'(t)|\, dt$, gives the total distance traveled by the particle, i.e., the arc length of the curve.

ii. In the preceding discussion, we have implicitly assumed that the parametrization $\mathbf{x} = \mathbf{g}(t)$ is one-to-one. This is not always the case if we think of $\mathbf{g}(t)$ as the position of a particle at time t, for the particle can traverse a path more than once. For example, $\mathbf{g}(t) = (\cos t, \sin t)$ represents a particle moving around the unit circle with constant speed. If we restrict t to an interval of length $\leq 2\pi$, we get a one-to-one parametrization of part or all of the circle, but from the physical point of view there is no reason to make such a restriction. However, the interpretations in the preceding paragraph hold whether $\mathbf{g}$ is one-to-one or not: $\int_a^b \mathbf{g}'(t)\, dt$ is still $\mathbf{g}(b) - \mathbf{g}(a)$, and $\int_a^b |\mathbf{g}'(t)|\, dt$ is still the total distance traveled by the particle from time a to time b; it can be interpreted as arc length if the portions of the curve that are traversed more than once are counted with the appropriate multiplicity.

iii. While theoretically simple, calculation of arc length tends to be difficult in practice because the square root implicit in the definition of the norm $|\mathbf{g}'(t)|$ often leads to unpleasant integrands. This is just a fact of life.

Line Integrals of Scalar Functions. If f is a continuous function whose domain includes a smooth (or piecewise smooth) curve C in $\mathbb{R}^n$, we can integrate f over the curve, taking the differential in the integral to be the element of arc length ds. Thus, if C is parametrized by $\mathbf{x} = \mathbf{g}(t)$, $a \leq t \leq b$, we define

(5.7)
$$\int_C f\, ds = \int_a^b f(\mathbf{g}(t))|\mathbf{g}'(t)|\, dt.$$

This is independent of the parametrization and the orientation, by the same chain-rule calculation that we performed above for the case $f \equiv 1$.

As an example of an application of such integrals, we can define the average value of f over the curve C, just like the average value over a region:

$$\text{Average of } f \text{ over } C = \frac{\int_C f \, ds}{\text{Arc length of } C} = \frac{\int_C f \, ds}{\int_C ds}.$$

EXAMPLE 1. What is the centroid of the upper half of the unit circle, $C = \{(x, y) : x^2 + y^2 = 1, \, y \geq 0\}$?

Solution. The centroid of C is the point whose coordinates $(\overline{x}, \overline{y})$ are the averages of x and y over C. Clearly $\overline{x} = 0$ by symmetry. Just to get some practice, let's do the calculation of the arc length of C (which of course is π) and $\int_C y \, ds$ with two different parametrizations: (i) taking x as the parameter and $y = \sqrt{1 - x^2}$, and (ii) taking the polar angle θ as the parameter, $x = \cos\theta$, $y = \sin\theta$. (Note that these two parametrizations give opposite orientations on C; the first goes from left to right, the second from right to left.)

In the first parametrization, we have

$$dy = \frac{-x \, dx}{\sqrt{1 - x^2}}, \qquad ds = \sqrt{dx^2 + dy^2} = \sqrt{1 + \frac{x^2}{1 - x^2}} \, dx = \frac{dx}{\sqrt{1 - x^2}},$$

so

$$\int_C y \, ds = \int_{-1}^{1} \frac{\sqrt{1 - x^2}}{\sqrt{1 - x^2}} \, dx = x \Big|_{-1}^{1} = 2,$$

$$\int_C ds = \int_{-1}^{1} \frac{dx}{\sqrt{1 - x^2}} = \arcsin x \Big|_{-1}^{1} = \pi.$$

In the second one, we have

$$dx = -\sin\theta \, d\theta, \quad dy = \cos\theta \, d\theta; \qquad ds = \sqrt{dx^2 + dy^2} = d\theta,$$

so

$$\int_C y \, ds = \int_0^\pi \sin\theta \, d\theta = -\cos\theta \Big|_0^\pi = 2, \qquad \int_C ds = \int_0^\pi d\theta = \pi.$$

Either way, $\overline{y} = 2/\pi$.

Line Integrals of Vector Fields. We can define the integral of an $\mathbb{R}^m$-valued function over a curve in $\mathbb{R}^n$, simply by integrating each component separately; that is, if $\mathbf{F} = (F_1, \ldots, F_m)$, then $\int_C \mathbf{F} \, ds = (\int_C F_1 \, ds, \ldots, \int_C F_m \, ds)$. There is not much to be said about such integrals that does not follow immediately from the facts about scalar-valued integrals. One significant fact, however, does require a little extra proof, namely the analogue of Theorem 4.9d. We state it for ordinary integrals over $[a, b]$; the generalization to integrals over curves is easy (Exercise 7a).

5.8 Proposition. *If* $\mathbf{F}$ *is a continuous* $\mathbb{R}^m$*-valued function on* $[a, b]$*, then*

$$\left| \int_a^b \mathbf{F}(t)\, dt \right| \leq \int_a^b |\mathbf{F}(t)|\, dt.$$

Proof. For any unit vector $\mathbf{u}$, we have

$$\left| \left(\int_a^b \mathbf{F}(t)\, dt \right) \cdot \mathbf{u} \right| = \left| \int_a^b \mathbf{F}(t) \cdot \mathbf{u}\, dt \right| \leq \int_a^b |\mathbf{F}(t) \cdot \mathbf{u}|\, dt \leq \int_a^b |\mathbf{F}(t)|\, dt.$$

Here we have applied Theorem 4.9d to the scalar-valued function $\mathbf{F}(t) \cdot \mathbf{u}$ and then invoked Cauchy's inequality. The desired result is obtained by taking $\mathbf{u}$ to be the unit vector in the direction of $\int_a^b \mathbf{F}(t)\, dt$. $\qquad\square$

Of greater interest is a *scalar-valued* line integral for vector fields — that is, for $\mathbb{R}^n$-valued functions on $\mathbb{R}^n$. If C is a smooth (or piecewise smooth) curve in $\mathbb{R}^n$ and $\mathbf{F}$ is a continuous vector field defined on some neighborhood of C in $\mathbb{R}^n$, the **line integral** of $\mathbf{F}$ over C is

$$\int_C \mathbf{F} \cdot d\mathbf{x} = \int_C (F_1\, dx_1 + F_2\, dx_2 + \cdots + F_n\, dx_n).$$

That is, if C is described parametrically by $\mathbf{x} = \mathbf{g}(t)$, $a \leq t \leq b$, then

$$(5.9) \qquad \int_C \mathbf{F} \cdot d\mathbf{x} = \int_a^b \mathbf{F}(\mathbf{g}(t)) \cdot \mathbf{g}'(t)\, dt.$$

If we make a change of parameters, say $t = \varphi(u)$, the chain rule $\mathbf{g}'(t)\, dt = \mathbf{g}'(\varphi(u))\varphi'(u)\, du$ together with the change-of-variable formula for ordinary (single) integrals guarantees that the quantity on the right of (5.9) is unchanged, except that the new endpoints of integration may end up in the wrong order. Therefore:

The line integral $\int_C \mathbf{F} \cdot d\mathbf{x}$ *is independent of the parametrization as long as the orientation is unchanged, but it acquires a factor of* -1 *if the orientation is reversed. That is,* $\int_C \mathbf{F} \cdot d\mathbf{x}$ *is a well-defined quantity once the vector field* $\mathbf{F}$ *and the* oriented *curve* C *are specified.*

The line integral $\int_C \mathbf{F} \cdot d\mathbf{x}$ can be expressed as an integral of a scalar function over C. Indeed, let us choose a parametrization $\mathbf{x} = \mathbf{g}(t)$ and set

$$\mathbf{t}(\mathbf{g}(t)) = \frac{\mathbf{g}'(t)}{|\mathbf{g}'(t)|}, \qquad F_{\text{tang}}(\mathbf{x}) = \mathbf{F}(\mathbf{x}) \cdot \mathbf{t}(\mathbf{x}).$$

That is, $\mathbf{t}(\mathbf{x})$ is the unit tangent vector to the curve C in the *forward* direction at the point $\mathbf{x}$, and $F_{\text{tang}}(\mathbf{x})$ is the component of $\mathbf{F}(\mathbf{x})$ in the direction of $\mathbf{t}(\mathbf{x})$. Then

$$\mathbf{F}(\mathbf{g}(t)) \cdot \mathbf{g}'(t) = \mathbf{F}(\mathbf{g}(t)) \cdot \mathbf{t}(\mathbf{g}(t))|\mathbf{g}'(t)|\, dt = F_{\text{tang}}(\mathbf{g}(t))\, ds,$$

so

(5.10)
$$\int_C \mathbf{F} \cdot d\mathbf{x} = \int_C F_{\text{tang}} \, ds.$$

That is, $\int_C \mathbf{F} \cdot d\mathbf{x}$ is the integral of the tangential component of $\mathbf{F}$ with respect to arc length. The dependence on the orientation here comes through F_{tang}, which changes sign if the orientiation is reversed. (Any temptation to compute specific line integrals by using (5.10), however, should probably be resisted, because the element of arc length ds is often hard to compute with. It is almost always better to use the basic definition (5.9) instead.)

Remarks.
i. If $\mathbf{F}$ is a force field, then $\int_C \mathbf{F} \cdot d\mathbf{x}$ represents a quantity of energy; it is the **work** done by the force on a particle that traverses the curve C.
ii. The integrand $\mathbf{F} \cdot d\mathbf{x} = F_1 \, dx_1 + \cdots + F_n \, dx_n$ in a line integral, with the dx's included, is often called a **differential form**, and we speak of integrating a differential form over a curve. We shall return to this notion in §5.9.

What does all this boil down to when $n = 1$? In this case, vector fields and scalar functions are the same thing, and both the scalar and vector versions of line integrals are just ordinary one-variable integrals. The former, however, is independent of orientation, whereas the latter depends on orientation. The distinction is the same as the one between formulas (4.32) and (4.33) in §4.4; it is a question of

$$\int_{[a,b]} f(x) \, dx \quad \text{versus} \quad \int_a^b f(x) \, dx.$$

In the integral on the left we must have $a \le b$; but in the integral on the right a and b can occur in either order, and the sign of the integral depends on the order.

EXAMPLE 2. Let C be the ellipse formed by the intersection of the circular cylinder $x^2 + y^2 = 1$ and the plane $z = 2y + 1$, oriented counterclockwise as viewed from above, and let $\mathbf{F}(x, y, z) = (y, z, x)$. Calculate $\int_C \mathbf{F} \cdot d\mathbf{x} = \int_C (y \, dx + z \, dy + x \, dz)$.

    Solution. We can parametrize C by $x = \cos t$, $y = \sin t$, $z = 2 \sin t + 1$, with $0 \le t \le 2\pi$. Then $d\mathbf{x} = (-\sin t, \cos t, 2 \cos t) \, dt$, so

$$\mathbf{F} \cdot d\mathbf{x} = \left(-\sin^2 t + (2 \sin t + 1) \cos t + 2 \cos^2 t\right) dt$$

$$= (\cos 2t + \sin 2t + \cos t + \cos^2 t) \, dt.$$

The integral of the first three terms over $[0, 2\pi]$ vanishes, and the integral of the last one is π. So $\int_C \mathbf{F} \cdot d\mathbf{x} = \pi$.

FIGURE 5.3: Approximation of a curve by a piecewise linear curve.

Note that it doesn't matter which point on C we choose to start and end at. Instead of taking $t \in [0, 2\pi]$, we could take $t \in [a, a + 2\pi]$ for any $a \in \mathbb{R}$; the answer is the same since the integral of a trig function over a complete period is independent of the particular period chosen.

Rectifiable Curves. There is an alternative definition of arc length that requires no a priori hypotheses about the smoothness of the curve. One cuts the curve C up into a finite number of pieces by inserting subdivision points and approximates C by the piecewise linear curve obtained by connecting the dots, as indicated in Figure 5.3. The length of the piecewise linear approximation is obtained by adding up the lengths of its constituent line segments, and the arc length of C is defined to be the limit of this sum as the subdivision is made finer and finer.

To make this more precise, it is convenient to describe C parametrically. Thus, we assume that C is the range of a one-to-one continuous mapping $\mathbf{g} : [a, b] \to \mathbb{R}^n$. Given a partition $P = \{t_0, \ldots, t_J\}$ of $[a, b]$, the sum of the lengths of the line segments joining the points $\mathbf{g}(t_j)$ is

$$L_P(C) = \sum_{1}^{J} |\mathbf{g}(t_j) - \mathbf{g}(t_{j-1})|.$$

If the set of numbers

$$\mathcal{L} = \{L_P(C) : P \text{ is a partition of } [a, b]\}$$

is bounded, then C is called **rectifiable**, and the **arc length** $L(C)$ is defined to be the supremum of $\mathcal{L}$:

$$L(C) = \sup\{L_P(C) : P \text{ is a partition of } [a, b]\}.$$

Note that if P' is a refinement of P then $L_{P'}(C) \geq L_P(C)$, by the triangle inequality; hence the supremum is indeed the appropriate sort of limit. This estimate also implies that the supremum is unchanged if we consider only partitions containing a given $c \in (a, b)$ among their subdivision points, and from this it follows that arc length is additive: If C_1 and C_2 are the curves parametrized by $\mathbf{g}(t)$ for $t \in [a, c]$ and $t \in [c, b]$, then $L(C) = L(C_1) + L(C_2)$. See Exercise 8.

We now show that this definition coincides with our previous one for C^1 curves.

5.11 Theorem. *With notation as above, if $\mathbf{g}$ is of class C^1, then C is rectifiable, and*

$$L(C) = \int_a^b |\mathbf{g}'(t)|\, dt.$$

Proof. For any partition P of $[a, b]$, by (5.5) and Proposition 5.8 we have

$$L_P(C) = \sum_1^J \left| \int_{t_{j-1}}^{t_j} \mathbf{g}'(t)\, dt \right| \leq \sum_1^J \int_{t_{j-1}}^{t_j} |\mathbf{g}'(t)|\, dt = \int_a^b |\mathbf{g}'(t)|\, dt.$$

It follows that $L(C) \leq \int_a^b |\mathbf{g}'(t)|\, dt$, and in particular that C is rectifiable.

Next, for $r, s \in [a, b]$, let C_r^s be the curve parametrized by $\mathbf{g}(t)$ with $t \in [r, s]$, and let $\varphi(s) = L(C_a^s)$. (That is, we consider the length of the curve C, starting at $t = a$, as a function of the right endpoint of the parameter interval.) Suppose $h > 0$. Since arc length is additive, we have $L(C_s^{s+h}) = \varphi(s + h) - \varphi(s)$, so by the inequality we have just proved (applied to the curve C_s^{s+h}) and the mean value theorem for integrals,

$$L(C_s^{s+h}) = \varphi(s + h) - \varphi(s) \leq \int_s^{s+h} |\mathbf{g}'(t)|\, dt = h|\mathbf{g}'(\sigma)|,$$

where σ is some number between s and $s+h$. On the other hand, $|\mathbf{g}(s+h) - \mathbf{g}(s)|$ is $L_P(C_s^{s+h})$ where P is the trivial partition $\{s,\ s + h\}$, and hence it is no bigger than $L(C_s^{s+h})$. Combining these estimates and dividing by h, we see that

$$\left| \frac{\mathbf{g}(s + h) - \mathbf{g}(s)}{h} \right| \leq \frac{\varphi(s + h) - \varphi(s)}{h} \leq |\mathbf{g}'(\sigma)|.$$

As $h \to 0$, the quantities on the left and right approach $|\mathbf{g}'(s)|$, and hence so does the one in the middle. A slight modification of this argument works also for $h < 0$, so we conclude that φ is differentiable and that $\varphi'(s) = |\mathbf{g}'(s)|$. The desired result is now immediate:

$$L(C) = \varphi(b) = \varphi(b) - \varphi(a) = \int_a^b |\mathbf{g}'(s)|\, ds.$$

$\square$

EXERCISES

1. Find the arc length of the following parametrized curves:
    a. $\mathbf{g}(t) = (a\cos t,\, a\sin t,\, bt),\, t \in [0, 2\pi]$.
    b. $\mathbf{g}(t) = (\frac{1}{3}t^3 - t,\, t^2),\, t \in [0, 2]$.
    c. $\mathbf{g}(t) = (\log t,\, 2t,\, t^2),\, t \in [1, e]$.
    d. $\mathbf{g}(t) = (6t,\, 4t^{3/2},\, -4t^{3/2},\, 3t^2),\, t \in [0, 2]$.

2. Express the arc length of the following curves in terms of the integral

$$E(k) = \int_0^{\pi/2} \sqrt{1 - k^2 \sin^2 t}\, dt \qquad (0 < k < 1),$$

   for suitable values of k. ($E(k)$ is one of the standard **elliptic integrals**, so called because of their connection with the arc length of an ellipse.)
    a. An ellipse with semimajor axis a and semiminor axis b.
    b. The portion of the intersection of the sphere $x^2 + y^2 + z^2 = 4$ and the cylinder $x^2 + y^2 - 2y = 0$ lying in the first octant.

3. Find the centroid of the curve $y = \cosh x,\, -1 \le x \le 1$.

4. Compute $\int_C \sqrt{z}\, ds$ where C is parametrized by $\mathbf{g}(t) = (2\cos t,\, 2\sin t,\, t^2)$, $0 \le t \le 2\pi$.

5. Compute $\int_C \mathbf{F} \cdot d\mathbf{x}$ for the following $\mathbf{F}$ and C:
    a. $\mathbf{F}(x, y, z) = (yz,\, x^2,\, xz)$; C is the line segment from $(0, 0, 0)$ to $(1, 1, 1)$.
    b. $\mathbf{F}$ is as in (a); C is the portion of the curve $y = x^2,\, z = x^3$ from $(0, 0, 0)$ to $(1, 1, 1)$.
    c. $\mathbf{F}(x, y) = (x - y,\, x + y)$; C is the circle $x^2 + y^2 = 1$, oriented clockwise.
    d. $\mathbf{F}(x, y) = (x^2 y,\, x^3 y^2)$; C is the closed curve formed by portions of the line $y = 4$ and the parabola $y = x^2$, oriented counterclockwise.

6. Compute the following line integrals:
    a. $\int_C (xe^{-y}\, dx + \sin \pi x\, dy)$, where C is the portion of the parabola $y = x^2$ from $(0, 0)$ to $(1, 1)$.
    b. $\int_C (y\, dx + z\, dy + xy\, dz)$, where C is given by $x = \cos t,\, y = \sin t,\, z = t$ with $0 \le t \le 2\pi$.
    c. $\int_C (y^2\, dx - 2x\, dy)$, where C is the triangle with vertices $(0, 0)$, $(1, 0)$, and $(1, 1)$, oriented counterclockwise.

7. Let $\mathbf{F} : \mathbb{R}^n \to \mathbb{R}^m$ be a continuous map, and let C be a C^1 curve in $\mathbb{R}^n$.
    a. Deduce from Proposition 5.8 that $|\int_C \mathbf{F}\, ds| \le \int_C |\mathbf{F}|\, ds$.
    b. In the case $m = n$, show that $|\int_C \mathbf{F} \cdot d\mathbf{x}| \le \int_C |\mathbf{F}|\, ds$.

8. Prove in detail that arc length, as defined for rectifiable curves, is additive; that is, if C, C_1, and C_2 are the curves parametrized by $\mathbf{g}(t)$ for $t \in [a, b], t \in [a, c]$, and $t \in [c, b]$, then $L(C) = L(C_1) + L(C_2)$.

9. Let $\mathbf{g}(t) = (g(t), h(t))$ be a C^1 parametrization of a plane curve. Given a partition $P = \{t_0, \ldots, t_J\}$ of $[a, b]$, the distance between two neighboring points $\mathbf{g}(t_{j-1})$ and $\mathbf{g}(t_j)$ is

$$\sqrt{[g(t_j) - g(t_{j-1})]^2 + [h(t_j) - h(t_{j-1})]^2}.$$

Use the mean value theorem to express the differences inside the square root in terms of g' and h', and then use Exercise 9 in §4.1 to give an alternate proof of Theorem 5.11. (Exactly the same idea works for curves in $\mathbb{R}^n$.)

5.2   Green's Theorem

Green's theorem is the simplest of a group of theorems — actually, they're all special cases of one big theorem, as we shall indicate in §5.9 — that say that "the integral of something over the boundary of a region equals the integral of something else over the region itself." To state it, we need some terminology.

A **simple closed curve** in $\mathbb{R}^n$ is a curve whose starting and ending points coincide, but that does not intersect itself otherwise. More precisely, a simple closed curve is one that can be parametrized by a continuous map $\mathbf{x} = \mathbf{g}(t)$, $a \leq t \leq b$, such that $\mathbf{g}(a) = \mathbf{g}(b)$ but $\mathbf{g}(s) \neq \mathbf{g}(t)$ unless $\{s, t\} = \{a, b\}$.

We shall use the term **regular region** to mean a compact set in $\mathbb{R}^n$ that is the closure of its interior. Equivalently, a compact set $S \subset \mathbb{R}^n$ is a regular region if every neighborhood of every point on the boundary ∂S contains points in S^{int}. For example, a closed ball is a regular region, but a closed line segment in $\mathbb{R}^n$ $(n > 1)$ is not, because its interior is empty.

Now let $n = 2$. We say that a regular region $S \subset \mathbb{R}^2$ has a **piecewise smooth boundary** if the boundary ∂S consists of a finite union of disjoint, piecewise smooth simple closed curves, where "piecewise smooth" has the meaning assigned in the previous section. (We thus allow the possibility that S contains "holes," so that its boundary may be disconnected.) In this case, the **positive orientation** on ∂S is the orientation on each of the closed curves that make up the boundary such that the region S is on the *left* with respect to the positive direction on the curve. More precisely, if $\mathbf{x}$ is a point on ∂S at which ∂S is smooth, and $\mathbf{t} = (t_1, t_2)$ is the unit tangent vector in the positive direction at that point, then the vector $\mathbf{n} = (t_2, -t_1)$, obtained by rotating $\mathbf{t}$ by 90° *clockwise,* points *out* of S. (That is, $\mathbf{x} + \epsilon \mathbf{n} \notin S$ for small $\epsilon > 0$.) See Figure 5.4.

If $\mathbf{F} = (F_1, F_2)$ is a continuous vector field on $\mathbb{R}^2$, we denote by

$$\int_{\partial S} \mathbf{F} \cdot d\mathbf{x} \quad \text{or} \quad \int_{\partial S} F_1 \, dx_1 + F_2 \, dx_2$$

FIGURE 5.4: A region with piecewise smooth, positively oriented boundary.

the sum of the line integrals of $\mathbf{F}$ over the *positively oriented* closed curves that make up ∂S.

5.12 Theorem (Green's Theorem). *Suppose S is a regular region in $\mathbb{R}^2$ with piecewise smooth boundary ∂S. Suppose also that $\mathbf{F}$ is a vector field of class C^1 on $\overline{S}$. Then*

(5.13)
$$\int_{\partial S} \mathbf{F} \cdot d\mathbf{x} = \iint_S \left(\frac{\partial F_2}{\partial x_1} - \frac{\partial F_1}{\partial x_2} \right) dA.$$

In the more common notation, if we set $\mathbf{F} = (P, Q)$ and $\mathbf{x} = (x, y)$,

(5.14)
$$\int_{\partial S} P\, dx + Q\, dy = \iint_S \left(\frac{\partial Q}{\partial x} - \frac{\partial P}{\partial y} \right) dA.$$

Proof. First we consider a very restricted class of regions, for which the proof is quite simple. We shall say that the region S is x-**simple** if it is the region between the graphs of two functions of x, that is, if it has the form

(5.15)
$$S = \big\{(x, y) : a \leq x \leq b,\ \varphi_1(x) \leq y \leq \varphi_2(x)\big\},$$

where φ_1 and φ_2 are continuous, piecewise smooth functions on $[a, b]$. Likewise, we say that S is y-**simple** if it has the form

(5.16)
$$S = \big\{(x, y) : c \leq y \leq d,\ \psi_1(y) \leq x \leq \psi_2(y)\big\},$$

where ψ_1 and ψ_2 are continuous, piecewise smooth functions on $[a, b]$.

EXAMPLE 1. The region bounded by the curve $y = \frac{1}{8}x^3 - 1$, the line $x + 2y = 2$, and the y-axis is both x-simple and y simple. (See Figure 5.5.) It has the forms (5.15) and (5.16) with

$$a = 0, \quad b = 2, \quad \varphi_1(x) = \tfrac{1}{8}x^3 - 1, \quad \varphi_2(x) = 1 - \tfrac{1}{2}x,$$

$$c = -1, \quad d = 1, \quad \psi_1(y) = 0, \quad \psi_2(y) = \begin{cases} 2(y+1)^{1/3} & \text{if } -1 \leq y \leq 0, \\ 2 - 2y & \text{if } 0 \leq y \leq 1. \end{cases}$$

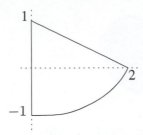

FIGURE 5.5: The region in Example 1.

Now let us suppose that S is both x-simple and y-simple. If we write S in the form (5.15), then ∂S consists of (i) the curve $y = \varphi_1(x)$, oriented from left to right, (ii) the curve $y = \varphi_2(x)$, oriented from right to left, and (iii) portions of the vertical lines $x = a$ and $x = b$, which may reduce to single points. The line integral $\int_{\partial S} P\, dx$ is the sum of the integrals over these pieces. On the vertical lines, x is constant and so $dx = 0$ (that is, $dx/dt = 0$ in any parametrization), so these pieces contribute nothing. On the curves $y = \varphi_1(x)$ and $y = \varphi_2(x)$ we can take x as the parameter, except that the orientation is wrong for $y = \varphi_2(x)$; hence

$$\int_{\partial S} P\, dx = \int_a^b P(x, \varphi_1(x))\, dx - \int_a^b P(x, \varphi_2(x))\, dx.$$

On the other hand, by the fundamental theorem of calculus,

$$\iint_S \frac{\partial P}{\partial y}\, dA = \int_a^b \int_{\varphi_1(x)}^{\varphi_2(x)} \frac{\partial P}{\partial y}\, dy\, dx = \int_a^b \left[P(x, \varphi_2(x)) - P(x, \varphi_1(x)) \right] dx.$$

Comparing these equalities, we obtain

$$\int_{\partial S} P\, dx = - \iint_S \frac{\partial P}{\partial y}\, dA.$$

In exactly the same way, using the representation (5.16) for S, we see that

$$\int_{\partial S} Q\, dy = \iint_S \frac{\partial Q}{\partial x}\, dA.$$

(There is no minus sign here, because if we take y as the parameter for the curves $x = \psi_1(y)$ and $\psi_2(y)$, the orientation is wrong for ψ_1 and right for ψ_2.) Adding these last two equalities, we obtain the desired result (5.14).

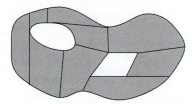

FIGURE 5.6: A decomposition of the region in Figure 5.4 into simple subregions.

Thus Green's theorem is established for regions that are both x-simple and y-simple. There is now an immediate generalization to a much larger class of regular regions. Namely, suppose the region S can be cut up into finitely many subregions, say $S = S_1 \cup \cdots \cup S_k$, where

a. the S_j's may intersect along common edges but have disjoint interiors;

b. each S_j has a piecewise smooth boundary and is both x-simple and y-simple.

(See Figure 5.6.) Since the S_j's overlap only in a set of zero content, by Corollary 4.23b we have

$$\iint_S \left(\frac{\partial Q}{\partial x} - \frac{\partial P}{\partial y} \right) dA = \sum_{j=1}^{k} \iint_{S_j} \left(\frac{\partial Q}{\partial x} - \frac{\partial P}{\partial y} \right) dA.$$

On the other hand, we also have

$$\int_{\partial S} (P\,dx + Q\,dy) = \sum_{j=1}^{k} \int_{\partial S_j} (P\,dx + Q\,dy),$$

because the integrals over the parts of the boundaries of the S_j's that are not parts of the boundary of S all cancel out. In more detail, if S_i and S_j have a common edge C, then C will have one orientation as part of ∂S_i and the opposite orientation as part of ∂S_j, so the two integrals over C that make up parts of $\int_{\partial S_i}$ and $\int_{\partial S_j}$ will cancel each other. Therefore, we obtain Green's theorem for the region S by applying Green's theorem to the simple regions S_j and adding up the results.

The result we have just obtained is sufficient for most practical purposes, but it is not definitive. The class of regular regions that can be cut up into simple subregions does not include all regions with C^1 boundary, much less all regions with piecewise smooth boundary, and it may be difficult to tell whether a given region has this property. For example, the region

$$\left\{ (x, y) : 0 \le x \le 1,\ 0 \le y \le 1 + x^3 \sin x^{-1} \right\}$$

is x-simple but cannot be cut up into finitely many y-simple subregions because the graph of $x^3 \sin x^{-1}$ has infinitely many "wiggles." The deduction of the general case from the special cases considered here requires some additional machinery that is of interest in its own right; we present it in Appendix B.7 (Theorem B.28).   □

EXAMPLE 2. Let C be the unit circle $x^2 + y^2 = 1$, oriented counterclockwise. The line integral

$$\int_C \left[\sqrt{1 + x^2} - ye^{xy} + 3y\right] dx + \left[x^2 - xe^{xy} + \log(1 + y^4)\right] dy$$

is difficult to evaluate directly, but it yields easily to Green's theorem. Indeed, C is the oriented boundary of the unit disc D, so the integral equals

$$\iint_D \left(\frac{\partial}{\partial x}\left[x^2 - xe^{xy} + \log(1 + y^4)\right] - \frac{\partial}{\partial y}\left[\sqrt{1 + x^2} - ye^{xy} + 3y\right]\right) dA$$

$$= \iint_D (2x - 3)\, dA = -6\pi.$$

(The integral of $2x$ over D vanishes by symmetry.)

EXAMPLE 3. It is an amusing and sometimes useful fact that the area of a regular region S in the plane can be expressed as a line integral over the boundary ∂S. This can be done in many different ways; for instance,

$$\text{Area of } S = \int_{\partial S} x\, dy = -\int_{\partial S} y\, dx = \int_{\partial S} \tfrac{1}{2}(x\, dy - y\, dx).$$

Indeed, Green's theorem shows that all of these integrals are equal to $\iint_S 1\, dA$.

The line integral $\int_{\partial S} \mathbf{F} \cdot d\mathbf{x}$ is the integral of the tangential component of $\mathbf{F}$ over ∂S. However, Green's theorem can also be interpreted as at statement about the integral of the normal component of a vector field.

To see this, recall that counterclockwise and clockwise rotations by $90°$ in the plane are given by the transformations $R_+(x, y) = (-y, x)$ and $R_-(x, y) = (y, -x)$, respectively. Thus, if $\mathbf{t} = (t_1, t_2)$ is the unit tangent vector to ∂S at a point on ∂S, pointing in the forward direction, then $\mathbf{n} = R_-(\mathbf{t}) = (t_2, -t_1)$ is the unit normal vector to ∂S pointing out of S. Given a vector field $\mathbf{F} = (F_1, F_2)$, let $\widetilde{\mathbf{F}} = R_+(\mathbf{F}) = (-F_2, F_1)$ be the vector field obtained by rotating the values of $\mathbf{F}$ by $90°$ counterclockwise. Then the normal component of $\mathbf{F}$ is the tangential component of $\widetilde{\mathbf{F}}$:

$$\mathbf{F} \cdot \mathbf{n} = F_1 t_2 - F_2 t_1 = \widetilde{\mathbf{F}} \cdot \mathbf{t}.$$

Hence, by applying Green's theorem to the rotated field $\widetilde{\mathbf{F}}$, we obtain the following result:

5.17 Corollary. *Suppose S is a regular region in $\mathbb{R}^2$ with piecewise smooth boundary ∂S, and let $\mathbf{n}(\mathbf{x})$ be the unit outward normal vector to ∂S at $\mathbf{x} \in \partial S$. Suppose also that $\mathbf{F}$ is a vector field of class C^1 on $\overline{S}$. Then*

$$(5.18) \qquad \int_{\partial S} \mathbf{F} \cdot \mathbf{n}\, ds = \iint_S \left(\frac{\partial F_1}{\partial x_1} + \frac{\partial F_2}{\partial x_2} \right) dA.$$

Let us see what Green's theorem says when $\mathbf{F}$ is the gradient of a C^2 function f, so that $F_1 = \partial_1 f$ and $F_2 = \partial_2 f$. Formula (5.13) gives

$$\int_{\partial S} \nabla f \cdot d\mathbf{x} = \iint_S (\partial_1 \partial_2 f - \partial_2 \partial_1 f)\, dA = \iint_S 0\, dA = 0.$$

This is no surprise; it is easy to see directly that the line integral of a gradient over any closed curve vanishes. Indeed, if the curve C is parametrized by $\mathbf{x} = \mathbf{g}(t)$ with $\mathbf{g}(a) = \mathbf{g}(b)$, then by the chain rule,

$$\int_C \nabla f \cdot d\mathbf{x} = \int_a^b \nabla f(\mathbf{g}(t)) \cdot \mathbf{g}'(t)\, dt = \int_a^b \frac{d}{dt} f(\mathbf{g}(t))\, dt$$
$$= f(\mathbf{g}(b)) - f(\mathbf{g}(a)) = 0.$$

The formula (5.18) gives a more interesting result. $\nabla f \cdot \mathbf{n}$ is the directional derivative of f in the outward normal direction to ∂S, or **normal derivative** of f on ∂S, often denoted by $\partial f / \partial n$; and (29) says that

$$\int_{\partial S} \frac{\partial f}{\partial n}\, ds = \iint_S \left(\frac{\partial^2 f}{\partial x_1^2} + \frac{\partial^2 f}{\partial x_2^2} \right) dA.$$

The integrand on the right is the Laplacian of f, which we encountered in §2.6 and which will play an important role in §5.6.

EXERCISES

1. Evaluate the following line integrals by using Green's theorem.
   a. The integral in Exercise 5c in §5.1.
   b. The integral in Exercise 6c in §5.1.
   c. $\int_C [(x^2 + 10xy + y^2)\, dx + (5x^2 + 5xy) dy]$, where C is the square with vertices $(0,0)$, $(2,0)$, $(0,2)$, and $(2,2)$, oriented counterclockwise.

    d. $\int_{\partial S}(3x^2\sin y^2\,dx + 2x^3y\cos y^2\,dy)$, where S is any regular region with piecewise smooth boundary.

2. Let S be the annulus $1 \leq x^2 + y^2 \leq 4$. Compute $\int_{\partial S}(xy^2\,dy - x^2y\,dx)$, both directly and by using Green's theorem.

3. Find the positively oriented simple closed curve C that maximizes the line integral $\int_C[y^3\,dx + (3x - x^3)\,dy]$.

4. Use Green's theorem as in Example 3 to calculate the area under one arch of the cycloid described parametrically by $x = R(t - \sin t)$, $y = R(1 - \cos t)$.

5. Let $S = \{(x,y) : a \leq x \leq b,\ 0 \leq y \leq f(x)\}$, where f is a nonnegative C^1 function on $[a, b]$. Explain how the formula $A = -\int_{\partial S}y\,dx$ for the area of S in Example 3 leads to the familiar formula $A = \int_a^b f(x)\,dx$.

6. Let S be a regular region in $\mathbb{R}^2$ with piecewise smooth boundary, and let f and g be functions of class C^2 on $\overline{S}$. Show that

$$\int_{\partial S} f\frac{\partial g}{\partial n}\,ds = \iint_S \left[f(\partial_x^2 g + \partial_y^2 g) + \nabla f\cdot\nabla g\right]dA.$$

7. The point of this exercise is to show how Green's theorem can be used to deduce a special case of Theorem 4.41. Let U, V be connected open sets in $\mathbb{R}^2$, and let $\mathbf{G} : U \to V$ be a one-to-one transformation of class C^1 whose derivative $D\mathbf{G}(\mathbf{u})$ is invertible for all $\mathbf{u} \in U$. Moreover, let S be a regular region in V with piecewise smooth boundary, let A be its area, and let $T = \mathbf{G}^{-1}(S)$.

    a. The Jacobian $\det D\mathbf{G}$ is either everywhere positive or everywhere negative on U; why?

    b. Suppose $\det D\mathbf{G}(\mathbf{u}) > 0$ for all $\mathbf{u} \in U$. Write $A = \int_{\partial S}y\,dx$ as in Example 3, make a change of variable to transform this line integral into a line integral over ∂T, and apply Green's theorem to deduce that $A = \iint_T \det D\mathbf{G}\,dA$.

    c. By a similar argument, show that if $\det D\mathbf{G}(\mathbf{u}) < 0$ for all $\mathbf{u} \in U$, then $A = -\iint_T \det D\mathbf{G}\,dA = \iint_T |\det D\mathbf{G}|\,dA$. Where does the minus sign come from?

5.3   Surface Area and Surface Integrals

In this section we discuss integrals of functions and vector fields over smooth surfaces in $\mathbb{R}^3$. Like line integrals, surface integrals come in two varieties, unoriented and oriented. On a curve the orientation is a matter of deciding which direction along a curve is "positive"; on a surface it is a matter of deciding which side of the surface is the "positive" side. The convenient way of specifying the orientation of

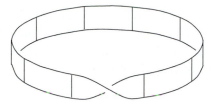

FIGURE 5.7: A Möbius band.

a smooth surface in $\mathbb{R}^3$ is to make a choice of one of the two unit normal vectors at each point of the surface, in such a way that the choice varies continuously with the point. The "positive" side of the surface is the one into which the normal arrow points.

It is important to note that not every surface can be oriented. The standard example of a nonorientable surface is the *Möbius band,* which can be constructed by taking a long strip of paper, giving it a half twist, and gluing the ends together. (That is, call the two sides of the original strip A and B; the ends are to be glued together so that side A of one end matches with side B of the other.) A sketch of a Möbius band is given in Figure 5.7, but the best way to appreciate the features of the Möbius band is to make one for yourself.

However, if a surface forms part of the boundary of a regular region in $\mathbb{R}^3$, it is always orientable, and the standard specification for the orientation is that the positive normal vector is the one pointing *out* of the region.

Surface Area. We begin by deriving a formula for the area of a region on a smooth surface S. We shall assume that S is represented parametrically as the image of a connected open set W in the uv-plane under a one-to-one C^1 map $\mathbf{G} : W \to \mathbb{R}^3$:

$$\mathbf{x} = (x, y, z) = \mathbf{G}(u, v), \quad (u, v) \in W.$$

For a given surface S, it may not be the case that all of S can be represented by a single parametrization. We shall assume, however, that S can be cut up into finitely many pieces which each admit a parametrization; it is then enough to consider the pieces separately. Also, it is usually sufficient to have a good parametrization for a subset of S whose complement is of lower dimension, such as the one provided by spherical coordinates on the unit sphere with the "international date line" removed.

To see how to compute surface area on S, consider a small rectangle in the uv-plane with vertices (u, v), $(u + \Delta u, v)$, $(u, v + \Delta v)$, and $(u + \Delta u, v + \Delta v)$. Its image under the map $\mathbf{G}$ is a small quadrilateral (with curved sides) on the surface

S whose vertices are $\mathbf{G}(u, v)$, $\mathbf{G}(u + \Delta u, v)$, etc. (See Figure 3.4 in §3.3.) In the limit in which the increments Δu and Δv become infinitesimals du and dv, this quadrilateral becomes a parallelogram whose sides from the vertex $\mathbf{x} = \mathbf{G}(u, v)$ to the two adjacent vertices are described by the vectors

$$\mathbf{G}(u + du, v) - \mathbf{G}(u, v) = \frac{\partial \mathbf{G}}{\partial u} du \quad \text{and} \quad \mathbf{G}(u, v + dv) - \mathbf{G}(v) = \frac{\partial \mathbf{G}}{\partial v} dv.$$

These two vectors are tangent to the surface S at $\mathbf{x}$, so their cross product is a vector normal to S at $\mathbf{x}$, whose magnitude is the area of the parallelogram they span. Therefore, the element of area on S is given in terms of the parametrization $\mathbf{x} = \mathbf{G}(u, v)$ by

$$(5.19) \qquad\qquad dA = \left| \frac{\partial \mathbf{G}}{\partial u} \times \frac{\partial \mathbf{G}}{\partial v} \right| du\, dv.$$

In other words, if R is a measurable subset of W in the uv-plane and $\mathbf{G}(R)$ is the corresponding region in the surface S,

$$(5.20) \qquad\qquad \text{Area of } \mathbf{G}(R) = \iint_R \left| \frac{\partial \mathbf{G}}{\partial u} \times \frac{\partial \mathbf{G}}{\partial v} \right| du\, dv.$$

Henceforth we shall take (5.20) as the *definition* of **area** for a parametrized surface. One might wonder if surface area can also be defined by considering polyhedral approximations to the surface, as polygonal approximations to a curve were used to define arc length in the appendix of §5.1. The answer is affirmative, but this matter is a good deal trickier than the theory of arc length, and we shall not pursue it further.

Let us be a little more explicit about the formula (5.19). With the notation $\mathbf{G}(u, v) = (x, y, z)$, we have

$$\frac{\partial \mathbf{G}}{\partial u} \times \frac{\partial \mathbf{G}}{\partial v} = \det \begin{pmatrix} \mathbf{i} & \mathbf{j} & \mathbf{k} \\ \partial_u x & \partial_u y & \partial_u z \\ \partial_v x & \partial_v y & \partial_v z \end{pmatrix} = \frac{\partial(y, z)}{\partial(u, v)} \mathbf{i} + \frac{\partial(z, x)}{\partial(u, v)} \mathbf{j} + \frac{\partial(x, y)}{\partial(u, v)} \mathbf{k}.$$

Thus,

$$(5.21) \qquad dA = \sqrt{\left[\frac{\partial(y, z)}{\partial(u, v)}\right]^2 + \left[\frac{\partial(z, x)}{\partial(u, v)}\right]^2 + \left[\frac{\partial(x, y)}{\partial(u, v)}\right]^2}\, du\, dv.$$

Computationally, this is usually a horrible mess. (But what did you expect? Arc length is already problematic; surface area must be worse!)

As with arc length, we must verify that our informally-derived formula for surface area really makes sense by checking that it is independent of the parametrization. Thus, suppose we make a change of variables $(u, v) = \Phi(s, t)$, where Φ is a one-to-one C^1 map from a region V in the st-plane to the region W in the uv-plane. The elements of area are then related by

$$du\, dv = \left| \frac{\partial(u, v)}{\partial(s, t)} \right| ds\, dt,$$

by Theorem 4.41. If we plug this into (5.21), we get

$$dA = \sqrt{\alpha^2 + \beta^2 + \gamma^2}\, ds\, dt, \text{ where } \alpha = \frac{\partial(y, z)}{\partial(u, v)} \frac{\partial(u, v)}{\partial(s, t)}, \text{ etc.}$$

But by the chain rule and the fact that the determinant of a product is the product of the determinants, we have

$$\frac{\partial(y, z)}{\partial(u, v)} \frac{\partial(u, v)}{\partial(s, t)} = \frac{\partial(y, z)}{\partial(s, t)},$$

and likewise for the other two terms. Hence, in the st-parametrization,

$$dA = \sqrt{\left[\frac{\partial(y, z)}{\partial(s, t)} \right]^2 + \left[\frac{\partial(z, x)}{\partial(s, t)} \right]^2 + \left[\frac{\partial(x, y)}{\partial(s, t)} \right]^2}\, ds\, dt.$$

This is of exactly the same form as (5.21), as we wished to show.

The formula for surface area becomes a little less hideous in the special case where the surface is the graph of a function, $z = \varphi(x, y)$. In this case we can take x and y as the parameters, that is,

$$\mathbf{G}(x, y) = \big(x,\ y,\ \varphi(x, y)\big).$$

Here $\partial_x \mathbf{G} = (1, 0, \partial_x \varphi)$ and $\partial_y \mathbf{G} = (0, 1, \partial_y \varphi)$, so

(5.22)

$$\frac{\partial \mathbf{G}}{\partial x} \times \frac{\partial \mathbf{G}}{\partial y} = -(\partial_x \varphi)\mathbf{i} - (\partial_y \varphi)\mathbf{j} + \mathbf{k},$$

$$dA = \sqrt{1 + (\partial_x \varphi)^2 + (\partial_y \varphi)^2}\, dx\, dy.$$

(Note that our surface is a level set of the function $\Phi(x, y, z) = z - \varphi(x, y)$ and that $-(\partial_x \varphi)\mathbf{i} - (\partial_y \varphi)\mathbf{j} + \mathbf{k} = \nabla\Phi$; we deduced that $\nabla\Phi$ is normal to the surface by other means in Theorem 2.37.)

EXAMPLE 1. Let us compute the surface area of the unit sphere $x^2 + y^2 + z^2 = 1$. We can proceed on two ways:

Solution I. The upper hemisphere is the graph of the function $\varphi(x, y) = \sqrt{1 - x^2 - y^2}$. A little calculation yields

$$\sqrt{1 + (\partial_x \varphi)^2 + (\partial_y \varphi)^2} = \frac{1}{\sqrt{1 - x^2 - y^2}},$$

and by (5.22), the area of the upper hemisphere is obtained by integrating this function over the unit disc. (Note that this integral is improper, as the integrand blows up along the boundary of the disc.) Switching to polar coordinates yields

$$\int_0^1 \int_0^{2\pi} \frac{r}{\sqrt{1 - r^2}} \, d\theta \, dr = -2\pi \sqrt{1 - r^2} \Big|_0^1 = 2\pi.$$

Hence the area of the whole sphere is 4π.

Solution II. We can parametrize the sphere by the spherical coordinates $x = \sin \varphi \cos \theta$, $y = \sin \varphi \sin \theta$, $z = \cos \varphi$. An easy calculation yields

$$\frac{\partial(y, z)}{\partial(\varphi, \theta)} = \sin^2 \varphi \cos \theta, \qquad \frac{\partial(z, x)}{\partial(\varphi, \theta)} = \sin^2 \varphi \sin \theta, \qquad \frac{\partial(x, y)}{\partial(\varphi, \theta)} = \cos \varphi \sin \varphi,$$

and the sum of the squares of these quantities is

$$\sin^4 \varphi (\cos^2 \theta + \sin^2 \theta) + \cos^2 \varphi \sin^2 \varphi = \sin^2 \varphi (\cos^2 \varphi + \sin^2 \varphi) = \sin^2 \varphi.$$

Hence, by (5.21), the area of the sphere is

$$\int_0^\pi \int_0^{2\pi} \sin \varphi \, d\theta \, d\varphi = -2\pi \cos \varphi \Big|_0^\pi = 4\pi.$$

Surface Integrals of Scalar Functions. Now that we know how to compute surface area, it is easy to define the integral of a real-valued continuous function over a surface: It is just $\iint_S f \, dA$, where dA is the element of surface area defined above. (To keep the notation simple, we shall take the region over which the integration is performed to be the whole surface S; the idea is exactly the same for integration over subsets of S.) More precisely, if S admits a parametrization $\mathbf{x} = \mathbf{G}(u, v)$ with $(u, v) \in W$, where W is tacitly assumed to be measurable,

$$\iint_S f \, dA = \iint_W f(\mathbf{G}(u, v)) \left| \frac{\partial \mathbf{G}}{\partial u} \times \frac{\partial \mathbf{G}}{\partial v} \right| du \, dv.$$

If S is the graph of a function $z = \varphi(x, y)$, $(x, y) \in W$, the result is

$$\iint_S f \, dA = \iint_W f(x, y, \varphi(x, y)) \sqrt{1 + (\partial_x \varphi)^2 + (\partial_y \varphi)^2} \, dx \, dy.$$

Surface Integrals of Vector Fields. The element of area dA on a surface S parametrized by $\mathbf{x} = \mathbf{G}(u, v)$ is the norm of the vector $(\partial_u \mathbf{G} \times \partial_v \mathbf{G}) \, du \, dv$. It is natural to regard the vector $(\partial_u \mathbf{G} \times \partial_v \mathbf{G}) \, du \, dv$ itself as a "vector element of area" for S: its magnitude gives the area of a small bit of S, and its direction, namely the normal direction to S, specifies how that bit is oriented in space. That is, we have

$$\left(\frac{\partial \mathbf{G}}{\partial u} \times \frac{\partial \mathbf{G}}{\partial v} \right) du \, dv = \mathbf{n} \, dA$$

where $\mathbf{n}$ is a unit normal vector to the surface S. We have already observed that dA is independent of the parametrization, and clearly so is $\mathbf{n}$ up to a factor of ± 1. However, using a different parametrization (for example, interchanging u and v) might result in replacing $\mathbf{n}$ by $-\mathbf{n}$. In other words, a parametrization for a surface S gives a definite *orientation* for the S, that is, a specification of which side of S is the "positive" side.

Now suppose S is a surface with a specified orientation, and $\mathbf{F}$ is a continuous vector field defined on a neighborhood of S. The **surface integral** of $\mathbf{F}$ over S is defined to be

$$\iint_S \mathbf{F} \cdot \mathbf{n} \, dA.$$

Thus, if S is parametrized by $\mathbf{x} = \mathbf{G}(u, v)$, $(u, v) \in W$, we have

$$\iint_S \mathbf{F} \cdot \mathbf{n} \, dA = \iint_W \mathbf{F}(\mathbf{G}(u, v)) \cdot \left(\frac{\partial \mathbf{G}}{\partial u} \times \frac{\partial \mathbf{G}}{\partial v} \right) du \, dv.$$

This integral is independent of the choice of parametrization as long as the parametrization induces the specified orientation of S; switching to the opposite orientation results in multiplying the integral by -1. (If S is a nonorientable surface such as a Möbius band, $\iint_S \mathbf{F} \cdot \mathbf{n} \, dA$ is not defined.)

A geometric-physical interpretation of this is easy to obtain. $\mathbf{F} \cdot \mathbf{n}$ is the **normal component** of $\mathbf{F}$ along S; it is positive or negative according as $\mathbf{F}$ points into the positive or negative side of S. We can think of $\mathbf{F}$ as representing the flow of some substance (air, for example, although there is no need to be specific at this point): the magnitude of $\mathbf{F}(\mathbf{x})$ is the rate of flow of the substance past $\mathbf{x}$ and its direction is the direction of flow. The integral $\iint \mathbf{F} \cdot \mathbf{n} \, dA$ then represents the net flow, or **flux**, of $\mathbf{F}$ across the surface S from the negative side to the positive side. We shall discuss this in more detail in §5.6.

As with line integrals, surface integrals of vector fields are often easier to compute than suface integrals of scalar functions because the inconvenient square root in the formula for dA does not appear in the vector $\mathbf{n} \, dA$. Let us see, for example, what $\iint_S \mathbf{F} \cdot \mathbf{n} \, dA$ becomes when S is the graph of a function with domain $W \subset \mathbb{R}^2$, say $z = \varphi(x, y)$. As in the preceding discussion of surface area, we take x and y as the parameters and find that

$$\mathbf{n} \cdot dA = [-(\partial_x \varphi)\mathbf{i} - (\partial_y \varphi)\mathbf{j} + \mathbf{k}]\, dx\, dy.$$

The orientation here is the one with the normal pointing upward, since its z component is positive. Thus, if $\mathbf{F} = F_1\mathbf{i} + F_2\mathbf{j} + F_3\mathbf{k}$ and $\mathbf{G}(x, y) = (x, y, \varphi(x, y))$,

(5.23)
$$\iint_S \mathbf{F} \cdot \mathbf{n}\, dA$$
$$= \iint_W \left[-F_1(\mathbf{G}(x, y))\frac{\partial \varphi}{\partial x} - F_2(\mathbf{G}(x, y))\frac{\partial \varphi}{\partial y} + F_3(\mathbf{G}(x, y)) \right] dx\, dy.$$

Here and in what follows, we adopt the common practice of denoting by $\mathbf{i}$, $\mathbf{j}$, and $\mathbf{k}$ the unit vectors in the positive coordinate directions and writing vector fields in $\mathbb{R}^3$ as $\mathbf{F} = F_1\mathbf{i} + F_2\mathbf{j} + F_3\mathbf{k}$ in preference to $\mathbf{F} = (F_1, F_2, F_3)$; this serves to emphasize the interpretation of $\mathbf{F}$ as a vector field rather than a transformation.

EXAMPLE 2. Let S be the portion of the cone $x^2 + y^2 = z^2$ with $0 \le z \le 1$, oriented so that the normal points upward, and let $\mathbf{F}(x, y, z) = x^2\mathbf{i} + yz\mathbf{j} + y\mathbf{k}$. Compute $\iint_S \mathbf{F} \cdot \mathbf{n}\, dA$.

Solution. One way is to use polar coordinates as parameters: $\mathbf{G}(r, \theta) = (r\cos\theta, r\sin\theta, r)$. Then we have $\partial_r\mathbf{G} = (\cos\theta)\mathbf{i} + (\sin\theta)\mathbf{j} + \mathbf{k}$ and $\partial_\theta\mathbf{G} = -(r\sin\theta)\mathbf{i} + (r\cos\theta)\mathbf{j}$, so

$$\partial_r\mathbf{G} \times \partial_\theta\mathbf{G} = -(r\cos\theta)\mathbf{i} - (r\sin\theta)\mathbf{j} + r\mathbf{k}.$$

This gives the right orientation since the z component, namely r, is positive. Thus,

$$\iint_S \mathbf{F} \cdot \mathbf{n}\, dA$$
$$= \int_0^{2\pi} \int_0^1 \left[(r\cos\theta)^2(-r\cos\theta) + (r\sin\theta)r(-r\sin\theta) + (r\sin\theta)r \right] dr\, d\theta,$$

whose value is easily found to be $-\frac{1}{4}\pi$. Alternatively, we could use the representation $z = \sqrt{x^2 + y^2}$ and use (5.23). The reader may verify that this leads to

$$\iint_S \mathbf{F} \cdot \mathbf{n}\, dA = \iint_{x^2+y^2 \le 1} \left[\frac{-x^3 - y^2\sqrt{x^2 + y^2}}{\sqrt{x^2 + y^2}} + y \right] dx\, dy,$$

and conversion of this integral to polar coordinates leads to the same $r\theta$-integral as before.

Finally, as a practical matter we need to extend the ideas in this section from smooth surfaces to piecewise smooth surfaces. Giving a satisfactory general definition of a "piecewise smooth surface" is a rather messy business, and we shall not attempt it. For our present purposes, it will suffice to assume that the surface S under consideration is the union of finitely many pieces $S_1, \ldots, S_k$ that satisfy the following conditions:

i. Each S_j admits a smooth parametrization as discussed above.

ii. The intersections $S_i \cap S_j$ are either empty or finite unions of smooth curves.

We then define integration over S in the obvious way:

$$\iint_S f \, dA = \sum_{j=1}^{k} \iint_{S_j} f \, dA.$$

Condition (ii) guarantees that the parts of S that are counted more than once on the right, namely the intersections $S_i \cap S_j$, contribute nothing to the integral, by Propositions 4.19 and 4.22.

EXAMPLE 3.
a. Let S be the surface of a cube; then we can take $S_1, \ldots, S_6$ to be the faces of the cube.
b. Let S be the surface of the cylindrical solid $\{(x, y, z) : x^2 + y^2 \leq 1, |z| \leq 1\}$. We can write $S = S_1 \cup S_2 \cup S_3$ where S_1 and S_2 are the discs forming the top and bottom and S_3 is the circular vertical side. S_1 and S_2 can be parametrized by $(x, y) \to (x, y, 1)$ and $(x, y) \to (x, y, -1)$ with $x^2 + y^2 \leq 1$, and S_3 can be parametrized by $(\theta, z) \to (\cos\theta, \sin\theta, z)$ with $0 \leq \theta < 2\pi$ and $|z| \leq 1$. If one wishes to use only one-to-one parametrizations with compact parameter domains, one can cut S_3 further into two pieces, say the left and right halves defined by $0 \leq \theta \leq \pi$ and $\pi \leq \theta \leq 2\pi$.

Remark. In condition (ii) above, we have in mind that the sets S_j will intersect each other only along their edges, although there is nothing to forbid them from crossing one another. For example, S could be the union of the two spheres $S_1 = \{\mathbf{x} : |\mathbf{x}| = 1\}$ and $S_2 = \{\mathbf{x} : |\mathbf{x} - \mathbf{i}| = 1\}$. This added generality is largely useless but also harmless.

EXERCISES

1. Find the area of the part of the surface $z = xy$ inside the cylinder $x^2 + y^2 = a^2$.

2. Find the area of the part of the surface $z = x^2 + y^2$ inside the cylinder $x^2 + y^2 = a^2$.

3. Suppose $0 < a < b$. Find the area of the torus obtained by revolving the circle $(x - b)^2 + z^2 = a^2$ in the xz-plane about the z axis. (*Hint:* The torus may be parametrized by $x = (b + a \cos \varphi) \cos \theta$, $y = (b + a \cos \varphi) \sin \theta$, $z = a \sin \varphi$, with $0 \le \varphi, \theta \le 2\pi$.)

4. Find the area of the ellipsoid $(x/a)^2 + (y/a)^2 + (z/b)^2 = 1$.

5. Find the centroid of the upper hemisphere of the unit sphere $x^2 + y^2 + z^2 = 1$.

6. Compute $\iint_S (x^2 + y^2)\, dA$ where S is the portion of the sphere $x^2 + y^2 + z^2 = 4$ with $z \ge 1$.

7. Compute $\iint_S (x^2 + y^2 - 2z^2)\, dA$ where S is the unit sphere. Can you find the answer by symmetry considerations without doing any calculations?

8. Calculate $\iint_S \mathbf{F} \cdot \mathbf{n}\, dA$ for the following $\mathbf{F}$ and S.
   a. $\mathbf{F}(x, y, z) = xz\mathbf{i} - xy\mathbf{k}$; S is the portion of the surface $z = xy$ with $0 \le x \le 1, 0 \le y \le 2$, oriented so that the normal points upward.
   b. $\mathbf{F}(x, y, z) = x^2\mathbf{i} + z\mathbf{j} - y\mathbf{k}$; S is the unit sphere $x^2 + y^2 + z^2 = 1$, oriented so that the normal points outward (away from the center).
   c. $\mathbf{F}(x, y, z) = xy\mathbf{i} + z\mathbf{j}$; S is the triangle with vertices $(2, 0, 0)$, $(0, 2, 0)$, $(0, 0, 2)$, oriented so that the normal points upward.
   d. $\mathbf{F}(x, y, z) = z^2\mathbf{k}$; S is the boundary of the region $x^2 + y^2 \le 1, a \le z \le b$, oriented so that the normal points out of the region. (You should be able to do this in your head.)
   e. $\mathbf{F}(x, y, z) = x\mathbf{i} + y\mathbf{j} + z\mathbf{k}$; S is the boundary of the region $x^2 + y^2 \le z \le \sqrt{2 - x^2 - y^2}$, oriented so that the normal points out of the region.

5.4   Vector Derivatives

Let ∇ denote the n-tuple of partial differential operators $\partial_j = \partial/\partial x_j$:

$$\nabla = (\partial_1, \ldots, \partial_n).$$

We are already familiar with this notation in connection with the gradient of a C^1 function on $\mathbb{R}^n$, which is the vector field defined by

$$\operatorname{grad} f = \nabla f = (\partial_1 f, \ldots, \partial_n f).$$

We can also use ∇ to form interesting combinations of the derivatives of a vector field, via the dot and cross product. If $\mathbf{F}$ is a C^1 vector field on an open subset of

$\mathbb{R}^n$, the **divergence** of $\mathbf{F}$ is the function defined by

$$\operatorname{div} \mathbf{F} = \nabla \cdot \mathbf{F} = \partial_1 F_1 + \cdots + \partial_n F_n.$$

The geometric (coordinate-invariant) meaning of $\nabla \cdot \mathbf{F}$ will be explained in §5.5.

Next, suppose $n = 3$. If $\mathbf{F}$ is a C^1 vector field on an open subset of $\mathbb{R}^3$, the **curl** of $\mathbf{F}$ is the vector field defined by

$$\operatorname{curl} \mathbf{F} = \nabla \times \mathbf{F} = (\partial_2 F_3 - \partial_3 F_2)\mathbf{i} + (\partial_3 F_1 - \partial_1 F_3)\mathbf{j} + (\partial_1 F_2 - \partial_2 F_1)\mathbf{k}.$$

(Some authors write $\operatorname{rot} \mathbf{F}$ instead of $\operatorname{curl} \mathbf{F}$; "rot" stands for "rotation.") Again, the curl has a geometric significance that will be explained later, in §5.7.

We shall employ the notations $\operatorname{div} \mathbf{F}$ and $\operatorname{curl} \mathbf{F}$ in preference to $\nabla\cdot\mathbf{F}$ and $\nabla\times\mathbf{F}$ because they seem to be more readable. In this section we shall also write $\operatorname{grad} f$ instead of ∇f for the sake of consistency; later we shall use these two notations interchangeably.

The operators grad, curl, and div satisfy product rules with respect to scalar multiplication and dot and cross products. As these rules are useful and some of them are not obvious, it is well to make a list for handy reference. In the following formulas, f and g are real-valued functions and $\mathbf{F}$ and $\mathbf{G}$ are vector fields, all of class C^1.

(5.24) $\operatorname{grad}(fg) = f \operatorname{grad} g + g \operatorname{grad} f$

(5.25) $\operatorname{grad}(\mathbf{F} \cdot \mathbf{G}) = (\mathbf{F} \cdot \nabla)\mathbf{G} + \mathbf{F} \times (\operatorname{curl} \mathbf{G}) + (\mathbf{G} \cdot \nabla)\mathbf{F} + \mathbf{G} \times (\operatorname{curl} \mathbf{F})$

(5.26) $\operatorname{curl}(f\mathbf{G}) = f \operatorname{curl} \mathbf{G} + (\operatorname{grad} f) \times \mathbf{G}$

(5.27) $\operatorname{curl}(\mathbf{F} \times \mathbf{G}) = (\mathbf{G} \cdot \nabla)\mathbf{F} + (\operatorname{div} \mathbf{G})\mathbf{F} - (\mathbf{F} \cdot \nabla)\mathbf{G} - (\operatorname{div} \mathbf{F})\mathbf{G}$

(5.28) $\operatorname{div}(f\mathbf{G}) = f \operatorname{div} \mathbf{G} + (\operatorname{grad} f) \cdot \mathbf{G}$

(5.29) $\operatorname{div}(\mathbf{F} \times \mathbf{G}) = \mathbf{G} \cdot (\operatorname{curl} \mathbf{F}) - \mathbf{F} \cdot (\operatorname{curl} \mathbf{G})$

In (5.25) and (5.27), $\mathbf{F} \cdot \nabla$ denotes the directional derivative $\sum F_j \partial_j$, that is,

$$(\mathbf{F} \cdot \nabla)\mathbf{G} = \sum F_j \frac{\partial \mathbf{G}}{\partial x_j}.$$

Equations (5.24) and (5.28) are valid in $\mathbb{R}^n$ for any n; the others, which involve cross products and curls, are restricted to $n = 3$. The proofs of all these formulas are just a matter of computation; we leave them to the reader as exercises.

We can combine the operations grad, curl, and div pairwise in several ways. That is, if f and $\mathbf{F}$ are of class C^2, we can form

$$\operatorname{curl}(\operatorname{grad} f), \quad \operatorname{div}(\operatorname{curl} \mathbf{F}), \quad \operatorname{div}(\operatorname{grad} f), \quad \operatorname{curl}(\operatorname{curl} \mathbf{F}), \quad \operatorname{grad}(\operatorname{div} \mathbf{F}).$$

It is an important fact that the first two of these always vanish, by the equality of mixed partials:

(5.30) $\mathrm{curl}(\mathrm{grad}\, f)$
$$= (\partial_2\partial_3 f - \partial_3\partial_2 f)\mathbf{i} + (\partial_3\partial_1 f - \partial_1\partial_3 f)\mathbf{j} + (\partial_1\partial_2 f - \partial_2\partial_1 f)\mathbf{k} = 0$$

and

(5.31) $\mathrm{div}(\mathrm{curl}\,\mathbf{F})$
$$= \partial_1(\partial_2 F_3 - \partial_3 F_2) + \partial_2(\partial_3 F_1 - \partial_1 F_3) + \partial_3(\partial_1 F_2 - \partial_2 F_1) = 0.$$

Schematically, we have

$$\begin{array}{ccccccc}
\text{scalar} & \xrightarrow{\text{grad}} & \text{vector} & \xrightarrow{\text{curl}} & \text{vector} & \xrightarrow{\text{div}} & \text{scalar} \\
\text{functions} & & \text{fields} & & \text{fields} & & \text{functions}
\end{array}$$

and (5.30) and (5.31) say that the composition of two successive mappings is zero.

The third combination, $\mathrm{div}(\mathrm{grad}\, f)$, which makes sense in any number of dimensions, is of fundamental importance for both physical and purely mathematical reasons. It is called the **Laplacian** of f and is usually denoted by $\nabla^2 f$ or Δf:

(5.32) $$\nabla^2 f = \Delta f = \mathrm{div}(\mathrm{grad}\, f) = \partial_1^2 f + \cdots + \partial_n^2 f.$$

The last two combinations are of less interest by themselves, but together they yield the Laplacian for vector fields in $\mathbb{R}^3$:

(5.33) $$\mathrm{grad}(\mathrm{div}\,\mathbf{F}) - \mathrm{curl}(\mathrm{curl}\,\mathbf{F}) = \nabla^2\mathbf{F} = (\nabla^2 F_1)\mathbf{i} + (\nabla^2 F_2)\mathbf{j} + (\nabla^2 F_3)\mathbf{k}.$$

The verification of (5.33) is a straightforward but somewhat tedious calculation that we leave to the reader.

EXERCISES

1. Compute the curl and divergence of the following vector fields.
   a. $\mathbf{F}(x, y, z) = xy^2\mathbf{i} + xy\mathbf{j} + xy\mathbf{k}$.
   b. $\mathbf{F}(x, y, z) = (\sin yz)\mathbf{i} + (xz\cos yz)\mathbf{j} + (xy\cos yz)\mathbf{k}$.
   c. $\mathbf{F}(x, y, z) = x^2 z\mathbf{i} + 4xyz\mathbf{j} + (y - 3xz^2)\mathbf{k}$.

2. Compute the Laplacians of the following functions.
   a. $f(x, y) = x^5 - 10x^3 y^2 + 5xy^4$.
   b. $f(x, y, z) = xy^2 - 4yz^3$.

   c. $f(\mathbf{x}) = |\mathbf{x}|^a$ ($\mathbf{x} \in \mathbb{R}^n \setminus \{\mathbf{0}\}$, $a \in \mathbb{R}$). (*Hint:* Use Exercise 9 in §2.6.)

   d. $f(x, y) = \log(x^2 + y^2)$ $((x, y) \neq (0, 0))$.

3. Let $\mathbf{F}(x, y, z) = x\mathbf{i} + y\mathbf{j} + z\mathbf{k}$. Show that for any $\mathbf{a} \in \mathbb{R}^3$, we have $\mathrm{curl}(\mathbf{a} \times \mathbf{F}) = 2\mathbf{a}$, $\mathrm{div}[(\mathbf{a} \cdot \mathbf{F})\mathbf{a}] = |\mathbf{a}|^2$ and $\mathrm{div}[(\mathbf{a} \times \mathbf{F}) \times \mathbf{a}] = 2|\mathbf{a}|^2$.

4. Prove (5.24) and (5.25).

5. Prove (5.26) and (5.27).

6. Prove (5.28) and (5.29).

7. Prove (5.33).

8. Why is the minus sign in (5.29) there? That is, on grounds of symmetry, without going through any calculations, why must the formula $\mathrm{div}(\mathbf{F} \times \mathbf{G}) = \mathbf{G} \cdot (\mathrm{curl}\,\mathbf{F}) + \mathbf{F} \cdot (\mathrm{curl}\,\mathbf{G})$ be wrong?

9. Show that for any C^2 functions f and g, $\mathrm{div}(\mathrm{grad}\,f \times \mathrm{grad}\,g) = 0$.

5.5 The Divergence Theorem

The divergence theorem, also known as **Gauss's theorem** or **Ostrogradski's theorem**, is the 3-dimensional analogue of the version (5.18) of Green's theorem; it relates surface integrals over the boundary of a regular region in $\mathbb{R}^3$ to volume integrals over the region itself. The divergence theorem is valid for regions with piecewise smooth boundaries, but we shall allow the meaning of "piecewise smooth" to remain a little vague; see the remarks at the end of §5.3. To formulate precise conditions that encompass all the cases of interest would involve a rather arduous excursion into technicalities, and the more retricted class of regions covered by the following argument suffices for most purposes.

5.34 Theorem (The Divergence Theorem). *Suppose R is a regular region in $\mathbb{R}^3$ with piecewise smooth boundary ∂R, oriented so that the positive normal points out of R. Suppose also that $\mathbf{F}$ is a vector field of class C^1 on R. Then*

$$(5.35) \qquad \iint_{\partial R} \mathbf{F} \cdot \mathbf{n}\, dA = \iiint_R \mathrm{div}\,\mathbf{F}\, dV.$$

Proof. As with Green's theorem, we begin by considering a class of simple regions. We say that R is xy-**simple** if it has the form

$$R = \{(x, y, z) : (x, y) \in W,\ \varphi_1(x, y) \leq z \leq \varphi_2(x, y)\},$$

where W is a regular region in the xy-plane and φ_1 and φ_2 are piecewise smooth functions on W. We define the notions of yz-simple and xz-simple similarly, and we say that R is **simple** if it is xy-simple, yz-simple, and xz-simple.

Suppose now that R is simple. We shall prove the divergence theorem for the region R by considering the components of $\mathbf{F}$ separately. That is, let $\mathbf{F} = F_1\mathbf{i} + F_2\mathbf{j} + F_3\mathbf{k}$; we shall show that

$$\iint_{\partial R} F_3\mathbf{k} \cdot \mathbf{n}\, dA = \iiint_R \partial_3 F_3\, dV,$$

and similarly for the other two components. Since R is xy-simple, the boundary ∂R consists of three pieces: the "top" and "bottom" surfaces $z = \varphi_2(x, y)$ and $z = \varphi_1(x, y)$ and the "sides" consisting of the union of the vertical line segments from $(x, y, \varphi_1(x, y))$ to $(x, y, \varphi_2(x, y))$ as (x, y) ranges over the boundary of W. The outward normal to R is horizontal on the sides, i.e., $\mathbf{k} \cdot \mathbf{n} = 0$ there, so the sides contribute nothing to the surface integral. For the top and bottom surfaces we use (5.23). The outward normal points upward on the top surface and downward on the bottom surface, so

$$\iint_{\partial R} F_3\mathbf{k} \cdot \mathbf{n}\, dA = \iint_W F_3(x, y, \varphi_2(x, y))\, dx\, dy - \iint_W F_3(x, y, \varphi_1(x, y))\, dx\, dy$$

$$= \iint_W \int_{\varphi_1(x,y)}^{\varphi_2(x,y)} \partial_3 F_3(x, y, z)\, dz\, dx\, dy$$

$$= \iiint_R \partial_3 F_3(x, y, z)\, dV,$$

as claimed. The proof for $F_1\mathbf{i}$ and $F_2\mathbf{j}$ is the same, using the assumptions that R is yz-simple and xz-simple.

It now follows that the divergence theorem is valid for regions that can be cut up into finitely many simple regions $R_1, \ldots, R_k$. The integrals of div $\mathbf{F}$ over the regions $R_1, \ldots, R_k$ add up to the integral over R, and the integrals of $\mathbf{F} \cdot \mathbf{n}$ over the boundaries $\partial R_1, \ldots, \partial R_k$ add up to the integral over ∂R because the integrals over the portions of the ∂R_j's that are not part of ∂R cancel out. (The reasoning is the same as in the proof of Green's theorem.)

The completion of the proof for general regular regions with smooth boundary, with indications of how to generalize it to the piecewise smooth case, is given in Appendix B.7 (Theorem B.30).   □

Armed with the divergence theorem, we can obtain a better understanding of the meaning of div $\mathbf{F}$. Suppose $\mathbf{F}$ is a vector field of class C^1 in some open set containing the point $\mathbf{a}$. For $r > 0$, let B_r be the ball of radius r about $\mathbf{a}$. If r is very small, the average value of div $\mathbf{F}(\mathbf{x})$ on the ball B_r is very nearly equal to div $\mathbf{F}(\mathbf{a})$. Therefore, by the divergence theorem,

$$\text{div } \mathbf{F}(\mathbf{a}) \approx \frac{3}{4\pi r^3} \iiint_{B_r} \text{div } \mathbf{F}\, dV = \frac{3}{4\pi r^3} \iint_{\partial B_r} \mathbf{F} \cdot \mathbf{n}\, dA.$$

This approximation becomes better and better as $r \to 0$, and hence

$$(5.36) \qquad \operatorname{div} \mathbf{F}(\mathbf{a}) = \lim_{r \to 0} \frac{3}{4\pi r^3} \iint_{|\mathbf{x}-\mathbf{a}|=r} \mathbf{F} \cdot \mathbf{n} \, dA.$$

The integral on the right is the flux of $\mathbf{F}$ across ∂B_r from the inside (B_r) to the outside (the complement of B_r). If we think of the vector field as representing the flow of some substance through space, the integral represents the amount of substance flowing out of B_r minus the amount of substance flowing in; thus, the condition $\operatorname{div} \mathbf{F}(\mathbf{a}) > 0$ means that there is a net outflow near $\mathbf{a}$, in other words, that $\mathbf{F}$ tends to "diverge" from $\mathbf{a}$. (The effect is subtle, though: One has to divide the flux by r^3 in (5.36) to get something that does not vanish in the limit.) In any case, the integral in (5.36) is a geometrically defined quantity that is independent of the choice of coordinates; this gives the promised coordinate-free interpretation of $\operatorname{div} \mathbf{F}$.

Among the important consequences of the divergence theorem are the following identities.

5.37 Corollary (Green's Formulas). *Suppose R is a regular region in $\mathbb{R}^3$ with piecewise smooth boundary, and f and g are functions of class C^2 on $\overline{R}$. Then*

$$(5.38) \qquad \iint_{\partial R} f \nabla g \cdot \mathbf{n} \, dA = \iiint_R (\nabla f \cdot \nabla g + f \nabla^2 g) \, dV,$$

$$(5.39) \qquad \iint_{\partial R} (f \nabla g - g \nabla f) \cdot \mathbf{n} \, dA = \iiint_R (f \nabla^2 g - g \nabla^2 f) \, dV.$$

Proof. An application of the product rule (5.28) shows that $\operatorname{div}(f \nabla g) = \nabla f \cdot \nabla g + f \cdot \nabla^2 g$, so the divergence theorem applied to $\mathbf{F} = f \nabla g$ yields (5.38). The corresponding equation with f and g switched also holds; by subtracting the latter equation from the former we obtain (5.39). $\qquad \square$

The directional derivative $\nabla f \cdot \mathbf{n}$ that occurs in these formulas is called the **outward normal derivative** of f on ∂R and is often denoted by $\partial f / \partial n$.

EXERCISES

In several of these exercises it will be useful to note that if S_r is the sphere of radius r about the origin, the unit outward normal to S_r at a point $\mathbf{x} \in S_r$ is just $r^{-1}\mathbf{x}$. This is geometrically obvious if you think about it a little. Alternatively, since S_r is a level set of the function $|\mathbf{x}|^2 = x^2 + y^2 + z^2$, we know that $\nabla(|\mathbf{x}|^2) = 2x\mathbf{i} + 2y\mathbf{j} + 2z\mathbf{k} = 2\mathbf{x}$ is normal to S_r, so the unit normal is $|\mathbf{x}|^{-1}\mathbf{x} = r^{-1}|\mathbf{x}|$ for $\mathbf{x} \in S_r$.

1. Use the divergence theorem to evaluate the surface integral $\iint_S \mathbf{F} \cdot \mathbf{n} \, dA$ for the following $\mathbf{F}$ and S, where S is oriented so that the positive normal points out of the region bounded by S.

   a. $\mathbf{F}$, S as in Exercise 8b in §5.3.

   b. $\mathbf{F}$, S as in Exercise 8e in §5.3.

   c. $\mathbf{F}(x, y, z) = x^2\mathbf{i} + y^2\mathbf{j} + z^2\mathbf{k}$; S is the surface of the cube $0 \le x, y, z \le a$.

   d. $\mathbf{F}(x, y, z) = (x/a^2)\mathbf{i} + (y/b^2)\mathbf{j} + (z/c^2)\mathbf{k}$; S is the ellipsoid $(x/a)^2 + (y/b)^2 + (z/c)^2 = 1$.

   e. $\mathbf{F}(x, y, z) = x^2\mathbf{i} - 2xy\mathbf{j} + z^2\mathbf{k}$; S is the surface of the cylindrical solid $\{(x, y, z) : (x, y) \in W, \, 1 \le z \le 2\}$ where W is a smoothly bounded regular region in the plane with area A.

2. Let $\mathbf{F}(x, y, z) = (x^2 + y^2 + z^2)(x\mathbf{i} + y\mathbf{j} + z\mathbf{k})$ and let S be the sphere of radius a about the origin. Compute $\iint_S \mathbf{F} \cdot \mathbf{n}$ both directly and by the divergence theorem.

3. Let R be a regular region in $\mathbb{R}^3$ with piecewise smooth boundary. Show that the volume of R is $\frac{1}{3} \iint_{\partial R} \mathbf{F} \cdot \mathbf{n} \, dA$ where $\mathbf{F}(x, y, z) = x\mathbf{i} + y\mathbf{j} + z\mathbf{k}$.

4. Prove the following integration-by-parts formula for triple integrals:

$$\iiint_R f\frac{\partial g}{\partial x} \, dV = -\iiint_R g\frac{\partial f}{\partial x} \, dV + \iint_{\partial R} fgn_x \, dA,$$

where n_x is the x-component of the unit outward normal to ∂R. (Of course, similar formulas also hold with x replaced by y and z.)

5. Suppose R is a regular region in $\mathbb{R}^3$ with piecewise smooth boundary, and f is a function of class C^2 on $\overline{R}$.

   a. Show that $\displaystyle\iint_{\partial R} \frac{\partial f}{\partial n} \, dA = \iiint_R \nabla^2 f \, dV.$

   b. Show that if $\nabla^2 f = 0$, then $\displaystyle\iint_{\partial R} f\frac{\partial f}{\partial n} \, dA = \iiint_R |\nabla f|^2 \, dV.$

6. Let $\mathbf{x} = (x, y, z)$ and $g(\mathbf{x}) = |\mathbf{x}|^{-1} = (x^2 + y^2 + z^2)^{-1/2}$.

   a. Compute $\nabla g(\mathbf{x})$ for $\mathbf{x} \ne \mathbf{0}$.

   b. Show that $\nabla^2 g(\mathbf{x}) = 0$ for $\mathbf{x} \ne \mathbf{0}$. (Cf. Exercise 9 in §2.6.)

   c. Show by direct calculation that $\iint_S (\partial g/\partial n) \, dA = -4\pi$ if S is any sphere centered at the origin.

   d. Since $\partial g/\partial n = \nabla g \cdot \mathbf{n}$ and $\nabla^2 g = \mathrm{div}(\nabla g)$, why do (b) and (c) not contradict the divergence theorem?

   e. Show that $\iint_{\partial R} (\partial g/\partial n) \, dA = -4\pi$ if R is any regular region with piecewise smooth boundary whose interior contains the origin. (*Hint:* Consider the region obtained by excising a small ball about the origin from R.)

7. Suppose that f is a C^2 function on $\mathbb{R}^3$ that satisfies Laplace's equation $\nabla^2 f = 0$.

    a. By applying (5.39) to f and g, with g as in Exercise 6 and $R = \{\mathbf{x} : \epsilon \leq |\mathbf{x}| \leq r\}$, show that the mean values of f on the spheres $|\mathbf{x}| = r$ and $|\mathbf{x}| = \epsilon$ are equal. (Use Exercises 5a and 6.)

    b. Conclude that the mean value of f on any sphere centered at the origin is equal to the value of f at the origin. (*Remark:* There is nothing special about the origin here. By applying this result to $\tilde{f}(\mathbf{x}) = f(\mathbf{x} + \mathbf{a})$, which also satisfies Laplace's equation, we see that the mean value of f on any sphere is the value of f at the center. The converse is also true; a function that has this mean value property must satisfy Laplace's equation.)

5.6 Some Applications to Physics

In this section we illustrate the uses of the divergence theorem by deriving some important differential equations of mathematical physics. We make a standing assumption that all unspecified mathematical functions that denote physical quantities are smooth enough to ensure the validity of the calculations.

    Flow of Material. We have previously alluded to an interpretation of a vector field in terms of material flowing through space. We now develop this idea in more detail.

    Suppose there is some substance moving through a region of space — it might be air, water, electric charge, or whatever. The distribution of the substance is given by a *density function* $\rho(\mathbf{x}, t)$; thus $\rho(\mathbf{x}, t)\,dV$ is the amount of substance at time t in a small box of volume dV located at the point $\mathbf{x} = (x, y, z)$. The substance is moving around, so we also have the *velocity field* $\mathbf{v}(\mathbf{x}, t)$ that gives the velocity of the substance at position $\mathbf{x}$ and time t.

    Now consider a small bit of oriented surface dS (imagined, not physical) with area dA and normal vector $\mathbf{n}$ located near the point $\mathbf{x}$. (We shall picture dS as a parallelogram, but its exact shape is unimportant.) At what rate does the substance flow through this bit of surface?

    First suppose that $\mathbf{n}$ is parallel to the velocity $\mathbf{v} = \mathbf{v}(\mathbf{x}, t)$. We picture a small box with vertical face dS and length $|\mathbf{v}|\,dt$, where dt is a small increment in time, as in Figure 5.8a. We assume that the box is sufficiently small so that that the velocity and density are essentially constant throughout the box during the time interval $(t, t + dt)$. Then the substance that flows through the surface dS in the time interval dt is just the contents of the box at time t. The volume of the box is

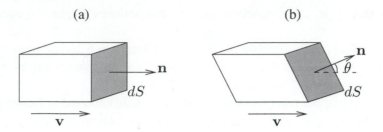

FIGURE 5.8: Flow of material through a surface element dS.

$|\mathbf{v}|\,dt\,dA$, so the amount of substance in the box is $\rho|\mathbf{v}|\,dt\,dA$. In short, the rate of flow of substance through dS is $\rho|\mathbf{v}|\,dA$.

Now suppose, more generally, that the angle from the velocity $\mathbf{v}$ to the normal $\mathbf{n}$ to dS is θ. We apply the same reasoning to the box in Figure 5.8b. The vertical height of the box is now $|\cos\theta|$ times the slant height of dS, so the volume of the box is $|\mathbf{v}|\,|\cos\theta|\,dt\,dA = |\mathbf{v}\cdot\mathbf{n}|\,dt\,dA$. Therefore, the rate of flow of substance through dS is $\rho\mathbf{v}\cdot\mathbf{n}\,dA$ if we take orientation into account, that is, if we count the flow as negative when it goes in across dS in the direction opposite to $\mathbf{n}$.

Passing from the infinitesimal level to the macroscopic level, we conclude that *the rate of flow of substance through a surface S is*

$$\iint_S \mathbf{J}\cdot\mathbf{n}\,dA, \quad where \quad \mathbf{J}(\mathbf{x},t) = \rho(\mathbf{x},t)\mathbf{v}(\mathbf{x},t).$$

The time-dependent vector field $\mathbf{J} = \rho\mathbf{v}$ that occurs here represents the momentum density if ρ is the mass density of the substance, and it represents the current density if the substance is electric charge and ρ is the charge density. Our earlier remarks about interpreting vector fields in terms of flows really mean thinking of the vector field as a momentum or current density.

A Conservation Law. Now we come to the application of the divergence theorem. In the context of the preceding discussion, suppose that the substance is *conserved*, i.e., that it is neither created nor destroyed. Consider a regular region R in space with smooth boundary ∂R. The total amount of substance in R at time t is $\iiint_R \rho(\mathbf{x},t)\,d^3\mathbf{x}$. Since the substance is conserved, the only way for this amount to change is for the substance to flow in or out through ∂R. Therefore,

$$\frac{d}{dt}\iiint_R \rho(\mathbf{x},t)\,d^3\mathbf{x} = -\iint_{\partial R} \mathbf{J}\cdot\mathbf{n}\,dA.$$

(The integral on the right is positive when the substance flows out of S, i.e., when the amount of substance in S is decreasing; hence the minus sign.) The quantity on the left is the integral over R of $\partial \rho / \partial t$, by Theorem 4.47. We can use the divergence theorem to convert the integral on the right to another integral over R, obtaining

$$(5.40) \qquad \iiint_R \frac{\partial \rho}{\partial t}(\mathbf{x}, t)\, dV = - \iiint_R \operatorname{div} \mathbf{J}\, dV.$$

Now, this relation holds for *any* region R. In particular, let us take $R = B_r$ to be the ball of radius r centered at the point $\mathbf{x}$. After division of both sides by the volume of B_r, (5.40) says that the mean values of $\partial \rho / \partial t$ and $- \operatorname{div} \mathbf{J}$ on B_r are equal. Letting $r \to 0$ and assuming that these functions are continuous, we see that their values at the center $\mathbf{x}$ are equal. In short, we have

$$(5.41) \qquad \frac{\partial \rho}{\partial t} + \operatorname{div} \mathbf{J} = 0,$$

the classic differential equation relating the charge and current densities (or mass and momentum densities, etc.).

This argument is reversible; that is, (5.41) implies that the substance is conserved. Indeed, suppose R is a regular region such that no substance flows in or out of R. Integrating (5.41) and using Theorem 4.47 and the divergence theorem, we obtain

$$\frac{d}{dt} \iiint_R \rho\, dV = \iiint_R \frac{\partial \rho}{\partial t}\, dV = - \iiint_R \operatorname{div} \mathbf{J}\, dV = - \iint_{\partial R} \mathbf{J} \cdot \mathbf{n}\, dA = 0,$$

so the amount of substance in R remains constant. Although (5.41) is equivalent to the conservation of the substance, it is more informative than the mere statement the substance is neither created or destroyed; it provides information about how the substance can move around.

The conservation law (5.41) has an important consequence for an incompressible fluid such as water. Incompressibility means that the density ρ is a constant, so that on the one hand, $\partial \rho / \partial t = 0$, and on the other, $\operatorname{div} \mathbf{J} = \operatorname{div}(\rho \mathbf{v}) = \rho \operatorname{div} \mathbf{v}$. Thus, (5.41) implies that *the velocity field $\mathbf{v}$ for an incompressible fluid satisfies* $\operatorname{div} \mathbf{v} = 0$.

The Heat Equation. We now derive a mathematical model for the transfer of heat through a substance by diffusion. (If the substance in question is a fluid like water or air, our model does *not* take convection effects into account; we must assume that the fluid is immobile on the macroscopic scale. But our model is valid for the diffusion of heat in solids as well as in fluids that cannot flow readily, such as air in a down jacket.) Our model will take the form of a differential equation for the temperature $u(\mathbf{x}, t)$ at position $\mathbf{x}$ and time t.

The first basic physical assumption (which may be a simplification of the real-life situation) is that the thermal energy density is proportional to the temperature. The constant of proportionality σ is the *specific heat density*; it is the product of the usual specific heat or heat capacity and the mass density of the substance. The total thermal energy (or "heat," for short) within a region R at time t is then

$$\iiint_R \sigma u(\mathbf{x}, t)\, d^3\mathbf{x}.$$

The next assumption is *Newton's law of cooling,* which says that heat flows from hotter to colder regions at a rate proportional to the difference in temperature. In our situation, the precise interpretation of this statement is that the flux of heat per unit area in the direction of the unit vector $\mathbf{n}$ at the point $\mathbf{x}$ is proportional to the directional derivative $\nabla u(\mathbf{x}) \cdot \mathbf{n}$ of the temperature in the direction $\mathbf{n}$, the constant of proportionality being negative since heat flows in the direction of *decreasing* temperature. Denoting the constant of proportionality by $-K$, then, we see that the flux of heat across an oriented surface S with normal vector $\mathbf{n}$ is

$$-\iint_S K\nabla u \cdot \mathbf{n}\, dA.$$

K is called the *thermal conductivity.*

Next, the amount of heat in a regular region R can change only by the flow of heat across the boundary ∂R or by the creation or destruction of heat within R (by a chemical or nuclear reaction, for example). Thus, if we denote by $F(\mathbf{x}, t)$ the rate per unit volume at which heat is being produced at position $\mathbf{x}$ at time t, we have

$$\frac{d}{dt}\iiint_R \sigma u(\mathbf{x}, t)\, d^3\mathbf{x} = \iint_{\partial R} K\nabla u(\mathbf{x}, t) \cdot \mathbf{n}\, dA + \iiint_R F(\mathbf{x}, t)\, d^3\mathbf{x}.$$

Here $\mathbf{n}$ denotes the unit *outward* normal to ∂R, as usual, and the minus sign on the surface integral has disappeared because a positive flow of heat *out* of R represents a *decrease* of heat in R.

As in the preceding subsection, we bring the d/dt inside the integral and apply the divergence theorem to obtain

$$\sigma \iiint_R \frac{\partial u}{\partial t}\, dV = K \iiint_R \nabla^2 u\, dV + \iiint_R F\, dV.$$

Since this holds for an arbitrary regular region R, we conclude as before that

(5.42)
$$\sigma \frac{\partial u}{\partial t}(\mathbf{x}, t) = K \nabla^2 u(\mathbf{x}, t) + F(\mathbf{x}, t).$$

This partial differential equation is known as the (inhomogeneous) **heat equation**; it is of fundamental importance in the study of all sorts of diffusion processes. The important special case $F = 0$ (the homogeneous equation) is what is usually called the heat equation.

We have implicitly assumed that the specific heat density σ and the thermal conductivity K are constants. However, the same arguments apply to the more general situation where they are allowed to depend on position, as will be the case where the material through which the heat is diffusing varies in some way from point to point. The reader may verify that the result is the following generalized heat equation:

$$\sigma(\mathbf{x}) \frac{\partial u}{\partial t}(\mathbf{x}, t) = \mathrm{div}\left[K(\mathbf{x}) \nabla u(\mathbf{x}, t) \right] + F(\mathbf{x}, t).$$

Potentials and Laplace's Equation. The **electric field** generated by a system of electric charges is the vector field $\mathbf{E}$ whose value at a point $\mathbf{x}$ is the force felt by a unit positive charge locted at $\mathbf{x}$ as the result of the electrostatic attraction or repulsion to the system of charges. If the system is just a single unit positive charge at the point $\mathbf{p}$, the field is given by the usual inverse square law force, $\mathbf{E}(\mathbf{x}) = (\mathbf{x} - \mathbf{p})/|\mathbf{x} - \mathbf{p}|^3$. (There should be a constant of proportionality, but we shall assume that units of measurement have been chosen so that the constant is 1.) For many purposes, it is more convenient to work with the electric **potential** $u(\mathbf{x}) = |\mathbf{p} - \mathbf{x}|^{-1}$, which is related to the electric field $\mathbf{E}$ by

$$\mathbf{E} = -\nabla u.$$

(For any points $\mathbf{x}_1$ and $\mathbf{x}_2$, $u(\mathbf{x}_2) - u(\mathbf{x}_1)$ is the work done in moving a unit positive charge from $\mathbf{x}_1$ to $\mathbf{x}_2$ through the field $\mathbf{E}$.)

If, instead of a single charge at one point, our system of charges consists of a number of charges located at different points, the electric field (resp. electric potential) generated by the system is just the sum of the fields (resp. potentials) generated by the individual charges. We wish to consider the case where there is a continuous distribution of charge (an idealization, but a useful one) in some bounded region of space. That is, we are given a charge density function $\rho(\mathbf{p})$, a continuous function that vanishes outside some bounded set R. The field generated by such a charge distribution is found in the usual way: Chop up the set R into tiny pieces, treat the charge coming from each piece as a point charge, and add up the resulting fields or potentials. We shall work primarily with the potentials, for which the result is

$$(5.43) \qquad u(\mathbf{x}) = \iiint_{\mathbb{R}^3} \frac{\rho(\mathbf{p})}{|\mathbf{p} - \mathbf{x}|}\, d^3\mathbf{p}.$$

It will be convenient to make the substitution $\mathbf{y} = \mathbf{p} - \mathbf{x}$. This is just a translation of coordinates, so its Jacobian is 1, and we obtain

$$(5.44) \qquad u(\mathbf{x}) = \iiint_{\mathbb{R}^3} \frac{\rho(\mathbf{x} + \mathbf{y})}{|\mathbf{y}|}\, d^3\mathbf{y}.$$

A couple of comments are in order about this integral. We have written it as an integral over $\mathbb{R}^3$, but it really extends only over the bounded region $R - \mathbf{x} = \{\mathbf{y} : \mathbf{x} + \mathbf{y} \in R\}$ on which $\rho(\mathbf{x} + \mathbf{y}) \neq 0$. The integral is improper because of the singularity of $|\mathbf{y}|^{-1}$ at the origin, but one can easily see that it is absolutely convergent by Proposition 4.65.

The main object of this subsection is to derive an important differential equation relating u and ρ. The key point is the fact that the Laplacian of $|\mathbf{y}|^{-1}$ vanishes except at the origin (where it is undefined):

$$(5.45) \qquad \nabla^2\big(|\mathbf{y}|^{-1}\big) = 0 \text{ for } \mathbf{y} \neq \mathbf{0}.$$

The proof is a straightforward calculation (Exercise 2c in §5.4 or Exercise 6b in §5.5).

5.46 Theorem. *Suppose ρ is a function of class C^2 on $\mathbb{R}^3$ that vanishes outside a bounded set, and let u be defined by (5.44). Then u is of class C^2 and $\nabla^2 u = -4\pi\rho$.*

Proof. We can differentiate u by passing the derivatives under the integral sign. They fall on ρ, which is assumed to be of class C^2, so u is of class C^2 and

$$\nabla^2 u(\mathbf{x}) = \iiint \frac{\nabla^2 \rho(\mathbf{x} + \mathbf{y})}{|\mathbf{y}|}\, d^3\mathbf{y}.$$

(Strictly speaking, Theorem 4.47 does not apply because of the singularity of the integrand at the origin, but this is a minor technicality. One can finesse the problem, for example, by switching to spherical coordinates, in which the $r^2 \sin\varphi$ coming from the volume element cancels the r^{-1} of the integrand with room to spare.) Here $\nabla^2\rho(\mathbf{x} + \mathbf{y})$ is obtained by differentiating ρ with respect to $\mathbf{x}$, but the same result is obtained by taking the derivatives with respect to $\mathbf{y}$, for $\partial_{x_j}[\rho(\mathbf{x} + \mathbf{y})] = (\partial_j \rho)(\mathbf{x} + \mathbf{y}) = \partial_{y_j}[\rho(\mathbf{x} + \mathbf{y})]$. We can therefore use Green's formula to transfer the derivatives to $|\mathbf{y}|^{-1}$. We need to take some care, however, since the singularity of $|\mathbf{y}|^{-1}$ does not remain harmless after being differentiated twice.

Let us fix the point $\mathbf{x}$ and choose positive numbers ϵ and K, with $\epsilon < 1$ and K large enough so that $\rho(\mathbf{x} + \mathbf{y}) = 0$ if $|\mathbf{y}| \geq K - 1$. Let $R_{\epsilon,K} = \{\mathbf{y} : \epsilon < |\mathbf{y}| < K\}$. We then have

$$\nabla^2 u(\mathbf{x}) = \lim_{\epsilon \to 0} \iiint_{R_{\epsilon,K}} \frac{\nabla^2 \rho(\mathbf{x} + \mathbf{y})}{|\mathbf{y}|} d^3\mathbf{y}.$$

The integrand has no singularities in the region $R_{\epsilon,K}$, so we can apply Green's formula (5.39) to obtain

$$\nabla^2 u(\mathbf{x}) = \lim_{\epsilon \to 0} \left[\iiint_{R_{\epsilon,K}} \rho(\mathbf{x} + \mathbf{y}) \nabla^2(|\mathbf{y}|^{-1}) d^3\mathbf{y} \right.$$
$$\left. + \iint_{\partial R_{\epsilon,K}} \left[\nabla\rho(\mathbf{x} + \mathbf{y})|\mathbf{y}|^{-1} - \rho(\mathbf{x} + \mathbf{y})\nabla(|\mathbf{y}|^{-1}) \right] \cdot \mathbf{n} \, dA \right].$$

The integral over $R_{\epsilon,K}$ on the right vanishes by (5.45). Also, the boundary of $R_{\epsilon,K}$ consists of two pieces, the sphere $|\mathbf{y}| = K$ and the sphere $|\mathbf{y}| = \epsilon$, and the integral over $|\mathbf{y}| = K$ is zero because $\rho(\mathbf{x} + \mathbf{y})$ and its derivatives vanish for $|\mathbf{y}| > K - 1$. Therefore,

$$(5.47) \quad \nabla^2 u(\mathbf{x}) = \lim_{\epsilon \to 0} \iint_{|\mathbf{y}|=\epsilon} \left[\nabla\rho(\mathbf{x} + \mathbf{y})|\mathbf{y}|^{-1} - \rho(\mathbf{x} + \mathbf{y})\nabla(|\mathbf{y}|^{-1}) \right] \cdot \mathbf{n} \, dA.$$

Here $\mathbf{n}$ denotes the unit normal to the sphere $|\mathbf{y}| = \epsilon$ that is *outward* with respect to $R_{\epsilon,K}$ and hence *inward* in the usual sense.

Since the first derivatives of ρ are continuous, $|\nabla\rho(\mathbf{x} + \mathbf{y})|$ is bounded by some constant C for $|\mathbf{y}| \leq 1$, and hence

$$(5.48) \quad \left| \iint_{|\mathbf{y}|=\epsilon} \frac{\nabla\rho(\mathbf{x} + \mathbf{y}) \cdot \mathbf{n}}{|\mathbf{y}|} dA \right| \leq \iint_{|\mathbf{y}|=\epsilon} \frac{C}{\epsilon} dA = \frac{C}{\epsilon} 4\pi\epsilon^2 = 4\pi C\epsilon,$$

which vanishes as $\epsilon \to 0$. To evaluate the second term in (5.47), we observe that $\mathbf{n} = -\epsilon^{-1}\mathbf{y}$. (See the remark preceding the exercises in §5.5.) An easy calculation gives $\nabla(|\mathbf{y}|^{-1}) = -\mathbf{y}/|\mathbf{y}|^3$, so $\nabla(|\mathbf{y}|^{-1}) \cdot \mathbf{n} = \epsilon^{-1}|\mathbf{y}|^2/|\mathbf{y}|^3 = \epsilon^{-2}$. Therefore, (5.47) and (5.48) show that

$$\nabla^2 u(\mathbf{x}) = -\lim_{\epsilon \to 0} \iint_{|\mathbf{y}|=\epsilon} \frac{\rho(\mathbf{x} + \mathbf{y})}{\epsilon^2} dA$$

$$= (-4\pi) \lim_{\epsilon \to 0} \left[\frac{1}{4\pi\epsilon^2} \iint_{|\mathbf{y}|=\epsilon} \rho(\mathbf{x} + \mathbf{y}) \, dA \right].$$

But the expression inside the brackets is just the mean value of $\rho(\mathbf{x} + \mathbf{y})$ on the sphere $|\mathbf{y}| = \epsilon$, which tends to $\rho(\mathbf{x})$ as $\epsilon \to 0$, so the proof is complete. $\square$

Remark. The hypothesis that ρ is of class C^2 can be weakened (C^1 is more than enough); we impose it simply to avoid technicalities in the proof. In fact, if ρ vanishes outside a bounded set and is integrable there, then the equation $\nabla^2 u = -4\pi\rho$ holds on any open set on which ρ is C^1. The key ideas of the proof are all present in the preceding argument.

5.49 Corollary. *The electric field* $\mathbf{E}$ *is related to the charge density* ρ *by* div $\mathbf{E} = 4\pi\rho$.

Proof. $4\pi\rho = \operatorname{div}(-\nabla u) = \operatorname{div} E$. $\square$

The differential equation $\nabla^2 u = -4\pi\rho$ is called the **inhomogeneous Laplace equation** or **Poisson equation**. The special case $\nabla^2 u = 0$, valid in regions where there are no charges, is the (homogeneous) **Laplace equation**. These equations have been extensively studied; solutions of $\nabla^2 u = 0$, in particular, have many interesting properties and applications in many areas.

Everything we have said applies also to gravitational potentials and fields generated by mass distributions with mass density ρ, except for some minus signs coming from the fact that masses attract whereas like charges repel. Specifically, the gravitational potential is given by $u(\mathbf{x}) = -\iiint \rho(\mathbf{x} + \mathbf{y})|\mathbf{y}|^{-1}\, d^3\mathbf{y}$, and it satisfies $\nabla^2 u = 4\pi\rho$.

It should be noted that the preceding discussion applies only to situations where the charge or mass density ρ is *static*, that is, unchanging in time. If the charges or masses move around, things become more complicated. The basic reason is that if a charge or mass at $\mathbf{p}$ is moved to a nearby point $\mathbf{p}'$, the potential it induces cannot change instantly from $|\mathbf{x} - \mathbf{p}|^{-1}$ to $|\mathbf{x} - \mathbf{p}'|^{-1}$ throughout all of space, because the news of the move can only travel with the speed of light. For electricity, the physics of time-varying fields is contained in Maxwell's equations, which we shall present below; for gravity, it is described by general relativity. (If the time dependence is not too rapid, however, the relativistic effects will be small and the preceding calculations can be used as a good approximation. This is more often the case with gravity than with electricity, because gravity is a much weaker interaction.)

Maxwell's Equations. Maxwell's equations are the fundamental differential equations that are the foundation for the classical (unquantized) theory of electicity and magnetism. They relate the electric field $\mathbf{E}$, the magnetic field $\mathbf{B}$, the charge density ρ, and the current density $\mathbf{J}$. In suitably normalized units, they are

(5.50)
$$\operatorname{div} \mathbf{E} = 4\pi\rho, \qquad \operatorname{curl} \mathbf{E} = -\frac{1}{c}\frac{\partial \mathbf{B}}{\partial t},$$
$$\operatorname{div} \mathbf{B} = 0, \qquad \operatorname{curl} \mathbf{B} = \frac{1}{c}\frac{\partial \mathbf{E}}{\partial t} + \frac{4\pi}{c}\mathbf{J},$$

where c is the speed of light. This is not the place for a thorough study of Maxwell's equations and their consequences for physics, but we wish to point out a couple of features of them in connection with the ideas we have been developing. In what follows we shall assume that all functions in question are of class C^2, so that the second derivatives make sense and the mixed partials are equal.

First, Maxwell's equations contain the law of conservation of charge. Indeed, by formula (5.30) we have

$$\frac{\partial \rho}{\partial t} = \frac{1}{4\pi}\operatorname{div}\frac{\partial \mathbf{E}}{\partial t} = \frac{c}{4\pi}\operatorname{div}(\operatorname{curl}\mathbf{B}) - \operatorname{div}\mathbf{J} = -\operatorname{div}\mathbf{J},$$

and this is the conservation law in the form (5.41). Second, in a region of space with no charges or currents ($\rho = 0$ and $\mathbf{J} = 0$), by formula (5.33) we have

$$\nabla^2\mathbf{E} = \nabla(\operatorname{div}\mathbf{E}) - \operatorname{curl}(\operatorname{curl}\mathbf{E}) = 0 + \frac{1}{c}\operatorname{curl}\frac{\partial \mathbf{B}}{\partial t} = \frac{1}{c^2}\frac{\partial^2 \mathbf{E}}{\partial t^2}$$

and

$$\nabla^2\mathbf{B} = \nabla(\operatorname{div}\mathbf{B}) - \operatorname{curl}(\operatorname{curl}\mathbf{B}) = 0 - \frac{1}{c}\operatorname{curl}\frac{\partial \mathbf{E}}{\partial t} = \frac{1}{c^2}\frac{\partial^2 \mathbf{B}}{\partial t^2}.$$

That is, the components of $\mathbf{E}$ and $\mathbf{B}$ all satisfy the differential equation

(5.51)
$$\nabla^2 f = \frac{1}{c^2}\frac{\partial^2 f}{\partial t^2}.$$

This is the **wave equation**, another of the fundamental equations of mathematical physics. It describes the propagation of waves in many different situations; here it concerns electromagnetic radiation — light, radio waves, X-rays, and so on.

EXERCISES

Besides distributions of charge or mass in 3-space, one can consider distributions on surfaces or curves (physically: thin plates or wires). The formula for the associated potential or field is similar to (5.43) except that the triple integral is replaced by a surface or line integral, and the density ρ represents charge or mass per unit area or unit length rather than unit volume. In the following exercises, "uniform" means "of constant density."

1. Consider a uniform distribution of mass on the sphere of radius r about the origin. Show that
   a. inside the sphere, the potential is constant and the gravitational field vanishes;
   b. outside the sphere; the potential and field are the same as if the entire mass were located at the origin.

2. Consider a uniform distribution of mass on the solid ball of radius R about the origin. Show that the gravitational field at a point $\mathbf{x}$ is the same as if the mass closer to the origin than $\mathbf{x}$ were all located at the origin and the mass farther from the origin than $\mathbf{x}$ (if any) were absent. (Use Exercise 1.)

3. Consider a uniform distribution of charge on the z-axis, with density ρ (charge per unit length).
   a. Compute the electric field generated by this distribution. (The relevant formula is similar to (5.43), but $1/|\mathbf{p} - \mathbf{x}|$ is replaced by the negative of its gradient with respect to $\mathbf{x}$, namely, $(\mathbf{x} - \mathbf{p})/|\mathbf{x} - \mathbf{p}|^3$.)
   b. Show that the modification of (5.43) that presumably gives the potential for this charge distribution is a divergent integral.
   c. To resolve the difficulty presented by (b), we make use of the fact that the defining property of the potential u, namely $\nabla u = -\mathbf{E}$, only determines u up to an additive constant, so we may subtract constants from u without affecting the physics. Consider instead a uniform distribution of charge on the interval $[-R, R]$ on the z-axis with density ρ. Compute the potential u_R generated by this distribution, and show that $u_R - 2\rho \log R$ converges as $R \to \infty$ to a function whose gradient is the negative of the field found in (a). (This sort of removal of divergences by "subtracting off infinite constants" is common in quantum field theory, where it is known as *renormalization*.)

4. Prove the following two-dimensional analogue of Theorem 5.46: Suppose ρ is a function of class C^2 on $\mathbb{R}^2$ that vanishes outside a bounded set, and let

$$u(\mathbf{x}) = \int \rho(\mathbf{x} + \mathbf{y}) \log |\mathbf{y}| \, d^2 \mathbf{y}.$$

Then u is of class C^2 and $\nabla^2 u = 2\pi \rho$. (The proof is very similar to that of Theorem 5.46; see Exercise 2d in §5.4.)

5.7   Stokes's Theorem

Stokes's theorem is the generalization of Green's theorem in which the plane is replaced by a curved surface. The precise setting is as follows.

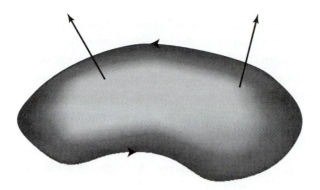

FIGURE 5.9: An oriented surface and its positively oriented boundary.

Let S_0 be a smooth surface in $\mathbb{R}^3$, and let S be a region in S_0 that is bounded by a piecewise smooth curve ∂S. By this we mean that ∂S is the boundary of S *within the surface* S_0.[1] (Of course, if we think of S as a subset of $\mathbb{R}^3$, it has no interior and so is its own boundary.) We assume that S is oriented by a choice of normal vector field $\mathbf{n}$, so we can speak of the positive and negative sides of S, and we give ∂S the orientation compatible with the orientation of S in the sense we used in Green's theorem. This means, informally speaking, that if you walk around ∂S in the positive direction, standing on the positive side of S, then S is on your left. In more mathematical terms, if $\mathbf{t}$ is the unit tangent to ∂S in the forward direction at a point $\mathbf{x} \in \partial S$, then $\mathbf{n} \times \mathbf{t}$, considered as an arrow emanating from $\mathbf{x}$, points into S. See Figure 5.9.

5.52 Theorem (Stokes's Theorem). *Let S and ∂S be as described above, and let $\mathbf{F}$ be a C^1 vector field defined on some neighborhood of S in $\mathbb{R}^3$. Then*

$$(5.53) \qquad \int_{\partial S} \mathbf{F} \cdot d\mathbf{x} = \iint_S (\mathrm{curl}\,\mathbf{F}) \cdot \mathbf{n}\, dA.$$

Proof. If S is a region in the xy-plane, then $\mathbf{n} = \mathbf{k} = (0,0,1)$; moreover, $\mathbf{F} \cdot d\mathbf{x}$ involves only the x- and y-components of $\mathbf{F}$, i.e., F_1 and F_2, and $(\mathrm{curl}\,\mathbf{F}) \cdot \mathbf{n}$ is the z-component of $\mathrm{curl}\,\mathbf{F}$, namely $\partial_1 F_2 - \partial_2 F_1$. Hence Stokes's theorem reduces to Green's theorem in this case.

---

[1] Here are the precise definitions: A point $\mathbf{x} \in S$ is in the interior of S relative to S_0 if it has a neighborhood U (in $\mathbb{R}^3$) such that $U \cap S_0 \subset S$; it is in the boundary of S relative to S_0 if all of its neighborhoods contain points in S and points in $S_0 \setminus S$. S is regular if it is compact and every neighborhood of every (relative) boundary point contains points in the (relative) interior.

Next, suppose that S admits a parametrization $\mathbf{x} = \mathbf{G}(u, v)$, so that S is the image under $\mathbf{G}$ of a regular region W in the uv-plane and ∂S is the image of ∂W. We assume that this parametrization yields the given orientation on S (otherwise, just switch u and v). We use the parametrization to pull back the integrals over S and ∂S to integrals over W and ∂W, and we apply Green's theorem to the latter. It is just a matter of seeing that this change of variables works out as it should.

As in the proofs of Green's theorem and the divergence theorem, we consider the components of $\mathbf{F}$ separately. Thus, if we write $\mathbf{F} = F\mathbf{i} + G\mathbf{j} + H\mathbf{k}$, it is enough to prove (5.53) for $F\mathbf{i}$, $G\mathbf{j}$, and $H\mathbf{k}$ separately. All three of them work the same way, so we shall just do $F\mathbf{i}$, for which (5.53) reduces to

$$(5.54) \qquad \int_{\partial S} F(x, y, z)\, dx = \iint_S [(\partial_z F)\mathbf{j} - (\partial_y F)\mathbf{k}] \cdot \mathbf{n}\, dA.$$

Now, using the parametrization $\mathbf{x} = \mathbf{G}(u, v)$, we have

(5.55)
$$\iint_S [(\partial_z F)\mathbf{j} - (\partial_y F)\mathbf{k}] \cdot \mathbf{n}\, dA = \iint_W [(\partial_z F)\mathbf{j} - (\partial_y F)\mathbf{k}] \cdot \left(\frac{\partial \mathbf{G}}{\partial u} \times \frac{\partial \mathbf{G}}{\partial v} \right) du\, dv$$

$$= \iint_W \left(\frac{\partial F}{\partial z} \frac{\partial(z, x)}{\partial(u, v)} - \frac{\partial F}{\partial y} \frac{\partial(x, y)}{\partial(u, v)} \right) du\, dv.$$

On the other hand, since the formalism of differentials automatically encodes the chain rule,

$$\int_{\partial S} F\, dx = \int_{\partial W} F \left(\frac{\partial x}{\partial u}\, du + \frac{\partial x}{\partial v}\, dv \right).$$

(In both of these equations, F and its derivatives are evaluated at $\mathbf{G}(u, v)$.) We apply Green's theorem to this last line integral:

$$\int_{\partial W} F \left(\frac{\partial x}{\partial u}\, du + \frac{\partial x}{\partial v}\, dv \right) = \iint_W \left(\frac{\partial}{\partial u} \left[F \frac{\partial x}{\partial v} \right] - \frac{\partial}{\partial v} \left[F \frac{\partial x}{\partial u} \right] \right) du\, dv.$$

By the product rule and the chain rule, the integrand on the right equals

$$\left[\frac{\partial F}{\partial x} \frac{\partial x}{\partial u} + \frac{\partial F}{\partial y} \frac{\partial y}{\partial u} + \frac{\partial F}{\partial z} \frac{\partial z}{\partial u} \right] \frac{\partial x}{\partial v} + F \frac{\partial^2 x}{\partial u \partial v}$$

$$- \left[\frac{\partial F}{\partial x} \frac{\partial x}{\partial v} + \frac{\partial F}{\partial y} \frac{\partial y}{\partial v} + \frac{\partial F}{\partial z} \frac{\partial z}{\partial v} \right] \frac{\partial x}{\partial u} - F \frac{\partial^2 x}{\partial v \partial u}$$

$$= \frac{\partial F}{\partial z} \frac{\partial(z, x)}{\partial(u, v)} - \frac{\partial F}{\partial y} \frac{\partial(x, y)}{\partial(u, v)}.$$

But this is the integrand on the right side of (5.55), so (5.54) is proved.

Finally, as in the proofs of Green's theorem and the divergence theorem, we obtain Stokes's theorem more generally for surfaces S that can be cut up into a finite number of pieces that each admit a parametrization by applying the preceding argument to the pieces and adding up the results. Alternatively, we can obtain Stokes's theorem for general surfaces by an adaptation of the proof of Green's theorem in Appendix B.7. □

EXAMPLE 1. Use Stokes's theorem to compute $\int_C \mathbf{F} \cdot d\mathbf{x}$ where $\mathbf{F}(x, y, z) = \sqrt{x^2 + 1}\,\mathbf{i} + x\mathbf{j} + 2y\mathbf{k}$ and C is the intersection of the surfaces $z = xy$ and $x^2 + y^2 = 1$, oriented counterclockwise as viewed from above.

Solution. C is the boundary of the portion of the surface $z = xy$ inside the cylinder $x^2 + y^2 = 1$, and its orientation is compatible with the orientation of S with the normal pointing upward. We have $\operatorname{curl}\mathbf{F}(x, y, z) = 2\mathbf{i} + \mathbf{k}$ and $\mathbf{n}\,dA = (-y\mathbf{i} - x\mathbf{j} + \mathbf{k})\,dx\,dy$, so

$$\int_C \mathbf{F} \cdot d\mathbf{x} = \iint_{x^2+y^2 \leq 1} (1 - 2y)\,dx\,dy = \pi.$$

(No computation is necessary here; the integral of 1 is the area of the disc and the integral of $-2y$ vanishes by symmetry.)

There is an interesting feature of Stokes's theorem that does not appear in its siblings. A closed curve in $\mathbb{R}^2$ is the boundary of just one regular region in $\mathbb{R}^2$, and a closed surface in $\mathbb{R}^3$ is the boundary of just one regular region in $\mathbb{R}^3$; but a closed curve in $\mathbb{R}^3$ is the boundary of infinitely many surfaces in $\mathbb{R}^3$! For example, the unit circle in the xy-plane is the boundary of the unit disc in the xy-plane, the upper and lower hemispheres of the unit sphere in $\mathbb{R}^3$, the portion of the paraboloid $z = 1 - x^2 - y^2$ lying above the unit disc, and so forth. Stokes's theorem says that if C is a closed curve in $\mathbb{R}^3$ and S is *any* oriented surface bounded by C, then

$$\int_C \mathbf{F} \cdot d\mathbf{x} = \iint_S (\operatorname{curl}\mathbf{F}) \cdot \mathbf{n}\,dA$$

for any C^1 vector field $\mathbf{F}$, provided that the orientations on C and S are compatible.

EXAMPLE 2. Let $\mathbf{F}(x, y, z) = [e^{xz} + e^{x+2y}]\mathbf{i} + [\log(2 + y + z) + 2e^{x+2y}]\mathbf{j} + 3xyz\mathbf{k}$. Compute $\iint_S \operatorname{curl}\mathbf{F} \cdot \mathbf{n}\,dA$, where S is the portion of the surface $z = 1 - x^2 - y^2$ above the xy-plane, oriented with the normal pointing upward.

Solution. We have $\operatorname{curl}\mathbf{F}(x, y, z) = [3xz - (2+y+z)^{-1}]\mathbf{i} + [xe^{xz} - 3yz]\mathbf{j}$ and $\mathbf{n}\,dA = (2x\mathbf{i} + 2y\mathbf{j} + \mathbf{k})\,dx\,dy$, so direct evaluation of the integral is quite unpleasant. By Stokes's theorem, the integral equals $\int_C \mathbf{F} \cdot d\mathbf{x}$ where C is the unit circle in the xy-plane; this is not much better. However, by Stokes's theorem again, the latter line integral is equal to $\iint_D \operatorname{curl} \mathbf{F} \cdot \mathbf{n}\, dA$ where D is the unit disc in the xy-plane. Here $\mathbf{n} = \mathbf{k}$, so $\operatorname{curl} \mathbf{F} \cdot \mathbf{n} = 0$ and the integral vanishes!

Here is an analogue of the fact that the integral of the gradient of a function over any closed curve vanishes:

5.56 Corollary. *If S is a closed surface (i.e., a surface with no boundary) in $\mathbb{R}^3$ with unit outward normal $\mathbf{n}$, and $\mathbf{F}$ is a C^1 vector field on S, then $\iint_S (\operatorname{curl} \mathbf{F}) \cdot \mathbf{n}\, dA = 0$.*

Proof. If $\mathbf{F}$ extends differentiably to the region R inside S, this follows from the divergence theorem, since $\operatorname{div}(\operatorname{curl} \mathbf{F}) = 0$ for any $\mathbf{F}$. However, it is true even if $\mathbf{F}$ has singularities inside S. To see this, draw a small simple closed curve C in S (say, the image of a small circle in the uv-plane under a parametrization $\mathbf{x} = \mathbf{G}(u, v)$). C divides S into two regular regions S_1 and S_2, and we have

$$(5.57) \qquad \iint_S (\operatorname{curl} \mathbf{F}) \cdot \mathbf{n}\, dA = \iint_{S_1} (\operatorname{curl} \mathbf{F}) \cdot \mathbf{n}\, dA + \iint_{S_2} (\operatorname{curl} \mathbf{F}) \cdot \mathbf{n}\, dA.$$

On the other hand, if we give C the orientation compatible with S_1, Stokes's theorem gives

$$\iint_{S_1} (\operatorname{curl} \mathbf{F}) \cdot \mathbf{n}\, dA = \int_C \mathbf{F} \cdot d\mathbf{x} = - \iint_{S_2} (\operatorname{curl} \mathbf{F}) \cdot \mathbf{n}\, dA,$$

because the orientation compatible with S_2 is the opposite one. Hence the terms on the right of (5.57) cancel.

(*Note:* We had to say that C is a "small" closed curve, because otherwise C might not divide S into two pieces. For example, take S to be a torus [the surface of a doughnut] and C to be a circle that goes completely around S in one direction.) $\qquad\square$

Stokes's theorem gives a geometric, coordinate-free interpretation of the curl of a vector field. Namely, suppose $\mathbf{F}$ is a C^1 vector field on some open set containing the point $\mathbf{a}$; here's how to find the component of $\operatorname{curl} \mathbf{F}(\mathbf{a})$ in the direction of any unit vector $\mathbf{u}$, that is, $(\operatorname{curl} \mathbf{F}(\mathbf{a})) \cdot \mathbf{u}$. Let D_ϵ be the disc of radius ϵ centered at $\mathbf{a}$ in the plane perpendicular to $\mathbf{u}$, oriented so that $\mathbf{u}$ is the positive normal for D_ϵ. As $\epsilon \to 0$, the average value of $(\operatorname{curl} \mathbf{F}) \cdot \mathbf{u}$ over D_ϵ approaches its value at $\mathbf{a}$:

$$(\operatorname{curl} \mathbf{F}(\mathbf{a})) \cdot \mathbf{u} = \lim_{\epsilon \to 0} \frac{1}{\pi \epsilon^2} \iint_{D_\epsilon} (\operatorname{curl} \mathbf{F}) \cdot \mathbf{u}\, dA.$$

Since **u** is the normal to D_ϵ, Stokes's theorem gives

$$(5.58) \qquad\qquad (\text{curl}\, \mathbf{F}(\mathbf{a})) \cdot \mathbf{u} = \lim_{\epsilon \to 0} \frac{1}{\pi \epsilon^2} \int_{C_\epsilon} \mathbf{F} \cdot d\mathbf{x},$$

where C_ϵ is the circle of radius ϵ about **a** in the plane perpendicular to **u**, traversed counterclockwise as viewed from the side on which **u** lies. This is the promised coordinate-free description of $\text{curl}\,\mathbf{F}$.

If we think of **F** as a force field, $\int_{C_\epsilon} \mathbf{F} \cdot d\mathbf{x}$ is the work done by **F** on a particle that moves around C_ϵ. Thus (5.58) says that $(\text{curl}\,\mathbf{F}(\mathbf{a})) \cdot \mathbf{u}$ represents the tendency of the force **F** to push the particle around C_ϵ, counterclockwise if $(\text{curl}\,\mathbf{F}(\mathbf{a})) \cdot \mathbf{u}$ is positive and clockwise if it is negative (as viewed from the **u**-side).

EXERCISES

1. Use Stokes's theorem to calculate $\int_C [(x - z)\, dx + (x + y)\, dy + (y + z)\, dz]$ where C is the ellipse where the plane $z = y$ intersects the cylinder $x^2 + y^2 = 1$, oriented counterclockwise as viewed from above.

2. Use Stokes's theorem to evaluate $\int_C [y\, dx + y^2\, dy + (x + 2z)\, dz]$ where C is the curve of intersection of the sphere $x^2 + y^2 + z^2 = a^2$ and the plane $y + z = a$, oriented counterclockwise as viewed from above.

3. Given any nonvertical plane P parallel to the x-axis, let C be the curve of intersection of P with the cylinder $x^2 + y^2 = a^2$. Show that $\int_C [(yz - y)\, dx + (xz + x)\, dy] = 2\pi a^2$.

4. Evaluate $\iint_S \text{curl}\,\mathbf{F} \cdot \mathbf{n}\, dA$ where $\mathbf{F}(x, y, z) = y\mathbf{i} + (x - 2x^3 z)\mathbf{j} + xy^3\mathbf{k}$ and S is the upper half of the sphere $x^2 + y^2 + z^2 = a^2$.

5. Let $\mathbf{F}(x, y, z) = 2x\mathbf{i} + 2y\mathbf{j} + (x^2 + y^2 + z^2)\mathbf{k}$ and let S be the lower half of the ellipsoid $(x^2/4) + (y^2/9) + (z^2/27) = 1$. Use Stokes's theorem to calculate the flux of $\text{curl}\,\mathbf{F}$ across S from the lower side to the upper side.

6. Define the vector field **F** on the complement of the z-axis by $\mathbf{F}(x, y, z) = (-y\mathbf{i} + x\mathbf{j})/(x^2 + y^2)$.
   a. Show that $\text{curl}\,\mathbf{F} = \mathbf{0}$.
   b. Show by direct calculation $\int_C \mathbf{F} \cdot d\mathbf{x} = 2\pi$ for any horizontal circle C centered at a point on the z-axis.
   c. Why do (a) and (b) not contradict Stokes's theorem?

7. Let C_r denote the circle of radius r about the origin in the xz-plane, oriented counterclockwise as viewed from the positive y-axis. Suppose **F** is a C^1 vector field on the complement of the y-axis in $\mathbb{R}^3$ such that $\int_{C_1} \mathbf{F} \cdot d\mathbf{x} = 5$ and $\text{curl}\,\mathbf{F}(x, y, z) = 3\mathbf{j} + (z\mathbf{i} - x\mathbf{k})/(x^2 + z^2)^2$. Compute $\int_{C_r} \mathbf{F} \cdot d\mathbf{x}$ for every r.

8. Let S be a smooth oriented surface in $\mathbb{R}^3$ with piecewise smooth, compatibly oriented boundary ∂S. Suppose f is C^1 and g is C^2 on some open set containing S. Show that

$$\int_{\partial S} f \nabla g \cdot d\mathbf{x} = \iint_S (\nabla f \times \nabla g) \cdot \mathbf{n}\, dA.$$

5.8   Integrating Vector Derivatives

In this section we study the question of solving the equations

$$\operatorname{grad} f = \mathbf{G}, \qquad \operatorname{curl} \mathbf{F} = \mathbf{G}, \qquad \operatorname{div} \mathbf{F} = g$$

for f or $\mathbf{F}$, given g or $\mathbf{G}$. We first consider the equation $\nabla f = \mathbf{G}$, and we begin with a simple and useful result:

5.59. Proposition. *Suppose $\mathbf{G}$ is a continuous vector field on an open set R in $\mathbb{R}^n$. The following two conditions are equivalent:*
a. *If C_1 and C_2 are any two oriented curves in R with the same initial point and the same final point, then $\int_{C_1} \mathbf{G} \cdot d\mathbf{x} = \int_{C_2} \mathbf{G} \cdot d\mathbf{x}$.*
b. *If C is any closed curve in R, $\int_C \mathbf{G} \cdot d\mathbf{x} = 0$.*

Proof. (a) implies (b): Suppose C starts and ends at $\mathbf{a}$. Then C has the same initial and final point as the "constant curve" C_2 described by $\mathbf{x}(t) \equiv \mathbf{a}$, and obviously $\int_{C_2} \mathbf{G} \cdot d\mathbf{x} = 0$ since $d\mathbf{x} = \mathbf{0}$ on C_2.

(b) implies (a): Suppose C_1 and C_2 start at $\mathbf{a}$ and end at $\mathbf{b}$. Let C be the closed curve obtained by following C_1 from $\mathbf{a}$ to $\mathbf{b}$ and then C_2 backwards from $\mathbf{b}$ to $\mathbf{a}$. Then $0 = \int_C \mathbf{G} \cdot d\mathbf{x} = \int_{C_1} \mathbf{G} \cdot d\mathbf{x} - \int_{C_2} \mathbf{G} \cdot d\mathbf{x}$. $\qquad\square$

A vector field $\mathbf{G}$ that satisfies (a) and (b) is called **conservative** in the region R. (The word "conservative" has to do with conservation of energy. If we interpret $\mathbf{G}$ as a force field, condition (b) says that the force does no net work on a particle that returns to its starting point.) A good deal of mathematical physics is based on the following characterization of conservative vector fields:

5.60. Proposition. *A continuous vector field $\mathbf{G}$ in an open set $R \subset \mathbb{R}^n$ is conservative if and only if $\mathbf{G}$ is the gradient of a C^1 function f on R.*

Proof. If $\mathbf{G} = \nabla f$ and C is a closed curve parametrized by $\mathbf{x} = \mathbf{g}(t)$, $a \le t \le b$, by the chain rule we have

$$\int_C \nabla f \cdot d\mathbf{x} = \int_a^b \nabla f(\mathbf{g}(t)) \cdot \mathbf{g}'(t)\, dt = \int_a^b \frac{d}{dt} f(\mathbf{g}(t))\, dt$$

$$= f(\mathbf{g}(b)) - f(\mathbf{g}(a)) = 0$$

because $\mathbf{g}(b) = \mathbf{g}(a)$, so condition (b) in Proposition 5.59 is satisfied.

Conversely, suppose $\mathbf{G}$ is conservative in R. To construct a function of which $\mathbf{G}$ is the gradient, we shall assume R is connected. (Otherwise we can consider each connected piece of R separately.) Pick a base point $\mathbf{a} \in R$. For any $\mathbf{x} \in R$, let C be a curve in R from $\mathbf{a}$ to $\mathbf{x}$ — such a curve always exists, by Theorem 1.30 — and define $f(\mathbf{x}) = \int_C \mathbf{G} \cdot d\mathbf{x}$. This definition makes sense by condition (a) in Proposition 5.59: It doesn't matter which curve we pick. We shall show that $\mathbf{G} = \nabla f$ by showing that $F_j = \partial_j f$ for each j; it is enough to do the case $j = 1$. Let $\mathbf{h} = (h, 0, \ldots, 0)$. Given $\mathbf{x} \in R$, suppose h is small enough so that the line segment L from $\mathbf{x}$ to $\mathbf{x} + \mathbf{h}$ lies entirely in R. We have $f(\mathbf{x}) = \int_C \mathbf{G} \cdot d\mathbf{x}$ where C is a curve from $\mathbf{a}$ to $\mathbf{x}$. We can make a curve from $\mathbf{a}$ to $\mathbf{x} + \mathbf{h}$ by joining L onto the end of C, so that $f(\mathbf{x} + \mathbf{h}) = \int_C \mathbf{G} \cdot d\mathbf{x} + \int_L \mathbf{G} \cdot d\mathbf{x}$. But then

$$\frac{f(\mathbf{x} + \mathbf{h}) - f(\mathbf{x})}{h} = \frac{1}{h} \int_L \mathbf{G} \cdot d\mathbf{x} = \frac{1}{h} \int_0^h G_1(x + t, x_2, \ldots, x_n) \, dt,$$

and by letting $h \to 0$ we obtain $\partial_1 f(\mathbf{x}) = G_1(\mathbf{x})$. $\qquad \square$

The function f in Proposition 5.60 is determined up to an additive constant, assuming that R is connected. It is called the **potential** associated to the conservative vector field $\mathbf{G}$.

It remains to find a good method for determining whether a vector field is conservative, i.e., whether it is the gradient of a function. Another way of phrasing this question: When is a differential form $G_1 \, dx_1 + \cdots + G_n \, dx_n$ the differential of a function? We shall assume that the vector field $\mathbf{G}$ is of class C^1 on an open set R. In this case, there is an obvious necessary condition for $\mathbf{G}$ to be a gradient of a function on R. Indeed, if $G_j = \partial_j f$, then $\partial_j G_k$ and $\partial_k G_j$ are both equal to the mixed partial $\partial_j \partial_k f$, so

(5.61) $$\frac{\partial G_j}{\partial x_k} - \frac{\partial G_k}{\partial x_j} = 0 \text{ for all } j \neq k.$$

We observe that when $n = 3$, the quantities in (5.61) are the components of curl $\mathbf{G}$, so that (5.61) is equivalent to the condition curl $\mathbf{G} = \mathbf{0}$.

The condition (5.61) is *almost* sufficient to guarantee that $\mathbf{G}$ is a gradient; the only possible problem arises from the geometry of R, as we shall explain in more detail below. When R is convex, the problem disappears, and we have the following result. Our proof will only be complete in dimensions 2 and 3 because it invokes Green's or Stokes's theorem, but the same idea works in higher dimensions.

5.62 Theorem. *Suppose R is a convex open set in $\mathbb{R}^n$ and $\mathbf{G}$ is a C^1 vector field on R. If $\mathbf{G}$ satisfies (5.61) in R (which means that $\operatorname{curl} \mathbf{G} = \mathbf{0}$ in R in the case $n = 3$), then $\mathbf{G}$ is the gradient of a C^2 function on R.*

Proof. The idea is similar to the proof of Proposition 5.60, but we do not know yet that condition (a) of Proposition 5.59 is satisfied, so we must be more careful. Pick a base point $\mathbf{a}$ in R, and define $f(\mathbf{x})$ for $\mathbf{x} \in R$ by $f(\mathbf{x}) = \int_{L(\mathbf{a},\mathbf{x})} \mathbf{G} \cdot d\mathbf{x}$, where $L(\mathbf{a}, \mathbf{x})$ is the line segment from $\mathbf{a}$ to $\mathbf{x}$. (We need the hypothesis of convexity so that this line segment lies in R.) To show that $\mathbf{G}(\mathbf{x}) = \nabla f(\mathbf{x})$, let $\mathbf{h} = (h, 0, \cdots, 0)$ be small enough so that $\mathbf{x} + \mathbf{h} \in R$. Let C be the triangular closed curve obtained by following $L(\mathbf{a}, \mathbf{x})$ from $\mathbf{a}$ to $\mathbf{x}$, $L(\mathbf{x}, \mathbf{x} + \mathbf{h})$ from $\mathbf{x}$ to $\mathbf{x} + \mathbf{h}$, and then $L(\mathbf{a}, \mathbf{x} + \mathbf{h})$ *backwards* from $\mathbf{x} + \mathbf{h}$ to $\mathbf{a}$. Green's theorem (if $n = 2$), Stokes's theorem (if $n = 3$), or the higher-dimensional version of Stokes's theorem (if $n > 3$; see §5.9) converts $\int_C \mathbf{G} \cdot d\mathbf{x}$ into a double integral over the solid triangle whose boundary is C, whose integrand vanishes by (5.61). Hence $\int_C \mathbf{G} \cdot d\mathbf{x} = 0$, or in other words,

$$f(\mathbf{x} + \mathbf{h}) - f(\mathbf{x}) = \int_{L(\mathbf{a},\mathbf{x})} \mathbf{G} \cdot d\mathbf{x} - \int_{L(\mathbf{a},\mathbf{x}+\mathbf{h})} \mathbf{G} \cdot d\mathbf{x} = \int_{L(\mathbf{x},\mathbf{x}+\mathbf{h})} \mathbf{G} \cdot d\mathbf{x}.$$

Now the same argument as in Proposition 5.60 shows that $\partial_1 f = G_1$, and likewise $\partial_j f = G_j$ for the other j. $\qquad\square$

The hypothesis of convexity in Theorem 5.62 is stronger than necessary; one can generalize the argument by using curves other than straight lines. What is crucial is that when one joins the points $\mathbf{a}$, $\mathbf{x}$, and $\mathbf{x} + \mathbf{h}$ by line segments or curves, the resulting "triangle" is the boundary of a piece of surface that lies entirely in R, so that the condition (5.61) and Stokes's theorem can be applied. This may not be the case if the region R has "holes." The following example shows what can go wrong in such a case.

EXAMPLE 1. Let R be the complement of the z-axis in $\mathbb{R}^3$, and let

$$\mathbf{G}(x, y, z) = \frac{-y\mathbf{i} + x\mathbf{j}}{x^2 + y^2}.$$

It is easily verified that the condition $\operatorname{curl} \mathbf{G} = \mathbf{0}$ is satisfied on R, but that $\mathbf{G}$ is not conservative on R; in fact, $\int_C \mathbf{G} \cdot d\mathbf{x} = 2\pi$ when C is the unit circle. (See Exercise 6 in §5.7.) The key to the mystery is as follows: $\mathbf{G}$ is really the gradient of the angular variable θ in cylindrical coordinates, *but θ is not a well-defined function on R.* It is defined only up to multiples of 2π.

However, if we choose a convex subregion of $S \subset R$ (for example, the half-space $y > 0$), we can choose a well-defined "branch" of the angle θ on S (for example, $0 < \theta < \pi$), and then $\mathbf{G}$ is the gradient of this function on S. The same example can be used in $\mathbb{R}^2$, taking R to be the complement of the origin.

The hypothesis on R that should replace convexity in Theorem 5.62 to give the best result is that *every simple closed curve in R is the boundary of a surface lying entirely in R*. (The proof requires more advanced techniques.) The region R in Example 1 does not have this property; no closed curve that encircles the z-axis can be the boundary of a surface in R.

In practice, if R is a rectangular box, to find a function whose gradient is $\mathbf{G}$ one can proceed in a more simple-minded way than is indicated in the proof of Theorem 5.62. Consider the 2-dimensional case, where $R = [a, b] \times [\alpha, \beta]$ and $\mathbf{G}(x, y) = P(x, y)\mathbf{i} + Q(x, y)\mathbf{j}$. Assuming that $\partial_x Q = \partial_y P$, we begin by integrating P with respect to x, including a "constant" of integration that can depend on the other variable y:

$$f(x, y) = \int_c^x P(t, y)\, dt + \varphi(y).$$

Here c can be any point in the interval $[a, b]$. Any such f will satisfy $\partial_x f = P$. To obtain $\partial_y f = Q$, differentiate the formula for f with respect to y and use Theorem 4.47:

$$\partial_y f(x, y) = \int_a^x \partial_y P(t, y)\, dt + \varphi'(y) = \int_a^x \partial_x Q(t, y)\, dt + \varphi'(y)$$

$$= Q(x, y) - Q(a, y) + \varphi'(y).$$

Thus we obtain the desired f by taking φ to be an antiderivative of $Q(a, y)$.

The same idea works in n variables. If $\mathbf{G}$ is a vector field on $\mathbb{R}^n$ that satisfies (5.61), we integrate G_1 with respect to x_1 to get

$$f(x_1, \ldots, x_n) = \int_a^{x_1} G_1(t, x_2, \ldots, x_n)\, dt + \varphi(x_2, \ldots, x_n).$$

Then $\partial_1 f = G_1$. Differentiating this formula with respect to $x_2, \ldots, x_n$ and using the facts that $\partial_j G_1 = \partial_1 G_j$, we obtain formulas for $\partial_2 \varphi, \ldots, \partial_n \varphi$. The problem is thereby reduced to a similar problem (finding a function with a given gradient) in one less variable, so we can proceed inductively.

EXAMPLE 2. Let $\mathbf{G}(x, y) = [y^2 e^{xy}]\mathbf{i} + [(xy + 1)e^{xy} + \cos y]\mathbf{j}$. We have $\partial_1 G_2 = \partial_2 G_1 = (2x + x^2 y)e^{xy}$, so (5.61) is satisfied. To find a function f

such that $\nabla f = \mathbf{G}$, we set

$$f(x, y) = \int y^2 e^{xy}\, dx = y e^{xy} + \varphi(y).$$

Then $\partial_y f = (xy + 1)e^{xy} + \varphi'(y)$; matching this up with the second component yields $\varphi'(y) = \cos y$, so we can take $\varphi(y) = \sin y$. The general solution is $f(x, y) = y e^{xy} + \sin y + C$.

EXAMPLE 3. Let $\mathbf{G}(x, y, z) = yz\mathbf{i} + (xz + y)\mathbf{j} + (xy - z)\mathbf{k}$. An easy calculation shows that curl $\mathbf{G} = 0$. To find a function f such that $\nabla f = \mathbf{G}$, we integrate the first component with respect to x, obtaining $f(x, y, z) = xyz + \varphi(y, z)$. Differentiating this in y and z yields $\partial_y f = xz + \partial_y \varphi$ and $\partial_z f = xy + \partial_z \varphi$. Therefore, we must have $\partial_y \varphi = y$ and $\partial_z \varphi = -z$. Integrating the first of these equations with respect to y gives $\varphi(y, z) = \frac{1}{2}y^2 + \psi(z)$, so $\partial_z \varphi = \psi'(z) = -z$ and $\psi(z) = -\frac{1}{2}z^2 + C$. Putting this all together,

$$f(x, y, z) = xyz + \tfrac{1}{2}y^2 - \tfrac{1}{2}z^2 + C.$$

Next, we turn to the question of solving the equation curl $\mathbf{F} = \mathbf{G}$, where $\mathbf{G}$ is a C^1 vector field on some open set $R \subset \mathbb{R}^3$. There is an obvious necessary condition for solvability: Since $\mathrm{div}(\mathrm{curl}\,\mathbf{F}) = 0$ for any $\mathbf{F}$ (formula (5.31)), we must have $\mathrm{div}\,\mathbf{G} = 0$ on R. Again, this condition turns out to be sufficient provided that R has "no holes," but here the meaning of "no holes" is somewhat different. Instead of requiring that every closed curve in R be the boundary of a surface that lies entirely in R, we require that every closed surface in R be the boundary of a 3-dimensional region that lies entirely in R. For example, the complement of the z-axis in $\mathbb{R}^3$ satisfies the second condition but not the first; the complement of the origin satisfies the first condition but not the second. An example of a vector field $\mathbf{G}$ that satisfies $\mathrm{div}\,\mathbf{G} = 0$ on the complement of the origin but is not the curl of any vector field there is provided by $\mathbf{G}(\mathbf{x}) = \mathbf{x}/|\mathbf{x}|^3$, the "inverse square law force." This $\mathbf{G}$ cannot be a curl because its integral over a sphere about the origin is nonzero, and this contradicts Corollary 5.56. (See Exercise 6 in §5.5; our $\mathbf{G}$ is the negative of the gradient of the g there.)

Convex regions have no holes, no matter what one means by "holes," and the following analogue of Theorem 5.62 is valid.

5.63 Theorem. *Suppose R is a convex open set in $\mathbb{R}^3$ and $\mathbf{G}$ is a C^1 vector field on R. If $\mathbf{G}$ satisfies $\mathrm{div}\,\mathbf{G} = 0$ on R, then $\mathbf{G}$ is the curl of a C^2 vector field on R.*

Proof. We shall not give the general proof but shall content ourselves with presenting an algorithm for solving curl $\mathbf{F} = \mathbf{G}$ when R is a rectangular box, similar to the one given above for solving $\nabla f = \mathbf{G}$. Suppose that $R = [a_1, b_1] \times [a_2, b_2] \times [a_3, b_3]$ and $\mathbf{G}$ is a C^1 vector field satisfying div $\mathbf{G} = 0$ on R. Unlike the problem of finding a function with a given gradient, whose solution is unique up to an additive constant, there is lots of freedom in choosing an $\mathbf{F}$ such that curl $\mathbf{F} = \mathbf{G}$, for if curl $\mathbf{F} = \mathbf{G}$ then also curl$(\mathbf{F} + \nabla f) = \mathbf{G}$ for any smooth function f. This gives enough leeway to allow us to assume that the z-component of $\mathbf{F}$ is zero. Thus, let us write $\mathbf{G} = G_1\mathbf{i} + G_2\mathbf{j} + G_3\mathbf{k}$ and $\mathbf{F} = F_1\mathbf{i} + F_2\mathbf{j}$; we then want

$$\text{curl } \mathbf{F} = -\partial_z F_2\mathbf{i} + \partial_z F_1\mathbf{j} + (\partial_x F_2 - \partial_y F_1)\mathbf{k} = G_1\mathbf{i} + G_2\mathbf{j} + G_3\mathbf{k}.$$

We solve the first two equations by taking

$$F_2 = -\int_c^z G_1(x, y, t)\, dt + \varphi(x, y), \qquad F_1 = \int_c^z G_2(x, y, t)\, dt + \psi(x, y),$$

where c is some chosen point in $[a_3, b_3]$. We then have

$$\partial_x F_2 - \partial_y F_1 = -\int_c^z \left[\partial_y G_2(x, y, t) + \partial_x G_1(x, y, t)\right] dt$$

$$+ \partial_x \varphi(x, y) + \partial_y \psi(x, y).$$

Since div $\mathbf{G} = 0$, this equals

$$\int_c^z \partial_z G_3(x, y, t)\, dt + \partial_x \varphi(x, y) + \partial_y \psi(x, y)$$

$$= G_3(x, y, z) - G_3(x, y, c) + \partial_x \varphi(x, y) + \partial_y \psi(x, y).$$

We therefore achieve our goal by choosing φ and ψ to satisfy

$$\partial_x \varphi(x, y) + \partial_y \psi(x, y) = G_3(x, y, c).$$

There is still lots of freedom here; for example, we could take

$$\varphi(x, y) = \int_a^x G_3(t, y, c)\, dt, \qquad \psi(x, y) = 0 \qquad (a \in [a_1, b_1]).$$

$\square$

If div $\mathbf{G} = 0$, a vector field $\mathbf{F}$ such that curl $\mathbf{F} = \mathbf{G}$ is called a **vector potential** for $\mathbf{G}$.

EXAMPLE 4.   Find   a   vector   potential   for   the   vector   field
$\mathbf{G}(x, y, z) = (6xz + x^3)\mathbf{i} - (3x^2y + y^2)\mathbf{j} + (4x + 2yz - 3z^2)\mathbf{k}$.

Solution. First one should verify that div $\mathbf{G} = 0$ so as not to go on a fool's errand. Having done so, one can take $\mathbf{F} = F_1\mathbf{i} + F_2\mathbf{j}$ where

$$\partial_z F_1 = -(3x^2y + y^2), \qquad \partial_z F_2 = -(6xz + x^3),$$
$$\partial_x F_2 - \partial_y F_1 = 4x + 2yz - 3z^2.$$

Solving the first two equations yields

$$F_1 = -3x^2yz - y^2z + \psi(x, y), \qquad F_2 = -3xz^2 - x^3z + \varphi(x, y),$$

and plugging these results into the third equation yields $\partial_x\varphi - \partial_y\psi = 4x$. Therefore, one solution (with $\varphi = 2x^2$ and $\psi = 0$) is

$$\mathbf{F}_0 = -(3x^2yz + y^2z)\mathbf{i} + (2x^2 - 3xz^3 - x^3z)\mathbf{j};$$

the general solution is $\mathbf{F} = \mathbf{F}_0 + \nabla f$ where f is an arbitrary C^1 function.

Now, what about the equation div $\mathbf{F} = g$? Here there are *no* obstructions to solvability, and there is an enormous amount of freedom in finding a solution. For example, if we wish to solve div $\mathbf{F} = g$ in a rectangular box in $\mathbb{R}^n$, we could take

$$\mathbf{F} = (F, 0, \dots, 0), \qquad F(\mathbf{x}) = \int_c^{x_1} g(t, x_2, \dots, x_n)\, dt,$$

or similar expressions with the variables permuted; there are many other possibilities. In fact, this problem is so easy that it seems reasonable to make it more interesting by imposing additional conditions on $\mathbf{F}$. We restrict attention to the three-dimensional situation, but there are similar results in higher dimensions.

The key result here is Theorem 5.46, which shows that we can solve the equation div $\mathbf{F} = g$ subject to the restriction that curl $\mathbf{F} = \mathbf{0}$. More precisely, suppose R is a bounded open set in $\mathbb{R}^3$ and g is of class C^1 on $\overline{R}$. (In Theorem 5.46 g was assumed to be C^2, but see the remarks following the proof.) Smoothness on $\overline{R}$ means that g can be extended as a C^1 function to an open set containing $\overline{R}$, and it can be modified outside $\overline{R}$ so as to vanish outside some bounded set while remaining of class C^1. (One multiplies g by a C^1 function that is identically 1 on $\overline{R}$ and vanishes outside some slightly larger region; we omit the details, which are of little importance for this argument.) Hence we may assume that g is C^1 on $\mathbb{R}^3$ and vanishes outside a bounded set. Then, by Theorem 5.46, the function

$$u(\mathbf{x}) = -\frac{1}{4\pi}\iiint_{\mathbb{R}^3} \frac{g(\mathbf{x} + \mathbf{y})}{|\mathbf{y}|}\, d^3\mathbf{y}$$

satisfies $\nabla^2 u = g$, and so the vector field $\mathbf{F} = \nabla u$ satisfies both $\operatorname{div} \mathbf{F} = g$ and $\operatorname{curl} \mathbf{F} = \mathbf{0}$ on R.

With this result in hand, we show that the equations $\operatorname{curl} \mathbf{F} = \mathbf{G}$ and $\operatorname{div} \mathbf{F} = g$ can be solved *simultaneously* (for the same $\mathbf{F}$).

5.64 Theorem. *Let R be a bounded convex open set in $\mathbb{R}^3$. For any C^1 function g on $\overline{R}$ and any C^2 vector field $\mathbf{G}$ on $\overline{R}$ such that $\operatorname{div} \mathbf{G} = 0$, there is a C^2 vector field $\mathbf{F}$ on $\overline{R}$ such that $\operatorname{curl} \mathbf{F} = \mathbf{G}$ and $\operatorname{div} \mathbf{F} = g$ on R.*

Proof. Let $\mathbf{H}$ be a solution of $\operatorname{curl} \mathbf{H} = \mathbf{G}$, as in Theorem 5.63, and let u be a solution of $\nabla^2 u = g - \operatorname{div} \mathbf{H}$, as explained above. Let $\mathbf{F} = \nabla u + \mathbf{H}$; then $\operatorname{curl} \mathbf{F} = \operatorname{curl}(\nabla u) + \mathbf{G} = \mathbf{G}$ and $\operatorname{div} \mathbf{F} = \nabla^2 u + \operatorname{div} \mathbf{H} = g$. $\qquad\square$

There is a companion result to Theorem 5.64: Not every vector field is a gradient, and not every vector field is a curl, but every vector field is the sum of a gradient and a curl. The proof is left to the reader as Exercise 3, where a more precise statement is given.

One might also ask about uniqueness in Theorem 5.64; that is, to what extent is a vector field determined by its curl and divergence? Clearly, if $\mathbf{F}$ satisfies $\operatorname{curl} \mathbf{F} = \mathbf{G}$ and $\operatorname{div} \mathbf{F} = g$, then so does $\mathbf{F} + \mathbf{H}$ whenever $\operatorname{curl} \mathbf{H} = \mathbf{0}$ and $\operatorname{div} \mathbf{H} = 0$. Solutions of the latter pair of equations can be obtained simply by taking $\mathbf{H} = \nabla \varphi$ where φ is any solution of Laplace's equation $\nabla^2 \varphi = 0$. Such solutions exist in great abundance, so the $\mathbf{F}$ in Theorem 5.64 is far from unique. However, one can pin down a unique solution by imposing suitable boundary conditions.

5.65 Proposition. *Let R be a bounded convex open set in $\mathbb{R}^3$ with piecewise smooth boundary. Suppose $\mathbf{H}$ is a C^1 vector field on $\overline{R}$ such that $\operatorname{curl} \mathbf{F} = \mathbf{0}$ and $\operatorname{div} \mathbf{F} = 0$ on R and $\mathbf{F} \cdot \mathbf{n} = 0$ on ∂R. Then $\mathbf{H}$ vanishes identically on R.*

Proof. By Theorem 5.62, $\mathbf{F}$ is the gradient of a function u on R, and $\nabla^2 u = \operatorname{div} \mathbf{F} = 0$. Since $\mathbf{F} \cdot \mathbf{n} = \partial u / \partial n$, by Green's formula (5.38) we have

$$0 = \iint_{\partial R} u \frac{\partial u}{\partial n}\, dA = \iiint_R (|\nabla u|^2 + u \nabla^2 u)\, dV = \iiint_R (|\mathbf{H}|^2 + 0)\, dV.$$

But $|\mathbf{H}|^2$ is a nonnegative continuous function, so its integral over R can be zero only if $|\mathbf{H}|^2$ (and hence $\mathbf{H}$) vanishes identically on R. $\qquad\square$

By applying Proposition 5.65 to the difference of two solutions of the problem in Theorem 5.64, we see that *if $\mathbf{F}$ and $\mathbf{F}'$ are vector fields with the same curl and divergence on R and the same normal component on ∂R, then $\mathbf{F} = \mathbf{F}'$ on R.*

We conclude with a few remarks about the application of the results of this section to Maxwell's equations (5.50). First, we observe that the curl of the electric field $\mathbf{E}$ vanishes only when there are no time-varying magnetic fields present. Only in this case is $\mathbf{E}$ the gradient of a potential function. However, $\operatorname{div} \mathbf{B} = 0$ always (this expresses the fact that there are no "magnetic charges"), so $\mathbf{B}$ is the curl of a vector potential $\mathbf{A}$. We then have

$$\operatorname{curl} \left(\mathbf{E} + \frac{1}{c} \frac{\partial \mathbf{A}}{\partial t} \right) = \operatorname{curl} \mathbf{E} + \frac{1}{c} \frac{\partial \mathbf{B}}{\partial t} = \mathbf{0},$$

so $\mathbf{E} + c^{-1} \partial_t \mathbf{A}$ *is the gradient of a function* $-\varphi$. The four-component quantity $(\varphi, \mathbf{A}) = (\varphi, A_1, A_2, A_3)$ is called the *electromagnetic 4-potential*. It is best regarded as a vector in 4-dimensional space-time, with φ being the time component, in the context of special relativity.

EXERCISES

1. Determine whether each of the following vector fields is the gradient of a function f, and if so, find f. The vector fields in (a)–(c) are on $\mathbb{R}^2$; those in (d)–(f) are on $\mathbb{R}^3$, and the one in (g) is on $\mathbb{R}^4$. In all cases $\mathbf{i}$, $\mathbf{j}$, $\mathbf{k}$, and $\mathbf{l}$ denote unit vectors along the positive x-, y-, z-, and w-axes.
   a. $\mathbf{G}(x, y) = (2xy + x^2)\mathbf{i} + (x^2 - y^2)\mathbf{j}$.
   b. $\mathbf{G}(x, y) = (3y^2 + 5x^4 y)\mathbf{i} + (x^5 - 6xy)\mathbf{j}$.
   c. $\mathbf{G}(x, y) = (2e^{2x} \sin y - 3y + 5)\mathbf{i} + (e^{2x} \cos y - 3x)\mathbf{j}$
   d. $\mathbf{G}(x, y, z) = (yz - y \sin xy)\mathbf{i} + (xz - x \sin xy + z \cos yz)\mathbf{j} + (xy + y \cos yz)\mathbf{k}$.
   e. $\mathbf{G}(x, y, z) = (y - z)\mathbf{i} + (x - z)\mathbf{j} + (x - y)\mathbf{k}$
   f. $\mathbf{G}(x, y, z) = 2xy\mathbf{i} + (x^2 + \log z)\mathbf{j} + ((y + 2)/z)\mathbf{k}$ $(z > 0)$.
   g. $\mathbf{G}(x, y, z, w) = (xw^2 + yzw)\mathbf{i} + (xzw + yz^2 - 2e^{2y+z})\mathbf{j} + (xyw + y^2 z - e^{2y+z} - w \sin zw)\mathbf{k} + (xyz + x^2 w - z \sin zw)\mathbf{l}$.

2. Determine whether each of the following vector fields is the curl of a vector field $\mathbf{F}$, and if so, find such an $\mathbf{F}$.
   a. $\mathbf{G}(x, y, z) = (x^3 + yz)\mathbf{i} + (y - 3x^2 y)\mathbf{j} + 4y^2 \mathbf{k}$.
   b. $\mathbf{G}(x, y, z) = (xy + z)\mathbf{i} + xz\mathbf{j} - (yz + x)\mathbf{k}$.
   c. $\mathbf{G}(x, y, z) = (xe^{-x^2 z^2} - 6x)\mathbf{i} + (5y + 2z)\mathbf{j} + (z - ze^{-x^2 z^2})\mathbf{k}$.

3. Let R be a bounded convex open set in $\mathbb{R}^3$. Show that for any C^2 vector field $\mathbf{H}$ on $\overline{R}$ there exist a C^2 function f and a C^2 vector field $\mathbf{G}$ such that $\mathbf{H} = \operatorname{grad} f + \operatorname{curl} \mathbf{G}$. (*Hint:* Solve $\nabla^2 f = \operatorname{div} \mathbf{H}$.)

4. Let $\mathbf{F} = F_1 \mathbf{i} + F_2 \mathbf{j}$ be a C^1 vector field on $S = \mathbb{R}^2 \setminus \{(0, 0)\}$ such that $\partial_1 F_2 = \partial_2 F_1$ on S (but $\mathbf{F}$ may be singular at the origin).

a. Let C_r be the circle of radius r about the origin, oriented counterclockwise. Show that $\int_{C_r} \mathbf{F} \cdot d\mathbf{x}$ is a constant α that does not depend on r. (*Hint:* Consider the region between two circles.)

b. Show that $\int_C \mathbf{F} \cdot d\mathbf{x} = \alpha$ for *any* simple closed curve C, oriented counterclockwise, that encircles the origin.

c. Let $\mathbf{F}_0 = (x\mathbf{j} - y\mathbf{i})/(x^2 + y^2)$ as in Example 1. Show that $\mathbf{F} - (\alpha/2\pi)\mathbf{F}_0$ is the gradient of a function on S. (Thus, all curl-free vector fields on S that are not gradients can be obtained from $\mathbf{F}_0$ by adding gradients.)

5.9 Higher Dimensions and Differential Forms

Green's theorem has to do with integrals of vector fields in the plane, and the divergence theorem and Stokes's theorem have do do with integrals of vector fields in 3-space. What happens in dimension n? There are a couple of things we can say without too much additional explanation.

First, the obvious analogue of the divergence theorem holds in $\mathbb{R}^n$ for any $n > 1$. To wit, if R is a regular region in $\mathbb{R}^n$ bounded by a piecewise smooth hypersurface ∂R, and $\mathbf{F}$ is a C^1 vector field on R, then

$$\int \cdots \int_{\partial R} \mathbf{F} \cdot \mathbf{n} \, dV^{n-1} = \int \int \cdots \int_R \operatorname{div} \mathbf{F} \, dV^n.$$

Here dV^n is the n-dimensional volume element in $\mathbb{R}^n$ and dV^{n-1} is the $(n-1)$-dimensional "area" element on ∂R. The "vector area element" $\mathbf{n} \, dV^{n-1}$ is given by a formula analogous to the one in $\mathbb{R}^3$. Namely, if (part of) ∂R is parametrized by $\mathbf{x} = \mathbf{G}(u_1, \ldots, u_{n-1})$, then

$$\mathbf{n} \, dV^{n-1} = \det \begin{pmatrix} \mathbf{e}_1 & \cdots & \mathbf{e}_n \\ \partial_1 G_1 & \cdots & \partial_1 G_n \\ \vdots & & \vdots \\ \partial_1 G_{n-1} & \cdots & \partial_{n-1} G_n \end{pmatrix} du_1 \cdots du_{n-1},$$

where $\mathbf{e}_1, \ldots, \mathbf{e}_n$ are the standard basis vectors for $\mathbb{R}^n$. (The reader may verify that in the case $n = 2$, these formulas yield Green's theorem in the form (5.18).)

Second, the analogue of the divergence theorem in dimension 1 is just the fundamental theorem of calculus:

$$f(b) - f(a) = \int_{[a,b]} f'(t) \, dt.$$

On the real line, vector fields are the same thing as functions, and the divergence of a vector field is just the derivative of a function. A regular region in $\mathbb{R}$ is an interval

$[a, b]$, whose boundary is the two-element set $\{a, b\}$. Since the boundary is finite, "integration" over the boundary is just summation, and the minus sign on $f(a)$ comes from assigning the proper "orientation" to the two points in the boundary.

There are also analogues of Stokes's theorem in higher dimensions, which say that the integral of some gadget G over the boundary of a k-dimensional submanifold of $\mathbb{R}^n$ equals the integral of another gadget formed from the first derivatives of G over the submanifold itself. However, to formulate things properly in this general setting, it is necessary to develop some additional algebraic machinery, the theory of *differential forms*. To do so is beyond the scope of this book; what follows is intended to provide an informal introduction to the ideas involved. For a detailed treatment of differential forms, we refer the reader to Hubbard and Hubbard [7] and Weintraub [19].

Roughly speaking, a differential k-form is an object whose mission in life is to be integrated over k-dimensional sets; thus, 1-forms are designed to be integrated over curves, 2-forms are designed to be integrated over surfaces, and so on. Here is how the ideas of vector analysis that we have been studying can be reformulated in terms of differential forms.

1-Forms. A **differential 1-form** on $\mathbb{R}^n$ is an expression of the form

$$\omega = F_1(x_1, \ldots, x_n) \, dx_1 + \cdots + F_n(x_1, \ldots, x_n) \, dx_n,$$

where the F_j's are continuous functions. There is an obvious correspondence between the 1-form ω and the vector field $\mathbf{F} = (F_1, \ldots, F_n)$. In particular, in 3 dimensions the correspondence between 1-forms and vector fields takes the form

$$(5.66) \qquad \omega = F \, dx + G \, dy + H \, dz \quad \longleftrightarrow \quad \mathbf{F} = F\mathbf{i} + G\mathbf{j} + H\mathbf{k}.$$

One type of 1-form that we have already encountered is the differential of a C^1 function,

$$df = (\partial_1 f) \, dx_1 + \cdots + (\partial_n f) \, dx_n.$$

However, not every 1-form is the differential of a function; the necessary condition for ω to be of the form df is (5.61).

We note that the set of 1-forms on $\mathbb{R}^n$ is a vector space. That is, it makes sense to add 1-forms to each other and to multiply them by scalars. In fact, the "scalars" here can be taken to be not just constants but arbitrary continuous functions on $\mathbb{R}^n$. Thus, if $\alpha = A_1 \, dx_1 + \cdots + A_n \, dx_n$ and $\beta = B_1 \, dx_1 + \cdots + B_n \, dx_n$ are 1-forms and f is a continuous function,

$$\alpha + \beta = (A_1 + B_1) \, dx_1 + \cdots + (A_n + B_n) \, dx_n,$$
$$f\alpha = (fA_1) \, dx_1 + \cdots + (fA_n) \, dx_n.$$

Any smooth mapping $\mathbf{T} : \mathbb{R}^k \to \mathbb{R}^n$ induces a mapping of 1-forms *in the opposite direction*, that is, an operation $\mathbf{T}^*$ which takes 1-forms on $\mathbb{R}^n$ to 1-forms on $\mathbb{R}^k$. Schematically:

$$\mathbb{R}^k \xrightarrow{\mathbf{T}} \mathbb{R}^n$$
$$\text{1-forms on } \mathbb{R}^k \xleftarrow{\mathbf{T}^*} \text{1-forms on } \mathbb{R}^n$$

This operation is just the "built-in chain rule" for differentials of functions, extended to arbitrary 1-forms. To wit, let $x_1, \ldots, x_n$ and $u_1, \ldots, u_k$ be the coordinates on $\mathbb{R}^n$ and $\mathbb{R}^k$, respectively. If $\omega = F_1 \, dx_1 + \cdots + F_n \, dx_n$ is a 1-form on $\mathbb{R}^n$, its **pullback** via $\mathbf{T}$ is the 1-form $\mathbf{T}^*\omega$ on $\mathbb{R}^k$ defined by substituting into ω the expressions for the x's in terms of the u's and the dx's in terms of the du's:

(5.67)
$$\mathbf{T}^*\omega = \tilde{A}_1 \left[\frac{\partial x_1}{\partial u_1} du_1 + \cdots + \frac{\partial x_1}{\partial u_k} du_k \right] + \cdots + \tilde{A}_n \left[\frac{\partial x_n}{\partial u_1} du_1 + \cdots + \frac{\partial x_n}{\partial u_k} du_k \right]$$
$$= \left[\tilde{A}_1 \frac{\partial x_1}{\partial u_1} + \cdots + \tilde{A}_n \frac{\partial x_n}{\partial u_1} \right] du_1 + \cdots + \left[\tilde{A}_1 \frac{\partial x_1}{\partial u_k} + \cdots + \tilde{A}_n \frac{\partial x_n}{\partial u_k} \right] du_k,$$

where

$$\tilde{A}_m(u_1, \ldots, u_k) = A_m\big(\mathbf{T}(u_1, \ldots, u_k)\big).$$

Two special cases are of particular interest. First, the chain rule says that when $\omega = df$, $\mathbf{T}^*\omega = d(f \circ \mathbf{T})$. Second, when $k = 1$ so that $\mathbf{T} : \mathbb{R} \to \mathbb{R}^n$ defines a curve in $\mathbb{R}^n$, (5.67) becomes

$$\mathbf{T}^*\omega = \left[(A_1 \circ \mathbf{T}) \frac{dx_1}{du} + \cdots + (A_n \circ \mathbf{T}) \frac{dx_n}{du} \right] du.$$

1-forms can be integrated over curves. To begin with, a 1-form on $\mathbb{R}$ is merely something of the form $\omega = g(t) \, dt$, and its integral over an interval $[a, b]$ is just what you think it is:

$$\int_{[a,b]} \omega = \int_a^b g(t) \, dt.$$

Now, if $\omega = A_1 \, dx_1 + \cdots + A_n \, dx_n$ is a 1-form on $\mathbb{R}^n$ and C is a smooth curve parametrized by $\mathbf{x} = \mathbf{g}(t)$, $\int_C \omega$ is defined by pulling ω back to $\mathbb{R}$ via $\mathbf{g}$ and integrating the result as before:

$$\int_C \omega = \int_{[a,b]} \mathbf{g}^*\omega = \int_a^b \left[F_1(\mathbf{g}(t)) \frac{dx_1}{dt} + \cdots + F_n(\mathbf{g}(t)) \frac{dx_n}{dt} \right] dt.$$

In other words, if we identify ω with the vector field $\mathbf{F}$ as before,

$$\int_C \omega = \int_C \mathbf{F} \cdot d\mathbf{x}.$$

2-Forms and the Exterior Product. We now define a notion of a "product of two 1-forms" that is related to the cross product of vector fields in $\mathbb{R}^3$ but works in any number of dimensions. This product is called the **exterior product**; the exterior product of two 1-forms α and β is denoted by $\alpha \wedge \beta$. The novel feature of this is that $\alpha \wedge \beta$ is no longer a 1-form but a *new type of object* called a 2-form.

Without specifying what a 2-form is just yet, we list the basic properties that the exterior product is to have. First, it distributes over addition and scalar multiplication in the usual way. That is, if α_1, α_2, and β are 1-forms on $\mathbb{R}^n$ and f_1 and f_2 are continuous functions on $\mathbb{R}^n$,

(5.68)
$$(f_1\alpha_1 + f_2\alpha_2) \wedge \beta = f_1(\alpha_1 \wedge \beta) + f_2(\alpha_2 \wedge \beta),$$
$$\beta \wedge (f_1\alpha_1 + f_2\alpha_2) = f_1(\beta \wedge \alpha_1) + f_2(\beta \wedge \alpha_2).$$

Second, the exterior product is *anticommutative*:

(5.69)
$$\beta \wedge \alpha = -\alpha \wedge \beta.$$

Thus, if $\alpha = A_1\,dx_1 + \cdots + A_n\,dx_n$ and $\beta = B_1\,dx_1 + \cdots + B_n\,dx_n$, we can expand $\alpha \wedge \beta$ according to (5.68) to obtain

(5.70)
$$\alpha \wedge \beta = \sum_{i=1}^{n}\sum_{j=1}^{n} A_i B_j\,dx_i \wedge dx_j.$$

But according to (5.69), $dx_j \wedge dx_i = -dx_i \wedge dx_j$ and $dx_i \wedge dx_i = 0$. Thus the terms with $i = j$ in (5.70) drop out, and for $i \neq j$ we can combine the ijth and jith terms into one:

$$A_i B_j\,dx_i \wedge dx_j + A_j B_i\,dx_j \wedge dx_i = (A_i B_j - A_j B_i)\,dx_i \wedge dx_j$$
$$= (A_j B_i - A_i B_j)\,dx_j \wedge dx_i.$$

We have the option of using either of the two expressions on the right, and the usual choice is to use the one where the first index is smaller than the second one. (In $\mathbb{R}^3$ a different choice is sometimes convenient, as we shall soon see.) Thus, we finally obtain

$$\alpha \wedge \beta = \sum_{1 \leq i < j \leq n} (A_i B_j - A_j B_i)\,dx_i \wedge dx_j.$$

In general, a **differential 2-form** on $\mathbb{R}^n$ is an expression of the type

(5.71)
$$\omega = \sum_{1 \leq i < j \leq n} C_{ij}(x_1,\ldots,x_n)\,dx_i \wedge dx_j,$$

where the C_{ij} are continuous functions on $\mathbb{R}^n$. We note that the number of terms in this sum, that is, the number of pairs (i, j) with $1 \le i < j \le n$, is $\frac{1}{2}n(n-1)$. In (5.71) we also have the option of rewriting $dx_i \wedge dx_j$ as $-dx_j \wedge dx_i$ if we so choose.

What does this really mean? We have been proceeding purely formally, without saying what meaning is to be attached to the expressions $dx_i \wedge dx_j$. In the full-dress treatment of this subject, 2-forms are defined to be alternating rank-2 tensor fields over $\mathbb{R}^n$, but this is somewhat beside the point. For now it is probably best to think of a 2-form on $\mathbb{R}^n$ simply as a $\frac{1}{2}n(n-1)$-tuple of functions, namely the functions C_{ij} in (5.71), and the expressions $dx_i \wedge dx_j$ simply as a convenient set of signposts to mark the various components, just as $\mathbf{i}$, $\mathbf{j}$, and $\mathbf{k}$ are used to mark the components of vector fields in $\mathbb{R}^3$. The important features of 2-forms are not their precise algebraic definition but the way they transform under changes of variables and the way they integrate over surfaces.

Before proceeding to these matters, however, let us see how things look in the 3-dimensional case. When $n = 3$ we also have $\frac{1}{2}n(n-1) = 3$, so 2-forms have 3 components just as vector fields and 1-forms do: This is the "accident" that makes $n = 3$ special! The general 2-form on $\mathbb{R}^3$ can be written as

$$\omega = F(x, y, z)\, dy \wedge dz + G(x, y, z)\, dz \wedge dx + H(x, y, z)\, dx \wedge dy,$$

so there is a one-to-one correspondence between 2-forms and vector fields:

$$(5.72) \quad \omega = F\, dy \wedge dz + G\, dz \wedge dx + H\, dx \wedge dy \quad \longleftrightarrow \quad \mathbf{F} = F\mathbf{i} + G\mathbf{j} + H\mathbf{k}.$$

Observe carefully how we have set this correspondence up: we have written the basis elements $dx_i \wedge dx_j$ with the variables in *cyclic order*,

$$dx \text{ before } dy \text{ before } dz \text{ before } dx,$$

rather than the "$i < j$" order we used above, so that the middle term is $dz \wedge dx$ rather than $dx \wedge dz$. Also, we identify the unit vector $\mathbf{i}$ in the x direction with the 2-form $dy \wedge dz$ from which dx is missing, and likewise for $\mathbf{j}$ and $\mathbf{k}$.

The exterior product in 3 dimensions looks like this: If

$$\alpha = A_1\, dx + A_2\, dy + A_3\, dz, \qquad \beta = B_1\, dx + B_2\, dy + B_3\, dz,$$

then

$$\alpha \wedge \beta = (A_2 B_3 - A_3 B_2)\, dy \wedge dz + (A_3 B_1 - A_1 B_3)\, dz \wedge dx$$
$$+ (A_1 B_2 - A_2 B_1)\, dx \wedge dy.$$

Thus, if we identify α and β with vector fields according to (5.66) and $\alpha \wedge \beta$ with a vector field according to (5.72), the exterior product turns into the cross product:

$$\alpha \quad \longleftrightarrow \quad \mathbf{F}, \qquad \beta \quad \longleftrightarrow \quad \mathbf{G}, \qquad \alpha \wedge \beta \quad \longleftrightarrow \quad \mathbf{F} \times \mathbf{G}.$$

Pullbacks and Integrals of 2-Forms. We have seen that a smooth mapping $\mathbf{T} : \mathbb{R}^k \to \mathbb{R}^n$ induces a "pullback" mapping $\mathbf{T}^*$ that takes 1-forms on $\mathbb{R}^n$ to 1-forms on $\mathbb{R}^k$. It also induces a pullback mapping, still denoted by $\mathbf{T}^*$, from 2-forms on $\mathbb{R}^n$ to 2-forms on $\mathbb{R}^k$, in exactly the same way: We simply substitute $\mathbf{T}(u)$ for x and $\sum_j (\partial x_m / \partial u_j) \, du_j$ for dx_j. Thus,

$$\mathbf{T}^*(dx_l \wedge dx_m) = \left[\frac{\partial x_l}{\partial u_1} \, du_1 + \cdots + \frac{\partial x_l}{\partial u_k} \, du_k \right] \wedge \left[\frac{\partial x_m}{\partial u_1} \, du_1 + \cdots + \frac{\partial x_m}{\partial u_k} \, du_k \right]$$
$$= \sum_{i<j} \frac{\partial(x_l, x_m)}{\partial(u_i, u_j)} \, du_i \wedge du_j,$$

so in general, if

$$\omega = \sum_{l<m} C_{lm}(x) \, dx_l \wedge dx_m,$$

then

$$\mathbf{T}^*\omega = \sum_{l<m} \sum_{i<j} C_{lm}(\mathbf{T}(u)) \frac{\partial(x_l, x_m)}{\partial(u_i, u_j)} \, du_i \wedge du_j.$$

It is a consequence of the chain rule that the pullback operation behaves properly under composition of mappings, namely, $(\mathbf{T}_1 \circ \mathbf{T}_2)^*\omega = \mathbf{T}_2^*(\mathbf{T}_1^*\omega)$.

We can now show how to integrate 2-forms over surfaces. First consider the simplest case, where the surface is simply a region D in $\mathbb{R}^2$. If we name the coordinates on $\mathbb{R}^2$ x and y, the general 2-form on $\mathbb{R}^2$ has the form $\omega = f(x, y) \, dx \wedge dy$, and its integral over D is the obvious thing:

$$(5.73) \qquad \iint_D f(x, y) \, dx \wedge dy = \iint_D f(x, y) \, dx \, dy,$$

the integral on the right being the ordinary double integral of f over D. The only subtle point is that the integral on the left is an *oriented* integral, the orientation being carried in the fact that dx comes before dy in $dx \wedge dy$. If we wrote $dy \wedge dx$ instead, we would introduce a minus sign.

The nice thing about (5.73) is that the change-of-variable formula for double integrals is more or less built into it. Namely, suppose $\mathbf{T} : \mathbb{R}^2 \to \mathbb{R}^2$ is an invertible $C^{(1)}$ transformation, say $\mathbf{T}(u, v) = (x, y)$. If $\omega = f(x, y) \, dx \wedge dy$, then

$$\mathbf{T}^*\omega = f(\mathbf{T}(u, v)) \frac{\partial(x, y)}{\partial(u, v)} \, du \wedge dv = f(\mathbf{T}(u, v))(\det D\mathbf{T}) \, du \, dv,$$

so the change-of-variable formula simply says that

(5.74)
$$\iint_{\mathbf{T}(D)} \omega = \iint_D \mathbf{T}^* \omega.$$

In other words, the formalism of differential forms produces the necessary Jacobian factor automatically. The change-of-variable formula as we have seen it before involved $|\det D\mathbf{T}|$ rather than $\det D\mathbf{T}$, but this discrepancy is accounted for by the difference between ordinary integrals and oriented integrals.

Now we turn to the case of integrals over a surface S in $\mathbb{R}^n$. The idea is the same as for line integrals: If ω is a 2-form on $\mathbb{R}^3$ and S is a surface parametrized by $\mathbf{x} = \mathbf{G}(u, v)$, $(u, v) \in D \subset \mathbb{R}^2$, we define $\iint_S \omega$ by pulling ω back to D via $\mathbf{G}$ and using (5.73) to define the resulting integral:

$$\iint_S \omega = \iint_D \mathbf{G}^* \omega.$$

This is independent of the parametrization, in the following sense: If $\mathbf{G} = \tilde{\mathbf{G}} \circ \mathbf{T}$ where $\mathbf{T} : \mathbb{R}^2 \to \mathbb{R}^2$ is a $C^{(1)}$ transformation, then by (5.74),

$$\iint_D \mathbf{G}^* \omega = \iint_D \mathbf{T}^* \tilde{\mathbf{G}}^* \omega = \iint_{\mathbf{T}(D)} \tilde{\mathbf{G}}^* \omega.$$

Let us see how this looks in the case $n = 3$. If

$$\omega = A\, dy \wedge dz + B\, dz \wedge dx + C\, dx \wedge dy \quad \text{and} \quad (x, y, z) = \mathbf{G}(u, v),$$

then $\mathbf{G}^* \omega$ equals

$$\left[A(\mathbf{G}(u, v)) \frac{\partial(y, z)}{\partial(u, v)} + B(\mathbf{G}(u, v)) \frac{\partial(z, x)}{\partial(u, v)} + C(\mathbf{G}(u, v)) \frac{\partial(x, y)}{\partial(u, v)} \right] du \wedge dv,$$

and hence $\iint_S \omega$ equals

$$\iint_D \left[A(\mathbf{G}(u, v)) \frac{\partial(y, z)}{\partial(u, v)} + B(\mathbf{G}(u, v)) \frac{\partial(z, x)}{\partial(u, v)} + C(\mathbf{G}(u, v)) \frac{\partial(x, y)}{\partial(u, v)} \right] du\, dv.$$

But this is something we have seen before. Indeed, we have

$$\frac{\partial(y, z)}{\partial(u, v)} \mathbf{i} + \frac{\partial(z, x)}{\partial(u, v)} \mathbf{j} + \frac{\partial(x, y)}{\partial(u, v)} \mathbf{k} = \frac{\partial \mathbf{G}}{\partial u} \times \frac{\partial \mathbf{G}}{\partial v},$$

so if we identify ω with the vector field $\mathbf{F} = A\mathbf{i} + B\mathbf{j} + C\mathbf{k}$ as in (5.72), we have

$$\iint_S \omega = \iint_S \mathbf{F} \cdot \mathbf{n}\, dA.$$

Hence the notion of surface integrals of vector fields in $\mathbb{R}^3$ also fits into the theory of differential forms.

3-Forms. A **differential 3-form** on $\mathbb{R}^n$ is an expression of the form

(5.75) $$\omega = \sum_{1 \le i < j < k \le n} C_{ijk}(x_1, \dots, x_n)\, dx_i \wedge dx_j \wedge dx_k.$$

Here, as in the case of 2-forms, one can think of the expressions $dx_i \wedge dx_j \wedge dx_k$ simply as formal basis elements, and one can put the indices i, j, k in an order other than $i < j < k$ with the understanding that whenever one interchanges two of the dx's one introduces a minus sign. The number of terms in the sum in (5.75) is the binomial coefficient $n!/3!(n-3)!$. When $n = 3$, this number is 1: All 3-forms on $\mathbb{R}^3$ have the form

$$\omega = f(x, y, z)\, dx \wedge dy \wedge dz$$

and hence can be identified with functions:

$$f(x, y, z)\, dx \wedge dy \wedge dz \quad \longleftrightarrow \quad f(x, y, z).$$

The notion of exterior product extends so as to yield a 3-form as the product of three 1-forms or as the product of a 1-form and a 2-form. The idea is pretty obvious: $dx_i \wedge dx_j \wedge dx_k$ is the exterior product of the three 1-forms dx_i, dx_j, and dx_k, or the 1-form dx_i and the 2-form $dx_j \wedge dx_k$, or the 2-form $dx_i \wedge dx_j$ and the 1-form dx_k. The exterior product distributes over sums and scalar multiples in the usual way, and the anticommutative law becomes

$$\alpha \wedge \beta = (-1)^{l+m-1}\beta \wedge \alpha \text{ if } \alpha \text{ is an } l\text{-form and } \beta \text{ is an } m\text{-form.}$$

Here is how it works when $n = 3$: If

$$\alpha = A_1\, dx + A_2\, dy + A_3\, dz,$$
$$\beta = B_1\, dx + B_2\, dy + B_3\, dz,$$
$$\gamma = C_1\, dx + C_2\, dy + C_3\, dz,$$
$$\omega = W_1\, dy \wedge dz + W_2\, dz \wedge dx + W_3\, dx \wedge dy,$$

then

$$\alpha \wedge (\beta \wedge \gamma) = (\alpha \wedge \beta) \wedge \gamma = \det \begin{pmatrix} A_1 & A_2 & A_3 \\ B_1 & B_2 & B_3 \\ C_1 & C_2 & C_3 \end{pmatrix} dx \wedge dy \wedge dz,$$

$$\alpha \wedge \omega = \omega \wedge \alpha = (A_1 W_1 + A_2 W_2 + A_3 W_3)\, dx \wedge dy \wedge dz.$$

Thus, if we identify α, β, γ with the vector fields $\mathbf{F}, \mathbf{G}, \mathbf{H}$ and ω with the vector field $\mathbf{V}$, the exterior product turns into the scalar triple product and dot product:

$$\alpha \wedge \beta \wedge \gamma \quad \longleftrightarrow \quad \mathbf{F} \cdot (\mathbf{G} \times \mathbf{H}), \qquad \alpha \wedge \omega \quad \longleftrightarrow \quad \mathbf{F} \cdot \mathbf{V}.$$

Pullbacks and integrals of 3-forms work just as before; we restrict ourselves to the 3-dimensional case. Let $\omega = f(x, y, z)\, dx \wedge dy \wedge dz$. If $\mathbf{T} : \mathbb{R}^3 \to \mathbb{R}^3$ is a C^1 transformation, say $\mathbf{T}(u, v, w) = (x, y, z)$, we obtain $\mathbf{T}^*\omega$ by subsituting in the formulas for x, y, z, dx, dy, and dz in terms of u, v, w; the result is

$$\mathbf{T}^*\omega = f(\mathbf{T}(u, v, w)) \frac{\partial(x, y, z)}{\partial(u, v, w)}\, du \wedge dv \wedge dw.$$

The integral of ω over a region $D \subset \mathbb{R}^3$ is defined in the obvious way:

$$\iiint_D f(x, y, z)\, dx \wedge dy \wedge dz = \iiint_D f,$$

and the change-of variable formula (for *oriented* integrals) reads

$$\iiint_{\mathbf{T}(D)} \omega = \iiint_D \mathbf{T}^*\omega.$$

We have now sketched the whole idea of differential forms in dimension 3. In dimension n one needs to develop the theory of k-forms for all $k \leq n$, which requires the machinery of multilinear algebra.

The Exterior Derivative. When the operations of gradient, curl, and divergence are expressed in terms of differential forms, they are all instances of a *single* operation, denoted by d and called the **exterior derivative**, which maps k-forms on $\mathbb{R}^n$ into $(k + 1)$-forms on $\mathbb{R}^n$:

$$0\text{-forms} \xrightarrow{\ d\ } 1\text{-forms} \xrightarrow{\ d\ } 2\text{-forms} \xrightarrow{\ d\ } 3\text{-forms} \xrightarrow{\ d\ } \cdots .$$

Here's how it works.

First, a **0-form** is, by definition, a function; if f is a 0-form, then df is just the differential of f. If we identify 1-forms with vector fields, df becomes ∇f. That is, *the gradient is the exterior derivative on 0-forms.*

Now, any k-form ω with $k \geq 1$ is a sum of terms of the form $f\beta$ where f is a function and β is one of the basis elements (dx_i for 1-forms, $dx_i \wedge dx_j$ for 2-forms, etc.). $d\omega$ is defined to be the $(k+1)$-form obtained by replacing each such term $f\beta$ by $df \wedge \beta$.

This is what it looks like when $\omega = A_1\, dx_1 + A_2\, dx_2 + \cdots + A_n\, dx_n$ is a 1-form:

$$d\omega = dA_1 \wedge dx_1 + \cdots + dA_n \wedge dx_n$$

$$= \left[\frac{\partial A_1}{\partial x_1} dx_1 + \cdots + \frac{\partial A_1}{\partial x_n} dx_n \right] \wedge dx_1 + \cdots + \left[\frac{\partial A_n}{\partial x_1} dx_1 + \cdots + \frac{\partial A_n}{\partial x_n} dx_n \right] \wedge dx_n$$

$$= \sum_{i<j} \left[\frac{\partial A_j}{\partial x_i} - \frac{\partial A_i}{\partial x_j} \right] dx_i \wedge dx_j.$$

When $n = 3$ and we write x, y, z instead of x_1, x_2, x_3, we obtain

$$dw = \left[\frac{\partial A_3}{\partial y} - \frac{\partial A_2}{\partial z}\right] dy \wedge dz + \left[\frac{\partial A_1}{\partial z} - \frac{\partial A_3}{\partial x}\right] dz \wedge dx$$
$$+ \left[\frac{\partial A_2}{\partial x} - \frac{\partial A_1}{\partial y}\right] dx \wedge dy.$$

But this is just the curl! That is, if we identify the 1-form ω and the 2-form $d\omega$ with vector fields $\mathbf{F}$ and $\mathbf{G}$ in the standard way, then $\mathbf{G} = \text{curl}\,\mathbf{F}$. *The curl is the exterior derivative on 1-forms in $\mathbb{R}^3$.*

Now suppose that $\omega = A\,dy \wedge dz + B\,dz \wedge dx + C\,dx \wedge dy$ is a 2-form. As the notation in higher dimensions gets messy, we shall write out only the 3-dimensional case:

$$dw = dA \wedge dy \wedge dz + dB \wedge dz \wedge dx + dC \wedge dx \wedge dy$$
$$= (\partial_x A\,dx + \partial_y A\,dy + \partial_z A\,dz) \wedge dy \wedge dz$$
$$+ (\partial_x B\,dx + \partial_y B\,dy + \partial_z B\,dz) \wedge dz \wedge dx$$
$$+ (\partial_x C\,dx + \partial_y C\,dy + \partial_z C\,dz) \wedge dx \wedge dy$$
$$= (\partial_x A + \partial_y B + \partial_z C)\,dx \wedge dy \wedge dz.$$

(For the last equality we have used the fact that an exterior product containing two identical factors vanishes and the fact that the product $dx \wedge dy \wedge dz$ is unchanged by cyclic permutation of its three terms.) If we identify ω with a vector field $\mathbf{F}$ and $d\omega$ with a function g as before, we see that $g = \text{div}\,\mathbf{F}$. *The divergence is the exterior derivative on 2-forms in $\mathbb{R}^3$.*

We observed earlier that $\text{curl}(\nabla f) = \mathbf{0}$ for any function f and $\text{div}(\text{curl}\,\mathbf{F}) = 0$ for any vector field $\mathbf{F}$. The interpretation of these identities in terms of differential forms is that $d(df) = 0$ for any 0-form (function) f and $d(d\omega) = 0$ for any 1-form ω. *It is true in general that*

$$(5.76) \qquad\qquad\qquad d(d\omega) = 0$$

for any k-form ω on $\mathbb{R}^n$. In all cases the proof of this fact boils down to the equality of mixed partials.

As an illustration of the exterior derivative, we give the relativistically covariant reformulation of Maxwell's equations (5.50). The key idea is to think of electromagnetism as a phenomenon in 4-dimensional space-time rather than a time-dependent phenomenon in 3-dimensional space. The electric and magnetic fields $\mathbf{E} = (E_x, E_y, E_z)$ and $\mathbf{B} = (B_x, B_y, B_z)$ are combined into a single entity, the *electromagnetic field tensor*, which we identify in two ways with a 2-form on $\mathbb{R}^4$:

$$\begin{aligned} \omega &= c(E_x\, dx \wedge dt + E_y\, dy \wedge dt + E_z\, dz \wedge dt) \\ &\quad + B_x\, dy \wedge dz + B_y\, dz \wedge dx + B_z\, dx \wedge dy, \\ \omega^* &= c(B_x\, dx \wedge dt + B_y\, dy \wedge dt + B_z\, dz \wedge dt) \\ &\quad - E_x\, dy \wedge dz - E_y\, dz \wedge dx - E_z\, dx \wedge dy, \end{aligned}$$

where c is the speed of light. Also, the current and charge densities ρ and $\mathbf{J} = (J_x, J_y, J_z)$ are combined into a single entity, the *4-current density*, which we identify with a 3-form on $\mathbb{R}^4$:

$$\gamma = c(J_x\, dy \wedge dz \wedge dt + J_y\, dz \wedge dx \wedge dt + J_z\, dx \wedge dy \wedge dt) - \rho\, dx \wedge dy \wedge dz.$$

The four Maxwell equations (5.50) then turn into the two equations

$$d\omega = 0, \qquad d\omega^* = 4\pi\gamma.$$

The verification of this is a good way for readers to see whether they have learned how to compute exterior derivatives!

Stokes's Theorem. We can now state the general theorem that encompasses the integral theorems of the preceding sections and their higher dimensional analogues:

5.77 Theorem (The General Stokes Theorem). *Let M be a smooth, oriented k-dimensional submanifold of $\mathbb{R}^n$ with a piecewise smooth boundary ∂M, and let ∂M carry the orientation that is (in a suitable sense) compatible with the one on M. If ω is a $(k-1)$-form of class C^1 on an open set containing M, then*

$$\int \cdots \int_{\partial M} \omega = \int\int \cdots \int_M d\omega.$$

We conclude with a final suggestive remark. The formal differential-algebraic identity $d(d\omega) = 0$ stated above has a geometric counterpart. The boundary of a region in the plane is a closed curve with no endpoints, and the boundary of a region in 3-space is a closed surface with no edge. *In general, the boundary of a (smoothly bounded) region M in a k-dimensional manifold is a $(k-1)$-dimensional manifold with no boundary,* that is,

$$(5.78) \qquad\qquad \partial(\partial M) = \varnothing.$$

The general Stokes theorem shows that (5.76) and (5.78) are in some sense equivalent. Indeed, if M is k-dimensional and ω is a $(k-2)$-form, the Stokes theorem gives

$$\int \cdots \int_{\partial(\partial M)} \omega = \int\int \cdots \int_{\partial M} d\omega = \int\int\int \cdots \int_M d(d\omega).$$

If we accept the geometric fact that $\partial(\partial M) = \varnothing$, then the integral on the left vanishes, and hence so does the integral on the right. But since this happens for *every* M, it follows that $d(d\omega) = 0$. Similarly, if we know that $d(d\omega) = 0$ for *every* ω, we can conclude that $\partial(\partial M) = \varnothing$. This sort of interplay of algebra, analysis, and geometry is a significant feature of much of modern mathematics.

Chapter 6

INFINITE SERIES

Infinite series are sums with infinitely many terms, of which the most familiar examples are the nonterminating decimal expansions. For instance, the equality $\pi = 3.14159\ldots$ is an abbreviation of the statement that π is the sum of the infinite series

$$3 + \frac{1}{10} + \frac{4}{10^2} + \frac{1}{10^3} + \frac{5}{10^4} + \frac{9}{10^5} + \cdots .$$

The procedure by which one makes sense out of such sums stands alongside differentiation and integration as one of the fundamental limiting processes of mathematical analysis. Just as decimal expansions provide a useful way of obtaining all real numbers from the finite decimal fractions, infinite series provide a flexible and powerful way of building complicated functions out of simple ones.

This chapter is devoted to the foundations of the theory of infinite series. In it we develop the basic facts about series of *numbers*; then in the next chapter we proceed to the study of series of *functions*.

6.1   Definitions and Examples

Informally speaking, an *infinite series* (or just a *series*, for short) is an expression of the form

$$\sum_{0}^{\infty} a_n = a_0 + a_1 + a_2 + \cdots .$$

Here the a_k's can be real numbers, complex numbers, vectors, and so on; for the present, we shall mainly consider the case where they are real numbers.

It is not immediately clear what precise meaning is to be attached an expression of the form $\sum_0^\infty a_n$ that involves a sum of infinitely many terms. The formal definition must be phrased in terms of limits of finite sums, as follows.

Given a sequence $\{a_n\}_0^\infty$ of real numbers (or complex numbers, vectors, etc.), we can form a new sequence $\{s_k\}_0^\infty$ by adding up the terms of the original sequence successively:

$$s_0 = a_0, \quad s_1 = a_0 + a_1, \quad s_2 = a_0 + a_1 + a_2, \quad \ldots ,$$
$$s_k = a_0 + a_1 + \cdots + a_k.$$

An **infinite series** is formally defined to be a pair of sequences $\{a_n\}$ and $\{s_k\}$ related by these equations, and the notation $\sum_0^\infty a_n$ is to be regarded as a convenient way of encoding this information. The a_n's are called the **terms** of the series, and the s_k's are called the **partial sums** of the series. If the sequence $\{s_k\}$ of partial sums converges to a limit S, then the series is said to be **convergent**, S is called its **sum**, and we write $\sum_0^\infty a_n = S$; otherwise, the series is said to be **divergent**, and no numerical meaning is attached to the expression $\sum_0^\infty a_n$. (However, if $s_k \to \infty$ as $k \to \infty$, we may say that $\sum_0^\infty a_n = \infty$.)

Remark. We have elected to start the numbering of the sequences $\{a_n\}$ and $\{s_k\}$ at $n = 0$ and $k = 0$, since this is perhaps the most common situation in practice. However, we could equally well start at some other point, for instance,

$$\sum_5^\infty a_n = a_5 + a_6 + a_7 + \cdots ,$$

for which we would write

$$s_5 = a_5, \ s_6 = a_5 + a_6, \ s_7 = a_5 + a_6 + a_7, \ \ldots .$$

Before proceeding further, let us record a couple of very simple but important facts about series.

6.1 Theorem.
a. *If the series $\sum_0^\infty a_n$ and $\sum_0^\infty b_n$ are convergent, with sums S and T, then $\sum_0^\infty (a_n + b_n)$ is convergent, with sum $S + T$.*
b. *If the series $\sum_0^\infty a_n$ is convergent, with sum S, then for any $c \in \mathbb{R}$ the series $\sum_0^\infty c a_n$ is convergent, with sum cS.*
c. *If the series $\sum_0^\infty a_n$ is convergent, then $\lim_{n \to \infty} a_n = 0$. Equivalently, if $a_n \not\to 0$ as $n \to \infty$, then the series $\sum_0^\infty a_n$ is divergent.*

Proof. Let $\{s_k\}$ and $\{t_k\}$ be the sequences of partial sums of the series $\sum_0^\infty a_n$ and $\sum_0^\infty b_n$, respectively. (a) and (b) follow from the fact that if $s_k \to S$ and $t_k \to T$, then $s_k + t_k \to S + T$ and $cs_k \to cS$. As for (c), we observe that $a_n = s_n - s_{n-1}$. If the series converges to the sum S, it follows that $\lim a_n = \lim s_n - \lim s_{n-1} = S - S = 0$. $\qquad\square$

At present we are thinking primarily of series whose terms are numbers, but most of the really significant applications of series come from situations where the terms a_n depend on a variable x. In this case the series $\sum_0^\infty a_n(x)$ may converge for some values of x and diverge for others, and it defines a function whose domain is the set of all x for which it converges. We shall explore this idea in more detail in the next chapter; at this point we recall some familiar examples.

One of the simplest and most useful infinite series is the **geometric series**, in which the ratio of two succeeding terms is a constant x. That is, the geometric series with initial term a and ratio x is

$$a + ax + ax^2 + ax^3 + \cdots = \sum_0^\infty ax^n.$$

The constant a can be factored out, according to Theorem 6.1b, so it suffices to consider the case $a = 1$.

The partial sums $s_k = \sum_0^k x^n$ of the series $\sum_0^\infty x^n$ are easily evaluated. If $x = 1$, then of course $s_k = 1 + 1 + \cdots + 1 = k + 1$. If $x \neq 1$, we observe that

$$s_k = 1 + x + \cdots + x^k,$$
$$x s_k = \quad\; x + \cdots + x^k + x^{k+1},$$

and subtracting the second equation from the first yields $(1 - x)s_k = 1 - x^{k+1}$. Therefore,

(6.2) $$s_k = \frac{1 - x^{k+1}}{1 - x} \text{ if } x \neq 1, \quad s_k = k + 1 \text{ if } x = 1.$$

If $|x| < 1$, then $x^{k+1} \to 0$ as $k \to \infty$, so $s_k \to (1 - x)^{-1}$. It also follows easily from (6.2), or from Theorem 6.1c, that $\{s_k\}$ diverges when $|x| \geq 1$. In short, we have:

6.3 Theorem. *The geometric series $\sum_0^\infty x^n$ converges if and only if $|x| < 1$, in which case its sum is $(1 - x)^{-1}$.*

Another familiar result that leads to infinite series is Taylor's theorem. We recall that if f is a function of class C^∞ (that is, possessing derivatives of all orders) on some interval $(-c, c)$ centered at the origin, for any positive integer k we have

$$(6.4) \qquad f(x) = f(0) + f'(0)x + \cdots + \frac{f^{(k)}(0)}{k!} x^k + R_k(x) \qquad (|x| < c).$$

If it happens that $R_k(x) \to 0$ as $k \to \infty$, we can let $k \to \infty$ in (6.4) to obtain an infinite series expansion of $f(x)$, the **Taylor series** of f (centered at $x = 0$):

$$(6.5) \qquad f(x) = \sum_0^\infty \frac{f^{(n)}(0)}{n!} x^n.$$

One simple sufficient condition to guarantee that $R_k(x) \to 0$ follows from the estimate for the Taylor remainder in Corollary 2.61:

$$|R_k(x)| \le \sup_{|t| \le |x|} |f^{(k+1)}(t)| \frac{|x|^{k+1}}{(k+1)!} \qquad (|x| < c).$$

6.6 Theorem. *Let f be a function of class C^∞ on the interval $(-c, c)$, where $0 < c \le \infty$.*

a. *If there exist constants $a, b > 0$ such that $|f^{(k)}(x)| \le ab^k k!$ for all $|x| < c$ and $k \ge 0$, then (6.5) holds for $|x| < \min(c, b^{-1})$.*

b. *If there exist constants $A, B > 0$ such that $|f^{(k)}(x)| \le AB^k$ for all $|x| < c$ and $k \ge 0$, then (6.5) holds for $|x| < c$.*

Proof. By Corollary 2.61, the estimate $|f^{(k)}(x)| \le ab^k k!$ implies the estimate $|R_{k-1}(x)| \le a|bx|^k$ for $|x| < c$. If also $|x| < b^{-1}$, then $|bx|^k \to 0$ as $k \to \infty$, so (6.4) yields the result (a). To deduce (b), we observe that the factorial function grows faster than exponentially (see Example 5 in §1.4), so that for any positive A, B, and b, the sequence $A(B/b)^k/k!$ tends to zero as $k \to \infty$. Letting a be the largest term in this sequence, we have

$$AB^k = \left[A \frac{(B/b)^k}{k!} \right] b^k k! \le ab^k k!,$$

so the estimate $|f^{(k)}(x)| \le AB^k$, for a given A and B, implies the estimate $|f^{(k)}(x)| \le ab^k k!$ for *every* $b > 0$ (with a depending on b). Hence (b) follows from (a). $\qquad\square$

Remark. The interval $(-c, c)$ might not be the whole set where the function f and its derivatives are defined. It may be necessary to restrict x to a proper subinterval of the domain of f to obtain the estimates on $f^{(k)}(x)$ in Theorem 6.6, as Example 2 will show.

EXAMPLE 1. Let $f(x) = \cos x$. The derivatives $f^{(k)}(x)$ are equal to $\pm \cos x$ or $\pm \sin x$, depending on k, so they all satisfy $|f^{(k)}(x)| \leq 1$ for all x. By Theorem 6.6b, it follows that $\cos x$ is the sum of its Taylor series, $\sum_0^\infty (-1)^n x^{2n}/(2n)!$, for all x. For exactly the same reason, $\sin x$ is the sum of its Taylor series, $\sum_0^\infty (-1)^n x^{2n+1}/(2n+1)!$, for all x.

EXAMPLE 2. Let $f(x) = e^x$. Here $f^{(k)}(x) = e^x$ for all k. We cannot obtain a good estimate on $f^{(k)}(x)$ that is valid for all x at once, but for $|x| < c$ we have $|f^{(k)}(x)| < e^c$. By Theorem 6.6b, it follows that e^x is the sum of its Taylor series, $\sum_0^\infty x^n/n!$, for $|x| < c$. But c is arbitrary, so in fact $e^x = \sum_0^\infty x^n/n!$ for all x.

Finally, we mention one other simple type of series that arises from time to time. Just as $\int_a^b f(x)\,dx$ is easy to compute when f is the derivative of a known function, the series $\sum_0^\infty a_n$ is easy to sum when the terms a_n are the differences of a known sequence $\{b_n\}$. That is, suppose $a_0 = b_0$ and $a_n = b_n - b_{n-1}$ for $n \geq 1$; then

$$s_k = a_0 + a_1 + \cdots + a_k = b_0 + (b_1 - b_0) + \cdots + (b_k - b_{k-1}) = b_k,$$

so the series $\sum_0^\infty a_n$ converges if and only if the sequence $\{b_n\}$ converges, in which case $\sum_0^\infty a_n = \lim b_n$. Such series are called **telescoping series**.

EXERCISES

1. Find the values of x for which each of the following series converges and compute its sum.
   a. $2(x+1) + 4(x+1)^4 + 8(x+1)^7 + \cdots + 2^{n+1}(x+1)^{3n+1} + \cdots$
   b. $10x^{-2} + 20x^{-4} + 40x^{-6} + \cdots + 10 \cdot 2^n x^{-2(n+1)} + \cdots$
   c. $1 + (1-x)/(1+x) + (1-x)^2/(1+x)^2 + \cdots + (1-x)^n/(1+x)^n + \cdots$
   d. $\log x + (\log x)^2 + (\log x)^3 + \cdots + (\log x)^n + \cdots$

2. Tell whether each of the following series converges; if it does, find its sum.
   a. $1 + \frac{3}{4} + \frac{5}{8} + \frac{9}{16} + \frac{17}{32} + \cdots$
   b. $\frac{1}{1\cdot2} + \frac{1}{2\cdot3} + \frac{1}{3\cdot4} + \frac{1}{4\cdot5} + \cdots$ (*Hint:* $[n(n+1)]^{-1} = n^{-1} - (n+1)^{-1}$).
   c. $(\sqrt{2} - \sqrt{1}) + (\sqrt{3} - \sqrt{2}) + (\sqrt{4} - \sqrt{3}) + \cdots$
   d. $1 - \frac{1}{2} + 1 - \frac{1}{3} + 1 - \frac{1}{4} + 1 - \frac{1}{5} + \cdots$

3. Let $f(x) = \log(1+x)$. Show that the Taylor remainder $R_{0,k}(x)$ (defined by (2.54)) tends to zero as $k \to \infty$ for $-1 < x \leq 1$, and conclude that

$$\log(1+x) = \sum_1^\infty (-1)^{n+1} \frac{x^n}{n} \quad \text{for } -1 < x \leq 1.$$

(*Hint:* Lagrange's formula for $R_{0,k}$ easily yields the desired result when $-\frac{1}{2} <$ $x \leq 1$ but not when $-1 < x \leq -\frac{1}{2}$. For $x < 0$, use the integral formula (2.56) for $R_{0,k}$ and the mean value theorem for integrals to show that $|R_{0,k}(x)| = |x|(x' - x)^n(x' + 1)^{-n-1}$ for some $x' \in (x, 0)$, and thence show that $|R_{0,k}(x)| < |x|^{n+1}/(1+x)$.)

4. Given a sequence $\{a_n\}$ of numbers, let $\prod_1^k a_n$ denote the product of the numbers $a_1, \ldots, a_k$. The **infinite product** $\prod_1^\infty a_n$ is said to converge to the number P if the sequence of partial products converges to P:

$$\prod_1^\infty a_n = \lim_{k \to \infty} \prod_1^k a_n = \lim_{k \to \infty} a_1 a_2 \cdots a_k.$$

(*Note:* In many books one finds a more complicated definition that takes account of the peculiar role of the number 0 with regard to multiplication.)

a. Show that if $\prod_1^\infty a_n$ converges to a nonzero number P, then $\lim_{n \to \infty} a_n = 1$. (This is the analogue of Theorem 6.1c for products.)

b. Show that if $\prod_1^\infty a_n$ converges to a nonzero number P, then $\sum_1^\infty \log a_n$ converges after omission of those terms for which $a_n < 0$. (By (a), there can only be finitely many such terms, and no a_n can be 0.) Conversely, show that if $a_n > 0$ for all n and $\sum_1^\infty \log a_n$ converges to S, then $\prod_1^\infty a_n$ converges to e^S. (See also Exercise 5 in §6.3.)

6.2   Series with Nonnegative Terms

In this section we begin the systematic study of the convergence of infinite series by considering series with nonnegative terms. If $a_n \geq 0$ for all n, the partial sums $s_k = a_0 + \cdots + a_k$ form an increasing sequence. By the monotone sequence theorem, therefore, the series $\sum_0^\infty a_n$ converges if and only if the partial sums s_k have a finite upper bound. This observation leads to a variety of *comparison tests*, in which the partial sums s_k are compared to more easily computable quantities that can be shown to be bounded or unbounded.

   The Integral Test. If $a_n = f(n)$ where f is a function of a real variable, a sum $\sum_{n=j}^k a_n$ can be compared to an integral $\int_j^k f(x)\, dx$. The virtue of this idea is that although integration is a more sophisticated concept than summation, integrals are often easier to compute than sums! The fundamental theorem, whose pictorial meaning is indicated in Figure 6.1, is as follows:

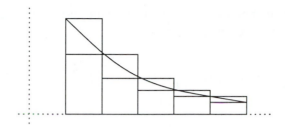

FIGURE 6.1: Comparison of $\int_j^k f(x)\,dx$ (the area under the curve) with $\sum_{n=j}^{k-1} f(n)$ and $\sum_{n=j+1}^{k} f(n)$ (its upper and lower Riemann sums).

6.7 Theorem. *Suppose f is a positive, decreasing function on the half-line $[a, \infty)$. Then for any integers j, k with $a \leq j < k$,*

$$\sum_{n=j}^{k-1} f(n) \geq \int_j^k f(x)\,dx \geq \sum_{n=j+1}^{k} f(n).$$

Proof. Since f is decreasing, for $n \leq x \leq n+1$ we have $f(n) \geq f(x) \geq f(n+1)$, and hence

$$f(n) = \int_n^{n+1} f(n)\,dx \geq \int_n^{n+1} f(x)\,dx \geq \int_n^{n+1} f(n+1)\,dx = f(n+1).$$

Adding up these inequalities from $n = j$ to $n = k - 1$, we obtain the asserted result. $\qquad\square$

An immediate corollary is the following test for convergence.

6.8 Corollary (The Integral Test). *Suppose f is a positive, decreasing function on the half-line $[1, \infty)$. Then the series $\sum_1^\infty f(n)$ converges if and only if the improper integral $\int_1^\infty f(x)\,dx$ converges.*

Proof. Let $s_k = \sum_{n=1}^{k} f(n)$. If $\int_1^\infty f(x)\,dx < \infty$, we have

$$s_k = f(1) + \sum_2^k f(n) \leq f(1) + \int_1^k f(x)\,dx \leq f(1) + \int_1^\infty f(x)\,dx,$$

so the partial sums are bounded above and hence the series converges. On the other hand, if $\int_1^\infty f(x)\,dx = \infty$, we have

$$s_k = \sum_1^{k-1} f(n) + f(k) \geq \int_1^k f(x)\,dx + f(k) \to \infty \text{ as } k \to \infty,$$

so the series diverges. $\qquad\qquad\qquad\qquad\qquad\qquad\qquad\qquad\qquad\qquad\square$

Of course, a similar result relates $\sum_J^\infty f(n)$ to $\int_J^\infty f(x)\,dx$, for any integer J. We chose $J = 1$ because it is appropriate for the following important application.

6.9 Theorem. *The series $\sum_1^\infty n^{-p}$ converges if $p > 1$ and diverges if $p \leq 1$.*

Proof. The same is true of the integrals $\int_1^\infty x^{-p}\,dx$, for

$$\int_1^\infty x^{-p}\,dx = \lim_{K\to\infty} \frac{x^{1-p}}{1-p}\Big|_1^K = \begin{cases} (p-1)^{-1} & \text{if } p > 1, \\ \infty & \text{if } p < 1, \end{cases}$$

and $\int_1^\infty x^{-1}\,dx = \lim_{K\to\infty} \log x \big|_1^K = \infty$. $\qquad\qquad\qquad\qquad\square$

Theorem 6.7 does more than provide a test for convergence; it also provides an approximation to the partial sums and the full sum of the series. In the convergent case, this can be used to provide a numerical approximation to the sum $\sum_1^\infty f(n)$ or an estimate of how many terms must be used for a partial sum to provide a good approximation; in the divergent case, it can be used to estimate how rapidly the partial sums grow.

Suppose, for example, that f is positive and decreasing, and that $\int_1^\infty f(x)\,dx < \infty$. By letting $k \to \infty$ in Theorem 6.7, we obtain

$$\sum_1^\infty f(n) \geq \int_1^\infty f(x)\,dx \geq \sum_2^\infty f(n),$$

and hence

$$\int_1^\infty f(x)\,dx \leq \sum_1^\infty f(n) \leq f(1) + \int_1^\infty f(x)\,dx.$$

This gives an approximation to the sum $\sum_1^\infty f(n)$ with an error of at most $f(1)$. A better approximation can be obtained by using this estimate not for the whole series but for its tail end:

$$\int_k^\infty f(x)\,dx \leq \sum_k^\infty f(n) \leq f(k) + \int_k^\infty f(x)\,dx.$$

Adding on the first $k - 1$ terms of the series, we see that

$$(6.10) \quad \sum_1^\infty f(n) = \sum_0^{k-1} f(n) + \int_k^\infty f(x)\, dx, \text{ with an error of at most } f(k).$$

The error $f(k)$ will be as small as we please provided k is sufficiently large.

EXAMPLE 1. To evaluate $\sum_1^\infty n^{-4}$ with an error of at most 0.0001, we take $k = 10$ in (6.10) to get

$$\sum_1^\infty n^{-4} \approx 1^{-4} + 2^{-4} + \cdots + 9^{-4} + \int_{10}^\infty x^{-4}\, dx$$

$$= 1^{-4} + 2^{-4} + \cdots + 9^{-4} + \tfrac{1}{3} 10^{-3}.$$

A bit of work with a pocket calculator yields the value of this last sum as $1.08226\ldots$, so we can conclude that $1.08226 < \sum_1^\infty n^{-4} < 1.08236$. (The exact value of $\sum_1^\infty n^{-4}$ is $\pi^4/90 = 1.0823232\ldots$; see Exercise 3 in §8.3 or Exercise 9a in §8.6.)

General Comparison Tests. One can often decide whether a series of nonnegative terms converges by comparing it to a series whose convergence or divergence is known. The general method is as follows.

6.11 Theorem. *Suppose* $0 \le a_n \le b_n$ *for* $n \ge 0$. *If* $\sum_0^\infty b_n$ *converges, then so does* $\sum_0^\infty a_n$. *If* $\sum_0^\infty a_n$ *diverges, then so does* $\sum_0^\infty b_n$.

Proof. Let $s_k = \sum_0^k a_n$ and $t_k = \sum_0^k b_n$; thus $0 \le s_k \le t_k$ for all k. If $\sum_0^\infty b_n$ converges, the numbers t_k form a bounded set; hence so do the numbers s_k, so the sequence $\{s_k\}$ converges by the monotone sequence theorem. This proves the first assertion, to which the second one is logically equivalent. $\qquad\square$

A couple of remarks are in order concerning this result. First, the convergence or divergence of a series is unaffected if finitely many terms are deleted from or added to the series. Hence, the comparison $a_n \le b_n$ only has to be valid for all $n \ge N$, where N is some (possibly large) positive integer. Second, the convergence or divergence of a series is unaffected if all the terms of the series are multiplied by a nonzero constant. Hence, the comparison $a_n \le b_n$ can be replaced by $a_n \le c b_n$, where c is any positive number.

When a_n is an algebraic function of n (obtained from n by applying various combinations of the arithmetic operations together with the operation of raising to a power, $x \to x^a$), one can usually decide the convergence of $\sum a_n$ by comparing it to one of the series $\sum_1^\infty n^{-p}$, discussed in Theorem 6.9. The rule of thumb, obtained by combining Theorems 6.9 and 6.11, is that if $a_n \geq cn^{-1}$ then $\sum a_n$ diverges, whereas if $a_n \leq cn^{-p}$ for some $p > 1$ then $\sum a_n$ converges.

EXAMPLE 2. The series $\sum_1^\infty (2n-1)^{-1} = 1 + \frac{1}{3} + \frac{1}{5} + \cdots$ diverges by comparison to $\sum_1^\infty n^{-1}$, for

$$\frac{1}{2n-1} > \frac{1}{2n} = \frac{1}{2} \cdot \frac{1}{n}.$$

EXAMPLE 3. The series $\sum_1^\infty (n^2 - 6n + 10)^{-1}$ converges by comparison to $\sum_1^\infty n^{-2}$, but here the comparison takes more work to establish. Since $6n > 10$ except for $n = 1$, it is not true that $(n^2 - 6n + 10)^{-1} \leq n^{-2}$. However, we can observe when $n > 12$ we have $6n < \frac{1}{2}n^2$, and hence

$$\frac{1}{n^2 - 6n + 10} < \frac{1}{(n^2/2) + 10} < \frac{1}{(n^2/2)} = \frac{2}{n^2} \qquad (n > 12),$$

which gives the desired comparison. However, there is also a simpler way to proceed. The key observation is that when n is large, $-6n + 10$ is negligibly small in comparison with n^2, so $(n^2 - 6n + 10)^{-1}$ is practically equal to n^{-2}. More precisely,

$$\frac{(n^2 - 6n + 10)^{-1}}{n^{-2}} = \frac{n^2}{n^2 - 6n + 10} = \frac{1}{1 - 6n^{-1} + 10n^{-2}} \to 1 \text{ as } n \to \infty,$$

which immediately gives the comparison $(n^2 - 6n + 10)^{-1} < 2n^{-2}$ when n is large.

The second method for solving Example 3 can be formulated quite generally; the result is often called the **limit comparison test**:

6.12 Theorem. *Suppose $\{a_n\}$ and $\{b_n\}$ are sequences of positive numbers and that a_n/b_n approaches a positive, finite limit as $n \to \infty$. Then the series $\sum_0^\infty a_n$ and $\sum_0^\infty b_n$ are either both convergent or both divergent.*

Proof. If $a_n/b_n \to l$ as $n \to \infty$, where $0 < l < \infty$, we have $\frac{1}{2}l < a_n/b_n < 2l$ when n is large; that is, $a_n < 2lb_n$ and $b_n < (2/l)a_n$. The result therefore follows from Theorem 6.11 and the remarks following it. $\qquad\square$

Theorem 6.12 can be extended a little. If $a_n/b_n \to 0$ as $n \to \infty$, then $a_n < b_n$ for large n, so the convergence of $\sum b_n$ will imply the convergence of $\sum a_n$. Likewise, if $a_n/b_n \to \infty$, then $a_n > b_n$ for large n, so the convergence of $\sum a_n$ will imply the convergence of $\sum b_n$. However, the reverse implications are not valid in these cases.

Comparisons to the Geometric Series. There are a couple of very useful convergence tests that are based on a comparison to the geometric series $\sum_0^\infty r^n$, where $r > 0$. We recall that this series converges for $r < 1$ and diverges for $r \geq 1$.

6.13 Theorem (The Ratio Test). *Suppose $\{a_n\}$ is a sequence of positive numbers.*

a. *If $a_{n+1}/a_n < r$ for all sufficiently large n, where $r < 1$, then the series $\sum_0^\infty a_n$ converges. On the other hand, if $a_{n+1}/a_n \geq 1$ for all sufficiently large n, then the series $\sum_0^\infty a_n$ diverges.*

b. *Suppose that $l = \lim_{n\to\infty} a_{n+1}/a_n$ exists. Then the series $\sum_0^\infty a_n$ converges if $l < 1$ and diverges if $l > 1$. No conclusion can be drawn if $l = 1$.*

Proof. Suppose $a_{n+1}/a_n < r < 1$ for all $n \geq N$. Then

$$a_{N+1} < r a_N, \quad a_{N+2} < r a_{N+1} < r^2 a_N, \quad a_{N+3} < r a_{N+2} < r^3 a_N, \quad \ldots,$$

so $a_{N+m} < r^m a_N$ for all $m \geq 0$. The series $\sum_0^\infty a_n$ therefore converges by comparison to the geometric series $\sum r^m$:

$$\sum_0^\infty a_n < a_0 + \cdots + a_{N-1} + a_N(1 + r + r^2 + \cdots) < \infty.$$

On the other hand, if $a_{n+1}/a_n \geq 1$ then $a_{n+1} \geq a_n$; if this is so for all $n \geq N$, then $a_n \not\to 0$, so $\sum a_n$ cannot converge. This proves (a).

Assertion (b) is a corollary of (a). If $l < 1$, choose r with $l < r < 1$. If $\lim a_{n+1}/a_n = l$, then $a_{n+1}/a_n < r$ for large n, so $\sum a_n$ converges. If $l > 1$, then $a_{n+1}/a_n \geq 1$ for large n, so $\sum a_n$ diverges. Finally, if we take $a_n = n^{-p}$, we know that $\sum_1^\infty a_n$ converges if $p > 1$ and diverges if $p \leq 1$; but $a_{n+1}/a_n = [n/(n+1)]^p \to 1$ no matter what p is. Hence the test is inconclusive if $l = 1$. □

6.14 Theorem (The Root Test). *Supppose $\{a_n\}$ is a sequence of positive numbers.*

a. *If $a_n^{1/n} < r$ for all sufficiently large n, where $r < 1$, then the series $\sum_0^\infty a_n$ converges. On the other hand, if $a_n^{1/n} \geq 1$ for all sufficiently large n, then the series $\sum_0^\infty a_n$ diverges.*

b. *Suppose that $l = \lim_{n\to\infty} a_n^{1/n}$ exists. Then the series $\sum_0^\infty a_n$ converges if $l < 1$ and diverges if $l > 1$. No conclusion can be drawn if $l = 1$.*

Proof. If $a_n^{1/n} < r$, we have $a_n < r^n$, so we have an immediate comparison to the geometric series $\sum r^n$ that gives the convergence of $\sum a_n$ when $r < 1$. If $a_n^{1/n} \geq 1$ then $a_n \geq 1$, so $a_n \not\to 0$ and $\sum a_n$ diverges. This proves (a).

Part (b) follows as in the proof of the ratio test. If $a_n^{1/n} \to l < 1$, we pick $r \in (l, 1)$ and obtain $a_n^{1/n} < r$ for large n, so $\sum a_n$ converges. If $a_n^{1/n} \to l > 1$, then $a_n^{1/n} \geq 1$ for large n, and $\sum a_n$ diverges. Finally, for $a_n = n^{-p}$ we have $a_n^{1/n} = n^{-p/n} \to 1$ for any p, so the test is inconclusive when $l = 1$. $\square$

Note: In the last line of this proof, and in Example 4 below, we use the fact that $\lim_{x\to\infty} x^{1/x} = 1$. To see, this, observe that $\log(x^{1/x}) = (\log x)/x$, and $\lim_{x\to\infty}(\log x)/x = 0$ by l'Hôpital's rule.

It can be shown that if a_{n+1}/a_n converges to a limit l, then $a_n^{1/n}$ also converges to the same limit; but the convergence of $a_n^{1/n}$ does not imply the convergence of a_{n+1}/a_n. (See Example 6.) Thus the root test is, in theory, more powerful than the ratio test. However, the ratio test is often more convenient to use in practice, especially for series whose terms involve factorials or similar sorts of products.

EXAMPLE 4. Let $a_n = n^2/2^n$. The ratio test and the root test can both be used to establish the convergence of $\sum_0^\infty a_n$:

$$\frac{a_{n+1}}{a_n} = \frac{(n+1)^2/2^{n+1}}{n^2/2^n} = \frac{1}{2}\left[\frac{n+1}{n}\right]^2 \to \frac{1}{2}, \qquad a_n^{1/n} = \frac{1}{2}n^{2/n} \to \frac{1}{2}.$$

EXAMPLE 5. Let $a_n = \dfrac{1 \cdot 4 \cdot 7 \cdots (3n+1)}{2^n n!}$. Here the root test is cumbersome, but the ratio test works easily:

$$\frac{a_{n+1}}{a_n} = \frac{1 \cdot 4 \cdots (3n+1)(3n+4)/2^{n+1}(n+1)!}{1 \cdot 4 \cdots (3n+1)/2^n n!} = \frac{3n+4}{2(n+1)} \to \frac{3}{2},$$

so $\sum_0^\infty a_n$ diverges.

EXAMPLE 6. Let $a_n = 2^{-n/2}$ if n is even and $a_n = 2^{-(n-1)/2}$ if n is odd; thus

$$\sum_0^\infty a_n = 1 + 1 + \tfrac{1}{2} + \tfrac{1}{2} + \tfrac{1}{4} + \tfrac{1}{4} + \tfrac{1}{8} + \tfrac{1}{8} + \cdots.$$

Here a_{n+1}/a_n equals 1 if n is even and $\frac{1}{2}$ if n is odd, so the ratio test (even the more general form in part (a) of Theorem 6.13) fails; its hypotheses are not satisfied. But the root test works: $a_n^{1/n}$ equals $2^{-1/2}$ if n is even and $2^{-(n-1)/2n}$ if n is odd; both of these expressions converge to $2^{-1/2}$ as $n \to \infty$, so the series converges. (Of course, this can also be proved more simply. By grouping the terms in pairs, one sees that $\sum_0^\infty a_n = 2 \sum_0^\infty 2^{-m} = 4$.)

Raabe's Test. The ratio test and the root test are, in a sense, rather crude, for the indecisive cases where $\lim a_{n+1}/a_n = 1$ or $\lim a_n^{1/n} = 1$ include many commonly encountered series such as $\sum_1^\infty n^{-p}$. The reason for this insensitivity is that the terms of the geometric series $\sum r^n$ either converge to zero exponentially fast (if $r < 1$) or not at all (if $r \geq 1$), so they do not furnish a useful comparison for quantities such as n^{-p} that tend to zero only polynomially fast. However, there is another test, Raabe's test, that is sometimes useful in the case where $\lim a_{n+1}/a_n = 1$. The class of problems for which Raabe's test is effective is rather limited, and there is another way of attacking the most important of them that we shall present in §7.6. Hence we view Raabe's test as an optional topic; however, the insight behind it is of interest in its own right.

The idea is to use the ratios a_{n+1}/a_n to compare the series $\sum a_n$ to one of the series $\sum n^{-p}$ rather than to the geometric series. For the series $\sum n^{-p}$, the ratio of two successive terms is $(n+1)^{-p}/n^{-p} = [1 + (1/n)]^{-p}$. To put this quantity in a form more amenable to comparison, we use the tangent line approximation to the function $f(x) = (1+x)^{-p}$ at $x = 0$. Since $f'(x) = -p(1+x)^{-p-1}$ and $f''(x) = p(p+1)(1+x)^{-p-2}$, Lagrange's formula for the error term gives

$$(1+x)^{-p} = 1 - px + E(x), \qquad 0 < E(x) < \frac{p(p+1)}{2}x^2 \text{ for } x > 0.$$

Hence,

(6.15)
$$\frac{(n+1)^{-p}}{n^{-p}} = \left[1 + \frac{1}{n}\right]^{-p} = 1 - \frac{p}{n} + E_n, \qquad 0 < E_n < \frac{p(p+1)}{2n^2}.$$

Thus, $n[1 - (n+1)^{-p}/n^{-p}]$ is approximately p when n is large. With this in mind, we are ready for the main result.

6.16 Theorem (Raabe's Test). *Let $\{a_n\}$ be a sequence of positive numbers. Suppose that*

$$\frac{a_{n+1}}{a_n} \to 1 \quad and \quad n\left[1 - \frac{a_{n+1}}{a_n}\right] \to L \text{ as } n \to \infty.$$

If $L > 1$, the series $\sum a_n$ converges, and if $L < 1$, the series $\sum a_n$ diverges. (If $L = 1$, no conclusion can be drawn.)

Proof. If $L > 1$, choose a number p with $1 < p < L$. Then, when n is large, we have $n[1 - (a_{n+1}/a_n)] > p$, that is, $a_{n+1}/a_n < 1 - (p/n)$. Thus, by (6.15),

$$\frac{a_{n+1}}{a_n} < 1 - \frac{p}{n} < \frac{(n+1)^{-p}}{n^{-p}}, \quad \text{or} \quad \frac{a_{n+1}}{(n+1)^{-p}} < \frac{a_n}{n^{-p}}.$$

292   *Chapter 6. Infinite Series*

Thus the sequence $\{a_n/n^{-p}\}$ is decreasing, so it is bounded above by a constant C. In other words, $a_n \leq Cn^{-p}$, so since $p > 1$, $\sum a_n$ converges by comparison to $\sum n^{-p}$.

On the other hand, if $L < 1$, choose numbers p and q with $L < q < p < 1$. Then, when n is large, we have $n[1 - (a_{n+1}/a_n)] < q$, that is, $(a_{n+1}/a_n) > 1 - (q/n)$. If also $n > p(p+1)/2(p-q)$, we have $p(p+1)/2n^2 < (p-q)/n$, so by (6.15),

$$\frac{a_{n+1}}{a_n} > 1 - \frac{q}{n} = 1 - \frac{p}{n} + \frac{p-q}{n} > 1 - \frac{p}{n} + E_n = \frac{(n+1)^{-p}}{n^{-p}}.$$

Thus $(n+1)^{-p}/a_{n+1} < n^{-p}/a_n$, so the sequence $\{n^{-p}/a_n\}$ is decreasing. As before, this gives $n^{-p} \leq Ca_n$, and $p < 1$ in this case, so $\sum a_n$ diverges by comparison to $\sum n^{-p}$.   □

The main applications of Raabe's test are to series whose terms involve quotients of factorial-like products. The following example is typical.

EXAMPLE 7. Let $a_n = \dfrac{1 \cdot 4 \cdot 7 \cdots (3n+1)}{n^2 3^n n!}$. We have

$$\frac{a_{n+1}}{a_n} = \frac{1 \cdot 4 \cdots (3n+1)(3n+4)/(n+1)^2 3^{n+1}(n+1)!}{1 \cdot 4 \cdots (3n+1)/n^2 3^n n!} = \frac{(3n+4)n^2}{3(n+1)^3}.$$

This tends to 1 as $n \to \infty$ (the dominant term on both top and bottom is $3n^3$), so the ratio test fails. But

$$n\left[1 - \frac{a_{n+1}}{a_n}\right] = n\left[1 - \frac{(3n+4)n^2}{3(n+1)^3}\right] = \frac{5n^3 + 9n^2 + 3}{3(n+1)^3} \to \frac{5}{3},$$

and $\frac{5}{3} > 1$, so the series $\sum a_n$ converges.

Concluding Remarks. Faced with an infinite series $\sum a_n$, how does one decide how to test it for convergence? Some series require more cleverness than others, but the following rules of thumb may be helpful.

- Does $a_n \to 0$ as $n \to \infty$? If not, $\sum a_n$ diverges.

- If a_n is an algebraic function of n (say, a rational function of n, or a similar expression involving fractional powers of n), try comparison with $\sum n^{-p}$ for a suitable value of p.

- If a_n involves expressions with n in the exponent, try the ratio test or the root test.

- If a_n involves factorial-like products, the ratio test is the best bet. If the ratio test fails because $\lim a_{n+1}/a_n = 1$, try Raabe's test.

- The integral test may be useful when numerical estimates are desired or when the series is near the borderline between convergence and divergence.

In any case, one should beware of confusing the various sequences that arise in the study of infinite series. For any infinite series $\sum a_n$, one has the sequence $\{a_n\}$ of *terms* and the sequence $\{s_k\}$ of *partial sums*. In the ratio test, one considers the sequence $\{a_{n+1}/a_n\}$ of ratios of successive terms of a series, whereas in the limit comparison test, one considers the sequence $\{a_n/b_n\}$ of ratios of corresponding terms of two different series. Don't mix these sequences up!

EXERCISES

In Exercises 1–18, test the series for convergence.

1. $\displaystyle\sum_0^\infty \frac{\sqrt{n+1}}{n^2-4n+5}.$

2. $\displaystyle\sum_1^\infty ne^{-n}.$

3. $\displaystyle\sum_1^\infty \frac{2n^2-n}{2n^{8/3}+n}.$

4. $\displaystyle\sum_1^\infty \frac{n+1}{n!}.$

5. $\displaystyle\sum_0^\infty \frac{(2n+1)^{3n}}{(3n+1)^{2n}}.$

6. $\displaystyle\sum_0^\infty \frac{1^2\cdot 3^2\cdots(2n+1)^2}{3^n(2n)!}.$

7. $\displaystyle\sum_1^\infty \frac{n!}{10^n}.$

8. $\displaystyle\sum_2^\infty (\log n)^{-100}.$

9. $\displaystyle\sum_0^\infty \frac{1\cdot 3\cdots(2n+1)}{2\cdot 5\cdots(3n+2)}.$

10. $\displaystyle\sum_0^\infty \frac{(n!)^2}{(2n)!}.$

11. $\displaystyle\sum_0^\infty \frac{3^n n!}{n^n}.$

12. $\displaystyle\sum_1^\infty \left(\frac{n}{n+1}\right)^{n^2}.$

13. $\displaystyle\sum_1^\infty [1 - \cos(1/n)].$

14. $\displaystyle\sum_1^\infty \frac{\sqrt{n+1} - \sqrt{n}}{\sqrt{n+2}}.$

15. $\displaystyle\sum_1^\infty \sin\frac{n}{n^2+3}.$

16. $\displaystyle\sum_1^\infty \frac{n^2[\pi + (-1)^n]^n}{5^n}.$

17. $\displaystyle\sum_1^\infty \frac{1 \cdot 3 \cdots (2n-1)}{4 \cdot 6 \cdots (2n+2)}.$

18. $\displaystyle\sum_1^\infty \frac{2 \cdot 4 \cdots (2n)}{3 \cdot 5 \cdots (2n+1)}.$

19. Suppose $a_n > 0$. Show that if $\sum a_n$ converges, then so does $\sum a_n^p$ for any $p > 1$.

20. Show that $\displaystyle\sum_2^\infty \frac{1}{n(\log n)^p}$ converges if $p > 1$ and diverges if $p \le 1$.

21. For which p does $\displaystyle\sum_4^\infty \frac{1}{n(\log n)(\log\log n)^p}$ converge?

22. By Exercise 20, $\sum_2^\infty 1/[n \log n]$ diverges while $\sum_2^\infty 1/[n(\log n)^2]$ converges. Use Theorem 6.7 to show that

$$4.88 < \sum_2^{10^{40}} \frac{1}{n \log n} < 5.61, \qquad \sum_{10^{40}}^\infty \frac{1}{n(\log n)^2} \approx 0.011.$$

The point is that for series such as these that are near the borderline between convergence and divergence, attempts at numerical approximation by adding up the first few terms aren't much use. If you add up the first 10^{40} terms of the first series, you get no clue that the series diverges; and if you add up the first 10^{40} terms of the second one, the answer you get still differs from the full sum in the second decimal place. (By way of comparison, the universe is around 10^{18} seconds old, and the earth contains around 10^{50} atoms.)

23. Verify that $x/(x^2 + 1)^2$ is decreasing for $x \geq 3^{-1/2}$, and thence show that $0.38 < \sum_1^\infty n/(n^2 + 1)^2 < 0.41$.

24. Let $c_k = 1 + \frac{1}{2} + \cdots + \frac{1}{k} - \log k$. Show that the sequence $\{c_k\}$ is positive and decreasing, and hence convergent. ($\lim_{k\to\infty} c_k$ is conventionally denoted by γ and is called **Euler's constant** or the **Euler-Mascheroni constant**. It is approximately equal to 0.57721; it is conjectured to be transcendental, but at present no one knows whether it is even irrational.)

25. Suppose $a_n > 0$ for all $n > 0$, and let $L = \limsup a_n^{1/n}$ (see Exercises 9–12 in §1.5). Show that $\sum_1^\infty a_n$ converges if $L < 1$ and diverges if $L > 1$.

6.3 Absolute and Conditional Convergence

We now consider the question of convergence of series whose terms may be either positive or negative. To a certain extent, this question may be reduced to the study of series with nonnegative terms, via the notion of absolute convergence.

A series $\sum_0^\infty a_n$ is called **absolutely convergent** if the series $\sum_0^\infty |a_n|$ converges. For series with nonnegative terms, absolute convergence is the same thing as convergence. For more general series, the basic result is as follows.

6.17 Theorem. *Every absolutely convergent series is convergent.*

Proof. Suppose $\sum_0^\infty |a_n|$ converges. Let $s_k = \sum_0^k a_n$ and $S_k = \sum_0^k |a_n|$. The sequence $\{S_k\}$ is convergent and hence Cauchy, so given $\epsilon > 0$, there exists an integer K such that

$$|a_{j+1}| + \cdots + |a_k| = S_k - S_j < \epsilon \text{ whenever } k > j \geq K.$$

But then

$$|s_k - s_j| = |a_{j+1} + \cdots + a_k| \leq |a_{j+1}| + \cdots + |a_j| < \epsilon \text{ whenever } k > j \geq K,$$

so the sequence $\{s_k\}$ is also Cauchy. By Theorem 1.20, the sequence $\{s_k\}$, and hence the series $\sum a_n$, is convergent. $\qquad\square$

Important Remark. We can consider series whose terms are complex numbers or n-dimensional vectors instead of real numbers. The definition of absolute convergence is the same, with $|a_n|$ denoting the norm of the vector a_n. *Theorem 6.17 remains valid in this more general setting, with exactly the same proof.*

The converse of Theorem 6.17 is false; a series that is not absolutely convergent may still converge because of cancellation between the positive and negative terms. A series that converges but does not converge absolutely is said to be **conditionally convergent**.

EXAMPLE 1. Let $a_n = 1/(n+1)$ if n is even, $a_n = -1/n$ if n is odd; thus,

$$\sum_0^\infty a_n = 1 - 1 + \tfrac{1}{3} - \tfrac{1}{3} + \tfrac{1}{5} - \tfrac{1}{5} + \cdots.$$

Clearly $s_k = 0$ if k is odd and $s_k = 1/(k+1)$ if k is even, so the series converges to the sum 0. However,

$$\sum_0^\infty |a_n| = 1 + 1 + \tfrac{1}{3} + \tfrac{1}{3} + \cdots = 2 \sum_0^\infty \frac{1}{2n+1},$$

which diverges by comparison to $\sum n^{-1}$.

EXAMPLE 2. Here is a more interesting example. The series

$$\sum_1^\infty \frac{(-1)^{n-1}}{n} = 1 - \tfrac{1}{2} + \tfrac{1}{3} - \tfrac{1}{4} + \cdots$$

is not absolutely convergent since $\sum_1^\infty n^{-1}$ diverges. However, it is the Taylor series for $f(x) = \log(1 + x)$ at $x = 1$. Indeed, for $n > 0$ we have $f^{(n)}(x) = (-1)^{n-1}(n-1)!(1+x)^{-n}$, so Taylor's formula gives

$$\log(1+x) = \sum_1^k \frac{(-1)^{n-1}(n-1)!}{n!} x^n + R_k(x)$$

$$= x - \frac{x^2}{2} + \frac{x^3}{3} + \cdots + \frac{(-1)^{k-1}x^k}{k} + R_k(x),$$

and by Corollary 2.61,

$$|R_k(1)| \le \frac{1}{(k+1)!} \sup_{0 \le t \le 1} \left| \frac{(-1)^k k!}{(1+t)^k} \right| = \frac{1}{k},$$

which tends to zero as $k \to \infty$. It follows that $\sum_1^\infty (-1)^{n-1}/n$ converges to $\log 2$.

It is to be emphasized that conditionally convergent series converge *only* because of cancellation between positive and negative terms. More precisely, let

$$a_n^+ = \max(a_n, 0) \qquad a_n^- = \max(-a_n, 0).$$

That is, $a_n^+ = a_n$ if a_n is positive and $a_n^+ = 0$ otherwise, and $a_n^- = |a_n|$ if a_n is negative and $a_n^- = 0$ otherwise; the nonzero a_n^+'s are the positive terms of the series $\sum a_n$, and the nonzero a_n^-'s are the absolute values of the negative terms. Observe that

$$a_n^+ - a_n^- = a_n, \qquad a_n^+ + a_n^- = |a_n|.$$

6.18 Theorem. *If $\sum a_n$ is absolutely convergent, the series $\sum a_n^+$ and $\sum a_n^-$ are both convergent. If $\sum a_n$ is conditionally convergent, the series $\sum a_n^+$ and $\sum a_n^-$ are both divergent.*

Proof. This theorem follows from the following three facts:

  i. The convergence of $\sum |a_n|$ implies the convergence of $\sum a_n^+$ and $\sum a_n^-$.
 ii. The divergence of $\sum |a_n|$ implies the divergence of at least one of $\sum a_n^+$ and $\sum a_n^-$.
iii. If $\sum a_n$ converges, it cannot happen that one of $\sum a_n^+$ and $\sum a_n^-$ converges while the other one diverges.

The first of these is clear since $0 \le a_n^+ \le |a_n|$ and $0 \le a_n^- \le |a_n|$, and the second is clear since $|a_n| = a_n^+ + a_n^-$. As for the third, let s_k and $s_k^{\pm}$ denote the kth partial sums of the series $\sum a_n$ and $\sum a_n^{\pm}$; thus $s_k = s_k^+ - s_k^-$. Suppose, to be definite, that $\sum a_n^+ = \infty$ while $\sum a_n^- = S < \infty$; then for any $C > 0$ (no matter how large), for sufficiently large k we will have $s_k^+ > C + S$, while $s_k^- \le S$, so that $s_k > C + S - S = C$. It follows that $s_k \to +\infty$, so $\sum a_n$ diverges. $\qquad\square$

Absolutely convergent series are much more pleasant to deal with than conditionally convergent ones. For one thing, they converge more rapidly; the partial sums s_k of conditionally convergent series tend to provide poor approximations to the full sum unless one takes k very large because the divergence of $\sum |a_n|$ implies that a_n cannot tend to zero very rapidly as $n \to \infty$. For another thing, the sum of an absolutely convergent series cannot be affected by rearranging the terms, but this is not the case for conditionally convergent series!

Let us explain this mysterious statement in more detail. The terms of a series $\sum_0^{\infty} a_n$ are presented in a definite order: $a_0, a_1, a_2, \ldots$. We might think of forming a new series by writing down these terms in a different order, such as

$$a_0, a_2, a_1, a_4, a_6, a_3, a_8, a_{10}, a_5, \ldots,$$

where we take the first two even-numbered terms, the first odd-numbered term, the next two even-numbered terms, the next odd-numbered term, and so forth. In general, if σ is any one-to-one mapping from the set of nonnegative integers onto itself, we can form the series $\sum_0^\infty a_{\sigma(n)}$, which we call a **rearrangement** of $\sum_0^\infty a_n$. (The reasons why we would want to do this are perhaps not so clear right now, but we will encounter situations in §6.5 where this issue must be addressed.) The sharp contrast between absolutely and conditionally convergent series with respect to rearrangements is explained in the following two theorems.

6.19 Theorem. *If $\sum_0^\infty a_n$ is absolutely convergent with sum S, then every rearrangement $\sum_0^\infty a_{\sigma(n)}$ is also absolutely convergent with sum S.*

Proof. First suppose $a_n \geq 0$ for all n. Every term of the rearranged series $\sum a_{\sigma(n)}$ is among the terms of the original series $\sum a_n$, and hence the partial sums of the rearranged series cannot exceed S. It follows that the full sum S' of the rearranged series satisfies $S' \leq S$. The same reasoning shows that $S \leq S'$, so $S' = S$.

Now we do the general case. If $\sum |a_n| < \infty$, we have $\sum |a_{\sigma(n)}| < \infty$ by what we have just proved. Hence, given $\epsilon > 0$, for k sufficiently large we have

$$\sum_{k+1}^\infty |a_n| < \epsilon \text{ and } \sum_{k+1}^\infty |a_{\sigma(n)}| < \epsilon.$$

Given such a k, let K be the largest of the numbers $\sigma(0), \ldots, \sigma(k)$, so that

$$\{\sigma(0), \sigma(1), \ldots, \sigma(k)\} \subset \{0, 1, \ldots, K\}.$$

The elements of $\{0, 1, \ldots, K\} \setminus \{\sigma(0), \sigma(1), \ldots, \sigma(k)\}$ are among the $\sigma(n)$'s with $n \geq k+1$, so

$$\left| \sum_0^K a_n - \sum_0^k a_{\sigma(n)} \right| \leq \sum_{k+1}^\infty |a_{\sigma(n)}| < \epsilon.$$

But then

$$\left| \sum_0^k a_{\sigma(n)} - S \right| \leq \left| \sum_0^k a_{\sigma(n)} - \sum_0^K a_n \right| + \left| \sum_0^K a_n - S \right| \leq \epsilon + \sum_{K+1}^\infty |a_n| < 2\epsilon.$$

As ϵ is arbitrary, we conclude that $\sum_0^\infty a_{\sigma(n)} = S$. $\square$

6.20 Theorem. *Suppose $\sum_0^\infty a_n$ is conditionally convergent. Given any real number S, there is a rearrangement $\sum_0^\infty a_{\sigma(n)}$ that converges to S.*

Proof. By Theorem 6.18, the series $\sum a_n^+$ and $\sum a_n^-$ of positive and negative terms from $\sum a_n$ both diverge; but since $\sum a_n$ converges, we have $a_n \to 0$ as $n \to \infty$. These pieces of information are all we need.

Suppose $S \geq 0$. (A similar argument works for $S < 0$.) We construct the desired rearrangement as follows:

1. Add up the positive terms from the series $\sum a_n$ (in their original order) until the sum exceeds S. This is possible since $\sum a_n^+ = \infty$. Stop as soon as the sum exceeds S.

2. Now start adding in the negative terms (in their original order) until the sum becomes less than S. Again, this is possible since $\sum a_n^- = \infty$. Stop as soon as the sum is less than S.

3. Repeat steps 1 and 2 ad infinitum. That is, add in positive terms until the sum is greater than S, then add in negative terms until the sum is less than S, and so forth. This process never terminates since the series $\sum a_n^+$ and $\sum a_n^-$ both diverge, and sooner or later every term from the original series will be added into the new series. The result is a rearrangement $\sum_0^\infty a_{\sigma(n)}$ of the original series.

We claim that this rearrangement converges to S. Indeed, given $\epsilon > 0$, there exists an integer N so that $|a_n| < \epsilon$ if $n > N$. If we choose K large enough so that all the terms $a_0, a_1, \ldots, a_N$ are included among the terms $a_{\sigma(0)}, a_{\sigma(1)}, \ldots a_{\sigma(K)}$, then $|a_{\sigma(n)}| < \epsilon$ if $n > K$. It follows that the partial sums $\sum_0^k a_{\sigma(n)}$ differ from S by less than ϵ if $k > K$, because the procedure specifies switching from positive to negative terms or vice versa *as soon as* the sum is greater than or less than S; if the sum became greater than $S + \epsilon$ or less than $S - \epsilon$, we would have added in too many terms of the same sign. Hence the sums $\sum_0^k a_{\sigma(n)}$ converge to S. $\qquad\square$

EXERCISES

1. Show that the following series are absolutely convergent.
   a. $\sum_0^\infty x^n \cos n\theta \quad (|x| < 1, \theta \in \mathbb{R})$.
   b. $\sum_1^\infty n^{-2} \sin n\theta \quad (\theta \in \mathbb{R})$.
   c. $\sum_1^\infty (-1)^n n^2 3^{1-n} x^n \quad (|x| < 3)$.

2. Suppose $\sum a_n$ is conditionally convergent. Show that there are rearrangements of $\sum a_n$ whose partial sums diverge to $+\infty$ or $-\infty$.

3. Consider the rearrangement of the series $\sum_1^\infty (-1)^{n-1}/n$ obtained by taking two positive terms, one negative term, two positive terms, one negative term, and so forth:

$$1 + \tfrac{1}{3} - \tfrac{1}{2} + \tfrac{1}{5} + \tfrac{1}{7} - \tfrac{1}{4} + \tfrac{1}{9} + \tfrac{1}{11} - \tfrac{1}{6} + \cdots.$$

Show that the sum of this series is $\frac{3}{2}\log 2$. (*Hint:* Deduce from Example 2 that $0 + \frac{1}{2} + 0 - \frac{1}{4} + 0 + \frac{1}{6} + 0 - \cdots = \frac{1}{2}\log 2$ and add this to the result of Example 2.)

4. Let $\sum_0^\infty a_n$ be a convergent series, and let $\sum_0^\infty b_n$ be its rearrangement obtained by interchanging each even-numbered term with the odd-numbered term immediately following it: $a_1 + a_0 + a_3 + a_2 + a_5 + a_4 + \cdots$. Show that $\sum_0^\infty b_n = \sum_0^\infty a_n$.

5. Suppose $a_n > -1$ for all n. By suitable applications of Taylor's theorem to the functions $\log(1 + x)$ or e^x, show the following:

   a. $\sum a_n$ is absolutely convergent if and only if $\sum \log(1 + a_n)$ is absolutely convergent. (This is of interest in connection with Exercise 4 of §6.1: If $\sum |a_n| < \infty$, then $\prod(1 + a_n)$ converges.)

   b. Let $a_n = (-1)^{n+1}/\sqrt{n}$. Then $\sum_1^\infty a_n$ is conditionally convergent (see Theorem 6.22 below), but $\sum_1^\infty \log(1 + a_n)$ diverges.

6.4 More Convergence Tests

The tests we developed in §6.2 for the convergence of series of nonnegative terms immediately yield tests for the *absolute* convergence of more general series. We sum up the most important results:

6.21 Theorem.

a. If $|a_n| \le Cn^{-1-\epsilon}$ for some $C, \epsilon > 0$, then $\sum a_n$ converges absolutely. If $|a_n| \ge Cn^{-1}$ for some $C > 0$, then $\sum a_n$ either converges conditionally or diverges.

b. (*The Ratio Test*) If $|a_{n+1}/a_n| \to l$ as $n \to \infty$, then $\sum a_n$ converges absolutely if $l < 1$ and diverges if $l > 1$.

c. (*The Root Test*) If $|a_n|^{1/n} \to l$ as $n \to \infty$, then $\sum a_n$ converges absolutely if $l < 1$ and diverges if $l > 1$.

In the ratio and root tests, the divergence (rather than conditional convergence) when $l > 1$ is guaranteed because $a_n \not\to 0$ in this case; see the proofs of Theorems 6.13 and 6.14. The statements of the ratio and root tests can be sharpened a bit as in Theorems 6.13a and 6.14a.

Warning. It is a common mistake to obtain incorrect results by forgetting the absolute values in Theorem 6.21. For example, the series $\sum_0^\infty (-2)^n$ satisfies $a_{n+1}/a_n = -2$, and $-2 < 1$, but the series diverges!

It remains to investigate criteria that will yield information about conditional convergence as well as absolute convergence. By far the most commonly used result of this kind pertains to **alternating series**, that is, series whose terms alternate in sign. Such a series can be written in the form $\sum(-1)^n a_n$ or $\sum(-1)^{n-1} a_n$ (depending on whether the even or odd numbered terms are positive), where $a_n > 0$; we shall consider the first form for the sake of definiteness.

6.22 Theorem (The Alternating Series Test). *Suppose the sequence $\{a_n\}$ is decreasing and $\lim_{n \to \infty} a_n = 0$. Then the series $\sum_0^\infty (-1)^n a_n$ is convergent. Moreover, if s_k and S denote the kth partial sum and the full sum of this series, we have*

$$s_k > S \text{ for } k \text{ even}, \quad s_k < S \text{ for } k \text{ odd}, \quad \text{and} \quad |s_k - S| < a_{k+1} \text{ for all } k.$$

Proof. Since $a_k \geq a_{k+1}$ for all k, we have

$$s_{2m+1} = s_{2m-1} + a_{2m} - a_{2m+1} \geq s_{2m-1},$$
$$s_{2m+2} = s_{2m} - a_{2m+1} + a_{2m+2} \leq s_{2m}.$$

Thus the sequence $\{s_{2m-1}\}$ of odd-numbered partial sums is increasing and the sequence $\{s_{2m}\}$ of even-numbered partial sums is decreasing. This monotonicity further yields

$$s_{2m-1} = s_{2m-2} - a_{2m-1} \leq s_{2m-2} \leq s_0,$$
$$s_{2m} = s_{2m-1} + a_{2m} \geq s_{2m-1} \geq s_1,$$

so $\{s_{2m-1}\}$ and $\{s_{2m}\}$ are bounded above and below, respectively. By the monotone sequence theorem, these sequences both converge, and since $s_{2m} - s_{2m-1} = a_{2m} \to 0$, their limits are equal. Thus the whole sequence $\{s_k\}$ converges, that is, the series $\sum(-1)^n a_n$ converges. The even-numbered partial sums decrease to the full sum S while the odd-numbered ones increase, so $S < s_{2m}$ and $S > s_{2m-1}$ for all m. In particular,

$$0 < S - s_{2m-1} < s_{2m} - s_{2m-1} = a_{2m},$$
$$0 < s_{2m} - S < s_{2m} - s_{2m+1} = a_{2m+1},$$

so $|s_k - S| < a_{k+1}$ whether k is even or odd. $\qquad\square$

EXAMPLE 1. The series $\sum_1^\infty (-1)^n (e^{1/n} - 1)$ converges by the alternating series test, because $e^{1/n}$ decreases to 1 as $n \to \infty$. The convergence is only conditional, however, since $e^{1/n} - 1 \approx 1/n$ when n is large. (More precisely, by Taylor's theorem we have $e^x = 1 + x + R(x)$ where $|R(x)| \leq Cx^2$ for $0 \leq x \leq 1$. Thus $\sum(e^{1/n} - 1) = \sum n^{-1} + \sum R(1/n)$; the first series on the right diverges, while the second converges by comparison to $\sum n^{-2}$.)

The alternating series test is a useful test for conditional convergence, but the fact that the difference between a partial sum and the full sum is less in absolute value than the first neglected term is also of interest in the absolutely convergent case. (This estimate for the error in replacing the full sum by a partial sum is, in most cases, accurate to within an order of magnitude.)

The alternating series test can be applied to a series $\sum(-1)^n a_n$ for which $\lim a_n = 0$ provided that the a_n's decrease *from some point onward*. (Of course, the inequalities for the partial sums are only valid from that point onward too.) However, the monotonicity condition cannot be dropped entirely, as the following example shows:

$$1 - \tfrac{1}{2} + \tfrac{1}{2} - \tfrac{1}{4} + \tfrac{1}{3} - \tfrac{1}{6} + \cdots + \tfrac{1}{m} - \tfrac{1}{2m} + \cdots.$$

Here $a_n \to 0$ as $n \to \infty$, but not monotonically, and the series diverges. (The sum of the first $2m$ terms is $\tfrac{1}{2}(1 + \tfrac{1}{2} + \tfrac{1}{3} + \cdots + \tfrac{1}{m})$, a partial sum of the divergent series $\tfrac{1}{2} \sum n^{-1}$.)

The tests we have developed can be used to analyze a wide variety of *power series*, that is, series of the form $\sum_0^\infty c_n(x - a)^n$ where x is a real variable. In typical cases, the ratio test or the root test will establish that there is some number r such that the series converges absolutely for $|x-a| < r$ and diverges for $|x-a| > r$. The convergence at the two remaining points $x = a \pm r$ can then be studied by one of the other tests.

EXAMPLE 2. Consider the series $\displaystyle\sum_0^\infty \frac{(-1)^n(x-3)^n}{(n+1)2^{2n+1}}$. We start with the ratio test:

$$\left| \frac{a_{n+1}}{a_n} \right| = \left| \frac{(-1)^{n+1}(x-3)^{n+1}/(n+2)2^{2n+3}}{(-1)^n(x-3)^n/(n+1)2^{2n+1}} \right| = \frac{n+1}{n+2} \frac{|x-3|}{4} \to \frac{|x-3|}{4}.$$

Thus the series converges absolutely for $|x-3| < 4$ and diverges for $|x-3| > 4$. (The root test would also yield this result.) The two remaining points are where $x - 3 = \pm 4$, that is, $x = -1$ and $x = 7$. At these two points the series becomes

$$\sum_0^\infty \frac{(-1)^n(-4)^n}{(n+1)2^{2n+1}} = \frac{1}{2} \sum_0^\infty \frac{1}{n+1} \quad \text{and} \quad \sum_0^\infty \frac{(-1)^n 4^n}{(n+1)2^{2n+1}} = \frac{1}{2} \sum_0^\infty \frac{(-1)^n}{n+1}.$$

The first of these diverges, while the second one converges by the alternating series test. The convergence is only conditional, by the divergence of the first series. Thus the original series converges absolutely for $-1 < x < 7$, converges conditionally at $x = 7$, and diverges elsewhere.

We conclude with another test for convergence (absolute or conditional) that generalizes the alternating series test and is sometimes useful for trigonometric series. Its proof is based on the following discrete analogue of the integration-by-parts formula, in which a sum $\sum_1^k a_n b_n$ is rewritten by "differentiating" the sequence $\{a_n\}$ and "integrating" the sequence $\{b_n\}$.

6.23 Lemma (Summation by Parts). *Given two numerical sequences $\{a_n\}$ and $\{b_n\}$, let*

$$a'_n = a_n - a_{n-1}, \qquad B_n = b_0 + \cdots + b_n.$$

Then

$$\sum_0^k a_n b_n = a_k B_k - \sum_1^k a'_n B_{n-1}.$$

Proof. We have $b_0 = B_0$, and $b_n = -B_{n-1} + B_n$ for $n \geq 1$, so

$$(6.24) \quad a_0 b_0 + a_1 b_1 + a_2 b_2 + \cdots + a_k b_k$$
$$= a_0 B_0 - a_1 B_0 + a_1 B_1 - a_2 B_1 + a_2 B_2 - \cdots - a_k B_{k-1} + a_k B_k$$
$$= -a'_1 B_0 - a'_2 B_1 - \cdots - a'_k B_{k-1} + a_k B_k.$$

$\square$

6.25 Theorem (Dirichlet's Test). *Let $\{a_n\}$ and $\{b_n\}$ be numerical sequences. Suppose that the sequence $\{a_n\}$ is decreasing and tends to 0 as $n \to \infty$, and that the sums $B_n = b_0 + \cdots + b_n$ are bounded in absolute value by a constant C independent of n. Then the series $\sum_0^\infty a_n b_n$ converges.*

Proof. With notation as in Lemma 6.23, $\sum_0^k a_n b_n = a_k B_k - \sum_1^k a'_n B_{n-1}$, so it is enough to show that $\lim_{k \to \infty} a_k B_k$ exists and that the series $\sum_1^\infty a'_n B_{n-1}$ converges. The first assertion is easy: Since $|B_k| \leq C$ and $a_k \to 0$, we have $|a_k B_k| \leq C a_k \to 0$. On the other hand, since $\{a_n\}$ is decreasing, we have $a'_n \leq 0$ for all n, so

$$\sum_1^k |a'_n B_{n-1}| \leq C \sum_1^k |a'_n|$$
$$= C\big[(a_0 - a_1) + (a_1 - a_2) + \cdots + (a_{k-1} - a_k)\big] = C(a_0 - a_k) \leq C a_0$$

for all k. It follows that the series $\sum_1^\infty a'_n B_{n-1}$ is absolutely convergent and hence convergent.

$\square$

Dirichlet's test includes the alternating series test as a special case, by taking $b_n = (-1)^n$, for which $B_n = 1$ or 0 according as n is even or odd. The other situations in which it is most commonly applied are those with $b_n = \sin n\theta$ or $b_n = \cos n\theta$, where θ is not an integer multiple of 2π. That the hypotheses on $\{b_n\}$ in Dirichlet's test are satisfied in these cases is shown by the following calculation.

6.26 Lemma. *If θ is not an integer multiple of 2π, then*

$$\sum_1^k \cos n\theta = \frac{\cos \frac{1}{2}(k+1)\theta \cdot \sin \frac{1}{2}k\theta}{\sin \frac{1}{2}\theta},$$

$$\sum_1^k \sin n\theta = \frac{\sin \frac{1}{2}(k+1)\theta \cdot \sin \frac{1}{2}k\theta}{\sin \frac{1}{2}\theta}.$$

Proof. These formulas can be established by using various trigonometric identities. The easiest method is to use Euler's formula $\cos x + i \sin x = e^{ix}$ (which we shall discuss in detail in §7.5). By the formula (6.2) for the sum of a finite geometric series,

$$\sum_1^k e^{in\theta} = e^{i\theta} \frac{e^{ik\theta} - 1}{e^{i\theta} - 1} = e^{i\theta} \frac{e^{ik\theta/2}[e^{ik\theta/2} - e^{-ik\theta/2}]}{e^{i\theta/2}[e^{i\theta/2} - e^{-i\theta/2}]}$$

$$= e^{i(k+1)\theta/2} \frac{e^{ik\theta/2} - e^{-ik\theta/2}}{e^{i\theta/2} - e^{-i\theta/2}}$$

$$= \left[\cos \frac{1}{2}(k+1)\theta + i \sin \frac{1}{2}(k+1)\theta\right] \frac{\sin \frac{1}{2}k\theta}{\sin \frac{1}{2}\theta}.$$

The asserted formulas follow by taking the real and imaginary parts of both sides.
□

6.27 Corollary. *Suppose that the sequence $\{a_n\}$ decreases to 0 as $n \to \infty$. Then the series $\sum_1^\infty a_n \cos n\theta$ converges for all θ except perhaps for integer multiples of 2π, and the series $\sum_1^\infty a_n \sin n\theta$ converges for all θ.*

Proof. The hypotheses of Dirichlet's test are satisfied for $\theta \neq 2\pi j$, for if b_n is either $\cos n\theta$ or $\sin n\theta$, the lemma implies that $|B_n| \leq |\csc \frac{1}{2}\theta|$ for all n. (If $\theta = 2\pi j$, the series $\sum a_n \sin n\theta$ converges trivially since $\sin n\theta = 0$ for all n.)
□

EXERCISES

In Exercises 1–9, determine the values of x at which the series converges absolutely or conditionally.

1. $\displaystyle\sum_0^\infty \frac{(x+2)^n}{n^2+1}$.

2. $\displaystyle\sum_1^\infty n^3(2x-1)^n$.

3. $\displaystyle\sum_0^\infty \frac{x^{2n}}{1\cdot 3\cdots(2n+1)}$.

4. $\displaystyle\sum_1^\infty \frac{nx^{n+2}}{5^n(n+1)^2}$.

5. $\displaystyle\sum_0^\infty \frac{(-1)^n(x-4)^n}{(2^n-3)\log(n+3)}$.

6. $\displaystyle\sum_1^\infty \frac{1}{\sqrt{n}}\left(\frac{x-1}{x+1}\right)^n$.

7. $\displaystyle\sum_1^\infty \frac{2\cdot 4\cdots(2n)}{1\cdot 3\cdots(2n-1)}(\tfrac{1}{2}x-3)^n$.

8. $\displaystyle\sum_0^\infty \frac{(-1)^n(x+1)^{2n}}{3n+2}$.

9. $\displaystyle\sum_0^\infty \frac{1\cdot 3\cdots(2n+1)}{2\cdot 5\cdots(3n+2)}x^n$.

In Exercises 10–14, determine whether the series converges absolutely, converges conditionally, or diverges.

10. $\displaystyle\sum_2^\infty (-1)^n \log\left(\frac{n+1}{n}\right)$.

11. $\displaystyle\sum_1^\infty (-1)^n \int_n^{n+1} \frac{\log(x+7)}{x}\,dx$.

12. $\displaystyle\sum_1^\infty \frac{(-1)^n}{n^{1/n}}$.

13. $\displaystyle\sum_1^\infty (-1)^{n-1} \log(n \sin n^{-1})$.

14. $\displaystyle\sum_1^\infty (-1)^{n-1} \left[e - \left(\frac{n+1}{n} \right)^n \right]$.

15. Use the alternating series test to show that $x^{-1} \sin x = 1 - \frac{1}{3!}x^2 + \frac{1}{5!}x^4 - \frac{1}{7!}x^6 + E(x)$ where $0 < E(x) < 0.027$ for $|x| \leq \pi$.

16. (*Abel's Test*) Suppose $\sum a_n$ is a convergent series and $\{b_n\}$ is a decreasing sequence of positive numbers. ($\lim b_n$ need not be zero.) Show that $\sum a_n b_n$ converges. (This can be done by using Dirichlet's test or by modifying the proof of Dirichlet's test.)

17. Show that if $\sum_1^\infty a_n$ converges, then so does $\sum_1^\infty n^{-p} a_n$ for any $p > 0$. For which p can you guarantee absolute convergence without knowing anything more about the a_n's?

18. For which x and θ does $\sum_1^\infty n^{-1} x^n \cos n\theta$ converge?

6.5   Double Series; Products of Series

A double infinite series, informally speaking, is an expression of the form

$$(6.28) \qquad\qquad \sum_{m,n=0}^\infty a_{mn},$$

that is, a series whose terms are indexed by ordered pairs of nonnegative integers. The difficulty in making precise sense out of such an expression is that it is not clear what one should mean by a "partial sum." Two obvious candidates are the "square" partial sums and the "triangular" partial sums

$$s_k^\square = \sum_{m,n=0}^k a_{mn}, \qquad s_k^\triangle = \sum_{m+n \leq k} a_{mn},$$

which are defined by adding up all the terms a_{mn} for which (m, n) lies in the outlined regions in Figure 6.2. (Note that passing from $s_k^\square$ or $s_k^\triangle$ to $s_{k+1}^\square$ or $s_{k+1}^\triangle$ involves adding not just a single term but a finite set of terms to the sum. It is not necessary to specify the order in which these terms are added, as finite addition is commutative.) Clearly there are many other possibilities. Indeed, there are infinitely many ways to enumerate the set of ordered pairs of nonnegative integers, each of which leads to a different notion of "partial sums."

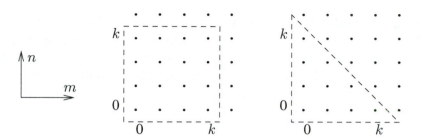

FIGURE 6.2: Schematic representation of square and triangular partial sums of a double series.

There is yet another possibility: One can consider the double series (6.28) as an iterated series, just as one can regard double integrals as iterated integrals. That is, one could interpret (6.28) as

$$\sum_{m=0}^{\infty}\left(\sum_{n=0}^{\infty} a_{mn}\right) \quad \text{or} \quad \sum_{n=0}^{\infty}\left(\sum_{m=0}^{\infty} a_{mn}\right),$$

in which one forms the ordinary series $\sigma_m = \sum_{n=0}^{\infty} a_{mn}$ for each m and then adds up the sums to obtain $\sum_{m=0}^{\infty} \sigma_m$, or similarly with m and n switched. This is different from the partial-sum procedures discussed above because the intermediate steps involve infinite sums rather than finite ones.

How is one to make sense out of all these ways of interpreting (6.28)? The answer, in a nutshell, is that the situation is similar to that for improper double integrals discussed in §4.7: *For series of positive terms, or for absolutely conver-gent series, there is no problem, as all interpretations lead to the same answer. Otherwise, one must proceed with great caution.*

Let us explain this in more detail. Given any one-to-one correspondence $j \leftrightarrow (m, n)$ between the set of nonnegative integers and the set of ordered pairs of non-negative integers, we can set $b_j = a_{mn}$ and form the ordinary infinite series $\sum_0^{\infty} b_j$; we call such a series an **ordering** of the double series $\sum_{m,n=0}^{\infty} a_{mn}$. The essential point is that *the orderings of $\sum a_{mn}$ are all rearrangements of one another,* and we can apply Theorem 6.19.

First, if $a_{mn} \geq 0$, then either all orderings of $\sum a_{mn}$ diverge or all orderings converge, and in the latter case their sums are all equal. Thus, the sum of the series $\sum a_{mn}$ is well defined as a positive number or $+\infty$, independent of the choice of ordering.

Second, without the assumption of positivity, if $\sum |b_j|$ is convergent for one ordering of $\sum a_{mn}$, then the same is true for every ordering. In this case the series

$\sum a_{mn}$ is called **absolutely convergent,** and by Theorem 6.19 again, all order-ings of $\sum a_{mn}$ have the same sum, which we call the sum of the double series $\sum a_{mn}$. Moreover, an argument similar to the proof of Theorem 6.19 shows that the double series $\sum a_{mn}$ is absolutely convergent if and only if the iterated series $\sum_m (\sum_n |a_{mn}|)$ is convergent, in which case $\sum_{m,n} a_{mn} = \sum_m (\sum_n a_{mn})$. (See Exercises 5 and 6.)

Given a double series $\sum a_{mn}$, we can therefore proceed as follows. First we evaluate the series $\sum |a_{mn}|$ by ordering it in some fashion or treating it as an iter-ated series; if it turns out to be finite, we can then evaluate $\sum a_{mn}$ by ordering it in any fashion or treating it as an iterated series.

What if $\sum a_{mn}$ is not absolutely convergent? Let us separate out the positive and negative terms as we did in Theorem 6.18. The argument in the proof of Theo-rem 6.18 shows that if $\sum a_{mn}^+ = \infty$ but $\sum a_{mn}^- < \infty$, then all orderings of $\sum a_{mn}$ diverge to $+\infty$; likewise, if $\sum a_{mn}^+ < \infty$ but $\sum a_{mn}^- = \infty$, then all orderings of $\sum a_{mn}$ diverge to $-\infty$. On the other hand, if $\sum a_{mn}^+ = \sum a_{mn}^- = \infty$ but $a_{mn} \to 0$ as $m, n \to \infty$, the proof of Theorem 6.20 shows that various orderings of $\sum a_{mn}$ can converge to any real number. In this case, therefore, we simply cannot make numerical sense out of the expression $\sum a_{mn}$ without specifying more precisely how the summation is to be performed.

An important situation in which double series occur is in multiplying two series together. The basic result is as follows.

6.29 Theorem. *Suppose that $\sum_0^\infty a_m$ and $\sum_0^\infty b_n$ are both absolutely convergent, with sums A and B. Then the double series $\sum_{m,n=0}^\infty a_m b_n$ is absolutely convergent, and its sum is AB.*

Proof. We consider the square partial sums of $\sum a_m b_n$, which are just the products of the partial sums of $\sum a_m$ and $\sum b_n$:

$$(6.30) \qquad \sum_{m,n=0}^k a_m b_n = \left(\sum_0^k a_m \right) \left(\sum_0^k b_n \right).$$

If we replace a_m and b_n by $|a_m|$ and $|b_n|$ in (6.30), the right side is bounded by the finite quantity $(\sum_0^\infty |a_m|)(\sum_0^\infty |b_n|)$, which shows that the double series $\sum a_m b_n$ is absolutely convergent. Then, letting $k \to \infty$ in (6.30), we obtain $\sum a_m b_n = AB$. $\qquad \square$

Under the conditions of Theorem 6.29, we are free to use any ordering of $\sum a_m b_n$ that we choose, and in particular, we can use the triangular partial sums rather than the square ones. This is the natural thing to do when considering power series. Indeed, if $\sum a_n x^n$ and $\sum b_n x^n$ are absolutely convergent for a particular value of x, their product is $\sum a_m b_n x^{m+n}$, which can also be expressed as a power series if we group together all the terms involving a given power of x. The terms involving x^j are those with $m + n = j$, i.e., those with $m = 0, 1, \ldots, j$ and $n = j - m$. Collecting these terms together yields

$$\left(\sum_0^\infty a_n x^n \right) \left(\sum_0^\infty b_n x^n \right) = \sum_{j=0}^\infty \left[\sum_{m+n=j} a_n b_m \right] x^j.$$

The expression on the right is a power series whose jth coefficient is a finite sum of products of the original coefficients; its partial sums are precisely the triangular partial sums of the double series $\sum a_m b_n x^{m+n}$.

The same procedure can also be used for series without an x (by taking $x = 1$, if you like). That is, given two convergent series $\sum_0^\infty a_m$ and $\sum_0^\infty b_n$, we can form the series

$$\sum_{j=0}^\infty \left(\sum_{m+n=j} a_n b_m \right) = \sum_{j=0}^\infty (a_0 b_j + a_1 b_{j-1} + \cdots + a_{j-1} b_1 + a_j b_0),$$

whose partial sums are the triangular partial sums of the double series $\sum a_m b_n$; it is called the **Cauchy product** of $\sum a_m$ and $\sum b_n$. As we have seen, if $\sum a_m$ and $\sum b_n$ are absolutely convergent, their Cauchy product is too, and its sum is $(\sum a_m)(\sum b_n)$. In fact, the Cauchy product converges to $(\sum a_m)(\sum b_n)$ provided that at least one of $\sum a_m$ and $\sum b_n$ is absolutely convergent (see Krantz [12, pp. 109–10], or Rudin [18, p. 74]). However, if $\sum a_m$ and $\sum b_n$ are both conditionally convergent, their Cauchy product may diverge. (See Exercise 4.)

EXERCISES

1. By multiplying the geometric series by itself, show that for $|x| < 1$,
   a. $(1 - x)^{-2} = \sum_0^\infty (n + 1) x^n$;
   b. $(1 - x)^{-3} = \frac{1}{2} \sum_0^\infty (n + 1)(n + 2) x^n$.

2. Let $f(x) = \sum_0^\infty x^n / n!$. Show directly from this formula that $f(x) f(y) = f(x + y)$.

3. Verify that the Taylor series of $(1 - 4x)^{-1/2}$ about $x = 0$ is $\sum_0^\infty (2n)! x^n / (n!)^2$ and that this series converges absolutely for $|x| < \frac{1}{4}$. Then, taking for granted that the sum of this series actually is $(1 - 4x)^{-1/2}$ (which we shall prove in

§7.3), multiply the series by itself and conclude that for any positive integer j,

$$\sum_{n=0}^{j} \frac{(2n)!(2j-2n)!}{(n!)^2((j-n)!)^2} = 4^j.$$

4. Show that the series $\sum_{0}^{\infty}(-1)^n(n+1)^{-1/2}$ is conditionally convergent and that the Cauchy product of this series with itself diverges. (*Hint:* The maximum of the function $f(x) = (x+1)(j-x+1)$ occurs at $x = \frac{1}{2}j$, and hence $(n+1)(j-n+1) \le (\frac{1}{2}j+1)^2$ for $n = 0, \ldots, j$.)

5. Show that $\sum_{m,n=0}^{\infty} a_{mn} = \sum_{m=0}^{\infty}(\sum_{n=0}^{\infty} a_{mn})$ whenever $a_{mn} \ge 0$ for all $m, n \ge 0$.

6. Suppose $\sum_{m,n=0}^{\infty} a_{mn}$ is absolutely convergent. Show that the iterated series $\sum_{m=0}^{\infty}(\sum_{n=0}^{\infty} a_{mn})$ converges to the sum $\sum_{m,n=0}^{\infty} a_{mn}$. (Use Exercise 5.)

7. Show that $\sum_{m,n=1}^{\infty}(m+n)^{-p}$ converges if and only if $p > 2$. (*Hint:* Use triangular partial sums.)

8. Let $a_{mn} = 1$ if $m = n$, $a_{mn} = -1$ if $m - n = 1$, and $a_{mn} = 0$ otherwise. Show that the iterated series $\sum_{n=0}^{\infty}\sum_{m=0}^{\infty} a_{mn}$ and $\sum_{m=0}^{\infty}\sum_{n=0}^{\infty} a_{mn}$ both converge, but their sums are unequal.

Chapter 7

FUNCTIONS DEFINED BY SERIES AND INTEGRALS

In this chapter we study the convergence of sequences and series whose terms are functions of a variable x and improper integrals whose integrand contains x as a free variable. In all these situations, the study of the resulting function of x may reveal unpleasant surprises unless we have some control over the way the rate of convergence varies along with x; the most commonly encountered form of such control, *uniform convergence*, is a major theme of this chapter.

7.1 Sequences and Series of Functions

We recall that a *sequence* $\{f_k\}_0^\infty$ of functions is a map that assigns to each non-negative integer k a function f_k. It is implicitly assumed that the functions f_k are all defined on some common domain S (usually a subset of $\mathbb{R}$ or $\mathbb{R}^n$) and all take values in the same space ($\mathbb{R}$, $\mathbb{C}$, or $\mathbb{R}^m$).

What does it mean for a sequence of functions $\{f_k\}$ defined on a set $S \subset \mathbb{R}^n$ to converge to a function f on S? The most obvious interpretation is that

(7.1) $$f_k(\mathbf{x}) \to f(\mathbf{x}) \text{ for every } \mathbf{x} \in S.$$

This is, indeed, what is usually meant by the statement "$f_k \to f$ on S" when no further qualification is added; when we wish to be very clear about it, we shall say that $f_k \to f$ **pointwise** on S when (7.1) holds.

Unfortunately, pointwise convergence is a rather badly behaved operation in the sense that it does not interact well with other limiting operations, such as differentiation and integration. Consider the following group of examples:

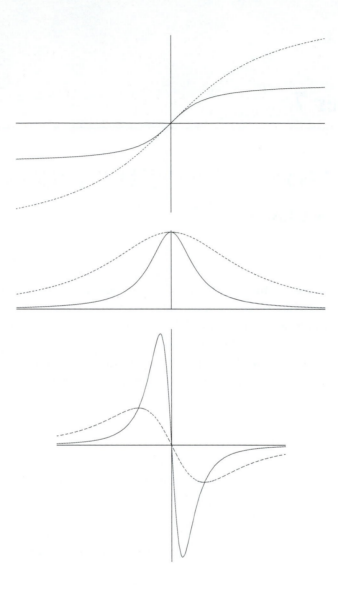

FIGURE 7.1: Some of the functions defined in (7.2). Top: f_1 (dashed) and f_3 (solid). Middle: g_1 (dashed) and g_3 (solid). Bottom: h_1 (dashed) and h_3 (solid).

EXAMPLE 1. Let

(7.2)
$$f_k(x) = \frac{1}{k} \arctan kx, \qquad g_k(x) = f_k'(x) = \frac{1}{k^2x^2 + 1},$$

$$h_k(x) = g_k'(x) = \frac{-2k^2x}{(k^2x^2 + 1)^2}.$$

Observe that $f_k(x) = k^{-1}f_1(kx)$, $g_k(x) = g_1(kx)$, and $h_k(x) = kh_1(kx)$. In graphical terms, as shown in Figure 7.1, this means that the graph of f_k is obtained from the graph of f_1 by shrinking the x and y scales by a factor of k; the graph of g_k is obtained from the graph of g_1 by shrinking the x scale by a factor of k and leaving the y scale unchanged; and the graph of h_k is obtained from the graph of h_1 by shrinking the x scale and expanding the y scale by a factor of k. We have:

i. $f_k(x) \to 0$ for all x, since $|f_k(x)| \le \pi/2k$.
ii. $g_k(x) \to 0$ for all $x \ne 0$, but $g_k(0) = 1$ for all k. That is,

$$\lim_{k\to\infty} g_k(x) = g(x) \equiv \begin{cases} 1 & \text{if } x = 0, \\ 0 & \text{otherwise.} \end{cases}$$

iii. $h_k(x) \to 0$ for all x. ($h_k(0) = 0$ for all k, and if $x \ne 0$, $h_k(x) \approx -2/k^2x^3$ for large k.)

Therefore, g is discontinuous even though the g_k's are all continuous; moreover, since g_k is the derivative of f_k and an antiderivative of h_k,

$$\lim_{k\to\infty} f_k'(0) = 1 \ne 0 = \left(\lim_{k\to\infty} f_k\right)'(0);$$

$$\lim_{k\to\infty} \lim_{x\to 0} g_k(x) = 1 \ne 0 = \lim_{x\to 0} \lim_{k\to\infty} g_k(x);$$

$$\lim_{k\to\infty} \int_0^1 h_k(x)\,dx = -1 \ne 0 = \int_0^1 \left[\lim_{k\to\infty} h_k(x)\right] dx.$$

Clearly, if we want some theorems to the effect that "the integral of the limit is the limit of the integrals," or "the derivative of a limit is the limit of the derivatives," pointwise convergence is the wrong condition to impose. We now develop a more stringent notion of convergence that removes some of the pathologies.

The real trouble with pointwise convergence is as follows. The statement "$f_k(x) \to f(x)$ for all $x \in S$" means that, for each x, $f_k(x)$ will be close to $f(x)$ provided k is sufficiently large, but the rate of convergence of $f_k(x)$ to $f(x)$ can be very different for different values of x. For example, if g_k is as in (7.2), for all $x \ne 0$ we have $g_k(x) \to 0$, so $|g_k(x)| < 10^{-4}$ (say) provided k is sufficiently large; for $x = 10$, "sufficiently large" means $k \geq 10$, but for $x = 0.1$, it means $k \geq 1000$. If, however, we have some control over the rate of convergence that is *independent of the particular point* x, then many of the pathologies disappear.

The precise definition is as follows. A sequence $\{f_k\}$ of functions defined on a set $S \subset \mathbb{R}^n$ is said to **converge uniformly on** S to the function f if for every $\epsilon > 0$ there is an integer K such that

$$(7.3) \qquad |f_k(\mathbf{x}) - f(\mathbf{x})| < \epsilon \text{ whenever } k > K \text{ and } \mathbf{x} \in S.$$

The point here is that the *same* K will work for *every* $\mathbf{x} \in S$. Another way of writing (7.3) is

$$(7.4) \qquad \sup_{\mathbf{x} \in S} |f_k(\mathbf{x}) - f(\mathbf{x})| \leq \epsilon \text{ whenever } k > K.$$

The geometry of this inequality is indicated in Figure 7.2. Yet another way of expressing uniform convergence is the following, which is sufficiently useful to be displayed as a theorem.

7.5 Theorem. *The sequence $\{f_k\}$ converges to f uniformly on S if and only if there is a sequence $\{C_k\}$ of positive constants such that $|f_k(\mathbf{x}) - f(\mathbf{x})| \leq C_k$ for all $\mathbf{x} \in S$ and $\lim_{k \to \infty} C_k = 0$.*

Proof. If $f_k \to f$ uniformly, by (7.4) we can take $C_k = \sup_{\mathbf{x} \in S} |f_k(\mathbf{x}) - f(\mathbf{x})|$. Conversely, if $C_k \to 0$, for any $\epsilon > 0$ there exists K such that $C_k < \epsilon$ whenever $k > K$, and hence $|f_k(\mathbf{x}) - f(\mathbf{x})| \leq C_k < \epsilon$ for all $\mathbf{x} \in S$ whenever $k > K$; that is, (7.3) holds. $\qquad \square$

Let us take another look at the examples in (7.2) with regard to uniform convergence. First, the sequence $\{f_k\}$ defined by $f_k(x) = k^{-1} \arctan kx$ converges uniformly to 0 on $\mathbb{R}$, since we can take $C_k = \pi/2k$ in Theorem 7.5. Second, the sequence $\{g_k\}$ defined by $g_k(x) = (k^2 x^2 + 1)^{-1}$ does not converge uniformly to its limit g on $\mathbb{R}$; indeed,

$$\sup_{x \in \mathbb{R}} |g_k(x) - g(x)| = \sup_{x \neq 0} \frac{1}{k^2 x^2 + 1} = 1 \text{ for all } k.$$

(Notice that the supremum is not actually achieved; the maximum of $(k^2 x^2 + 1)^{-1}$ occurs at $x = 0$, but $g(0) = 1$, so $g_k(0) - g(0) = 0$. See Figure 7.2.) Finally, the sequence $\{h_k\}$ defined by $h_k(x) = -2k^2 x (k^2 x^2 + 1)^{-2}$ does not converge uniformly to its limit 0 on $\mathbb{R}$. Indeed, a bit of calculus shows that the minimum and maximum values of $h_k(x)$, achieved at $x = \pm 1/2k$, are $\mp 16k/25$, so $\sup_x |h_k(x) - 0|$ actually tends to ∞ rather than 0.

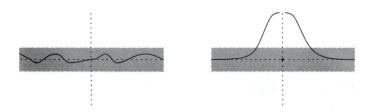

FIGURE 7.2: Left: Uniform convergence. For k large, the graph of $f_k - f$ is contained in the shaded strip $|y| < \epsilon$. Right: Nonuniform convergence of the sequence $\{g_k\}$ in (7.2). The spike of $g_k - g$ around the origin becomes narrower as $k \to \infty$ but is never wholly within the shaded strip.

On the other hand, the bad behavior in these examples is all at $x = 0$. The sequences $\{g_k\}$ and $\{h_k\}$ do converge uniformly to 0 on the intervals $[\delta, \infty)$ and $(-\infty, -\delta]$ for any $\delta > 0$. For g_k this is clear:

$$|g_k(x) - 0| = \frac{1}{k^2 x^2 + 1} \le \frac{1}{\delta^2 k^2 + 1} \qquad (x \le -\delta \text{ or } x \ge \delta),$$

and $(\delta^2 k^2 + 1)^{-1} \to 0$ as $k \to \infty$. For h_k we do not get a good estimate for the first few values of k, but (by the same bit of calculus as in the preceding paragraph) when $k > 1/2\delta$ the function h_k is positive and increasing on $(-\infty, -\delta]$ and negative and increasing on $[\delta, \infty)$, so the maximum of $|h_k|$ on these intervals occurs at the endpoints $\pm\delta$:

$$|h_k(x) - 0| \le \frac{2\delta k^2}{(\delta^2 k^2 + 1)^2} \qquad \left(x \le -\delta \text{ or } x \ge \delta, \ k > \frac{1}{2\delta} \right).$$

The phenomenon exhibited here is quite common. That is, one has a sequence $\{f_k\}$ of functions that converge pointwise to f on a set S; the convergence is not uniform on all of S but is uniform on many "slightly smaller" subsets of S. The situation we shall encounter most often is where S is an open interval (a, b), and the "bad behavior" occurs near the endpoints, so that the convergence is uniform on $[a + \delta, b - \delta]$ for any $\delta > 0$. In this case, the sequence of constants C_k in Theorem 39 will generally depend on δ — as they do in the preceding examples.

The notion of Cauchy sequence has an obvious adaptation to the context of uniform convergence. Namely, a sequence $\{f_k\}$ of functions on a set S is **uniformly Cauchy** if for every $\epsilon > 0$ there is an integer K so that

(7.6) $\qquad |f_j(\mathbf{x}) - f_k(\mathbf{x})| < \epsilon$ whenever $j, k > K$ and $\mathbf{x} \in S$, or in other words,

$$\sup_{\mathbf{x} \in S} |f_j(\mathbf{x}) - f_k(\mathbf{x})| < \epsilon \text{ whenever } j, k > K.$$

We have the following analogue of Theorem 1.20:

7.7 Theorem. *The sequence $\{f_k\}$ is uniformly Cauchy on S if and only if there is a function f on S such that $f_k \to f$ uniformly on S.*

Proof. If $\{f_k\}$ is uniformly Cauchy, then for each $\mathbf{x} \in S$ the numerical sequence $\{f_k(\mathbf{x})\}$ is Cauchy. By Theorem 1.20, it has a limit, which we call $f(\mathbf{x})$. Letting $j \to \infty$ in (7.6), we see that $|f_k(\mathbf{x}) - f(\mathbf{x})| \le \epsilon$ whenever $k > K$ and $x \in S$, so that $f_k \to f$ uniformly on S. Conversely, if $f_k \to f$ uniformly on S, we have $|f_k(\mathbf{x}) - f(\mathbf{x})| \le C_k$ for all $\mathbf{x} \in S$, where $C_k \to 0$ as $k \to \infty$, and

$$|f_j(\mathbf{x}) - f_k(\mathbf{x})| \le |f_j(\mathbf{x}) - f(\mathbf{x})| + |f(\mathbf{x}) - f_k(\mathbf{x})| \le C_j + C_k,$$

and $C_j + C_k < \epsilon$ when j and k are sufficiently large, so (7.6) holds.   □

One of the most important properties of uniform convergence is that it preserves continuity, as mere pointwise convergence does not (see the example $\{g_k\}$ in (7.2)).

7.8 Theorem. *Suppose $f_k \to f$ uniformly on S. If each f_k is continuous on S, then so is f.*

Proof. Given a point $\mathbf{a} \in S$, we show that f is continuous at $\mathbf{a}$. Given $\epsilon > 0$, we can choose k large enough so that $|f_k(\mathbf{x}) - f(\mathbf{x})| < \epsilon/3$ for all $\mathbf{x} \in S$. Since f_k is continuous, there exists $\delta > 0$ so that $|f_k(\mathbf{x}) - f_k(\mathbf{a})| < \epsilon/3$ whenever $|\mathbf{x} - \mathbf{a}| < \delta$ and $\mathbf{x} \in S$. But then, under these same conditions on $\mathbf{x}$, we have

$$|f(\mathbf{x}) - f(\mathbf{a})| \le |f(\mathbf{x}) - f_k(\mathbf{x})| + |f_k(\mathbf{x}) - f_k(\mathbf{a})| + |f_k(\mathbf{a}) - f(\mathbf{a})|$$
$$< \frac{\epsilon}{3} + \frac{\epsilon}{3} + \frac{\epsilon}{3} = \epsilon,$$

which shows that f is continuous at $\mathbf{a}$.   □

Theorem 7.8 can be strengthened somewhat, because the continuity of a function f at a point $\mathbf{a}$ depends only on the behavior of f at points close to $\mathbf{a}$. Hence, if f_k is continuous on S and $f_k \to f$ pointwise on S, it is not necessary to have uniform convergence on all of S to guarantee continuity of the limit function f; it is enough to have uniform convergence on some neighborhood of each point in S. For example, if S is the interval (a, b) and $f_k \to f$ uniformly on $[a + \delta, b - \delta]$ for each $\delta > 0$, we conclude that f is continuous on $[a + \delta, b - \delta]$ for each δ and hence that f is continuous on all of (a, b).

The preceding discussion of sequences of functions leads immediately to results about series of functions. Namely, given a sequence of functions $\{f_n\}_0^\infty$ defined on a set S, we can form the infinite series $\sum_0^\infty f_n(\mathbf{x})$ for each $\mathbf{x} \in S$. If this series converges for each $\mathbf{x} \in S$, we say that the series $\sum_0^\infty f_n$ is **(pointwise) convergent on** S; in this case, its sum defines a function on S, which we also denote by $\sum_0^\infty f_n$. The series $\sum_0^\infty f_n$ is said to be **uniformly convergent on** S if the sequence of partial sums, $s_k = \sum_0^k f_n$, is uniformly convergent on S.

EXAMPLE 2. The geometric series $\sum_0^\infty x^n$ converges pointwise on $(-1, 1)$ to $(1-x)^{-1}$. Denoting the kth partial sum by $s_k(x)$, we have

$$s_k(x) = \frac{1 - x^{k+1}}{1 - x}, \quad \text{so} \quad \left| s_k(x) - \frac{1}{1-x} \right| = \frac{|x|^{k+1}}{1-x}.$$

The latter quantity tends to ∞ as $x \to 1$ and to $\frac{1}{2}$ as $x \to -1$ no matter what k is, so the convergence is not uniform on $(-1, 1)$. (This is hardly surprising, since the series diverges at both endpoints.) But it is uniform on $[-r, r]$ for any $r < 1$, for

$$\frac{|x|^{k+1}}{1-x} \le \frac{r^{k+1}}{1-r} \text{ for } |x| \le r,$$

and this quantity vanishes as $k \to \infty$.

The following is the most commonly used test for uniform convergence of series:

7.9 Theorem (The Weierstrass M-Test). *Let* $\{f_n\}_0^\infty$ *be a sequence of functions on the set* S. *Suppose there is a sequence* $\{M_n\}_0^\infty$ *of positive constants such that (i)* $|f_n(x)| \le M_n$ *for all* $x \in S$ *and all* n, *and (ii)* $\sum_0^\infty M_n < \infty$. *Then the series* $\sum_0^\infty f_n$ *is absolutely and uniformly convergent on* S.

Proof. The series $\sum_0^\infty f_n(x)$ is absolutely convergent for each $x \in S$ by comparison to the series $\sum_0^\infty M_n$. Let us denote its sum by $s(x)$, the kth partial sum $\sum_0^k f_n(x)$ by $s_k(x)$, and $\sum_{k+1}^\infty M_n$ by C_k; then

$$|s(x) - s_k(x)| \le \sum_{k+1}^\infty |f_n(x)| \le \sum_{k+1}^\infty M_n = C_k \qquad (x \in S).$$

But $C_k \to 0$ as $k \to \infty$ since the series $\sum M_n$ is convergent, so it follows from Theorem 7.5 that the sequence $\{s_k\}$, i.e., the series $\sum f_n$, is uniformly convergent on S. $\qquad\square$

The tribute to Weierstrass in the name of this theorem is appropriate, since Weierstrass was one of the pioneers in the rigorous theory of infinite series; but the term "M-test" signifies nothing more than the fact that the sequence of constants in the theorem is traditionally denoted by $\{M_n\}$.

It is quite possible for a series of functions to be uniformly convergent on S without being absolutely convergent. (See Exercises 5 and 6.) Therefore, the Weierstrass M-test, unlike its cousin Theorem 7.5, gives a sufficient condition for uniform convergence but not a necessary one.

EXAMPLE 3. The M-test gives an easy verification that the geometric series $\sum_0^\infty x^n$ converges uniformly on $[-r, r]$ for any $r < 1$, by taking $M_n = r^n$. ($|x^n| \le r^n$ for $|x| \le r$, and $\sum r^n < \infty$.)

EXAMPLE 4. The Taylor series for $\log(1 + x)$, $\sum_1^\infty (-1)^{n-1} x^n / n$, converges absolutely for $x \in (-1, 1)$ (by the ratio test) and conditionally at $x = 1$ (by the alternating series test). Since $|(-1)^{n-1} x^n / n| \le r^n / n$ when $|x| \le r$, the M-test (with $M_n = r^n / n$) shows that this series converges uniformly on $[-r, r]$ for any $r < 1$. It actually converges uniformly on $[-r, 1]$ for any $r < 1$, but the M-test will not yield this result because the convergence at 1 is only conditional. (The result needed here is a theorem of Abel that we shall present in §7.3.)

Theorem 7.8, concerning the continuity of limits of sequences, translates immediately into a theorem about continuity of sums of series, as follows:

7.10 Theorem. *Suppose $\{f_n\}$ is a sequence of continuous functions on a set S. If the series $\sum f_n$ converges uniformly on S, its sum is a continuous function on S.*

Proof. Apply Theorem 7.8 to the sequence of partial sums.   □

The remarks following Theorem 7.8, to the effect that *local* uniform convergence is enough to yield continuity, apply to this situation also.

EXERCISES

1. For each of the following sequences $\{f_k\}$ of functions, compute $\lim_{k \to \infty} f_k$ on the given interval and tell whether the convergence is uniform on that interval. If not, is the convergence uniform on some slightly smaller sets?
   a. $f_k(x) = x^k$, $x \in [0, 1]$.
   b. $f_k(x) = x^{1/k}$, $x \in [0, 1]$.
   c. $f_k(x) = \sin^k x$, $x \in [0, \pi]$.
   d. $f_k(x) = k^{-1} e^{-x^2/k}$, $x \in \mathbb{R}$.

    e. $f_k(x) = kxe^{-kx}$, $x \in [0, \infty)$.

    f. $f_k(x) = (x/k)e^{-x/k}$, $x \in [0, \infty)$.

    g. $f_k(x) = x^k/(1 + x^{2k})$, $x \in [0, \infty)$.

2. Test the following series for absolute and uniform convergence; state the interval(s) on which you obtain such convergence. What can you conclude about the continuity of the sum of the series?

    a. $\displaystyle\sum_0^\infty e^{-nx}$.

    b. $\displaystyle\sum_0^\infty \frac{x^n}{n^2 + n + 1}$.

    c. $\displaystyle\sum_1^\infty \frac{nx^n}{2^{n+3}}$.

    d. $\displaystyle\sum_1^\infty \frac{\cos nx}{n^3}$.

    e. $\displaystyle\sum_1^\infty \frac{1}{x^2 + n^2}$.

    f. $\displaystyle\sum_1^\infty n^{-x}$.

3. Let $f_k(x) = g(x)x^k$, where g is continuous on $[0, 1]$ and $g(1) = 0$. Show that $f_k \to 0$ uniformly on $[0, 1]$. (Cf. Exercise 1a.)

4. Show that the series $\displaystyle\sum_1^\infty \frac{1}{x^2 - n^2}$ converges uniformly on any compact interval that does not contain a nonzero integer, and conclude that the sum of the series is a continuous function on $\mathbb{R} \setminus \{\pm 1, \pm 2, \ldots\}$.

5. Show that the series $\displaystyle\sum_1^\infty \frac{(-1)^{n-1}}{x^2 + n}$ converges uniformly on $\mathbb{R}$, although the convergence is conditional at every point.

6. Given a sequence $\{c_n\}$ of real numbers such that $\sum_1^\infty c_n$ converges, consider the series $\displaystyle\sum_1^\infty c_n \frac{x^n}{1 - x^n}$ $(x \neq \pm 1)$. (Such a series is called a **Lambert series**.)

    a. Show that the series converges absolutely and uniformly on $[-a, a]$ for any $a < 1$.

    b. Show that the series converges uniformly on $(-\infty, -b]$ and on $[b, \infty)$ for any $b > 1$, and that the convergence is absolute if and only if $\sum_1^\infty |c_n| <$

∞. (Hint: $x^n(1 - x^n)^{-1} = (1 - x^n)^{-1} - 1$.)

7. Let $\{f_k\}$ be a sequence of functions defined on a set S, and let $S_1, \ldots, S_M$ be a finite collection of subsets of S. Show that if $\{f_k\}$ converges uniformly on each S_m, then it converges uniformly on $\bigcup_1^M S_m$.

8. Let $\{f_k\}$ be a sequence of continuous functions on $[a, b]$. Show that if $\{f_k\}$ converges uniformly on (a, b), then it converges uniformly on $[a, b]$.

9. Let $\{f_k\}$ be a sequence of continuous functions on a compact set $S \subset \mathbb{R}^n$. Suppose that (a) the sequence $\{f_k(\mathbf{x})\}$ is bounded and increasing (and hence has a limit) for each $\mathbf{x} \in S$, and (b) the function $f = \lim_{k \to \infty} f_k$ is continuous on S. Show that $f_k \to f$ uniformly on S. (Hint: Given $\epsilon > 0$, apply Exercise 5 in §1.6 to the sets $S_k = \{\mathbf{x} \in S : f(\mathbf{x}) - f_k(\mathbf{x}) \geq \epsilon\}$.)

7.2   Integrals and Derivatives of Sequences and Series

If $\{f_k\}$ is a sequence of functions on the interval $[a, b]$ and $f_k \to f$ on $[a, b]$, is it true that $\int_a^b f_k(x)\, dx \to \int_a^b f(x)\, dx$? The sequence $\{h_k\}$ in (7.2) shows that the answer is sometimes *no*. The best general affirmative result in the context of Riemann integration is the bounded convergence theorem that we stated in §4.5. As we indicated there, the proof of that theorem is beyond the scope of this book; however, uniform convergence yields a affirmative result with an easy proof. It works equally well for n-dimensional integrals, so we present it in that generality.

7.11 Theorem. *Suppose S is a measurable set in $\mathbb{R}^n$ and $\{f_k\}$ is a sequence of integrable functions on S that converges uniformly to an integrable function f on S. Then*

$$\int \cdots \int_S f(\mathbf{x})\, d^n\mathbf{x} = \lim_{k \to \infty} \int \cdots \int_S f_k(\mathbf{x})\, d^n\mathbf{x}.$$

Proof. By Theorem 7.5, there is a sequence $\{C_k\}$ of constants such that $C_k \to 0$ and $|f_k(\mathbf{x}) - f(\mathbf{x})| \leq C_k$ for $\mathbf{x} \in S$. But then

$$\left| \int \cdots \int_S f_k(\mathbf{x})\, d^n\mathbf{x} - \int \cdots \int_S f(\mathbf{x})\, d^n\mathbf{x} \right| \leq \int \cdots \int_S |f_k(\mathbf{x}) - f(\mathbf{x})|\, d^n\mathbf{x}$$

$$\leq \int \cdots \int_S C_k\, d^n\mathbf{x}.$$

This last quantity is the n-dimensional volume of S times C_k, which tends to zero as $k \to \infty$. $\square$

Returning to the one-dimensional situation, we now ask the corresponding question for derivatives: If $f_k \to f$, is it true that $f_k' \to f'$? Equivalently, setting $g_k = f_k - f$, if $g_k \to 0$, is it true that $g_k' \to 0$? Here the answer is clearly *no* in general; the function g_k can be very small but also very wiggly, so that g_k' is not small.

EXAMPLE 1. Let $g_k(x) = k^{-1} \sin kx$. Then $|g_k(x)| \le k^{-1}$ for all x, so $g_k \to 0$ uniformly on $\mathbb{R}$. On the other hand, $g_k'(x) = \cos kx$; the sequence $\{\cos kx\}$ does not converge at all for most values of x, and when it does — namely, when x is an even multiple of π — its limit is 1, not 0.

In this situation, the crucial uniformity hypothesis is not on the original sequence $\{f_k\}$ but on the differentiated sequence $\{f_k'\}$. Here is the result:

7.12 Theorem. *Let $\{f_k\}$ be a sequence of functions of class C^1 on the interval $[a, b]$. Suppose that $\{f_k\}$ converges pointwise to f and that $\{f_k'\}$ converges uniformly to g on $[a, b]$. Then f is of class C^1 on $[a, b]$, and $g = f'$.*

Proof. The function g is continuous on $[a, b]$ by Theorem 7.8, so it is integrable over any subinterval of $[a, b]$. By Theorem 7.11,

$$\int_a^x g(t)\, dt = \lim_{k \to \infty} \int_a^x f_k'(t)\, dt = \lim_{k \to \infty} \left[f_k(x) - f_k(a) \right] = f(x) - f(a).$$

Thus $f(x) = f(a) + \int_a^x g(t)\, dt$. But by the fundamental theorem of calculus, the function on the right is differentiable and its derivative is g. $\qquad \square$

The example $\{f_k\}$ in (7.2) shows that pointwise convergence of $\{f_k'\}$ is not sufficient to obtain $\lim(f_k') = (\lim f_k)'$. On the other hand, Theorem 7.12 can be extended somewhat. Since differentiability (like continuity) is a local property, it is enough for the convergence of $\{f_k'\}$ to be uniform on a neighborhood of each point, rather than on the whole interval in question. In many situations, the sequence $\{f_k\}$ is defined on an *open* interval (a, b) and one has uniform convergence of $\{f_k'\}$ on each compact subinterval $[a + \delta, b - \delta]$; this suffices to guarantee that $\lim(f_k') = (\lim f_k)'$ on (a, b).

The results on term-by-term integration and differentiation of series are immediate consequences of those for sequences. We have merely to apply Theorems 7.11 and 7.12 to the partial sums of the series to obtain the following theorem.

7.13 Theorem. *Suppose that $\{f_n\}$ is a sequence of continuous functions on the interval $[a, b]$ and that the series $\sum f_n$ converges pointwise on $[a, b]$.*

a. If $\sum f_n$ converges uniformly on $[a, b]$, then

$$\int_a^b \left[\sum f_n(x)\right] dx = \sum \int_a^b f_n(x) \, dx.$$

b. If the f_n's are of class C^1 and the series $\sum f_n'$ converges uniformly on $[a, b]$, then the sum $\sum f_n$ is of class C^1 on $[a, b]$ and

$$\frac{d}{dx} \left[\sum f_n(x)\right] = \sum f_n'(x) \qquad (x \in [a, b]).$$

EXERCISES

1. Let $f(x) = \sum_1^\infty n^{-2} \sin nx$. Show that f is a continuous function on $\mathbb{R}$ and that $\int_0^{\pi/2} f(x) \, dx = \sum_{n=1,3,5,\dots} n^{-3} + 2 \sum_{n=2,6,10,\dots} n^{-3}$.

2. Let $f(x) = \sum_1^\infty (x + n)^{-2}$. Show that f is a continuous function on $[0, \infty)$ and that $\int_0^1 f(x) \, dx = 1$.

3. Let $f_k(x) = x \arctan kx$.
   a. Show that $\lim_{k\to\infty} f_k(x) = \frac{1}{2}\pi|x|$.
   b. Show that $\lim_{k\to\infty} f_k'(x)$ exists for every x, including $x = 0$, but that the convergence is not uniform in any interval containing 0.

4. For each of the series (a–f) in Exercise 2, §7.1, show that the series can be differentiated term-by-term on its interval of convergence (except at the endpoints in (b)).

5. For $x \neq \pm1, \pm2, \dots$, let $f(x) = 2x \sum_1^\infty (x^2 - n^2)^{-1}$ (see Exercise 4, §7.1). Show that f is of class C^1 on its domain and that $f'(x) = -\sum_1^\infty [(x - n)^{-2} + (x + n)^{-2}]$.

6. Let f be a continuous function on $[0, \infty)$ such that $0 \leq f(x) \leq Cx^{-1-\epsilon}$ for some $C, \epsilon > 0$, and let $a = \int_0^\infty f(x) \, dx$. (The estimate on f implies the convergence of this integral.) Let $f_k(x) = kf(kx)$.
   a. Show that $\lim_{k\to\infty} f_k(x) = 0$ for all $x > 0$ and that the convergence is uniform on $[\delta, \infty)$ for any $\delta > 0$.
   b. Show that $\lim_{k\to\infty} \int_0^1 f_k(x) \, dx = a$.
   c. Show that $\lim_{k\to\infty} \int_0^1 f_k(x)g(x) \, dx = ag(0)$ for any integrable function g on $[0, 1]$ that is continuous at 0. (Hint: Write $\int_0^1 = \int_0^\delta + \int_\delta^1$.)

7.3 Power Series

A **power series** is an infinite series of the form

$$(7.14) \qquad \sum_{0}^{\infty} a_n(x-b)^n = a_0 + a_1(x-b) + a_2(x-b)^2 + \cdots,$$

where x is a real or complex variable. The lower limit of summation is always $n = 0$ in principle, although the first few terms might vanish ($a_0 = \cdots = a_k = 0$); the crucial point is that only *nonnegative integer* powers of $x-b$ are allowed. (Thus, one might think of a power series as a "polynomial of infinite degree in $x - b$.") The study of series of the general form (7.14) can be reduced to the special case $b = 0$ by the change of variable $x \to x + b$, and we do so henceforth.

The first order of business in studying power series is to determine the range of values of the variable x for which they converge. The key observation is as follows.

7.15 Lemma. *If the power series $\sum_{0}^{\infty} a_n x^n$ converges for $x = x_0$, then it converges absolutely for all x such that $|x| < |x_0|$.*

Proof. The convergence of $\sum a_n x_0^n$ implies that $a_n x_0^n \to 0$, and in particular that $|a_n x_0^n| \le C$ for some constant C independent of n. Since

$$|a_n x^n| = |a_n x_0^n| \left| \frac{x^n}{x_0^n} \right| \le C \left| \frac{x}{x_0} \right|^n,$$

for $|x| < |x_0|$ the series $\sum a_n x^n$ converges absolutely by comparison with the geometric series $\sum |x/x_0|^n$. $\qquad \square$

7.16 Theorem. *For any power series $\sum_{0}^{\infty} a_n x^n$, there is a number $R \in [0, \infty]$, called the **radius of convergence** of the series, such that the series converges absolutely for $|x| < R$ and diverges for $|x| > R$. (When $R = 0$, this means that the series converges only for $x = 0$; when $R = \infty$, it means that the series converges absolutely for all x.)*

Proof. Let $R = \sup\{|x_0| : \sum a_n x_0^n \text{ converges}\}$. ($R \ge 0$ since the series always converges at $x_0 = 0$.) Thus $\sum a_n x^n$ diverges if $|x| > R$. On the other hand, if $|x| < R$, there exists x_0 such that $|x_0| > |x|$ and $\sum a_n x_0^n$ converges, and then $\sum a_n x^n$ converges absolutely by Lemma 7.15. $\qquad \square$

Important Remark. The reader has probably been thinking of a_n and x as real numbers, but Theorem 7.16 is valid, with exactly the same proof, when a_n and x are complex numbers.

Theorem 7.16 says that the set of all real x such that $\sum a_n x^n$ converges is an open interval centered at 0, possibly together with one or both endpoints, and the set of all complex x such that $\sum a_n x^n$ converges is an open *disc* centered at 0 in the complex plane, possibly together with some or all of its boundary points. The behavior of the series on the boundary of the region of convergence must be decided on a case-by-case basis.

EXAMPLE 1. Consider the series

$$\sum_{1}^{\infty} \frac{x^n}{n^2}, \qquad \sum_{0}^{\infty} x^n, \qquad \sum_{1}^{\infty} \frac{x^n}{n}.$$

An easy application of the ratio test shows that each of these series converges absolutely for $|x| < 1$ and diverges for $|x| > 1$, so their radius of convergence is 1. The first one is absolutely convergent when $|x| = 1$ by comparison with $\sum n^{-2}$, whereas the second is divergent when $|x| = 1$ because $x^n \not\to 0$ as $n \to \infty$ in that case. The third one is divergent when $x = 1$ but is conditionally convergent at $x = -1$ by the alternating series test. It is also conditionally convergent at all other complex numbers x such that $|x| = 1$, by Dirichlet's test. (Indeed, take $a_n = n^{-1}$ and $b_n = x^n$. Then $b_1 + \cdots + b_n$ is a finite geometric series whose sum equals $x(1 - x^n)/(1 - x)$, and this is bounded by $2|x|/(|1 - x|)$ as $n \to \infty$.)

The standard tools for determining the radius of convergence of a power series are the ratio test and the root test. We have already seen how this works in §6.4 (especially Example 2 and Exercises 1–9), so we shall not belabor the point here. However, see Exercise 1. In fact, a slight extension of the root test yields a formula for the radius of convergence of an arbitrary power series; see Exercise 4.

Theorem 7.16 shows that any power series converges *absolutely* within the region $|x| < R$. Equally important is that it converges *uniformly* on compact subsets of this region.

7.17 Theorem. *Let R be the radius of convergence of $\sum_0^\infty a_n x^n$. For any $r < R$, the series $\sum_0^\infty a_n x^n$ converges uniformly on the set $\{x : |x| \le r\}$, and its sum is a continuous function on the set $\{x : |x| < R\}$.*

Proof. For $|x| \le r$ we have $|a_n x^n| \le |a_n| r^n$, and the series $\sum |a_n| r^n$ is convergent since $\sum a_n x^n$ is absolutely convergent at $x = r$. The first assertion therefore follows from the Weierstrass M-test, and the second follows from the first by Theorem 7.8. $\qquad\square$

We now turn to the question of integrating power series. In this discussion we take x to be a real variable.

7.18 Theorem. *Suppose the series* $f(x) = \sum_0^\infty a_n x^n$ *has radius of convergence* $R > 0$.

a. *If* $-R < a < b < R$, *then* $\displaystyle \int_a^b f(x)\,dx = \sum_0^\infty a_n \frac{b^{n+1} - a^{n+1}}{n+1}$.

b. *If* F *is any antiderivative of* f, *then* $\displaystyle F(x) = F(0) + \sum_0^\infty \frac{a_n}{n+1} x^{n+1}$ *for* $|x| < R$.

Proof. Assertion (a) follows immediately from Theorems 7.13a and 7.17. The fundamental theorem of calculus then shows that $\sum_0^\infty a_n x^{n+1}/(n+1)$ is an antiderivative of f on $(-R, R)$ — specifically, the one whose value at $x = 0$ is zero — and any other antiderivative differs from this one by a constant. $\square$

Theorem 7.18 gives a way of generating new series expansions from old ones.

EXAMPLE 2. If we integrate the geometric series $\sum_0^\infty (-x)^n = (1 + x)^{-1}$ ($|x| < 1$), we obtain

$$\log(1 + x) = \int_0^x \frac{dt}{1 + t} = \sum_0^\infty \frac{(-1)^n}{n+1} x^{n+1} = \sum_1^\infty \frac{(-1)^{n-1}}{n} x^n \qquad (|x| < 1).$$

(The last equality is obtained by the change of variable $n \to n - 1$.) Similarly, integration of the geometric series $\sum_0^\infty (-x^2)^n = (1 + x^2)^{-1}$ leads to

$$\arctan x = \int_0^x \frac{dt}{1 + t^2} = \sum_0^\infty \frac{(-1)^n x^{2n+1}}{2n + 1} \qquad (|x| < 1).$$

The series for $\log(1+x)$ is easily obtained from Taylor's theorem (see Exercise 3 in §6.1), but not the series for $\arctan x$; the computation of the high-order derivatives of the latter function is very cumbersome. (*Remark:* The expansion of $\log(1+x)$ is also valid at $x = 1$, and that of $\arctan x$ is also valid at $x = \pm 1$. However, these facts do not follow from Theorem 7.18. The extra result needed here is Abel's theorem, which we shall present below.)

Theorem 7.18 also offers a technique for expressing definite or indefinite integrals of functions that have no elementary antiderivatives in a computable form.

EXAMPLE 3. The function $f(x) = x^{-1} \sin x$ has no elementary antiderivative, but

$$\int_0^x \frac{\sin t}{t} \, dt = \int_0^x \sum_0^\infty \frac{(-1)^m t^{2m}}{(2m+1)!} \, dt = \sum_0^\infty \frac{(-1)^m x^{2m+1}}{(2m+1) \cdot (2m+1)!}.$$

This gives a precise analytic expression for $\int_0^x t^{-1} \sin t \, dt$ that is valid for all x, and the first few terms, $x - \frac{1}{18}x^3 + \frac{1}{600}x^5 + \cdots$, furnish a good numerical approximation to the integral when x is not too large.

Next, what about term-by-term differentiation of a power series $\sum_0^\infty a_n x^n$? According to Theorem 7.13b, we must examine the convergence of the series $\sum_0^\infty n a_n x^{n-1}$ obtained by termwise differentiation, which we shall call the **derived series**. At first glance, the latter series seems less likely to converge than the original series, since the nth term of the derived series is much larger than the corresponding term of the original series when n is large (by a factor of $n/|x|$). But in fact, the only values of x for which this really matters are those on the boundary of the interval (or disc) of convergence; elsewhere, the exponential behavior of x^n as $n \to \infty$ swamps the extra factor of n, as will be seen in the following proof.

7.19 Theorem. *The radius of convergence of any power series $\sum_0^\infty a_n x^n$ is equal to the radius of convergence of the derived series $\sum_0^\infty n a_n x^{n-1}$.*

Proof. Let R and R' be the radii of convergence of $\sum_0^\infty a_n x^n$ and $\sum_0^\infty n a_n x^{n-1}$, respectively. Suppose $|x| < R'$. Then $\sum n a_n x^{n-1}$ is absolutely convergent, and

$$|a_n x^n| = \frac{|x|}{n} |n a_n x^{n-1}| \leq |n a_n x^{n-1}| \text{ for large } n,$$

so $\sum a_n x^n$ is absolutely convergent by comparison. Thus, if $|x| < R'$ then $|x| \leq R$, and it follows that $R' \leq R$.

On the other hand, if $|x| < R$, we can pick a number r such that $|x| < r < R$. Then the series $\sum a_n r^n$ is absolutely convergent, and

$$|n a_n x^{n-1}| = \frac{1}{|x|} \left(n \left| \frac{x}{r} \right|^n \right) |a_n| r^n.$$

Since $|x/r| < 1$, the sequence $|x/r|^n$ tends to 0 exponentially fast as $n \to \infty$, and hence $n|x/r|^n \to 0$ also. In particular, we have $|n a_n x^{n-1}| \leq |a_n| r^n$ for n large, so $\sum n a_n x^{n-1}$ converges (absolutely) by comparison to $\sum |a_n r^n|$. In short, if $|x| < R$ then $|x| \leq R'$, and it follows that $R \leq R'$. Combining this inequality with the one in the preceding paragraph, we conclude that $R = R'$. $\square$

Combining this result with Theorem 7.13b, we obtain the fundamental theorem on term-by-term differentiation of a power series.

7.20 Theorem. *Suppose the radius of convergence of the series $f(x) = \sum a_n x^n$ is $R > 0$. Then the function f is of class C^∞ on the interval $(-R, R)$, and its kth derivative may be computed on $(-R, R)$ by differentiating the series $\sum_0^\infty a_n x^n$ termwise k times.*

Proof. In view of Theorem 7.19, Theorem 7.13b shows that $f'(x) = \sum n a_n x^{n-1}$ for $|x| < R$. It now follows by induction on k that, for any positive integer k, f is of class C^k on $(-R, R)$ and that $f(k)$ is the sum of the k-times derived series. $\quad\square$

7.21 Corollary. *Every power series $\sum_0^\infty a_n x^n$ with a positive radius of convergence is the Taylor series of its sum; that is, if $f(x) = \sum_0^\infty a_n x^n$ for $|x| < R$ $(R > 0)$, then*

$$a_n = \frac{f^{(n)}(0)}{n!}.$$

Proof. Since $(d/dx)^n x^k = 0$ when $k < n$ and $(d/dx)^n x^n \equiv n!$, we have

$$f^{(n)}(x) = \frac{d^n}{dx^n}\left(a_0 + a_1 x + \cdots + a_n x^n + \cdots\right) = n! a_n + \cdots,$$

where the last set of dots denotes terms containing positive powers of x. Setting $x = 0$, we obtain $f^{(n)}(0) = n! a_n$. $\quad\square$

7.22 Corollary. *If $\sum_0^\infty a_n x^n = \sum_0^\infty b_n x^n$ for $|x| < R$ $(R > 0)$, then $a_n = b_n$ for all n.*

Proof. We have $a_n = f^{(n)}(0)/n! = b_n$ where $f(x)$ is the common sum of the two series. $\quad\square$

The following examples will illustrate the use of Theorem 7.20. The second one contains a result of importance in its own right, the binomial formula for fractional and negative exponents.

EXAMPLE 4. Suppose we wish to express the sum of the series $\sum_1^\infty x^n/n^2$ in terms of familiar elementary functions. The key is to recognize that this series is related to the geometric series $\sum_0^\infty x^n$, and that the factors of $1/n$ should arise from integrating the latter series. With this in mind, we proceed as follows. Setting $f(x) = \sum_1^\infty x^n/n^2$, we obtain successively

$$f'(x) = \sum_1^\infty \frac{x^{n-1}}{n}, \quad xf'(x) = \sum_1^\infty \frac{x^n}{n}, \quad (xf')'(x) = \sum_1^\infty x^{n-1} = \frac{1}{1-x}.$$

Undoing these transformations in turn yields

$$xf'(x) = -\log(1-x), \qquad f'(x) = -\frac{\log(1-x)}{x},$$

and, finally,

$$f(x) = -\int_0^x \frac{\log(1-t)}{t}\, dt.$$

EXAMPLE 5. Let α be a real number. Since

$$\frac{d^n}{dx^n}(1+x)^\alpha = \alpha(\alpha-1)\cdots(\alpha-n+1)(1+x)^{\alpha-n},$$

the Taylor series of $(1+x)^\alpha$ is

$$(7.23) \quad f_\alpha(x) = \sum_{n=0}^\infty \binom{\alpha}{n} x^n, \quad \text{where} \quad \binom{\alpha}{n} = \frac{\alpha(\alpha-1)\cdots(\alpha-n+1)}{n!}$$

(with the understanding that $\binom{\alpha}{0} = 1$). This series is called the **binomial series** of order α. When α is a nonnegative integer k, the terms with $n > k$ all vanish since they contain a factor of $(\alpha - k)$, and we obtain the familiar binomial expansion formula for $(1+x)^k$. For other values of α, the Taylor series is a genuine infinite series, and one can easily check by the ratio test that its radius of convergence is 1. Our aim is to verify that the sum of this series is actually $(1+x)^\alpha$ for $|x| < 1$.

We need the following formulas concerning the generalized binomial coefficients $\binom{\alpha}{n}$:

$$(7.24) \qquad n\binom{\alpha}{n} = \frac{\alpha(\alpha-1)\cdots(\alpha-n+1)}{(n-1)!} = \alpha\binom{\alpha-1}{n-1};$$

$$(7.25)$$
$$\binom{\alpha}{n} = \frac{[(\alpha-n)+n](\alpha-1)\cdots(\alpha-n+1)}{n!} = \binom{\alpha-1}{n} + \binom{\alpha-1}{n-1}.$$

Now, if $f_\alpha(x)$ is defined by (7.23) for $|x| < 1$, by (7.24) we have

$$f'_\alpha(x) = \sum_1^\infty n\binom{\alpha}{n} x^{n-1} = \sum_1^\infty \alpha\binom{\alpha-1}{n-1} x^{n-1} = \alpha \sum_0^\infty \binom{\alpha-1}{n} x^n$$

$$= \alpha f_{\alpha-1}(x).$$

(For the third equality we have made the change of variable $n \to n + 1$.) On the other hand,

$$(1 + x)f_{\alpha-1}(x) = \sum_0^\infty \binom{\alpha - 1}{n} x^n + \sum_0^\infty \binom{\alpha - 1}{n} x^{n+1}$$

$$= \sum_0^\infty \left[\binom{\alpha - 1}{n} + \binom{\alpha - 1}{n - 1} \right] x^n = \sum_0^\infty \binom{\alpha}{n} x^n = f_\alpha(x).$$

In the second equality, we substituted $n - 1$ for n in the second sum, and the third equality comes from (7.25). Combining these results, we see that $(1 + x)f'_\alpha(x) = \alpha f_\alpha(x)$. Multiplying through by $(1 + x)^{-\alpha-1}$ yields

$$0 = (1 + x)^{-\alpha} f'_\alpha(x) - \alpha(1 + x)^{-\alpha-1} f_\alpha(x) = \frac{d}{dx} \left[(1 + x)^{-\alpha} f_\alpha(x) \right].$$

Thus $(1 + x)^{-\alpha} f_\alpha(x)$ is a constant C, and setting $x = 0$, we see that $C = f_\alpha(0) = 1$. In short, $f_\alpha(x) = (1 + x)^\alpha$, as claimed.

EXAMPLE 6. The series $\sum_0^\infty (-1)^n x^{2n}$ is a geometric series with ratio $-x^2$, so it converges to $(1 + x^2)^{-1}$ for $|x| < 1$ and diverges elsewhere. By Corollary 7.21, this series is the Taylor series of the function $f(x) = (1 + x^2)^{-1}$ about $x = 0$. Now, the function f is C^∞ on the whole real line, so it seems rather mysterious that its Taylor series converges only on a finite interval. Why should the series behave badly as $x \to \pm 1$ when the function itself does not? The mystery is dispelled by considering *complex* values of x and recalling that the region of convergence of a power series in the complex plane is always a disc. The function $f(x)$ does blow up at $x = \pm i$, so the largest disc about the origin in the complex plane on which f is smooth is the disc $|x| < 1$.

Abel's Theorem. Suppose $f(x) = \sum_0^\infty a_n x^n$ is a power series whose radius of convergence R is positive and finite. We have seen that the convergence is uniform on any compact subinterval of $(-R, R)$ and hence that f is continuous on $(-R, R)$. But now suppose that the series converges at one of the endpoints, say $x = R$. Does the uniformity of convergence and the continuity of the sum persist up to this point?

If the series converges absolutely at $x = R$, then the M-test (with $M_n = |a_n| R^n$) shows that the series converges absolutely and uniformly on $[-R, R]$, so its sum is continuous there. But when the convergence is only conditional, a more subtle argument is needed. The necessary tool is the summation-by-parts formula that we used to obtain Dirichlet's test; since we need a slightly different version of that formula than the one given in Lemma 6.23 (namely, formula (7.27)), we shall simply derive it as we proceed.

7.26 Theorem (Abel's Theorem). *If the series $\sum_0^\infty a_n x^n$ converges at $x = R$ (resp. $x = -R$), then it converges uniformly on the interval $[0, R]$ (resp. $[-R, 0]$) and hence defines a continuous function on that interval.*

Proof. Convergence at $x = -R$ (and uniform convergence on $[-R, 0]$) of $f(x) = \sum a_n x^n$ is the same as convergence at $x = R$ (and uniform convergence on $[0, R]$) of $f(-x) = \sum (-1)^n a_n x^n$, so it is enough to consider convergence at $x = R$. Moreover, convergence at $x = R$ (and uniform convergence on $[0, R]$) of $f(x) = \sum a_n x^n$ is the same as convergence at $x = 1$ (and uniform convergence on $[0, 1]$) of $f(Rx) = \sum a_n R^n x^n$. In short, it is enough to assume that $\sum_0^\infty a_n$ converges and to prove that $\sum_0^\infty a_n x^n$ converges uniformly on $[0, 1]$. To do this we must show that the tail end $\sum_k^\infty a_n x^n$ of the series converges uniformly to zero on $[0, 1]$ as $k \to \infty$.

For $k \geq 1$, let $A_k = \sum_k^\infty a_n$ be the kth tail end of the series $\sum_0^\infty a_n$, so that $a_k = A_k - A_{k+1}$. For $l > k$ and $x \in [0, 1]$ we have

$$a_k x^k + \cdots + a_l x^l = (A_k - A_{k+1})x^k + \cdots + (A_l - A_{l+1})x^l$$
$$= A_k x^k + A_{k+1}(x^{k+1} - x^k) + \cdots + A_l(x^l - x^{l-1}) - A_{l+1}x^l.$$

Let $l \to \infty$: then $A_{l+1} \to 0$ and x^l remains bounded, so the last term on the right disappears and we obtain

$$(7.27) \qquad \sum_k^\infty a_n x^n = A_k x^k + \sum_k^\infty A_{n+1}(x^{n+1} - x^n).$$

Now, given $\epsilon > 0$, we can choose k so large that $|A_n| < \frac{1}{2}\epsilon$ whenever $n \geq k$. Since $x \in [0, 1]$, we have $x^{n+1} - x^n \leq 0$, so (7.27) yields

$$\left| \sum_k^\infty a_n x^n \right| \leq |A_k| x^k + \sum_k^\infty |A_{n+1}|(x^n - x^{x+1})$$
$$\leq \tfrac{1}{2}\epsilon x^k + \tfrac{1}{2}\epsilon \sum_k^\infty (x^n - x^{n+1}).$$

If $x = 1$, the series on the right vanishes; if $0 \leq x < 1$, it is a telescoping series whose sum is x^k. In either case, we obtain

$$\left| \sum_k^\infty a_n x^n \right| \leq \epsilon x^k \leq \epsilon$$

for all $x \in [0, 1]$ when k is sufficiently large, which establishes the desired uniform convergence. $\qquad \square$

Remark. If $\sum a_n R^n$ converges, we already know (Theorem 7.17) that $\sum a_n x^n$ converges uniformly on $[-r, r]$ for any $r < R$. Combining this with Abel's theorem, we see that $\sum a_n x^n$ converges uniformly on $[-r, R]$. (See Exercise 7 in §7.1.)

The continuity of the series at the endpoint can be restated in the following way. Recall that $\lim_{x \to a-} f(x)$ denotes the limit of $f(x)$ as x approaches a from the left.

7.28 Corollary. *If $\sum_0^\infty a_n$ converges, then $\lim_{x \to 1-} \sum_0^\infty a_n x^n = \sum_0^\infty a_n$.*

EXAMPLE 7. The expansion $\arctan x = \sum_0^\infty (-1)^n x^{2n+1}/(2n+1)$ was established in Example 2 for $|x| < 1$. Since the series also converges at $x = 1$ (by the alternating series test), we obtain a neat series formula for π:

$$\tfrac{1}{4}\pi = \lim_{x \to 1-} \arctan x = \sum_0^\infty \frac{(-1)^n}{2n+1} = 1 - \tfrac{1}{3} + \tfrac{1}{5} - \tfrac{1}{7} + \cdots .$$

The converse of Corollary 7.28 is false: The limit $S = \lim_{x \to 1-} \sum_0^\infty a_n x^n$ may exist even when $\sum_0^\infty a_n$ diverges. (Example: Take $a_n = (-1)^n$; then $\sum_0^\infty a_n x^n = (1+x)^{-1}$ for $|x| < 1$, so $S = \tfrac{1}{2}$.) In this case the series $\sum a_n$ is said to be **Abel summable** to the sum S. Abel summation provides a way of making sense out of certain divergent series that is useful in some situations, one of which we shall discuss in §8.2.

EXERCISES

1. Let $\{a_n\}_0^\infty$ be a sequence of real or complex numbers.
   a. Suppose that $|a_{n+1}/a_n|$ converges to a limit L as $n \to \infty$. Show that the radius of convergence of $\sum_0^\infty a_n x^n$ is L^{-1}.
   b. Suppose that $|a_n|^{1/n}$ converges to a limit L as $n \to \infty$. Show that the radius of convergence of $\sum_0^\infty a_n x^n$ is L^{-1}.

2. Show that if the sequence $\{a_n\}_0^\infty$ is bounded, the radius of convergence of $\sum_0^\infty a_n x^n$ is at least 1.

3. Suppose the radius of convergence of $\sum_0^\infty a_n x^n$ is R. What is the radius of convergence of $\sum_0^\infty a_n x^{kn}$ $(k = 2, 3, 4, \ldots)$?

4. Show that for any sequence $\{a_n\}_0^\infty$, the radius of convergence of $\sum_0^\infty a_n x^n$ is the reciprocal of $\limsup_{n \to \infty} |a_n|^{1/n}$. (See Exercises 9–12 in §1.5 and Exercise 25 in §6.2.)

5. Show that each of the following functions of x admits a power series expansion on some interval centered at the origin. Find the expansion and give its interval of validity.

   a. $\int_0^x e^{-t^2} dt$.

   b. $\int_0^x \cos t^2 \, dt$.

   c. $\int_0^x t^{-1} \log(1 + 2t) \, dt$.

6. Use the series expansions in Exercise 5 to calculate the following integrals to three decimal places, and prove the accuracy of your answer.

   a. $\int_0^1 e^{-t^2} dt$.

   b. $\int_0^1 \cos t^2 \, dt$.

   c. $\int_0^{1/2} t^{-1} \log(1 + 2t) \, dt$.

7. Let $f(x) = \sum_0^\infty a_n x^n$ be a power series with positive radius of convergence. Show that $f(-x) = f(x)$ (resp. $f(-x) = -f(x)$) for all x in the interval of convergence if and only if $a_n = 0$ for all odd n (resp. all even n).

8. Let k be a nonnegative integer. The **Bessel function of order** k is the function J_k defined by

$$J_k(x) = \sum_0^\infty \frac{(-1)^n}{n!(n+k)!} \left[\frac{x}{2}\right]^{2n+k}.$$

   a. Verify that the series defining $J_k(x)$ converges for all x.

   b. Show that $(d/dx)[x^k J_k(x)] = x^k J_{k-1}(x)$.

   c. Show that $(d/dx)[x^{-k} J_k(x)] = -x^{-k} J_{k+1}(x)$.

   d. Show that $u = J_k(x)$ satisfies the differential equation $x^2 u'' + xu' + (x^2 - k^2)u = 0$.

9. Show that the series

$$1 + \frac{x^3}{2 \cdot 3} + \frac{x^6}{2 \cdot 3 \cdot 5 \cdot 6} + \cdots + \frac{x^{3n}}{2 \cdot 3 \cdot 5 \cdot 6 \cdots (3n-1)(3n)} + \cdots$$

converges for all x and that its sum $f(x)$ satisfies $f''(x) = xf(x)$.

10. Express the sums of the following series in terms of elementary functions and (perhaps) their antiderivatives in the manner of Example 4.

   a. $\displaystyle\sum_1^\infty \frac{nx^n}{(n+1)!}$.

   b. $\displaystyle\sum_0^\infty \frac{(-1)^n x^{2n+1}}{(2n+1) \cdot (2n+2)!}$.

   c. $\displaystyle\sum_0^\infty \frac{x^n}{(n+1)^2 n!}$.

   d. $\displaystyle\sum_0^\infty \frac{(-1)^n (2n+1) x^{2n}}{(2n)!}$.

11. Consider the function $f(x) = \int_0^x \arctan t \, dt$.
   a. Perform the integration to evaluate f in terms of elementary functions.
   b. Using the result of Example 2, compute the Taylor series of $f(x)$ (centered at the origin) and show that it converges to $f(x)$ for $x \in [-1, 1]$. (The endpoints require special attention.)
   c. Deduce that

$$1 - \tfrac{1}{2} - \tfrac{1}{3} + \tfrac{1}{4} + \tfrac{1}{5} - \tfrac{1}{6} - \tfrac{1}{7} + \cdots = \tfrac{1}{4}\pi - \tfrac{1}{2}\log 2.$$

7.4   The Complex Exponential and Trig Functions

The series $\sum_0^\infty z^n/n!$ converges absolutely for every *complex* number z, by the ratio test, so we can use it to *define* the exponential function for a complex variable:

$$\exp(z) = e^z = \sum_0^\infty \frac{z^n}{n!} \qquad (z \in \mathbb{C}).$$

This extended exponential function still obeys the basic law of exponents. Indeed, by Theorem 6.29,

$$(7.29) \quad e^z e^w = \sum_{m,n=0}^\infty \frac{z^m w^n}{m!n!} = \sum_{k=0}^\infty \sum_{m+n=k} \frac{z^m w^n}{m!n!} = \sum_{k=0}^\infty \frac{(z+w)^k}{k!} = e^{z+w}.$$

(In the third equality we have used the binomial theorem.)

Let $i = \sqrt{-1}$ be the imaginary unit. Since $i^2 = -1$, we have $i^3 = -i$ and $i^4 = 1$, so

$$i^{4n} = 1, \quad i^{4n+1} = i, \quad i^{4n+2} = -1, \quad i^{4n+3} = -i \qquad (n = 0, 1, 2, \ldots).$$

Therefore, when $z = ix$ is purely imaginary,

$$e^{ix} = \sum_0^\infty \frac{i^n x^n}{n!} = \left(1 - \frac{x^2}{2!} + \frac{x^4}{4!} - \cdots\right) + i\left(x - \frac{x^3}{3!} + \frac{x^5}{5!} - \cdots\right).$$

The series on the right are the Taylor series of $\cos x$ and $\sin x$, so we have arrived at Euler's formula

$$(7.30) \qquad\qquad\qquad e^{ix} = \cos x + i\sin x.$$

This is the appropriate place to raise the issue of the *definition* of $\cos x$ and $\sin x$. These functions are so familiar that we take them entirely for granted, but the definitions presented in elementary trigonometry — as ratios of sides of right triangles, or as the coordinates of the point where the unit circle intersects the ray that makes an angle x with the positive horizontal axis — are quite unsatisfactory, for they provide neither a precise formula nor a computationally effective algorithm. (Think for a minute: How could you possibly use these definitions to calculate $\cos(1)$ to four decimal places?)[1] In fact, the best procedure is to use Taylor series as a definition! That is, we *define* $\cos x$ and $\sin x$ for all real (or, for that matter, complex) numbers x by

$$(7.31) \qquad \cos x = \sum_0^\infty \frac{(-1)^n x^{2n}}{(2n)!}, \qquad \sin x = \sum_0^\infty \frac{(-1)^n x^{2n+1}}{(2n+1)!}.$$

We now indicate how to derive all the familiar properties of the trig functions from these definitions. First, it is clear from (7.31) that

$$(7.32) \qquad \cos(-x) = \cos x, \qquad \sin(-x) = -\sin x,$$

so that $e^{-ix} = \cos x - i\sin x$. Second, termwise differentiation of (7.31) immediately yields

$$(7.33) \qquad \cos' = -\sin, \qquad \sin' = \cos.$$

Third, the addition formulas for sine and cosine follow easily from the law of exponents:

$$\cos(x \pm y) + i\sin(x \pm y) = e^{i(x\pm y)} = e^{ix}e^{\pm iy}$$
$$= (\cos x + i\sin x)(\cos y \pm i\sin y)$$
$$= (\cos x \cos y \mp \sin x \sin y) + i(\sin x \cos y \pm \cos x \sin y).$$

Taking the real and imaginary parts of both sides, we obtain

$$(7.34) \qquad \begin{aligned} \cos(x \pm y) &= \cos x \cos y \mp \sin x \sin y, \\ \sin(x \pm y) &= \sin x \cos y \pm \cos x \sin y. \end{aligned}$$

In particular, we have the Pythagorean identity

$$(7.35) \qquad \cos^2 x + \sin^2 x = \cos(x - x) = \cos 0 = 1.$$

---

[1] A similar problem arises if one tries to define e^x directly. However, here there is an alternative: Define $\log x$ to be $\int_1^x t^{-1}\, dt$ and then define exp to be the inverse function of log. The analogous procedure for developing trig functions, taking the equation $\arcsin x = \int_0^x (1 - t^2)^{-1/2}\, dt$ as a starting point, is less satisfactory, because the inverse function of arcsin is not the whole sine function but only its restriction to the interval $[-\pi/2, \pi/2]$.

Next, we have to bring the number π into play somehow. We can proceed as follows. The series $\sum_0^\infty (-1)^n 2^{2n}/(2n)!$ for $\cos 2$ is an alternating series whose terms decrease in magnitude starting with $n = 1$, so by the alternating series test,

$$\cos 2 = 1 - \frac{2^2}{2!} = -1 \text{ with error less than } \frac{2^4}{4!} = \frac{2}{3}.$$

In particular, $\cos 2 < 0$, and of course $\cos 0 = 1 > 0$, so by the intermediate value theorem there is at least one number $a \in (0, 2)$ such that $\cos a = 0$. Therefore, the set $Z = \{x \geq 0 : \cos x = 0\}$ is nonempty; it is closed since $\cos$ is continuous; hence it contains its greatest lower bound, which is positive since $\cos 0 = 1$. We denote this smallest positive zero of $\cos$ by $\frac{1}{2}\pi$. (Again, this may be taken as a *definition* of the number π, from which its other familiar properties can be derived.)

Now, by (7.33), $(d/dx) \sin x = \cos x > 0$ for $0 \leq x < \frac{1}{2}\pi$, so $\sin$ is increasing on $[0, \frac{1}{2}\pi]$, and $\sin 0 = 0$; hence $\sin \frac{1}{2}\pi > 0$. But by (7.35), $\sin^2 \frac{1}{2}\pi = \sin^2 \frac{1}{2}\pi + \cos^2 \frac{1}{2}\pi = 1$; hence, $\sin \frac{1}{2}\pi = 1$. In summary,

$$(7.36) \qquad \cos 0 = \sin \tfrac{1}{2}\pi = 1, \qquad \sin 0 = \cos \tfrac{1}{2}\pi = 0.$$

All of the familiar formulas of (precalculus) trigonometry can be derived from the even-odd relations (7.32), the addition formulas (7.34), and the special values (7.36), and these together with (7.33) yield all the formulas for integration and differentiation of trigonometric functions. For example, (7.34) and (7.36) yield the complementarity relations

$$(7.37) \qquad \begin{aligned} \cos(\tfrac{1}{2}\pi - x) &= \cos \tfrac{1}{2}\pi \cos x + \sin \tfrac{1}{2}\pi \sin x = \sin x, \\ \sin(\tfrac{1}{2}\pi - x) &= \sin \tfrac{1}{2}\pi \cos x - \cos \tfrac{1}{2}\pi \sin x = \cos x. \end{aligned}$$

These, in turn, yield the 2π-periodicity of sine and cosine. Indeed, replacing x by $-x$ in (7.37) and using (7.32), we see that $\cos(x + \frac{1}{2}\pi) = -\sin x$ and $\sin(x + \frac{1}{2}\pi) = \cos x$, whence

$$\begin{aligned} \cos(x + \pi) &= \cos(x + \tfrac{1}{2}\pi + \tfrac{1}{2}\pi) = -\sin(x + \tfrac{1}{2}\pi) = -\cos x, \\ \sin(x + \pi) &= \sin(x + \tfrac{1}{2}\pi + \tfrac{1}{2}\pi) = \cos(x + \tfrac{1}{2}\pi) = -\sin x, \end{aligned}$$

and therefore

$$\cos(x + 2\pi) = -\cos(x + \pi) = \cos x, \qquad \sin(x + 2\pi) = -\sin(x + \pi) = \sin x.$$

EXERCISES

1. Recall that the hyperbolic sine and cosine functions are defined by $\sinh z = \frac{1}{2}(e^z - e^{-z})$ and $\cosh z = \frac{1}{2}(e^z + e^{-z})$. Here, z may now be taken to be a complex number.

   a. Show that $\sinh ix = i \sin x$ and $\cosh ix = \cos x$.

   b. Show that $\sinh(z+w) = \sinh z \cosh w + \cosh z \sinh w$ and $\cosh(z+w) = \cosh z \cosh w + \sinh z \sinh w$.

   c. Express $\sinh(x + iy)$ and $\cosh(x + iy)$ in terms of real functions of the real variables x and y.

2. Verify that the formula $(d/dx)e^{cx} = ce^{cx}$ remains valid when c is a complex number. (However, x is still a real variable, since we have not discussed differentiation of functions of a complex variable.)

3. Let a and b be real numbers. Compute $\int e^{(a+ib)x}\,dx$ by using the result of Exercise 2; then, by taking real and imaginary parts, deduce the formulas

$$\int e^{ax} \cos bx \, dx = \frac{e^{ax}(a \cos bx + b \sin bx)}{a^2 + b^2},$$

$$\int e^{ax} \sin bx \, dx = \frac{e^{ax}(a \sin bx - b \cos bx)}{a^2 + b^2}.$$

7.5 Functions Defined by Improper Integrals

In the preceding sections we have considered infinite series of functions. The analogue for integrals is an improper integral $\int_c^d f(x,t)\,dt$, where the integrand contains a free variable x as well as the variable of integration and the resulting integral defines a function of x. The integral may be improper because $c = -\infty$ or $d = \infty$ or because of singularities of the function f. To keep the notation simple, we shall restrict our discussion to the case where $d = \infty$ and f has no singularities on $[c, \infty)$, but everything we say extends to the other cases with the obvious modifications.

In this situation, the notion of uniform convergence is as follows: We say that the integral $\int_c^\infty f(x,t)\,dt$ **converges uniformly for** $x \in I$ (where I is an interval in $\mathbb{R}$) if the difference between the "partial integral" $\int_c^d$ and the full integral $\int_c^\infty$ — that is, the "tail end" $\int_d^\infty$ — tends to zero uniformly for $x \in I$ as $d \to \infty$. Precisely, this means that

$$\sup_{x \in I} \left| \int_d^\infty f(x,t)\,dt \right| \to 0 \text{ as } d \to \infty.$$

The most useful test for uniform convergence is the following analogue of the Weierstrass M-test. The proof is essentially identical to that of the M-test, and we leave the details to the reader (Exercise 1).

7.38 Theorem. *Suppose there is a function $g(t) \geq 0$ on $[c, \infty)$ such that (i) $|f(x, t)| \leq g(t)$ for all $x \in I$ and $t \geq c$, and (ii) $\int_c^\infty g(t)\, dt < \infty$. Then $\int_c^\infty f(x, t)\, dt$ converges absolutely and uniformly for $x \in I$.*

The consequences of uniform convergence for continuity, integration, and differentiation of the function $F(x) = \int_c^\infty f(x, t)\, dt$ are much the same as for series. The following two theorems provide analogues of Theorems 7.10 and 7.13 in the present setting.

7.39 Theorem. *Suppose that $f(x, t)$ is a continuous function on the set $\{(x, t) : x \in I,\ t \geq c\}$ and that the integral $\int_c^\infty f(x, t)\, dt$ is uniformly convergent for $x \in I$. Then:*

a. The function $F(x) = \int_c^\infty f(x, t)\, dt$ is continuous on I.
b. If $[a, b] \subset I$, then

$$\int_a^b \int_c^\infty f(x, t)\, dt\, dx = \int_c^\infty \int_a^b f(x, t)\, dx\, dt.$$

Proof. The conclusions are true if $\int_c^\infty$ is replaced by $\int_c^d$ where $d < \infty$, by Theorems 4.46 and 4.26. (a) then follows because the uniform limit of continuous functions is continuous, and (b) follows by the argument in the proof of Theorem 7.11. $\qquad\square$

7.40 Theorem. *Suppose that $f(x, t)$ and its partial derivative $\partial_x f(x, t)$ are continuous functions on the set $\{(x, t) : x \in I,\ t \geq c\}$. Suppose also that the integral $\int_c^\infty f(x, t)\, dt$ converges for $x \in I$ and the integral $\int_c^\infty \partial_x f(x, t)\, dt$ converges uniformly for $x \in I$. Then the former integral is differentiable on I as a function of x, and*

$$\frac{d}{dx} \int_c^\infty f(x, t)\, dt = \int_c^\infty \frac{\partial f}{\partial x}(x, t)\, dt.$$

Theorem 7.40 may be deduced from Theorem 7.39 in much the same way as Theorem 7.12 was deduced from Theorem 7.11 (Exercise 2).

Let us state explicitly the result of combining Theorems 7.39 and 7.40 with Theorem 7.38:

7.41 Theorem. *The conclusions of Theorem 7.39 are valid whenever $|f(x, t)| \leq g(t)$ for all $x \in I$ and $t \geq c$, where $\int_c^\infty g(t)\, dt < \infty$. The conclusions of Theorem 7.40 are valid whenever $\int_c^\infty f(x, t)\, dt$ converges for $x \in I$ and $|\partial_x f(x, t)| \leq g(t)$ for all $x \in I$ and $t \geq c$, where $\int_c^\infty g(t)\, dt < \infty$.*

The manipulation of improper integrals by the foregoing theorems can be quite an entertaining exercise, and it leads to a number of interesting and useful results. Let us look at some examples.

EXAMPLE 1. Evaluate $\int_0^\infty \dfrac{\arctan(bt) - \arctan(at)}{t}\, dt$ where $0 < a < b$.

Solution: We recognize that the integrand is $\int_a^b (x^2t^2 + 1)^{-1} dx$. For $x \ge a$ and $t \ge 0$ we have $(x^2t^2 + 1)^{-1} \le (a^2t^2 + 1)^{-1}$, and $\int_0^\infty (a^2t^2 + 1)^{-1}\, dt < \infty$. Thus, by Theorem 7.38, the integral $\int_0^\infty (x^2t^2+1)^{-1}\, dt$ is uniformly convergent for $x \ge a$, so we can apply Theorem 7.39 to obtain

$$\int_0^\infty \frac{\arctan(bt) - \arctan(at)}{t}\, dt = \int_0^\infty \int_a^b \frac{1}{x^2t^2 + 1}\, dx\, dt$$

$$= \int_a^b \int_0^\infty \frac{1}{x^2t^2 + 1}\, dt\, dx = \int_a^b x^{-1} \arctan xt \Big|_0^\infty = \int_a^b \frac{\pi}{2x}\, dx$$

$$= \frac{\pi}{2} \log\left(\frac{b}{a}\right).$$

EXAMPLE 2. Let

$$F(x) = \int_0^\infty e^{-xt^2}\, dt, \qquad x > 0.$$

Since $(\partial^k/\partial x^k)e^{-xt^2} = (-t^2)^k e^{-xt^2}$, by Theorem 7.40 we can conclude that

$$F^{(k)}(x) = (-1)^k \int_0^\infty t^{2k} e^{-xt^2}\, dt \qquad (x > 0),$$

provided that we establish the uniform convergence of the integral on the right. In fact, the convergence is not uniform on the whole interval $(0, \infty)$, but it is uniform on $[\delta, \infty)$ for any $\delta > 0$, which is sufficient. This follows easily from Theorem 7.38, since $t^{2k}e^{-xt^2} \le t^{2k}e^{-\delta t^2}$ for $x \ge \delta$.

On the other hand, we can evaluate $F(x)$ explicitly by making the substitution $u = x^{1/2}t$ and invoking Proposition 4.66:

$$F(x) = \int_0^\infty e^{-u^2} x^{-1/2}\, du = \frac{\sqrt{\pi}}{2} x^{-1/2},$$

and therefore

$$F^{(k)}(x) = \frac{\sqrt{\pi}}{2} (-\tfrac{1}{2})(-\tfrac{3}{2}) \cdots (-k + \tfrac{1}{2}) x^{-k-(1/2)}.$$

Comparing the two formulas for $F^{(k)}(x)$, we conclude that

$$\int_0^\infty t^{2k} e^{-xt^2}\, dt = (\tfrac{1}{2})(\tfrac{3}{2}) \cdots (k - \tfrac{1}{2}) \frac{\sqrt{\pi}}{2x^{k+(1/2)}}.$$

This result can also be obtained by a laborious k-fold integration by parts ($u = t^{2k-1}$, $dv = te^{-xt^2}\, dt$, etc.), but differentiation under the integral gives a rather painless derivation.

EXAMPLE 3. We now derive one of the most important of all integral formulas:

(7.42)
$$\int_0^\infty \frac{\sin t}{t}\, dt = \frac{\pi}{2}.$$

This is a bit tricky, since the integral is not absolutely convergent. (Note that since $t^{-1} \sin t \to 1$ as $t \to 0$, the integral over $[0, 1]$ is an ordinary proper integral. The convergence of the integral over $[1, \infty)$ was proved in §4.6 [Example 3].) Our strategy will be to consider an improper integral with *two* parameters:

(7.43)
$$F(x, y) = \int_0^\infty \frac{e^{-xt} \sin yt}{t}\, dt \qquad (x > 0,\ y \in \mathbb{R}).$$

Again, this integral is proper at $t = 0$, and for $x > 0$ it is absolutely convergent.

First, we fix $x > 0$ and consider the integral as a function of y. Formal differentiation of (7.43) with respect to y leads to

$$\frac{\partial F}{\partial y} = \int_0^\infty e^{-xt} \cos yt\, dt.$$

By Theorem 7.41, this formula is indeed valid, since $|e^{-xt} \cos yt| \le e^{-xt}$ for all y and $\int_0^\infty e^{-xt}\, dt < \infty$. The integral on the right can be evaluated by elementary calculus (integrate by parts twice, or use Exercise 3 in §7.4), and the result is

$$\frac{\partial F}{\partial y} = e^{-xt} \frac{y \sin yt - x \cos yt}{x^2 + y^2} \Big|_0^\infty = \frac{x}{x^2 + y^2}.$$

Now we can recover F by integrating in y. Obviously $F(x, 0) = 0$, so we get the right constant of integration by starting the integration at 0:

$$F(x, y) = \int_0^y \frac{x}{x^2 + s^2}\, ds = \arctan(y/x).$$

The variable y has now served its purpose, and we henceforth set it equal to 1. We have shown that

(7.44) $$\int_0^\infty \frac{e^{-xt}\sin t}{t}\,dt = \arctan(1/x) \qquad (x > 0).$$

We now wish to let $x \to 0$. In order to pass the limit under the integral sign in (7.44), it is enough to show that the integral in (7.44) is uniformly convergent for $x \geq 0$. Unfortunately, Theorem 7.38 does not apply here, since the integral is not absolutely convergent at $x = 0$. (Theorem 7.38 easily yields the uniform convergence for $x \geq \delta$ for any $\delta > 0$, but that isn't good enough!) Recall the meaning of uniform convergence: What we need to show is that

$$\sup_{x \geq 0} \left| \int_b^\infty \frac{e^{-xt}\sin t}{t}\,dt \right| \to 0 \text{ as } b \to \infty.$$

To this end, we use integration by parts,[2] taking $u = t^{-1}$ and $dv = e^{-xt}\sin t\,dt$; the result is

$$\int_b^\infty \frac{e^{-xt}\sin t}{t}\,dt = \frac{e^{-bx}(x\sin b + \cos b)}{(x^2+1)b} - \int_b^\infty \frac{e^{-xt}(x\sin t + \cos t)}{(x^2+1)t^2}\,dt.$$

Now,

$$\left| \frac{e^{-xt}(x\sin t + \cos t)}{(x^2+1)} \right| \leq \frac{x+1}{x^2+1}.$$

The quantity on the right is continuous on $\mathbb{R}$ and tends to zero as $x \to \infty$, so it is bounded by a constant C for $x \geq 0$. Therefore,

$$\sup_{x \geq 0} \left| \int_b^\infty \frac{e^{-xt}\sin t}{t}\,dt \right| \leq \frac{C}{b} + C\int_b^\infty \frac{dt}{t^2} = \frac{2C}{b},$$

which tends to zero as $b \to \infty$, as desired. Thus the convergence is uniform in (7.44), and it follows that

$$\int_0^\infty \frac{\sin t}{t}\,dt = \lim_{x \to 0+} \int_0^\infty \frac{e^{-xt}\sin t}{t}\,dt = \lim_{x \to 0+} \arctan(1/x) = \frac{\pi}{2}.$$

---

[2]The idea is much the same as the use of summation by parts in the proof of Abel's theorem.

EXERCISES

1. Prove Theorem 7.38.

2. Prove Theorem 7.40.

3. Suppose $x > 0$. Verify that $\int_0^\infty e^{-xt}\, dt = x^{-1}$, justify differentiating under the integral sign, and deduce that $\int_0^\infty t^n e^{-xt}\, dt = n! x^{-n-1}$.

4. Verify that $\int_0^\infty (t^2 + x)^{-1}\, dt = \frac{1}{2}\pi x^{-1/2}$, justify differentiating under the integral sign, and thence evaluate $\int_0^\infty (t^2 + x)^{-n}\, dt$.

5. Show that $\displaystyle\int_0^\infty \frac{e^{-bx} - e^{-ax}}{x}\, dx = \log\frac{a}{b}$ for $a, b > 0$.

6. Show that $\displaystyle\int_0^\infty \frac{e^{-bx} - e^{-ax}}{x} \cos x\, dx = \frac{1}{2}\log\frac{1+a^2}{1+b^2}$ for $a, b > 0$.

7. Show that $\displaystyle\int_0^\infty e^{-x}\frac{1 - \cos ax}{x}\, dx = \frac{1}{2}\log(1+a^2)$ for all $a \in \mathbb{R}$.

8. Deduce from (7.42) that

$$\int_0^\infty \frac{\sin xt}{t}\, dt = \begin{cases} \frac{1}{2}\pi & \text{if } x > 0, \\ 0 & \text{if } x = 0, \\ -\frac{1}{2}\pi & \text{if } x < 0. \end{cases}$$

Show that the convergence is uniform for $x \in I$ if I is any compact interval with $0 \notin I$, but not if $0 \in I$.

9. Use Exercise 8 to show that $\displaystyle\int_0^\infty \frac{\sin^2 xt}{t^2}\, dt = \frac{1}{2}\pi x$ for $x > 0$.

10. Let $I(a, b) = \displaystyle\int_0^\infty \frac{\cos bx - \cos ax}{x^2}\, dx$.
    a. Show that $I(a, b)$ is convergent for all $a, b \in \mathbb{R}$ and that the convergence is uniform for a in any finite interval when b is fixed (or vice versa).
    b. Use Exercise 8 to show that $I(a, b) = \frac{1}{2}\pi(a - b)$ if $a, b > 0$.
    c. Show that $I(a, b) = \frac{1}{2}\pi(|a| - |b|)$ for all $a, b \in \mathbb{R}$.

11. Let $F(x) = \int_0^\infty e^{-t^2}\cos xt\, dt$ for $x \in \mathbb{R}$.
    a. Justify differentiating under the integral sign and thence show that $F'(x) = -\frac{1}{2}xF(x)$.
    b. Show that $F(x) = \frac{1}{2}\sqrt{\pi}e^{-x^2/4}$.

12. Let $G(x) = \int_0^\infty e^{-t^2}\sin xt\, dt$ for $x \in \mathbb{R}$. Proceeding as in Exercise 11, show that $G(x) = e^{-x^2/4}\int_0^x e^{t^2/4}\, dt$.

13. Show that $\displaystyle\int_0^\infty \frac{1 - e^{-xt^2}}{t^2}\, dt = \sqrt{\pi x}$ for $x \geq 0$.

14. Let $F(x) = \int_0^\infty e^{-t^2-(x^2/t^2)}\, dt$.

    a. Show that F is a continuous function on $\mathbb{R}$ that satisfies $F'(x) = -2F(x)$ for $x > 0$ and $F'(x) = 2F(x)$ for $x < 0$.

    b. Show that $F(x) = \frac{1}{2}\sqrt{\pi}\, e^{-2|x|}$.

    c. Evaluate $\int_0^\infty e^{-pt^2-(q/t^2)}\, dt$ for $p, q > 0$.

15. Let f be a continuous function on $[0, \infty)$ that satisfies $|f(x)| \le a(1+x)^N e^{bx}$ for some $a, b, N \ge 0$. The **Laplace transform** of f is the function $L[f]$ defined on (b, ∞) by

$$L[f](s) = \int_0^\infty e^{-sx} f(x)\, dx.$$

    a. Show that $L[f]$ is of class C^∞ on (b, ∞) and $(d/ds)^n L[f] = (-1)^n L[f_n]$ where $f_n(x) = x^n f(x)$.

    b. Suppose that f is of class C^1 on $[0, \infty)$ and that f' satisfies the same sort of exponential growth condition as f. Show that $L[f'](s) = sL[f](s) - f(0)$.

7.6 The Gamma Function

Perhaps the most important of all functions defined by improper integrals is the **gamma function** $\Gamma(x)$ defined for $x > 0$ by

$$(7.45) \qquad\qquad \Gamma(x) = \int_0^\infty t^{x-1} e^{-t}\, dt,$$

which has a way of turning up in many unexpected places. Let us analyze the integrals over $[0, 1]$ and $[1, \infty)$ separately. The integral over $[0, 1]$ is proper for $x \ge 1$ and improper but convergent for $0 < x < 1$. In fact, by Theorem 7.38 it is uniformly convergent for $x \ge \delta$, for any $\delta > 0$, since $0 < t^{x-1} e^{-t} \le t^{\delta-1}$ for $x \ge \delta$ and $0 \le t \le 1$. The integral over $[1, \infty)$ is convergent for all x and uniformly convergent for $x \le C$, for any constant C, since $0 < t^{x-1} e^{-t} \le t^{C-1} e^{-t}$ for $x \le C$ and $t \ge 1$. Therefore, the integral defining $\Gamma(x)$ is convergent for $x > 0$ and uniformly convergent on $\delta \le x \le C$ for any $\delta > 0$ and $C > 0$.

    It follows that Γ is a continuous function on $(0, \infty)$. In fact, Γ is of class C^∞ on $(0, \infty)$, and its derivatives can be calculated by differentiating under the integral:

$$(7.46) \qquad\qquad \Gamma^{(k)}(x) = \int_0^\infty (\log t)^k t^{x-1} e^{-t}\, dt.$$

Since $|\log t|$ grows more slowly than any power of t as $t \to 0$ or $t \to \infty$, the argument of the preceding paragraph shows that the integral on the right is absolutely and uniformly convergent for $\delta \le x \le C$ for any positive δ and C, so Theorem 7.40 guarantees the validity of (7.46).

The most important property of Γ is that it satisfies the **functional equation**

$$(7.47) \qquad \Gamma(x+1) = x\Gamma(x).$$

The proof is a simple integration by parts ($u = t^x$, $dv = e^{-t}\,dt$):

$$\Gamma(x+1) = \int_0^\infty t^x e^{-t}\,dt = -t^x e^{-t}\big|_0^\infty + \int_0^\infty x t^{x-1} e^{-t}\,dt = 0 + x\Gamma(x).$$

There are two values of Γ that can be calculated easily by hand:

$$\Gamma(1) = \int_0^\infty e^{-t}\,dt = -e^{-t}\big|_0^\infty = 1,$$

$$\Gamma(\tfrac{1}{2}) = \int_0^\infty t^{-1/2} e^{-t}\,dt = 2\int_0^\infty e^{-u^2}\,du = \sqrt{\pi}.$$

(For the second one we set $u = \sqrt{t}$ and used Proposition 4.66.) The functional equation (7.47) now yields the values of Γ at all positive integers and half-integers:

$$\Gamma(2) = 1\Gamma(1) = 1, \quad \Gamma(3) = 2\Gamma(2) = 2!, \quad \Gamma(4) = 3\Gamma(3) = 3!, \ldots$$
$$\Gamma(\tfrac{3}{2}) = \tfrac{1}{2}\Gamma(\tfrac{1}{2}) = \tfrac{1}{2}\sqrt{\pi}, \quad \Gamma(\tfrac{5}{2}) = \tfrac{3}{2}\Gamma(\tfrac{3}{2}) = \tfrac{3}{2}\cdot\tfrac{1}{2}\sqrt{\pi},$$

and so by induction,

$$(7.48) \qquad \Gamma(n) = (n-1)!, \qquad \Gamma(n+\tfrac{1}{2}) = (n-\tfrac{1}{2})\cdots\tfrac{3}{2}\cdot\tfrac{1}{2}\sqrt{\pi}.$$

Thus the gamma function provides an extension of the factorial function to non-integers: $x! = \Gamma(x+1)$, if you like. It is the *natural* extension of the factorial function, not just because it gives the right values at the integers, but because the functional equation $\Gamma(x+1) = x\Gamma(x)$ is the natural generalization of the recursive formula $n! = n\cdot(n-1)!$ that defines factorials.

Other factorial-like products — more precisely, products of numbers in an arithmetic progression — can also be expressed in terms of the gamma function. Indeed, since

$$\Gamma(c+n+1) = (c+n)\Gamma(c+n) = \cdots = (c+n)(c+n-1)\cdots c\Gamma(c),$$

for $a, b > 0$ we have
(7.49)

$$a(a+b)\cdots(a+nb) = b^{n+1}\left(\frac{a}{b}\right)\left(\frac{a}{b}+1\right)\cdots\left(\frac{a}{b}+n\right) = b^{n+1}\frac{\Gamma(\frac{a}{b}+n+1)}{\Gamma(\frac{a}{b})}.$$

The functional equation, written in the form

$$\Gamma(x) = \frac{\Gamma(x+1)}{x},$$

shows that $\Gamma(x)$ blows up like x^{-1} as $x \to 0$. It also provides a way of extending the gamma function to negative values of x. Indeed, the expression on the right is defined for all $x > -1$ except $x = 0$, and it can be taken as a *definition* of $\Gamma(x)$ for $-1 < x < 0$. Once this has been done, $\Gamma(x+1)/x$ is defined for all $x > -2$ except $x = 0, -1$, and it can be taken as a definition of $\Gamma(x)$ for $-2 < x < -1$. Proceeding inductively, we eventually obtain a definition of $\Gamma(x)$ for all x except the nonpositive integers, where $\Gamma(x)$ blows up. In more explicit form, it is

$$(7.50) \qquad \Gamma(x) = \frac{\Gamma(x+n)}{x(x+1)\cdots(x+n-1)} \qquad (x > -n).$$

This extended gamma function still satisfies the functional equation (7.47), more or less by definition, and (7.49) remains valid provided that a/b is not a nonpositive integer.

The qualitative behavior of the gamma function for $x > 0$ can be analyzed as follows: Since $\Gamma(1) = \Gamma(2) = 1$, there is a critical point x_0 in the interval $(1, 2)$ by Rolle's theorem. On the other hand, from (7.46) it is clear that $\Gamma''(x) > 0$ for $x > 0$, so that $\Gamma'(x)$ is strictly increasing. It follows that Γ is decreasing for $0 < x < x_0$ and increasing for $x > x_0$; in particular, it has a minimum at x_0. Also, it tends to ∞ as $x \to 0$ or $x \to \infty$, so its graph is roughly U-shaped. The behavior for $x < 0$ can then be deduced from (7.50). The graph of Γ is sketched in Figure 7.3.

A number of useful integrals can be transformed into the integral defining $\Gamma(x)$ by a change of variables. We single out two particularly useful ones, obtained by setting $u = bt$ and $v = t^2$, respectively:

$$(7.51) \qquad \int_0^\infty t^{x-1} e^{-bt}\, dt = \int_0^\infty \left[\frac{u}{b}\right]^{x-1} e^{-u} \frac{du}{b} = b^{-x}\Gamma(x) \qquad (b > 0),$$

$$(7.52) \qquad \int_0^\infty t^{2x-1} e^{-t^2}\, dt = \int_0^\infty v^{(2x-1)/2} e^{-v} \frac{dv}{2v^{1/2}} = \tfrac{1}{2}\Gamma(x).$$

There is another important integral related to the gamma function, the so-called **beta function**

$$(7.53) \qquad B(x, y) = \int_0^1 t^{x-1}(1-t)^{y-1}\, dt \qquad (x, y > 0).$$

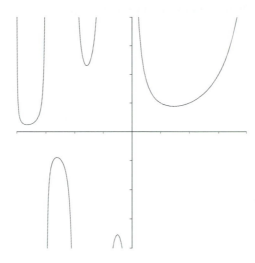

FIGURE 7.3: Graph of the equation $y = \Gamma(x)$, $-4 < x < 4$. (The lines $x = -k$, $k = 0, 1, 2, \ldots$, are vertical asymptotes.)

Since the integrand is approximately equal to t^{x-1} for t near 0 and to $(1 - t)^{y-1}$ for t near 1, the integral is proper when $x, y \geq 1$ and convergent for $x, y > 0$. Like the gamma function, the beta function can be expressed in a number of different forms by changes of variable in the integral. Other than (7.53), the most important of these is obtained by the substitution $t = \sin^2 \theta$, which makes $1 - t = \cos^2 \theta$ and $dt = 2 \sin \theta \cos \theta \, d\theta$, so that

$$(7.54) \qquad B(x, y) = 2 \int_0^{\pi/2} \sin^{2x-1} \theta \cos^{2y-1} \theta \, d\theta.$$

The relation between the gamma and beta functions is as follows:

7.55 Theorem. *For $x, y > 0$, $B(x, y) = \dfrac{\Gamma(x)\Gamma(y)}{\Gamma(x + y)}$.*

Proof. We employ the same device that we used to calculate $\int_{-\infty}^{\infty} e^{-x^2} \, dx$ in §4.7: We express $\Gamma(x)$ and $\Gamma(y)$ by (7.52), write $\Gamma(x)\Gamma(y)$ as an iterated integral, convert the latter to a double integral, and switch to polar coordinates:

$$\Gamma(x)\Gamma(y) = 4 \int_0^\infty t^{2x-1} e^{-t^2} \, dt \int_0^\infty s^{2y-1} e^{-s^2} \, ds$$

$$= 4 \int_0^\infty \int_0^\infty s^{2y-1} t^{2x-1} e^{-s^2-t^2} \, ds \, dt$$

$$= 4 \int_0^{\pi/2} \int_0^\infty (r\cos\theta)^{2y-1} (r\sin\theta)^{2x-1} e^{-r^2} r \, dr \, d\theta$$

$$= 4 \int_0^{\pi/2} \cos^{2y-1}\theta \sin^{2x-1}\theta \, d\theta \int_0^\infty r^{2x+2y-1} e^{-r^2} \, dr$$

$$= B(x,y)\Gamma(x+y).$$

In the last step we have used (7.52) and (7.54).   □

   We draw two useful consequences from Theorem 7.55. The first one is another functional equation for the gamma function; the second one compares the growth of $\Gamma(x)$ and $\Gamma(x+a)$ as $x \to \infty$.

7.56 Theorem (The Duplication Formula). $\Gamma(2x) = \pi^{-1/2} 2^{2x-1} \Gamma(x)\Gamma(x + \frac{1}{2})$.

Proof. Assume that $x > 0$. By taking $y = x$ in Theorem 7.55 and observing that the function $t(1-t)$ is symmetric about $t = \frac{1}{2}$, we see that

$$\frac{\Gamma(x)^2}{\Gamma(2x)} = \int_0^1 [t(1-t)]^{x-1} \, dt = 2 \int_0^{1/2} [t(1-t)]^{x-1} \, dt.$$

By the substitution

$$t = \tfrac{1}{2}(1 - s^{1/2}), \qquad dt = -\tfrac{1}{4} s^{-1/2} \, ds, \qquad t(1-t) = \tfrac{1}{4}(1-s)$$

and another application of Theorem 7.55, we obtain

$$\frac{\Gamma(x)^2}{\Gamma(2x)} = 2^{1-2x} \int_0^1 s^{-1/2}(1-s)^{x-1} \, ds = 2^{1-2x} \frac{\Gamma(\frac{1}{2})\Gamma(x)}{\Gamma(x + \frac{1}{2})}.$$

Since $\Gamma(\frac{1}{2}) = \pi^{1/2}$, the result follows. The extension to negative values of x is left to the reader (Exercise 6).   □

7.57 Theorem. *For $a > 0$,* $\displaystyle\lim_{x\to\infty} \frac{\Gamma(x+a)}{x^a \Gamma(x)} = 1.$

Proof. By Theorem 7.55, the substitution $t = e^{-u}$, and formula (7.51),

$$\frac{\Gamma(x)\Gamma(a)}{\Gamma(x+a)} = \int_0^1 t^{x-1}(1-t)^{a-1}\,dt = \int_0^\infty (1-e^{-u})^{a-1}e^{-xu}\,du.$$

When x is large, e^{-xu} is very small unless u is close to 0, and in that case $1-e^{-u}$ is approximately u. Hence, the integral on the right should be approximately equal to $\int_0^\infty u^{a-1}e^{-xu}\,du = x^{-a}\Gamma(a)$, which is what we are trying to show. More precisely, we have

$$\frac{\Gamma(x)\Gamma(a)}{\Gamma(x+a)} = \int_0^\infty u^{a-1}e^{-xu}\,du + \int_0^\infty \left[(1-e^{-u})^{a-1} - u^{a-1}\right]e^{-xu}\,du$$

$$= x^{-a}\Gamma(a) + \int_0^\infty \left[(1-e^{-u})^{a-1} - u^{a-1}\right]e^{-xu}\,du.$$

Multiplying both sides by $x^a/\Gamma(a)$, we obtain

(7.58) $$\frac{x^a\Gamma(x)}{\Gamma(x+a)} - 1 = \frac{x^a}{\Gamma(a)}\int_0^\infty \left[\left(\frac{1-e^{-u}}{u}\right)^{a-1} - 1\right]u^{a-1}e^{-xu}\,du.$$

It remains to show that the quantity on the right tends to zero as $x \to \infty$.

The function defined by $f(u) = (1 - e^{-u})/u$ for $u \neq 0$ and $f(0) = 1$ is everywhere positive and of class C^∞ (even at $u = 0$, for it is the sum of the power series $\sum_1^\infty (-1)^{n-1}u^{n-1}/n!$). Hence the same is true of $f(u)^{a-1}$, so the function $g(u) = f(u)^{a-1} - 1$ is smooth and vanishes at $u = 0$. By the mean value theorem, then, for $0 \leq u \leq 1$ we have $|g(u)| = |g(u) - g(0)| \leq Cu$ where C is the maximum value of $|g'(u)|$ on $[0,1]$. On the other hand, for $u > 1$ we clearly have $0 < f(u) < 1$ and hence $-1 < g(u) < 0$. Therefore, the quantity on the right of (7.58) is bounded in absolute value by

$$\frac{x^a}{\Gamma(a)}\left[\int_0^1 Cu^a e^{-xu}\,du + \int_1^\infty u^{a-1}e^{-xu}\,du\right]$$

$$\leq \frac{x^a}{\Gamma(a)}\left[C\int_0^1 u^a e^{-xu}\,du + \int_1^\infty u^a e^{-xu}\,du\right] \leq (C+1)\frac{x^a}{\Gamma(a)}\int_0^\infty u^a e^{-xu}\,du$$

$$= (C+1)\frac{\Gamma(a+1)}{x\Gamma(a)} = (C+1)\frac{a}{x},$$

where we have used (7.51) again in the last step. In short, the right side of (7.58) is dominated by x^{-1} as $x \to \infty$, so we are done. $\square$

Theorem 7.57 can be used as an effective alternative to Raabe's test to decide the convergence of series involving quotients of factorial-like products, for such quotients can be expressed as quotients of gamma functions by (7.49).

EXAMPLE 1. Let us reconsider Example 7 from §6.2, namely,

$$\sum_{1}^{\infty} \frac{1 \cdot 4 \cdot 7 \cdots (3n+1)}{n^2 3^n n!}.$$

Since

$$1 \cdot 4 \cdot 7 \cdots (3n+1) = 3^n \left[\tfrac{4}{3} \cdot \tfrac{7}{3} \cdots (n+\tfrac{1}{3})\right] = 3^n \frac{\Gamma(n+\tfrac{4}{3})}{\Gamma(\tfrac{4}{3})},$$

and $n! = \Gamma(n+1)$, the nth term of the series is

$$\frac{\Gamma(n+\tfrac{4}{3})}{n^2 \Gamma(\tfrac{4}{3}) \Gamma(n+1)}.$$

By Theorem 7.57, $\Gamma(n+\tfrac{4}{3})/\Gamma(n+1)$ is approximately $n^{1/3}$ when n is large, so the series converges by comparison to $\sum n^{-5/3}$.

EXERCISES

1. Prove the duplication formula for the case where x is a positive integer simply by using (7.48).

2. Show that for $a, b > 0$,

$$\int_0^1 \left(\log \frac{1}{t}\right)^{a-1} dt = \Gamma(a), \qquad \int_0^1 \left(\log \frac{1}{t}\right)^{a-1} t^{b-1} dt = b^{-a} \Gamma(a).$$

3. Evaluate the following integrals:
   a. $\int_0^\infty x^4 e^{-x^2} \, dx$.
   b. $\int_0^\infty e^{-3x} \sqrt{x} \, dx$.
   c. $\int_0^\infty x^9 e^{-x^4} \, dx$.

4. Prove the following identities directly from the definition (7.53) (without using Theorem 7.55):
   a. $B(x, y) = B(y, x)$.
   b. $B(x, 1) = x^{-1}$.

    c. $B(x+1, y) + B(x, y+1) = B(x, y)$.

    d. $B(x, y) = \int_0^\infty (1+t)^{-x-y} t^{y-1}\, dt$.

5. Given $a, b, c > 0$, evaluate $\int_0^1 x^a (1 - x^b)^c\, dx$ in terms of the gamma function.

6. Use the functional equation (7.47) to show that if the duplication formula (7.56) is valid for a particular value of x, then it is also true for $x - 1$. Thence show how to deduce its validity for all x from its validity for $x > 0$. (In case x is a nonpositive integer or half-integer, the formula is valid in the sense that both sides are infinite.)

7. Use (7.54), Theorem 7.55, and (7.48) to evaluate $\int_0^{\pi/2} \sin^k x\, dx$. (The form of the answer is different depending on whether k is even or odd.)

8. Prove *Wallis's formula*:

$$\frac{\pi}{2} = \lim_{n\to\infty} \frac{2 \cdot 2 \cdot 4 \cdot 4 \cdot 6 \cdot 6 \cdots (2n)(2n)}{1 \cdot 3 \cdot 3 \cdot 5 \cdot 5 \cdot 7 \cdots (2n-1)(2n+1)}.$$

(Hint: Denote the fraction on the right by c_n. Use Exercise 7 and the fact that $\sin^{2n+1} x < \sin^{2n} x < \sin^{2n-1} x$ for $0 < x < \frac{1}{2}\pi$ to show that $c_n < \frac{1}{2}\pi < (2n+1)c_n/2n$.)

9. Suppose f is a continuous function on $[0, \infty)$. For $\alpha > 0$, define the function $I_\alpha[f]$ on $[0, \infty)$ by

$$I_\alpha[f](x) = \frac{1}{\Gamma(\alpha)} \int_0^x (x - t)^{\alpha-1} f(t)\, dt.$$

$I_\alpha[f]$ is called the αth-order **fractional integral** of f.

    a. Show that the derivative of $I_{\alpha+1}[f]$ is $I_\alpha[f]$ for $\alpha > 0$, and the derivative of $I_1[f]$ is f. (This generalizes Exercise 6 in §4.5.)

    b. Show that $I_\alpha[I_\beta[f]] = I_{\alpha+\beta}[f]$ for all $\alpha, \beta > 0$.

10. Test the following series for convergence in the manner of Example 1.

    a. $\displaystyle\sum_0^\infty \frac{1 \cdot 4 \cdots (3n+1)}{2 \cdot 5 \cdots (3n+2)}$

    b. $\displaystyle\sum_0^\infty \frac{4^n n!}{5 \cdot 9 \cdots (4n+5)}$

11. Show that $\displaystyle\sum_1^\infty \left[\frac{1 \cdot 3 \cdots (2n-1)}{2 \cdot 4 \cdots (2n)}\right]^p$ converges if and only if $p > 2$. (Try both Theorem 7.57 and Raabe's test; you'll find that the latter doesn't work in the borderline case $p = 2$.)

12. Suppose $a, b, c > 0$. Show that $\displaystyle\sum_{0}^{\infty} \frac{\Gamma(a+n)\Gamma(b+n)}{\Gamma(c+n)n!}$ converges if and only if $a + b < c$.

7.7 Stirling's Formula

Stirling's formula is a simple and useful approximation to $\Gamma(x)$ for large x; in particular, it tells precisely how rapidly $\Gamma(x)$ grows as $x \to \infty$.

We begin by analyzing the case where x is an integer $n + 1$, for which $\Gamma(x) = n!$. First, observe that

$$\log(n!) = \log 1 + \log 2 + \cdots + \log n.$$

The sum on the right suggests a Riemann sum for $\int \log x \, dx$. Indeed, it is the midpoint Riemann sum for $\int_{1/2}^{n+(1/2)} \log x \, dx$ corresponding to a division into n equal subintervals, so the latter integral provides an approximation to $\log(n!)$. In more detail, using this Riemann sum means taking $\log k$ as an approximation to

$$\int_{k-(1/2)}^{k+(1/2)} \log x \, dx = \int_{-1/2}^{1/2} \log(k + x) \, dx.$$

To see how good this approximation is, we approximate $\log(k + x)$ by its tangent line at $x = 0$ and use Taylor's theorem to estimate the error:

$$\log(k + x) = \log k + \frac{x}{k} + E_k(x), \qquad |E_k(x)| \le \sup_{|t| \le |x|} \frac{1}{(k + t)^2} \frac{x^2}{2!}.$$

(Here $(k+t)^{-2}$ is the absolute value of the second derivative of $\log(k+t)$.) Clearly, for $|x| \le \frac{1}{2}$ and $k \ge 1$ we have

$$|E_k(x)| \le \frac{1}{8(k - \frac{1}{2})^2} \le \frac{1}{8(\frac{1}{2}k)^2} \le \frac{1}{2k^2}.$$

Hence,

$$\int_{-1/2}^{1/2} \log(k + x) \, dx = \int_{-1/2}^{1/2} [\log k + k^{-1}x + E_k(x)] \, dx = \log k + c_k,$$

where

(7.59) $$|c_k| = \left| \int_{-1/2}^{1/2} E_k(x) \, dx \right| \le \frac{1}{2k^2}.$$

Adding these equalities up from $k = 1$ to $k = n$, we obtain

$$\int_{1/2}^{n+(1/2)} \log x \, dx = \log(n!) + \sum_{1}^{n} c_k.$$

On the other hand,

$$\int_{1/2}^{n+(1/2)} \log x \, dx = [x \log x - x]_{1/2}^{n+(1/2)} = (n + \tfrac{1}{2}) \log(n + \tfrac{1}{2}) - n - \tfrac{1}{2} \log \tfrac{1}{2}$$

$$= (n + \tfrac{1}{2}) \log n - n + (n + \tfrac{1}{2}) \log \left(\frac{n + \frac{1}{2}}{n} \right) - \tfrac{1}{2} \log \tfrac{1}{2}.$$

Therefore,

$$\log(n!) - (n + \tfrac{1}{2}) \log n + n = (n + \tfrac{1}{2}) \log(1 + (2n)^{-1}) - \tfrac{1}{2} \log \tfrac{1}{2} - \sum_{1}^{n} c_k.$$

Since $\log(1 + x) \approx x$ for x near 0, as $n \to \infty$ the quantity on the right approaches the constant $\tfrac{1}{2} - \tfrac{1}{2} \log \tfrac{1}{2} - \sum_{1}^{\infty} c_k$, where the series converges by the estimate (7.59). Exponentiating both sides, we obtain a preliminary version of Stirling's formula:

7.60 Lemma. *As $n \to \infty$, $\dfrac{n!}{n^{n+(1/2)} e^{-n}}$ approaches a limit $L \in (0, \infty)$.*

Since $\Gamma(n) = n!/n$, Lemma 7.60 says that $\Gamma(n)/(n^{n-(1/2)} e^{-n}) \to L$ as $n \to \infty$. We now extend this result from integers n to real numbers x. To do so we need a slight strengthening of Theorem 7.57, namely, the uniformity of the convergence with respect to the parameter a.

7.61 Lemma. *For any $A > 0$, $\displaystyle\sup_{0 \le a \le A} \left| \frac{x^a \Gamma(x)}{\Gamma(x + a)} - 1 \right| \to 0$ as $x \to \infty$.*

Proof. With $g(u) = f(u)^{a-1} - 1$ as in the proof of Theorem 7.57, the function $|g'(u)| = |(a - 1)f(u)^{a-2} f'(u)|$ is jointly continuous in a and u in the compact region $a \in [0, A]$, $u \in [0, 1]$, so its maximum on this region is finite. The constant C in that proof can be taken to be this maximum when $a \in [0, A]$, and the conclusion of the proof shows that

$$\sup_{0 \le a \le A} \left| \frac{x^a \Gamma(x)}{\Gamma(x + a)} - 1 \right| \le \frac{(C + 1)A}{x},$$

which yields the desired result. $\qquad\square$

7.62 Lemma. $\lim\limits_{x\to\infty}\dfrac{\Gamma(x)}{x^{x-(1/2)}e^{-x}}=L$, *where L is as in Lemma 7.60.*

Proof. Any number $x\geq 1$ can be written as $x=n+a$ where n is a positive integer and $0\leq a<1$, so that

$$\frac{\Gamma(x)}{x^{x-(1/2)}e^{-x}}=\frac{\Gamma(n+a)}{(n+a)^{n+a-(1/2)}e^{-n-a}}$$

$$=\left[\frac{\Gamma(n)}{n^{n-(1/2)}e^{-n}}\right]\left[\frac{\Gamma(n+a)}{n^a\Gamma(n)}\right]\left[e^a\left(\frac{n+a}{n}\right)^{-n-a+(1/2)}\right].$$

By Lemma 7.61, the first factor in this last expression will be as close to L as we please when n is sufficiently large. By Lemma 7.62, the second factor will be as close to 1 as we please when n is sufficiently large and $0\leq a\leq 1$. The same is also true of the third factor; indeed, by taking logarithms it is enough to verify that

$$\left|a-(n+a-\tfrac{1}{2})\log\left(1+\frac{a}{n}\right)\right|$$

will be as close to 0 as we please when n is sufficiently large and $0\leq a<1$, and this is easily accomplished by using the Taylor expansion of $\log(1+t)$ about $t=0$. (Details are left to the reader as Exercise 1.) Combining these results, we see that $\Gamma(x)/x^{x-(1/2)}e^{-x}$ becomes as close to L as we please when x is sufficiently large, as claimed. $\qquad\square$

7.63 Theorem (Stirling's Formula). $\lim\limits_{x\to\infty}\dfrac{\Gamma(x)}{x^{x-(1/2)}e^{-x}}=\sqrt{2\pi}.$

Proof. It remains only to identify the constant L in Lemma 7.62. According to that lemma, the quantities

$$\frac{\Gamma(x)}{x^{x-(1/2)}e^{-x}},\qquad\frac{\Gamma(x+\frac{1}{2})}{(x+\frac{1}{2})^x e^{-x-(1/2)}},\qquad\frac{\Gamma(2x)}{(2x)^{2x-(1/2)}e^{-2x}}$$

all approach L as $x\to\infty$. Dividing the product of the first two by the third and using the duplication formula

$$\frac{\Gamma(x)\Gamma(x+\frac{1}{2})}{\Gamma(2x)}=2^{1-2x}\sqrt{\pi},$$

we see that

$$L=\lim_{x\to\infty}\frac{\Gamma(x)}{x^{x-(1/2)}e^{-x}}\cdot\frac{\Gamma(x+\frac{1}{2})}{(x+\frac{1}{2})^x e^{-x-(1/2)}}\cdot\frac{(2x)^{2x-(1/2)}e^{-2x}}{\Gamma(2x)}$$

$$=\lim_{x\to\infty}\sqrt{2\pi e}\left(1+\frac{1}{2x}\right)^{-x}.$$

The last factor on the right tends to $e^{-1/2}$ as $x\to\infty$, so we are done. $\qquad\square$

Stirling's formula is often written as

$$\Gamma(x) \sim \sqrt{2\pi}\, x^{x-(1/2)} e^{-x},$$

where $\sim$ means that the ratio of the quantities on the left and right approaches 1 as $x \to \infty$. (The *difference* of these two quantites, however, tends to ∞ along with x.)

EXERCISES

1. Complete the proof of Lemma 7.62 by showing that for some constant $C > 0$ we have $\sup_{0 \le a \le 1} \left| a - (n + a - \frac{1}{2}) \log[1 + (a/n)] \right| \le C/n$.

2. If a fair coin is tossed $2n$ times, the probability that it will come up heads exactly n times is $(2n)!/(n!)^2 2^{2n}$. (The total number of possible outcomes is 2^{2n}, and the number of those with exactly n heads is the binomial coefficient $\binom{2n}{n} = (2n)!/(n!)^2$.) Use Stirling's formula to estimate this probability when n is large.

3. Stirling's formula for factorials,

$$\lim_{n \to \infty} \frac{n!}{n^{n+(1/2)} e^{-n}} = \sqrt{2\pi},$$

can be proved more simply than the general case. One begins, as we did, by proving Lemma 7.60, but it is then enough to evaluate the constant L there. To do this, show that the fraction on the right of Wallis's formula (Exercise 8 in §7.6) equals $[2^n n!]^4/[(2n)!]^2(2n+1)$, then use Lemma 7.60 to show that it approaches $\frac{1}{4}L^2$ as $n \to \infty$; conclude that $L = \sqrt{2\pi}$.

Chapter 8

FOURIER SERIES

Fourier series are infinite series that use the trigonometric functions $\cos n\theta$ and $\sin n\theta$, or, equivalently, $e^{in\theta}$ and $e^{-in\theta}$, as the basic building blocks, in the same way that power series use the monomials x^n. They are a basic tool for analyzing periodic functions, and they therefore have applications in the study of physical phenomena that are periodic in time (such as circular or oscillatory motion) or in space (such as crystal lattices). They can also be used to analyze functions defined on finite intervals in ways that are useful in solving differential equations, and this leads to many other applications in physics and engineering. The theory of Fourier series and its ramifications is an extensive subject that lies at the heart of much of modern mathematical analysis. Here we present only the basics; for further information we refer the reader to Folland [6], Kammler [10], and Körner [11].

8.1 Periodic Functions and Fourier Series

A function f on $\mathbb{R}$ is called **periodic** with period P, or P-periodic for short, if $f(x + P) = f(x)$ for all x. In this case, f is completely determined by its values on any interval $[a, a + P)$ of length P, including one but not both of the endpoints; conversely, any function f defined on an interval $[a, a + P)$ can be extended in a unique way to be a periodic function on $\mathbb{R}$, by declaring that $f(x + nP) = f(x)$ for all $x \in [a, a + P)$ and all integers n. This correspondence between functions on intervals and periodic functions on $\mathbb{R}$ will be useful later; for the time being, we focus our attention on periodic functions.

Unlike power series, Fourier series can be used to represent functions that are quite irregular. To keep the discussion reasonably simple, however, we shall make a standing assumption that *all functions under consideration are piecewise continu-*

ous. By this we mean, precisely, the following: A function f defined on an interval $[a, b]$ is **piecewise continuous** on $[a, b]$ if it is continuous except at finitely many points in $[a, b]$, and at each such point the one-sided limits

$$(8.1) \qquad f(x+) = \lim_{\epsilon \to 0+} f(x + \epsilon), \qquad f(x-) = \lim_{\epsilon \to 0+} f(x - \epsilon)$$

exist (and are finite). Moreover, we shall say that a P-periodic function f on $\mathbb{R}$ is piecewise continuous if it is piecewise continuous on each interval of length P. (If it is piecewise continuous on one such interval, of course, it is piecewise continuous on all of them.)

Note. It is sometimes convenient to allow a piecewise continuous function to be undefined at the points where it has jumps. This does not affect anything that follows in a significant way.

A piecewise continuous function is integrable over every bounded interval in its domain. In this connection, the following elementary fact is worth pointing out explicitly: If f is P-periodic and piecewise continuous, the integrals of f over all intervals of length P are equal:

$$(8.2) \qquad \int_a^{a+P} f(x) \, dx = \int_0^P f(x) \, dx \text{ for every } a \in \mathbb{R}.$$

The proof is left to the reader (Exercise 9).

By making the change of variable $\theta = 2\pi x/P$, we can convert any P-periodic function into a 2π-periodic function. Namely, if $f(x + P) = f(x)$ and we set $g(\theta) = f(x) = f(P\theta/2\pi)$, then $g(\theta + 2\pi) = g(\theta)$. We may therefore restrict attention to the case where the period is 2π, and we shall generally denote the independent variable by θ. There is no presumption that θ denotes an angle, however; it is just a convenient name for a real variable.

The basic idea of Fourier analysis is that an arbitrary piecewise continuous 2π-periodic function $f(\theta)$ can be expanded as an infinite linear combination of the functions $e^{in\theta}$ ($n = 0, \pm 1, \pm 2, \ldots$), or equivalently of the functions $\cos n\theta$ and $\sin n\theta$ ($n = 0, 1, 2, \ldots$). In terms of the functions $e^{in\theta}$, this expansion takes the form

$$(8.3) \qquad f(\theta) = \sum_{-\infty}^{\infty} c_n e^{in\theta}.$$

Here f may be either real-valued or complex-valued; the c_n's are complex numbers, and the series on the right is *always* to be interpreted as the limit of the symmetric partial sums in which the nth and $(-n)$th terms are added in together:

$$\sum_{-\infty}^{\infty} c_n e^{in\theta} = \lim_{k \to \infty} \sum_{-k}^{k} c_n e^{in\theta}.$$

Since $e^{\pm in\theta} = \cos n\theta \pm i \sin n\theta$, combining the nth and $(-n)$th terms gives

$$c_n e^{in\theta} + c_{-n} e^{-in\theta} = a_n \cos n\theta + b_n \sin n\theta,$$

where

$$a_n = c_n + c_{-n}, \qquad b_n = i(c_n - c_{-n}).$$

Therefore, (8.3) can be rewritten as

$$(8.4) \qquad f(\theta) = \tfrac{1}{2}a_0 + \sum_{1}^{\infty}(a_n \cos n\theta + b_n \sin n\theta).$$

The grouping of the nth and $(-n)$th terms in (8.4) corresponds to the grouping of the $\cos n\theta$ and $\sin n\theta$ terms in (8.4). (The factor of $\tfrac{1}{2}$ in front of a_0 is an artifact of the definition $a_0 = c_0 + c_{-0} = 2c_0$.)

The series (8.3) and (8.4) can be used interchangeably. The more traditional form is (8.4), but each of them has its advantages. The advantages of (8.4) derive from the fact that $\cos n\theta$ and $\sin n\theta$ are real-valued and are respectively even and odd; the advantages of (8.3) derive from the fact that exponentials tend to be easier to manipulate than trig functions. For developing the basic theory, the latter consideration is compelling, so we shall work mostly with (8.3).

The questions that face us are as follows: Given a 2π-periodic function f, can it be expanded in a series of the form (8.3)? If so, how do we find the coefficients c_n in this series? It turns out to be easier to tackle the second question first. That is, we first assume that f can be expressed in the form (8.3) and figure out what the coefficients c_n must be; then we show that with this choice of c_n, the expansion (8.3) is actually valid under suitable hypotheses on f.

Suppose, then, that the series $\sum_{-\infty}^{\infty} c_n e^{in\theta}$ converges pointwise to the function $f(\theta)$, and suppose also that the convergence is sufficiently well behaved that term-by-term integration is permissible. The coefficients c_n can then be evaluated by the following device. To compute c_k, we multiply both sides of (8.3) by $e^{-ik\theta}$ and integrate over $[-\pi, \pi]$:

$$\int_{-\pi}^{\pi} f(\theta)e^{-ik\theta}\, d\theta = \sum_{-\infty}^{\infty} c_n \int_{-\pi}^{\pi} e^{i(n-k)\theta}\, d\theta.$$

Now,

$$(8.5) \qquad \int_{-\pi}^{\pi} e^{i(n-k)\theta}\, d\theta = \begin{cases} [i(n-k)]^{-1} e^{i(n-k)\theta}\big|_{-\pi}^{\pi} = 0 & \text{if } n \neq k, \\ \theta\big|_{-\pi}^{\pi} = 2\pi & \text{if } n = k. \end{cases}$$

Thus all the terms on the right of the integrated series vanish except for the one with $n = k$, and we obtain

$$\int_{-\pi}^{\pi} f(\theta)e^{-ik\theta}\, d\theta = 2\pi c_k,$$

or, relabeling k as n,

(8.6)
$$c_n = \frac{1}{2\pi} \int_{-\pi}^{\pi} f(\theta)e^{-in\theta}\, d\theta.$$

This is the promised formula for the coefficients c_n. The corresponding formula for a_n and b_n in (8.4) follows immediately:
(8.7)

$$a_n = c_n + c_{-n} = \frac{1}{2\pi} \int_{-\pi}^{\pi} f(\theta)[e^{-in\theta} + e^{in\theta}]\, d\theta = \frac{1}{\pi} \int_{-\pi}^{\pi} f(\theta)\cos n\theta\, d\theta,$$

$$b_n = i(c_n - c_{-n}) = \frac{i}{2\pi} \int_{-\pi}^{\pi} f(\theta)[e^{-in\theta} - e^{in\theta}]\, d\theta = \frac{1}{\pi} \int_{-\pi}^{\pi} f(\theta)\sin n\theta\, d\theta.$$

Of course, according to (8.2), the integrals over $[-\pi, \pi]$ in (8.6) and (8.7) can be replaced by integrals over any interval of length 2π.

It is useful to keep in mind that in either (8.3) or (8.4), the constant term in the series is

(8.8)
$$c_0 = \tfrac{1}{2}a_0 = \frac{1}{2\pi} \int_{-\pi}^{\pi} f(\theta)\, d\theta,$$

the mean value of f on the interval $[-\pi, \pi]$ (or on any interval of length 2π).

What have we accomplished? We have shown that *if* $f(\theta)$ is the sum of a series $\sum_{-\infty}^{\infty} c_n e^{in\theta}$, and *if* term-by-term integration is legitimate, *then* the coefficients c_n must be given by (8.6), but as yet we know almost nothing about the class of functions that can be represented by such series. But now the formula (8.6) provides a starting point for studying this matter. Indeed, if f is *any* integrable 2π-periodic function, the quantities

$$a_n = \frac{1}{\pi} \int_{-\pi}^{\pi} f(\theta)\cos n\theta\, d\theta, \quad b_n = \frac{1}{\pi} \int_{-\pi}^{\pi} f(\theta)\sin n\theta\, d\theta,$$

$$c_n = \frac{1}{2\pi} \int_{-\pi}^{\pi} f(\theta)e^{-in\theta}\, d\theta$$

are well defined. We call them the **Fourier coefficients** of f, and we call the series

$$\sum_{-\infty}^{\infty} c_n e^{in\theta} = \tfrac{1}{2}a_0 + \sum_{1}^{\infty}(a_n \cos n\theta + b_n \sin n\theta)$$

the **Fourier series** of f.

The study of general Fourier series will be undertaken in the following sections. We conclude this one by working out two simple examples.

EXAMPLE 1. Let $f(\theta)$ be the 2π-periodic function determined by the formula

$$f(\theta) = \theta, \qquad (-\pi < \theta \le \pi).$$

That is, f is the **sawtooth wave** depicted in the top graph of Figure 8.1. The calculation of the Fourier coefficients c_n is an easy integration by parts for $n \ne 0$:

$$c_n = \frac{1}{2\pi} \int_{-\pi}^{\pi} \theta e^{-in\theta}\, d\theta = \frac{1}{2\pi} \left[\frac{\theta e^{-in\theta}}{-in} + \frac{e^{-in\theta}}{n^2} \right]_{-\pi}^{\pi} = \frac{(-1)^{n+1}}{in},$$

since $e^{\pm in\pi} = (-1)^n$. Moreover, $c_0 = 0$ since the mean value of f is clearly zero. Thus the Fourier series of f is

$$\sum_{n \ne 0} \frac{(-1)^{n+1}}{in} e^{in\theta}.$$

Grouping together the nth and $(-n)$th terms yields the equivalent form

(8.9)
$$2 \sum_{1}^{\infty} \frac{(-1)^{n+1}}{n} \sin n\theta.$$

(We could also have obtained this series directly by using (8.7); we have $a_n = 0$ for all n since f is odd, and a calculation similar to the one above shows that $b_n = 2(-1)^{n+1}/n$.)

The series (8.9) converges for all θ by Dirichlet's test. (See Corollary 6.27. The factor of $(-1)^{n+1}$ does not affect the result, since $(-1)^n \sin n\theta = \sin n(\theta + \pi)$.) The sketches of some of the partial sums in Figure 8.1 lend plausibility to the conjecture that (8.9) does indeed converge to the function $f(\theta)$, at least at the points where f is continuous. (At the points $\theta = (2k+1)\pi$ where f is discontinuous, every term in (8.9) vanishes.)

EXAMPLE 2. Let $g(\theta)$ be the 2π-periodic function determined by the formula

$$g(\theta) = |\theta|, \qquad (-\pi \le \theta \le \pi).$$

That is, g is the **triangle wave** depicted in the top graph of Figure 8.2. Here it is a bit easier to calculate the Fourier coefficients in terms of sines and cosines. Since g is an even function, we have $b_n = 0$ for all n and

$$a_n = \frac{1}{\pi} \int_{-\pi}^{\pi} g(\theta) \cos n\theta\, d\theta = \frac{2}{\pi} \int_{0}^{\pi} \theta \cos n\theta\, d\theta.$$

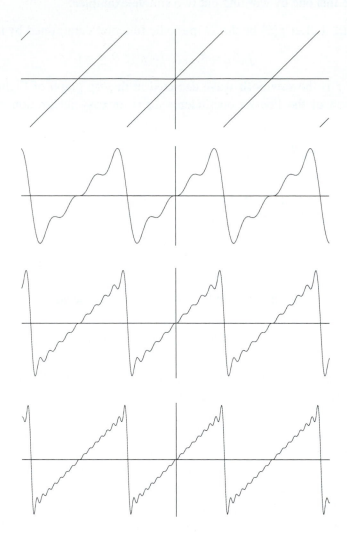

FIGURE 8.1: Top to bottom: The sawtooth wave of Example 1 and the partial sums S_4, S_{10}, and S_{16} of its Fourier series ($S_k = 2\sum_1^k (-1)^{n+1} n^{-1} \sin n\theta$).

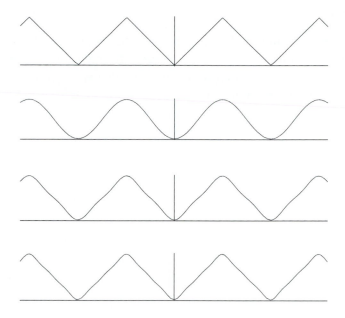

FIGURE 8.2: Top to bottom: The triangle wave of Example 2 and the partial sums S_1, S_2, and S_3 of its Fourier series ($S_k = (\pi/2) - (4/\pi) \sum_1^k (2m-1)^{-2} \cos(2m-1)\theta$).

For $n = 0$ we have $a_0 = (2/\pi) \int_0^\pi \theta \, d\theta = \pi$, and for $n > 0$ an integration by parts gives

$$a_n = \frac{2}{\pi} \left[\frac{\theta \sin n\theta}{n} + \frac{\cos n\theta}{n^2} \right]_0^\pi = \frac{2}{\pi} \frac{(-1)^n - 1}{n^2}.$$

In other words, $a_n = 0$ when n is even and $a_n = -4/\pi n^2$ when n is odd, so we obtain the Fourier series

(8.10) $$\frac{\pi}{2} - \frac{4}{\pi} \sum_{n=1,3,5,\dots} \frac{\cos n\theta}{n^2} = \frac{\pi}{2} - \frac{4}{\pi} \sum_1^\infty \frac{\cos(2m-1)\theta}{(2m-1)^2}.$$

Since $\sum_1^\infty n^{-2} < \infty$, this series converges absolutely and uniformly by the Weierstrass M-test. Again, a glance at its first few partial sums in Figure 8.2 supports the conjecture that its full sum is $g(\theta)$.

EXERCISES

In Exercises 1–8, find the Fourier series of the 2π-periodic function $f(\theta)$ that is given on the interval $(-\pi, \pi)$ by the indicated formula. (All but Exercise 5 are either even or odd, so their Fourier series are naturally expressed in terms of cosines or sines.) Sketches of these functions are given in Figure 8.3.

1. $f(\theta) = \begin{cases} -1 & (-\pi < \theta < 0) \\ 1 & (0 < \theta < \pi) \end{cases}$ (the **square wave**).

2. $f(\theta) = \sin^2 \theta$. (You don't need calculus if you look at this the right way.)

3. $f(\theta) = |\sin \theta|$. (*Hint:* $\sin a \cos b = \frac{1}{2}[\sin(a+b) + \sin(a-b)]$.)

4. $f(\theta) = \theta^2$.

5. $f(\theta) = e^{b\theta}$ $(b > 0)$.

6. $f(\theta) = \theta(\pi - |\theta|)$.

7. $f(\theta) = \begin{cases} 1/a & (|\theta| < a), \\ -1/(\pi - a) & (a < |\theta| < \pi), \end{cases}$ where $0 < a < \pi$. (The values of f

   are chosen to make the areas of the rectangles between the graph of f and the x-axis on the intervals $[0, a]$ and $[a, \pi]$ both equal to 1.)

8. $f(\theta) = \begin{cases} a^{-2}(a - |\theta|) & (|\theta| < a), \\ 0 & (a < |\theta| < \pi), \end{cases}$ where $0 < a < \pi$. (The constants are

   chosen to make the areas of the triangles under the graph of f equal to 1.)

9. Prove that (8.2) is valid for every piecewise continuous P-periodic function f. (This can be done either directly by changes of variable or by differentiating $\int_a^{a+P}$ with respect to a via Theorem 4.15a.)

8.2 Convergence of Fourier Series

Given a piecewise continuous 2π-periodic function f, we form its Fourier series:

$$(8.11) \qquad \sum_{-\infty}^{\infty} c_n e^{in\theta}, \qquad c_n = \frac{1}{2\pi} \int_{-\pi}^{\pi} f(\theta) e^{-in\theta} \, d\theta.$$

Does this series converge? If so, what is its sum?

These questions are rather delicate. In the first place, since $|e^{in\theta}| \equiv 1$, a *necessary* condition for the convergence of the Fourier series is that $c_n \to 0$ as $n \to \infty$, but the only estimate on the c_n's that is obvious from the definition is that they are

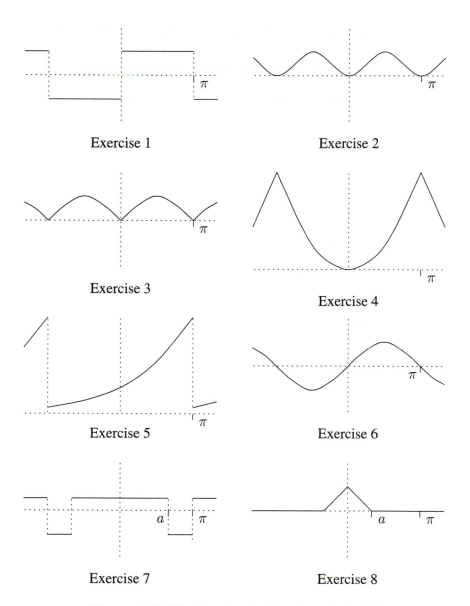

Exercise 1

Exercise 2

Exercise 3

Exercise 4

Exercise 5

Exercise 6

Exercise 7

Exercise 8

FIGURE 8.3: The functions in Exercises 1–8 of §8.1.

bounded by a constant:

$$|c_n| \le \frac{1}{2\pi} \int_{-\pi}^{\pi} |f(\theta)e^{-in\theta}| \, d\theta = \frac{1}{2\pi} \int_{-\pi}^{\pi} |f(\theta)| \, d\theta \le \sup_{\theta} |f(\theta)|.$$

However, it is actually true that $c_n \to 0$; in fact, we can say something more precise.

8.12 Theorem (Bessel's Inequality). *If f is 2π-periodic and piecewise continuous and c_n is defined by (8.11), then*

$$\sum_{-\infty}^{\infty} |c_n|^2 \le \frac{1}{2\pi} \int_{-\pi}^{\pi} |f(\theta)|^2 \, d\theta.$$

In particular, $\sum |c_n|^2 < \infty$, and hence $\lim_{n \to \pm\infty} c_n = 0$.

Proof. We examine the difference between f and a partial sum of its Fourier series. Since the absolute value of a complex number z is given by $|z|^2 = z\bar{z}$, we have

$$\left| f(\theta) - \sum_{-N}^{N} c_n e^{in\theta} \right|^2 = \left(f(\theta) - \sum_{-N}^{N} c_n e^{in\theta} \right) \left(\overline{f(\theta)} - \sum_{-N}^{N} \bar{c}_n e^{-in\theta} \right)$$

$$= |f(\theta)|^2 - \sum_{-N}^{N} \left[\bar{c}_n f(\theta) e^{-in\theta} + c_n \overline{f(\theta)} e^{in\theta} \right] + \sum_{m,n=-N}^{N} c_m \bar{c}_n e^{i(m-n)\theta}.$$

Next, integration of both sides over $[-\pi, \pi]$, using the definition of c_n and the relation (8.5), yields

$$\frac{1}{2\pi} \int_{-\pi}^{\pi} \left| f(\theta) - \sum_{-N}^{N} c_n e^{in\theta} \right|^2 d\theta = \frac{1}{2\pi} \int_{-\pi}^{\pi} |f(\theta)|^2 d\theta - \sum_{-N}^{N} [\bar{c}_n c_n + c_n \bar{c}_n] + \sum_{-N}^{N} c_n \bar{c}_n$$

$$= \frac{1}{2\pi} \int_{-\pi}^{\pi} |f(\theta)|^2 \, d\theta - \sum_{-N}^{N} |c_n|^2.$$

The integral on the left is clearly nonnegative, so

$$0 \le \frac{1}{2\pi} \int_{-\pi}^{\pi} |f(\theta)|^2 \, d\theta - \sum_{-N}^{N} |c_n|^2.$$

Letting $N \to \infty$, we obtain the desired result. $\square$

To proceed further in our study of the convergence of the Fourier series of a function f, we must take a closer look at the partial sums

$$(8.13) \qquad S_N^f(\theta) = \sum_{-N}^{N} c_n e^{in\theta}, \qquad c_n = \frac{1}{2\pi} \int_{-\pi}^{\pi} f(\psi) e^{-in\psi} \, d\psi.$$

Substitution of the formula for c_n into the sum yields

$$S_N^f(\theta) = \sum_{-N}^{N} \frac{1}{2\pi} \int_{-\pi}^{\pi} f(\psi) e^{in(\theta-\psi)} \, d\psi = \sum_{-N}^{N} \frac{1}{2\pi} \int_{-\pi}^{\pi} f(\psi) e^{in(\psi-\theta)} \, d\psi$$

$$= \sum_{-N}^{N} \frac{1}{2\pi} \int_{-\pi}^{\pi} f(\varphi + \theta) e^{in\varphi} \, d\varphi.$$

(The second equality is obtained by replacing n by $-n$, which leaves the sum from $-N$ to N unchanged, and the third one comes from the change of variable $\varphi = \psi - \theta$ with the help of (8.2).) In other words,

$$(8.14) \qquad S_N^f(\theta) = \int_{-\pi}^{\pi} f(\varphi + \theta) D_N(\varphi) \, d\varphi, \text{ where } D_N(\varphi) = \frac{1}{2\pi} \sum_{-N}^{N} e^{in\varphi}.$$

D_N is called the Nth **Dirichlet kernel**. Its essential properties are summarized in the following lemma.

8.15 Lemma. *Let $D_N(\varphi)$ be the function defined in (8.14). Then:*

a. $\displaystyle \int_{-\pi}^{0} D_N(\varphi) \, d\varphi = \int_{0}^{\pi} D_N(\varphi) \, d\varphi = \frac{1}{2}.$

b. $\displaystyle D_N(\varphi) = \frac{1}{2\pi} \frac{e^{i(N+1)\varphi} - e^{-iN\varphi}}{e^{i\varphi} - 1}.$

Proof. The validity of (a) is most easily seen by rewriting (8.14) as $D_N(\varphi) = (2\pi)^{-1} + \pi^{-1} \sum_{1}^{N} \cos n\varphi$ and integrating this sum term by term. Since $\sin 0 = \sin(\pm n\pi) = 0$, only the constant term gives a nonzero contribution. To prove (b), we use the formula (6.2) for the sum of a finite geometric progression:

$$2\pi D_N(\varphi) = e^{-iN\varphi} \sum_{0}^{2N} e^{in\varphi} = e^{-iN\varphi} \frac{e^{i(2N+1)\varphi} - 1}{e^{i\varphi} - 1} = \frac{e^{i(N+1)\varphi} - e^{-iN\varphi}}{e^{i\varphi} - 1}.$$

$\square$

Incidentally, if we multiply and divide the formula in Lemma 8.15b for $D_N(\varphi)$ by $e^{-i\varphi/2}$, we obtain

$$D_N(\varphi) = \frac{1}{2\pi} \frac{e^{i(N+(1/2))\varphi} - e^{-i(N+(1/2))\varphi}}{e^{i\varphi/2} - e^{-i\varphi/2}} = \frac{\sin(N + \frac{1}{2})\varphi}{2\pi \sin \frac{1}{2}\varphi}.$$

This shows that D_N is real-valued and gives an easy way to visualize it: Its graph is the rapidly oscillating sine wave $y = \sin(N + \frac{1}{2})\varphi$, amplitude-modulated to fit inside the envelope $y = \pm(2\pi \sin \frac{1}{2}\varphi)^{-1}$. (The reader may wish to generate graphs of D_N for various values of N on a computer.)

We are now ready to formulate and prove the basic convergence theorem for Fourier series. It turns out that piecewise continuity of a periodic function f is not enough to yield a good result. Instead we shall assume, in effect, that not only f but also its derivative f' is piecewise continuous. More precisely, we shall say that a periodic function f is **piecewise smooth** if, on any bounded interval, f is of class C^1 except at finitely many points, at which the one-sided limits $f(\theta+)$, $f(\theta-)$, $f'(\theta+)$, and $f'(\theta-)$ (as defined in (8.1)) exist and are finite. (Note that this definition of piecewise smoothness is more general than that given in §5.1, which required the function to be continuous.) Pictorially, f is piecewise smooth if its graph over any bounded interval is a smooth curve except at finitely many points where it has jumps (if f is discontinuous) or corners (if f is continuous but f' is discontinuous). In addition, the one-sided tangent lines at the jumps and corners are not allowed to be vertical.

8.16 Theorem. *Suppose f is 2π-periodic and piecewise smooth. Then the partial sums $S_N^f(\theta)$ of the Fourier series of f, defined by (8.13), converge pointwise to $\frac{1}{2}[f(\theta-) + f(\theta+)]$. In particular, they converge to $f(\theta)$ at each point θ where f is continuous.*

Proof. By Lemma 8.15a, we have

$$\tfrac{1}{2}f(\theta-) = f(\theta-)\int_{-\pi}^{0} D_N(\varphi)\,d\varphi, \qquad \tfrac{1}{2}f(\theta+) = f(\theta+)\int_{0}^{\pi} D_N(\varphi)\,d\varphi,$$

so by (8.14), the difference between $S_N^f(\theta)$ and its asserted limit is

$$S_N^f(\theta) - \tfrac{1}{2}[f(\theta-) + f(\theta+)]$$
$$= \int_{-\pi}^{0} [f(\varphi + \theta) - f(\theta-)] D_N(\varphi)\,d\varphi + \int_{0}^{\pi} [f(\varphi + \theta) - f(\theta+)] D_N(\varphi)\,d\varphi.$$

Our object is to show that this quantity vanishes as $N \to \infty$. By Lemma 8.15b, we can rewrite it as

(8.17)
$$\frac{1}{2\pi} \int_{-\pi}^{\pi} g(\varphi) \left[e^{i(N+1)\varphi} - e^{-iN\varphi} \right] d\varphi,$$

where

$$g(\varphi) = \begin{cases} \dfrac{f(\varphi+\theta) - f(\theta-)}{e^{i\varphi} - 1} & \text{if } -\pi \le \varphi < 0, \\[2ex] \dfrac{f(\varphi+\theta) - f(\theta+)}{e^{i\varphi} - 1} & \text{if } 0 < \varphi \le \pi. \end{cases}$$

(We could define $g(0)$ to be anything we please; altering the value at this one point does not affect (8.17), by Proposition 4.14.) On the interval $[-\pi, \pi]$, $g(\varphi)$ is continuous wherever $f(\varphi+\theta)$ is and has jump discontinuities wherever $f(\varphi+\theta)$ does, except for an additional singularity at $\varphi = 0$ caused by the vanishing of $e^{i\varphi} - 1$ there. But this singularity is also at worst a jump discontinuity; that is, the limits $g(0+)$ and $g(0-)$ both exist. Indeed, by l'Hôpital's rule,

$$g(0+) = \lim_{\varphi \to 0+} \frac{f(\varphi+\theta) - f(\theta+)}{e^{i\varphi} - 1} = \lim_{\varphi \to 0+} \frac{f'(\varphi+\theta)}{ie^{i\varphi}} = \frac{f'(\theta+)}{i},$$

and likewise $g(0-) = i^{-1} f'(\theta-)$. In short, g is piecewise continuous.

Now we are done. By Bessel's inequality, the Fourier coefficients of g,

$$C_n = \frac{1}{2\pi} \int_{-\pi}^{\pi} g(\varphi) e^{-in\varphi} \, d\varphi,$$

tend to zero as $n \to \pm\infty$. But the quantity (8.17) is simply $C_{-N-1} - C_N$, so it vanishes as $N \to \infty$, as desired. □

If f is piecewise continuous, there may be some question as to how to define f at its points of discontinuity; as we mentioned earlier, we may wish to allow f to remain undefined at these points. But Theorem 8.16 shows that for the purposes of Fourier analysis, the natural choice is the average of the left- and right-hand limits: $f(\theta) = \frac{1}{2}[f(\theta-) + f(\theta+)]$. We shall say that f is **standardized** if it satisfies this condition at all θ; thus, every standardized piecewise smooth 2π-periodic function is the sum of its Fourier series at every point.

8.18 Corollary. *If f and g are standardized piecewise smooth 2π-periodic functions with the same Fourier coefficients, then $f = g$.*

Proof. f and g are the sum of the same Fourier series. □

To illustrate Theorem 8.16, let us consider the two examples in §8.1.

EXAMPLE 1. The sawtooth wave $f(\theta)$ defined by $f(\theta) = \theta$ for $|\theta| < \pi$ is smooth except at the odd multiples of π, where its left- and right-hand limits are π and $-\pi$, respectively. Thus the Fourier series of f converges to f everywhere except at the odd multiples of π, where it converges to 0. On the interval $(-\pi, \pi)$, the result is

$$\sum_1^\infty \frac{(-1)^{n+1}}{n} \sin n\theta = \frac{\theta}{2} \text{ for } -\pi < \theta < \pi.$$

In particular, by taking $\theta = \frac{1}{2}\pi$, we obtain the interesting formula

$$\sum_1^\infty \frac{(-1)^{m-1}}{2m-1} = 1 - \frac{1}{3} + \frac{1}{5} - \frac{1}{7} + \cdots = \frac{\pi}{4},$$

which we derived by other methods in Example 5 of §7.3.

EXAMPLE 2. The triangle wave $g(\theta)$ defined by $f(\theta) = |\theta|$ for $|\theta| < \pi$ is piecewise smooth and everywhere continuous, so it is the sum of its Fourier series at every point. Thus,

$$\frac{\pi}{2} - \frac{4}{\pi} \sum_1^\infty \frac{\cos(2m-1)\theta}{(2m-1)^2} = |\theta| \text{ for } -\pi \le \theta \le \pi.$$

By taking $\theta = 0$ (or $\theta = \pm\pi$), we obtain another interesting formula:

$$\sum_1^\infty \frac{1}{(2m-1)^2} = 1 + \frac{1}{3^2} + \frac{1}{5^2} + \frac{1}{7^2} + \cdots = \frac{\pi^2}{8}.$$

From this it is also easy to obtain the sum

$$S = \sum_1^\infty \frac{1}{n^2} = 1 + \frac{1}{2^2} + \frac{1}{3^2} + \frac{1}{4^2} + \cdots$$

by separating out the odd and even terms:

$$S = \left(1 + \frac{1}{3^2} + \frac{1}{5^2} + \cdots\right) + \left(\frac{1}{2^2} + \frac{1}{4^2} + \frac{1}{6^2} + \cdots\right)$$

$$= \frac{\pi^2}{8} + \frac{1}{4}\left(1 + \frac{1}{2^2} + \frac{1}{3^2} + \cdots\right) = \frac{\pi^2}{8} + \frac{S}{4},$$

so that $3S/4 = \pi^2/8$, or $S = \pi^2/6$.

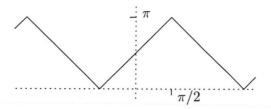

FIGURE 8.4: The function h of Example 3.

We conclude by remarking that one can often use simple changes of variable to generate new Fourier expansions from old ones without recalculating the coefficients from scratch.

EXAMPLE 3. Consider the modified triangle wave h whose graph is given in Figure 8.4. It is related to the triangle wave g in Example 2 by $h(\theta) = g(\theta + \frac{1}{2}\pi)$, and $\cos(2m - 1)(\theta + \frac{1}{2}\pi) = (-1)^m \sin(2m - 1)\theta$, so

$$h(\theta) = \frac{\pi}{2} + \frac{4}{\pi} \sum_{1}^{\infty} \frac{(-1)^{m-1} \sin(2m - 1)\theta}{(2m - 1)^2}.$$

Abel Summability of Fourier Series. The Fourier coefficients of a periodic function f are defined whenever f is piecewise continuous, but we have proved the convergence of the Fourier series only when f is piecewise smooth. In fact, it has been known since 1876 that there are continuous periodic functions whose Fourier series fail to converge at some points. (The examples are all quite complicated.) However, if f is merely piecewise continuous, we can still recover f from its Fourier series $\sum_{-\infty}^{\infty} c_n e^{in\theta}$ by the method of Abel summation that we discussed at the end of §7.3. Namely, for $0 < r < 1$ we consider the series

(8.19) $$A_r f(\theta) = \sum_{-\infty}^{\infty} r^{|n|} c_n e^{in\theta}$$

and its limit as $r \to 1-$ (i.e., as r approaches 1 from the left).

Since the coefficients c_n are bounded, the series (8.19) converges absolutely by comparison to the geometric series $\sum r^{|n|}$. Moreover, substitution of the formula (8.6) for c_n into (8.19) gives

$$A_r f(\theta) = \frac{1}{2\pi} \sum_{-\infty}^{\infty} \int_{-\pi}^{\pi} r^{|n|} f(\psi) e^{in(\theta - \psi)} \, d\psi.$$

Since f is bounded, the Weierstrass M-test (comparison to $\sum r^{|n|}$ again) gives the uniform convergence to justify interchange of the summation and integration, and a couple of manipulations like those that lead to (8.14) then show that

$$(8.20) \quad A_r f(\theta) = \int_{-\pi}^{\pi} f(\theta + \varphi) P_r(\varphi)\, d\varphi, \text{ where } P_r(\varphi) = \frac{1}{2\pi} \sum_{-\infty}^{\infty} r^{|n|} e^{in\varphi}.$$

The function P_r is called the **Poisson kernel**. Like the Dirichlet kernel, it satisfies

$$(8.21) \qquad \int_{-\pi}^{0} P_r(\varphi)\, d\varphi = \int_{0}^{\pi} P_r(\varphi)\, d\varphi = \frac{1}{2}$$

(write $P_r(\varphi) = (2\pi)^{-1} + \pi^{-1} \sum_{1}^{\infty} r^n \cos n\varphi$ and integrate term by term), and it is easily expressed in closed form since it is the sum of two geometric series:

$$(8.22)$$

$$P_r(\varphi) = \frac{1}{2\pi}\left[\sum_{0}^{\infty} r^n e^{in\varphi} + \sum_{1}^{\infty} r^n e^{-in\varphi}\right] = \frac{1}{2\pi}\left[\frac{1}{1 - re^{i\varphi}} + \frac{re^{-i\varphi}}{1 - re^{-i\varphi}}\right]$$

$$= \frac{1 - r^2}{2\pi(1 - re^{i\varphi})(1 - re^{-i\varphi})} = \frac{1 - r^2}{2\pi(1 + r^2 - 2r\cos\varphi)}.$$

However, the Poisson kernel has one additional crucial property that is not shared by the Dirichlet kernel:

$$(8.23)$$

　　For any $\delta > 0$, $P_r(\varphi) \to 0$ uniformly on $[-\pi, -\delta]$ and on $[\delta, \pi]$ as $r \to 1-$.

Indeed, by (8.22), for $\delta \le |\varphi| \le \pi$ we have

$$0 \le P_r(\varphi) \le \frac{1 - r^2}{2\pi(1 + r^2 - 2r\cos\delta)},$$

and the expression on the right tends to zero as $r \to 1-$. With these results in hand, we come to the main theorem.

8.24 Theorem. *Suppose that f is 2π-periodic. If f is piecewise continuous, then*

$$\lim_{r \to 1-} A_r f(\theta) = \tfrac{1}{2}[f(\theta-) + f(\theta+)]$$

for every θ. If f is continuous, then $A_r f \to f$ uniformly on $\mathbb{R}$ as $r \to 1$.

Proof. We sketch the ideas and leave the details to the reader as Exercises 5 and 6. Given $\theta \in \mathbb{R}$ and $\epsilon > 0$, we choose $\delta > 0$ small enough so that $|f(\theta+\varphi)-f(\theta+)| < \epsilon$ when $0 < \varphi < \delta$ and $|f(\theta + \varphi) - f(\theta-)| < \epsilon$ when $-\delta < \varphi < 0$. We then write the formula (8.20) for $A_r f(\theta)$ as

$$A_r f(\theta) = \left[\int_{-\pi}^{-\delta} + \int_{-\delta}^{0} + \int_{0}^{\delta} + \int_{\delta}^{\pi} \right] f(\theta + \varphi) P_r(\varphi) \, d\varphi.$$

The first and last integrals tend to zero as $r \to 1-$ by (8.23). In the second and third integrals, $f(\theta + \varphi)$ is within ϵ of $f(\theta-)$ and $f(\theta+)$, respectively, and (8.21) and (8.23) together show that the integrals of $P_r(\varphi)$ over $[-\delta, 0]$ and $[0, \delta]$ tend to $\frac{1}{2}$ as $r \to 1-$. The upshot is that $A_r f(\theta)$ is within 2ϵ of $\frac{1}{2}[f(\theta-) + f(\theta+)]$ when r is sufficiently close to 1, and since ϵ is arbitrary, the first assertion is proved.

If f is continuous, it is uniformly continuous on $[-\pi, \pi]$ by Theorem 1.33 and hence uniformly continuous on $\mathbb{R}$ by periodicity. This means that the δ in the preceding paragraph can be chosen independent of θ, and the argument given there then yields uniform convergence. $\qquad\square$

EXERCISES

1. Find the Fourier series of the sawtooth waves depicted below by modifying the series in Example 1.

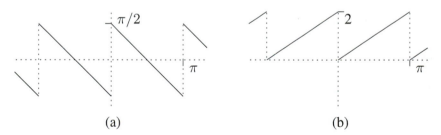

(a)   (b)

2. Find the Fourier series of the 2π-periodic function $f(\theta)$ defined by $f(\theta) = (\theta - \frac{1}{4}\pi)^2$ on the interval $[-\frac{3}{4}\pi, \frac{5}{4}\pi]$. Use the result of Exercise 4 in §8.1.

3. Find the Fourier series of the 2π-periodic functions defined on the interval $(-\pi, \pi)$ by the indicated formulas by modifying the series in the exercises of §8.1.

   a. $f(\theta) = \begin{cases} 0 & (-\pi < \theta < 0), \\ 1 & (0 < \theta < \pi). \end{cases}$

b. $f(\theta) = \begin{cases} 0 & (-\pi < \theta < 0), \\ \sin\theta & (0 < \theta < \pi). \end{cases}$ (*Hint:* $\max(x, 0) = \frac{1}{2}(x + |x|).$)

c. $f(\theta) = \begin{cases} (2a)^{-1} & (|\theta| < a), \\ 0 & (a < |\theta| < \pi), \end{cases}$ where $0 < a < \pi$.

d. $f(\theta) = \sinh\theta$.

4. Find the sums of the following series by applying Theorem 8.16 to the series obtained in the indicated exercises from §8.1 and choosing appropriate values of θ.

a. $\displaystyle\sum_1^\infty \frac{1}{4n^2 - 1}$ and $\displaystyle\sum_1^\infty \frac{(-1)^{n+1}}{4n^2 - 1}$ (Exercise 3). Can you sum the first series in a more elementary way by rewriting it as a telescoping series?

b. $\displaystyle\sum_1^\infty \frac{1}{n^2}$ and $\displaystyle\sum_1^\infty \frac{(-1)^{n+1}}{n^2}$ (Exercise 4).

c. $\displaystyle\sum_1^\infty \frac{(-1)^n}{n^2 + b^2}$ and $\displaystyle\sum_1^\infty \frac{1}{n^2 + b^2}$, where $b > 0$ (Exercise 5).

d. $\displaystyle\sum_1^\infty \frac{(-1)^{n+1}}{(2n - 1)^3}$ (Exercise 6).

5. Fill in the details of the proof of the first assertion of Theorem 8.24.

6. Fill in the details of the proof of the second assertion of Theorem 8.24.

8.3   Derivatives, Integrals, and Uniform Convergence

We next study the differentiation and integration of Fourier series. As a first step, we point out that by the fundamental theorem of calculus as stated in §4.1, the formula

(8.25)
$$f(b) - f(a) = \int_a^b f'(\theta)\, d\theta$$

is valid when f is continuous and piecewise smooth, even though f' may be un-defined at finitely many points. (However, it is generally false if f itself has jump discontinuities.) In particular, if f and g are both continuous and piecewise smooth, then so is fg, and an application of (8.25) to the latter function yields the integration-by-parts formula

$$\int_a^b f'(x)g(x)\, dx = f(x)g(x)\Big|_a^b - \int_a^b f(x)g'(x)\, dx.$$

The first main result is that there is a very simple relation between the Fourier coefficients of f and those of f'.

8.26 Theorem. *Suppose f is 2π-periodic, continuous, and piecewise smooth, and let c_n and c'_n be the Fourier coefficients of f and f', given by (8.6). Then*

$$c'_n = inc_n.$$

Equivalently, if a_n, b_n and a'_n, b'_n are the Fourier coefficients of f and f' given by (8.7), then $a'_n = nb_n$ and $b'_n = -na_n$.

Proof. Simply integrate by parts:

$$c'_n = \frac{1}{2\pi} \int_{-\pi}^{\pi} f'(\theta) e^{-in\theta}\, d\theta = \frac{1}{2\pi} f(\theta) e^{-in\theta} \Big|_{-\pi}^{\pi} - \frac{1}{2\pi} \int_{-\pi}^{\pi} f(\theta)(-ine^{-in\theta})\, d\theta.$$

The first term on the right vanishes because $f(\theta)e^{-in\theta}$ is 2π-periodic, and the second one is inc_n. The argument for a_n and b_n is similar (Exercise 1). $\square$

Note that Theorem 8.26 makes no claim about the Fourier *series* of f'; it is valid whether or not that series actually converges. If we add more conditions on f to ensure that it does, we obtain the following result:

8.27 Corollary. *Suppose that f is 2π-periodic, continuous, and piecewise smooth, and that f' is also piecewise smooth. If*

$$\sum_{-\infty}^{\infty} c_n e^{in\theta} = \tfrac{1}{2}a_0 + \sum_{1}^{\infty} (a_n \cos n\theta + b_n \sin n\theta)$$

is the Fourier series of f, then $f'(\theta)$ is the sum of the derived series

$$\sum_{-\infty}^{\infty} inc_n e^{in\theta} = \sum_{1}^{\infty} (nb_n \cos n\theta - na_n \sin n\theta)$$

at every θ at which $f'(\theta)$ exists. At the exceptional points where f' has jumps, the series converges to $\tfrac{1}{2}[f'(\theta-) + f'(\theta+)]$.

Proof. By Theorem 8.16, f' is the sum of its Fourier series everywhere except where it has jumps, and the coefficients in that series are given by Theorem 8.26. $\square$

EXAMPLE 1. The triangle wave (Example 2 in §8.1) is continuous and piece-wise smooth, and its derivative is the square wave (Exercise 1 in §8.1). We can therefore recover the result of the latter exercise by differentiating the series (8.10):

$$\frac{4}{\pi} \sum_1^\infty \frac{\sin(2m-1)\theta}{2m-1} = \begin{cases} -1 & (-\pi < \theta < 0), \\ 1 & (0 < \theta < \pi). \end{cases}$$

Next, we consider integration of Fourier series. There is one annoying point that must be kept in mind: If f is a piecewise continuous 2π-periodic function, its indefinite integral $F(\theta) = \int_0^\theta f(\varphi)\, d\varphi$ will be periodic only when

$$\int_{-\pi}^\pi f(\varphi)\, d\varphi = F(\pi) - F(-\pi) = 0,$$

that is, when the mean value of f over an interval of length 2π is zero, or, equiv-alently, when the constant term in the Fourier series of f vanishes. We make this assumption in the following theorem; if it is not valid, we may wish to subtract off the constant term and deal with it separately.

8.28 Theorem. *Suppose f is 2π-periodic and piecewise continuous, with Fourier coefficients c_n given by (8.6) or a_n, b_n given by (8.7). Assume that $c_0 = \frac{1}{2}a_0 = 0$. If F is a continuous, piecewise smooth function such that $F' = f$ (except at the points where f has jumps), then*

$$F(\theta) = C_0 + \sum_{n \neq 0} \frac{c_n}{in} e^{in\theta} = C_0 + \sum_1^\infty \left(\frac{a_n}{n} \sin n\theta - \frac{b_n}{n} \cos n\theta \right)$$

for all θ, where C_0 is the mean value of F on $[-\pi, \pi]$.

Proof. F is 2π-periodic by (8.2), for

$$F(\theta + 2\pi) - F(\theta) = \int_\theta^{\theta+2\pi} f(\varphi)\, d\varphi = \int_{-\pi}^\pi f(\varphi)\, d\varphi = 2\pi c_0 = 0.$$

By Theorem 8.16, F is the sum of its Fourier series at every point, and by Theorem 8.26, its Fourier coefficients C_n are given for $n \neq 0$ by $inC_n = c_n$ (and likewise for the cosine and sine coefficients). The constant term C_0 is, as always, the mean value of F.   □

Observe that the series in Theorem 8.28 is obtained by formally integrating the Fourier series of f term-by-term, whether the latter series converges or not.

EXAMPLE 2. Subtraction of the mean value from the triangle wave (Example 2 in §8.2) and multiplication by -2 gives

$$\pi - 2|\theta| = \frac{8}{\pi} \sum_{1}^{\infty} \frac{\cos(2m-1)\theta}{(2m-1)^2} \qquad (|\theta| \leq \pi),$$

and integration of both sides from 0 to θ then yields

$$\pi\theta - \theta|\theta| = \frac{8}{\pi} \sum_{1}^{\infty} \frac{\sin(2m-1)\theta}{(2m-1)^3} \qquad (|\theta| \leq \pi),$$

which is the result of Exercise 6 in §8.1.

Theorem 8.28 and the Corollary 8.27 exhibit situations where we can integrate or differentiate a series termwise without worrying about uniform convergence. However, uniform and absolute convergence are still highly desirable things, so we present a simple criterion for the Fourier series of a function to have these properties.

8.29 Theorem. *If f is 2π-periodic, continuous, and piecewise smooth, then the Fourier series of f is absolutely and uniformly convergent.*

Proof. Let c_n and c_n' be the Fourier coefficients of f and f'. Since $|c_n e^{in\theta}| = |c_n|$, the absolute convergence of $\sum c_n e^{in\theta}$ is equivalent to the convergence of $\sum |c_n|$, and by the Weierstrass M-test, this also implies the uniform convergence of $\sum c_n e^{in\theta}$. But by Theorem 8.26, $c_n = c_n'/in$ for $n \neq 0$, so

$$|c_n| \leq \tfrac{1}{2}(|c_n'|^2 + |n|^{-2}) \qquad (n \neq 0).$$

(The inequality $\alpha\beta \leq \frac{1}{2}(\alpha^2 + \beta^2)$ is valid for all $\alpha, \beta \in \mathbb{R}$ since $\alpha^2 + \beta^2 - 2\alpha\beta = (\alpha - \beta)^2 \geq 0$.) But the series $\sum |c_n'|^2$ and $\sum n^{-2}$ are both convergent — by Bessel's inequality in the former case, since f' is piecewise continuous — and hence so is $\sum |c_n|$. $\qquad \square$

We conclude this section with an important feature of Fourier series, which we state as a general principle rather than as a precise theorem:

The degree of smoothness of a periodic function is closely related to the rate of decay of its Fourier coefficients, that is, to the rate of convergence of its Fourier series.

Indeed, let f be a 2π-periodic function f with Fourier coefficients c_n. If f is of class C^k, then $f^{(k)}$ is a continuous 2π-periodic function whose Fourier coefficients are $(in)^k c_n$, by Theorem 8.26. By Bessel's inequality, $\lim_{n\to\infty} |n^k c_n| = 0$, so $|c_n|$ tends to zero faster than $|n|^{-k}$ as $n \to \pm\infty$. Conversely, suppose $|c_n| \le C|n|^{-k-\epsilon}$ for some $C, \epsilon > 0$. Then $\sum |n^j c_n| < \infty$ for $j < k$, so the series $\sum c_n e^{in\theta}$ can be differentiated termwise $k - 1$ times with the differentiated series being absolutely and uniformly convergent, and hence f is of class C^{k-1}. (A number of other variations can be played on this theme.)

We can see this phenomenon in the examples of §8.1. The sawtooth wave has discontinuities, and its Fourier coefficients decay like n^{-1}; the triangle wave is continuous but its first derivative is not, and its Fourier coefficients decay like n^{-2}. Figures 8.1 and 8.2 show clearly that the Fourier series of the triangle wave converges more rapidly than that of the sawtooth wave.

EXERCISES

1. Verify the assertion about a_n and b_n in Theorem 8.26.

2. Given $a \in (0, \pi)$, let f be the 2π-periodic function defined by $f(\theta) = a^{-1}$ for $|\theta| < a$ and $f(\theta) = (a - \pi)^{-1}$ for $a < |\theta| < \pi$.
   a. Find the formula for $g(\theta) = \int_0^\theta f(\varphi)\, d\varphi$ on $[-\pi, \pi]$ and sketch its graph.
   b. Use the Fourier series of f found in Exercise 7 of §8.1 to compute the Fourier series of g.

3. By applying Theorem 8.28 to the result of Exercise 4 of §8.1, show that:
   a. $\theta^3 - \pi^2\theta = 12 \sum_1^\infty \dfrac{(-1)^n \sin n\theta}{n^3}$ $(|\theta| \le \pi)$.

   b. $\theta^4 - 2\pi^2\theta^2 = \dfrac{-7\pi^4}{15} + 48 \sum_1^\infty \dfrac{(-1)^{n+1} \cos n\theta}{n^4}$ $(|\theta| \le \pi)$.

   c. $\sum_1^\infty \dfrac{1}{n^4} = \dfrac{\pi^4}{90}$.

4. From Exercise 3 of §8.1, we know that
$$\sin\theta = \frac{2}{\pi} - \frac{4}{\pi} \sum_1^\infty \frac{\cos 2n\theta}{4n^2 - 1} \text{ for } 0 \le \theta \le \pi.$$

Show that this series can be differentiated or integrated termwise to yield two apparently different series expansions of $\cos\theta$ for $0 < \theta < \pi$, and reconcile these two expansions. (*Hint:* Example 1 of §8.2 is useful.)

5. Let $f(\theta)$ be the 2π-periodic function such that $f(\theta) = e^{\theta}$ for $|\theta| < \pi$, and let $\sum_{-\infty}^{\infty} c_n e^{in\theta}$ be its Fourier series. If we formally differentiate this equation, we obtain $e^{\theta} = \sum_{-\infty}^{\infty} in c_n e^{in\theta}$ for $|\theta| < \pi$. But then c_n and $in c_n$ are both equal to $(2\pi)^{-1} \int_{-\pi}^{\pi} e^{\theta} e^{-in\theta}\, d\theta$, so $c_n = in c_n$ and hence $c_n = 0$ for all n. Clearly this is wrong; where is the mistake?

6. How smooth are the following functions? That is, for which k can you show that the function is of class C^k?

   a. $\displaystyle\sum_{n\neq 0} \frac{e^{in\theta}}{n^{6/5}(1+n^6)}.$
   b. $\displaystyle\sum_{0}^{\infty} \frac{\cos n\theta}{2^n}.$
   c. $\displaystyle\sum_{0}^{\infty} \frac{\cos 2^n\theta}{2^n}$

8.4 Fourier Series on Intervals

A 2π-periodic function is completely determined by its values on any interval of length 2π. Conversely, if one is given a function f defined on an interval of length 2π, say $[-\pi, \pi]$, one can extend f to be a 2π-periodic function on $\mathbb{R}$ by declaring that $f(\theta + 2k\pi) = f(\theta)$ for all $\theta \in [-\pi, \pi]$ and $k \in \mathbb{Z}$. (Actually, this definition is not consistent at the points $\theta = (2k+1)\pi$ unless $f(-\pi) = f(\pi)$, but one can redefine f to be any given number at these points, such as $\frac{1}{2}[f(-\pi) + f(\pi)]$.) If the original f on $[-\pi, \pi]$ is piecewise continuous or piecewise smooth, the same will be true of its periodic extension. However, even if f is perfectly smooth on $[-\pi, \pi]$, there will usually be discontinuities in the periodic extension or its derivatives at the points $(2k+1)\pi$ where the translates of f are joined together. (For example, the periodic extension of $f(\theta) = \theta$ on $[-\pi, \pi]$ is the sawtooth wave.)

By considering the periodic extension, then, one can use Fourier series to expand a piecewise smooth function on $[-\pi, \pi]$ in terms of trig functions. All of the results in the preceding sections apply, except that in using the results of §8.3 one must remember to take into account the possible extra discontinuities in the periodic extension or its derivatives at the points $(2k+1)\pi$.

There is an extra twist we can add to this construction that is useful in many situations. Suppose that we are considering functions on $[0, \pi]$ rather than $[-\pi, \pi]$. Given a piecewise continuous function f on $[0, \pi]$, we first extend it to $[-\pi, \pi]$ by declaring it to be *even* (see Figure 8.5), and then extend it to be 2π-periodic on $\mathbb{R}$. That is, we define the even extension f_{even} of f on $[-\pi, \pi]$ by

$$f_{\text{even}}(\theta) = \begin{cases} f(\theta) & \text{if } 0 \le \theta \le \pi, \\ f(-\theta) & \text{if } -\pi \le \theta \le 0. \end{cases}$$

For this extension the Fourier sine coefficients b_n all vanish because $f_{\text{even}}(\theta) \sin n\theta$

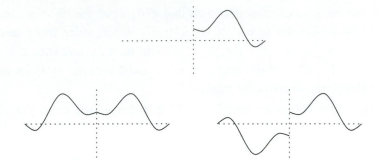

FIGURE 8.5: A function on $[0, \pi]$ (above) and its even and odd extensions to $[-\pi, \pi]$ (below, left and right).

is an odd function, and the cosine coefficients a_n are given by

$$a_n = \frac{1}{\pi} \int_{-\pi}^{\pi} f_{\text{even}}(\theta) \cos n\theta \, d\theta = \frac{2}{\pi} \int_0^{\pi} f(\theta) \cos n\theta \, d\theta.$$

The resulting Fourier series is $\frac{1}{2} a_0 + \sum_1^{\infty} a_n \cos n\theta$.

On the other hand, we could also consider the *odd* extension of f to $[-\pi, \pi]$ (see Figure 8.5):

$$f_{\text{odd}}(\theta) = \begin{cases} f(\theta) & \text{if } 0 < \theta < \pi, \\ -f(-\theta) & \text{if } -\pi < \theta < 0, \\ 0 & \text{if } \theta = 0, \pm\pi. \end{cases}$$

Here the Fourier cosine coefficients a_n all vanish, and the sine coefficients b_n are given by

$$b_n = \frac{1}{\pi} \int_{-\pi}^{\pi} f_{\text{odd}}(\theta) \sin n\theta \, d\theta = \frac{2}{\pi} \int_0^{\pi} f(\theta) \sin n\theta \, d\theta.$$

The resulting Fourier series is $\sum_1^{\infty} b_n \sin n\theta$.

We are thus led to the following definitions: If f is a piecewise continuous function on $[0, \pi]$, its **Fourier cosine series** is the series

$$\tfrac{1}{2} a_0 + \sum_1^{\infty} a_n \cos n\theta, \qquad a_n = \frac{2}{\pi} \int_0^{\pi} f(\theta) \cos n\theta \, d\theta,$$

and its **Fourier sine series** is the series

$$\sum_1^{\infty} b_n \sin n\theta, \qquad b_n = \frac{2}{\pi} \int_0^{\pi} f(\theta) \sin n\theta \, d\theta.$$

EXAMPLE 1. Let $f(\theta) = \theta$ on $[0, \pi]$. The even and odd periodic extensions of f are the triangle and sawtooth waves, respectively, and the Fourier cosine and sine series of f are

$$\frac{\pi}{2} - \frac{4}{\pi} \sum_1^\infty \frac{\cos(2m-1)\theta}{(2m-1)^2} \quad \text{and} \quad 2 \sum_1^\infty \frac{(-1)^{n+1} \sin n\theta}{n},$$

respectively.

If f is piecewise smooth on $[0, \pi]$, its even and odd periodic extensions will be piecewise smooth on $\mathbb{R}$. If $f(0) = f(0+)$ and $f(\pi) = f(\pi-)$, its even periodic extension will be continuous at both 0 and π, but its odd periodic extension will have jumps at 0 or π unless $f(0) = 0$ or $f(\pi) = 0$, respectively. In any case, an application of Theorem 8.16 to these extensions easily yields the following:

8.30 Theorem. *Suppose f is piecewise smooth on $[0, \pi]$. The Fourier cosine series and the Fourier sine series of f converge to $\frac{1}{2}[f(\theta-) + f(\theta+)]$ at every $\theta \in (0, \pi)$. The cosine series converges to $f(0+)$ at $\theta = 0$ and to $f(\pi-)$ at $\theta = \pi$; the sine series converges to 0 at both these points.*

We may wish to consider periodic functions with period other than 2π, or functions defined on intervals other than $[0, \pi]$. The general situation can be reduced to the one we have studied by a linear change of variable; we record the results for future reference.

Suppose $f(x)$ is a piecewise smooth $2l$-periodic function. We make the change of variables

$$\theta = \frac{\pi x}{l}, \qquad g(\theta) = f(x) = f\left(\frac{l\theta}{\pi}\right).$$

Then g is 2π-periodic, and we have

$$g(\theta) = \sum_{-\infty}^\infty c_n e^{in\theta}, \qquad c_n = \frac{1}{2\pi} \int_{-\pi}^\pi g(\theta) e^{-in\theta}\, d\theta.$$

The substitution $\theta = \pi x / l$ then yields the Fourier series for f.

$$(8.31) \qquad f(x) = \sum_{-\infty}^\infty c_n e^{in\pi x/l}, \qquad c_n = \frac{1}{2l} \int_{-l}^l f(x) e^{-in\pi x/l}\, dx.$$

The corresponding formula in terms of sines and cosines is

$$f(x) = \tfrac{1}{2}a_0 + \sum_1^\infty \left[a_n \cos \frac{n\pi x}{l} + b_n \sin \frac{n\pi x}{l} \right],$$

where

$$a_n = \frac{1}{l} \int_{-l}^{l} f(x) \cos \frac{n\pi x}{l}\, dx, \qquad b_n = \frac{1}{l} \int_{-l}^{l} f(x) \sin \frac{n\pi x}{l}\, dx.$$

It follows that the Fourier cosine and sine series of a piecewise smooth function f on the interval $[0, l]$ are

(8.32) $$f(x) = \tfrac{1}{2}a_0 + \sum_{1}^{\infty} a_n \cos \frac{n\pi x}{l}, \qquad a_n = \frac{2}{l} \int_{0}^{l} f(x) \cos \frac{n\pi x}{l}\, dx,$$

and

(8.33) $$f(x) = \sum_{0}^{\infty} b_n \sin \frac{n\pi x}{l}, \qquad b_n = \frac{2}{l} \int_{0}^{l} f(x) \sin \frac{n\pi x}{l}\, dx.$$

We conclude with a few remarks comparing Taylor series and Fourier series,

$$f(x) = \sum_{0}^{\infty} \frac{f^{(n)}(0)}{n!} x^n \quad \text{and} \quad f(x) = \sum_{-\infty}^{\infty} c_n e^{in\pi x/l},$$

as ways of expanding a function f on an interval centered at the origin. First, Taylor series are only defined for functions of class C^{∞}, whereas the smoothness requirements for Fourier series are quite minimal. The Taylor coefficients $f^{(n)}(0)/n!$ depend only on the values of f in an arbitrarily small neighborhood of the origin, whereas the Fourier coefficients c_n depend on the values of f over the whole interval $[-l, l]$. The partial sums of the Taylor series provide an excellent approximation to $f(x)$ when $|x|$ is small but are often quite useless when $|x|$ is large; the partial sums of the Fourier series tend to approximate f about equally well over the whole interval $[-l, l]$. (This last statement is a bit of an oversimplification!)

Despite their differences, there is a connection between Taylor and Fourier series that is of considerable importance in more advanced mathematics. Namely, let us consider a power series $f(z) = \sum_{0}^{\infty} a_n z^n$ as a function of the *complex* variable z. If we write z in polar coordinates as $z = re^{i\theta}$ and fix r, we obtain a function $g(\theta) = f(re^{i\theta})$ of the variable θ, and the power series for f becomes a Fourier series for g: $g(\theta) = \sum_{0}^{\infty}(a_n r^n)e^{in\theta}$. (It is a special kind of Fourier series, however, since the coefficient of $e^{in\theta}$ vanishes for all $n < 0$.)

EXERCISES

1. Find the Fourier cosine series and the Fourier sine series of the following functions on the interval $[0, \pi]$. All of these series can be derived from the results of the examples and exercises in §8.1 without computing the coefficients from scratch.
   a. $f(\theta) = 1$.
   b. $f(\theta) = \sin\theta$.
   c. $f(\theta) = \theta^2$. (For the sine series, use Example 1 and Exercise 6 in §8.1.)
   d. $f(\theta) = \theta$ for $0 \le \theta \le \frac{1}{2}\pi$, $f(\theta) = \pi - \theta$ for $\frac{1}{2}\pi \le \theta \le \pi$.

2. Expand the given function in a series of the given type. As in Exercise 1, use previously derived results as much as possible.
   a. $f(x) = 1$; sine series on $[0, 1]$.
   b. $f(x) = 1$ for $0 < x < 2$, $f(x) = -1$ for $2 < x < 4$; cosine series on $[0, 4]$.
   c. $f(x) = lx - x^2$; sine series on $[0, l]$.
   d. $f(x) = e^x$; series of the form $\sum_{-\infty}^{\infty} c_n e^{2\pi i n x}$ on $[0, 1]$.

3. Suppose f is a piecewise continuous function on $[0, 2l]$ that satisfies $f(x) = f(2l - x)$ (that is, the graph of f is symmetric about the line $x = l$). Let a_n and b_n be the Fourier cosine and sine coefficients of f (given by (8.32) and (8.33) with l replaced by $2l$). Show that $a_n = 0$ for n odd and $b_n = 0$ for n even.

4. Show that a piecewise smooth function f on $[0, l]$ can be expanded in a series as follows:
$$f(x) = \sum_{1}^{\infty} B_n \sin\frac{(n - \frac{1}{2})\pi x}{l}, \qquad B_n = \frac{2}{l}\int_0^l f(x)\sin\frac{(n - \frac{1}{2})\pi x}{l}\,dx.$$

   (*Hint:* Extend f to $[0, 2l]$ by making it even about $x = l$, i.e., $f(x) = f(2l - x)$ for $x \in [l, 2l]$, and use Exercise 3.)

8.5 Applications to Differential Equations

Fourier series were originally invented in order to solve some boundary value problems of mathematical physics. In this section we study a few basic examples.

Heat Flow in an Insulated Rod. Consider a rod occupying the interval $[0, l]$, insulated so that no heat can enter or leave it, and let $f(x)$ be the temperature at position x and time $t = 0$. How does the temperature distribution evolve with time? (*Note:* Instead of thinking of a thin rod, one can think of a thick cylindrical slab

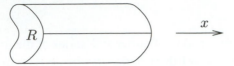

FIGURE 8.6: The cylindrical slab $\{(x, y, z) : 0 \le x \le l, \ (y, z) \in R\}$.

occupying the region where $0 \le x \le l$ and $(y, z) \in R$, where R is a bounded region in the yz-plane, as in Figure 8.6. The model of heat flow described here is valid under the hypothesis that the temperature depends only on x.)

Let $u(x, t)$ denote the temperature at position x and time t; thus u satisfies the *initial condition* $u(x, 0) = f(x)$. As we showed in §5.6, u obeys the *heat equation* $\partial_t u = k \partial_x^2 u$, where k is a positive constant (equal to K/σ in (5.42)). Since the rate of heat flow across the point x is proportional to $-\partial_x u(x, t)$ (Newton's law of cooling), the fact that no heat enters or leaves the ends of the rod means that u satisfies the *boundary conditions* $\partial_x u(0, t) = \partial_x u(l, t) = 0$. In summary,

$$(8.34) \qquad \frac{\partial u}{\partial t} = k \frac{\partial^2 u}{\partial x^2}, \qquad u(x, 0) = f(x), \qquad \frac{\partial u}{\partial x}(0, t) = \frac{\partial u}{\partial x}(l, t) = 0.$$

This is the problem we propose to solve.

The first step is to find a family of solutions of the heat equation satisfying the right boundary conditions by a device called **separation of variables**. The idea is to look for solutions of the form $u(x, t) = \varphi(x)\psi(t)$. For such a function, the heat equation becomes

$$\varphi(x)\psi'(t) = k\varphi''(x)\psi(t), \quad \text{or} \quad \frac{\psi'(t)}{k\psi(t)} = \frac{\varphi''(x)}{\varphi(x)}.$$

In this last equation, the quantities on the left and right depend only on t and x, respectively, so they must both be equal to a constant that we call $-\alpha$. Thus,

$$\psi'(t) = -k\alpha\psi(t), \qquad \varphi''(x) = -\alpha\varphi(x).$$

These are simple ordinary differential equations, and the general solutions are readily found:

$$\psi(t) = C_0 e^{-k\alpha t}, \qquad \varphi(x) = C_1 \cos \sqrt{\alpha}\, x + C_2 \sin \sqrt{\alpha}\, x.$$

We have thus found a large family of solutions of the heat equation of the form $\varphi(x)\psi(t)$. For these solutions, the boundary conditions $\partial_x u(0, t) = \partial_x u(l, t) = 0$

translate into the conditions $\varphi'(0) = \varphi'(l) = 0$. But

$$\varphi'(x) = \sqrt{\alpha}(-C_1 \sin \sqrt{\alpha}\, x + C_2 \cos \sqrt{\alpha}\, x),$$

so the condition $\varphi'(0) = 0$ forces $C_2 = 0$, and the condition $\varphi'(l) = 0$ then forces $\sqrt{\alpha}$ to be a multiple of π/l, or $\alpha = n^2\pi^2/l^2$ where n is an integer (which might as well be nonnegative). In short, we have obtained the following family of solutions of the heat equation together with the boundary conditions:

$$u_n(x,t) = \exp\left(\frac{-n^2\pi^2 kt}{l^2}\right) \cos \frac{n\pi x}{l} \qquad (n = 0, 1, 2, 3, \dots).$$

Since the heat equation and the boundary conditions are linear, we obtain more general solutions by taking linear combinations of these. In fact, we can pass to *infinite* linear combinations — that is, infinite series of the form

$$(8.35) \qquad u(x,t) = \sum_0^\infty a_n \exp\left(\frac{-n^2\pi^2 kt}{l^2}\right) \cos \frac{n\pi x}{l}.$$

Finally, we are ready to tackle the initial condition $u(x,0) = f(x)$. If we set $t = 0$ in (8.35), we obtain

$$u(x,0) = \sum_0^\infty a_n \cos \frac{n\pi x}{l},$$

so we can make $u(x,0)$ equal to $f(x)$ by taking the series on the right to be the Fourier cosine series of f, defined by (8.32)! (Note that the constant term, which we called $\frac{1}{2}a_0$ before, is called a_0 here.) In other words, to solve the problem (8.34), we take $u(x,t)$ to be defined by (8.35), where the coefficients a_n are given in terms of the initial data f by

$$a_0 = \frac{1}{l}\int_0^l f(x)\, dx, \qquad a_n = \frac{2}{l}\int_0^l f(x)\cos \frac{n\pi x}{l}\, dx \quad (n > 0).$$

At this point we should stop to verify that the proposed solution (8.35) of the problem (8.34) really works, as the passage from finite linear combinations to infinite series has the potential to cause difficulties. In fact, everything turns out quite nicely for this problem. In the first place, if the initial temperature distribution $f(x)$ is continuous and piecewise smooth (a reasonable physical assumption), the same will be true of its even $2l$-periodic extension, so by Theorem 8.29, its Fourier series is absolutely and uniformly convergent. In particular, $\sum_1^\infty |a_n| < \infty$. The absolute value of the nth term of the series in (8.35) is at most $|a_n|$, so the Weierstrass

M-test shows that this series converges absolutely and uniformly for $0 \le x \le l$ and $t \ge 0$ to define a continuous function $u(x,t)$ there. Moreover, for $t > 0$, the exponential factors in (8.35) decay rapidly as $n \to \infty$, which makes the convergence even better. In particular, repeated differentiation with respect to t or x introduces factors of n^k into the series, which are still overpowered by the decay of the exponential factors, so the differentiated series still converges absolutely and uniformly. If follows that $u(x,t)$ is of class C^∞ for $t > 0$ and that termwise differentiation is permissible; u therefore satisfies the heat equation and the boundary conditions because each term of the series does.

Two further remarks: First, as $t \to \infty$, the exponential factors in (8.35) all tend rapidly to zero except for the one with $n = 0$, and so $u(x,t)$ approaches the constant a_0, the mean value of f on the interval $[0,l]$. In physical terms this means that the rod approaches thermal equilibrium as time progresses. Second, the series (8.35) will usually diverge when $t < 0$, for then the exponential factors grow rather than decay! This corresponds to the physical fact that time is irreversible for diffusion processes governed by the heat equation.

The Vibrating String. We now study the vibrations of a string stretched across the interval $0 \le x \le l$ and fixed at the endpoints. (Think of a guitar string, and see Figure 8.7.) Here $u(x,t)$ will denote the displacement of the string (in a direction perpendicular to the x-axis) at position x and time t. The relevant differential equation is the *wave equation* $\partial_t^2 u = c^2 \partial_x^2 u$, where c is a positive constant that can be interpreted as the speed with which disturbances propagate down the string. (See Folland [6, pp. 388–90] or Kammler [10, pp. 526–7] for a derivation of the wave equation from physical principles.) Since the string is fixed at both ends, the *boundary conditions* for this problem are $u(0,t) = u(l,t) = 0$. As for initial conditions, since the wave equation is second-order in t we need to specify both the initial displacement $u(x,0)$ and the initial velocity $\partial_t u(x,0)$. Thus the problem we have to solve is

(8.36)
$$\frac{\partial^2 u}{\partial t^2} = c^2 \frac{\partial^2 u}{\partial x^2}, \quad u(x,0) = f(x), \quad \frac{\partial u}{\partial t}(x,0) = g(x), \quad u(0,t) = u(l,t) = 0,$$

where f and g are specified functions on $[0,l]$.

Again we employ the technique of separation of variables and look for solutions of the wave equation of the form $u(x,t) = \varphi(x)\psi(t)$. For such functions the wave equation becomes

$$\varphi(x)\psi''(t) = c^2 \varphi''(x)\psi(t), \quad \text{or} \quad \frac{\psi''(t)}{c^2\psi(t)} = \frac{\varphi''(x)}{\varphi(x)}.$$

FIGURE 8.7: A vibrating string fixed at its ends.

In the last equation, the quantities on the left and right depend only on t and x, respectively, so they are both equal to a constant $-\alpha$, and we obtain the ordinary differential equations

$$\psi''(t) + \alpha c^2 \psi(t) = 0, \qquad \varphi''(x) + \alpha \varphi(x) = 0.$$

The general solution of the second equation is

$$\varphi(x) = C_1 \cos \sqrt{\alpha}\, x + C_2 \sin \sqrt{\alpha}\, x.$$

The boundary condition $u(0,t) = 0$ forces C_1 to vanish, and then the boundary condition $u(l,t) = 0$ forces $\sqrt{\alpha}$ to be a multiple of π/l, so $\alpha = n^2\pi^2/l^2$ for some (positive) integer n. With this value of α, the general solution of the differential equation for ψ is

$$\psi(t) = b \cos \frac{n\pi ct}{l} + B \sin \frac{n\pi ct}{l}.$$

(The arbitrary constants are labeled b and B for reasons that will become clearer in a moment.)

For each positive integer n, we therefore have the solution

$$u_n(x,t) = \left(b_n \cos \frac{n\pi ct}{l} + B_n \sin \frac{n\pi ct}{l} \right) \sin \frac{n\pi x}{l}.$$

Taking linear combinations and passing to limits, we are led to the series solution

$$(8.37) \qquad u(x,t) = \sum_{1}^{\infty} \left(b_n \cos \frac{n\pi ct}{l} + B_n \sin \frac{n\pi ct}{l} \right) \sin \frac{n\pi x}{l}.$$

It remains to satisfy the initial conditions. Setting $t = 0$ in (8.37) yields

$$u(x,0) = \sum_{1}^{\infty} b_n \sin \frac{n\pi x}{l},$$

so we satisfy the condition $u(x,0) = f(x)$ by taking the b_n's to be the Fourier sine coefficients of f:

$$b_n = \frac{2}{l} \int_0^l f(x) \sin \frac{n\pi x}{l}\, dx.$$

Moreover, *formally* differentiating (8.37) with respect to t and then setting $t = 0$ yields

$$\frac{\partial u}{\partial t}(x,0) = \sum_{1}^{\infty} \frac{n\pi c}{l} B_n \sin \frac{n\pi x}{l},$$

so we should be able to satisfy the condition $\partial_t u(x,0) = g(x)$ by taking $n\pi c B_n/l$ to the nth Fourier sine coefficient of g:

$$B_n = \frac{2}{n\pi c} \int_0^l g(x) \sin \frac{n\pi x}{l} \, dx.$$

Again, we ask: Does this really work? It is physically reasonable to assume that the initial functions f and g are continuous and piecewise smooth and satisfy the boundary conditions $f(0) = f(l) = g(0) = g(l) = 0$. Their odd $2l$-periodic extensions will then have the same properties, so their Fourier series will be absolutely and uniformly convergent by Theorem 8.29. In particular, $\sum |b_n| < \infty$ and $\sum |nB_n| < \infty$, so by the Weierstrass M-test, the series (8.37) is absolutely and uniformly convergent for $0 \le x \le l$, $-\infty < t < \infty$. However, there is no reason for the twice-differentiated series that should represent $\partial_t^2 u$ or $c^2 \partial_x^2 u$, namely,

$$(8.38) \qquad -\frac{\pi^2 c^2}{l^2} \sum_{1}^{\infty} n^2 \left(b_n \cos \frac{n\pi ct}{l} + B_n \sin \frac{n\pi ct}{l} \right) \sin \frac{n\pi x}{l},$$

to converge. The extra factor of n^2 makes the terms larger, and there is no exponential decay anywhere to compensate. If we recall that the decay of Fourier coefficients is related to the degree of smoothness of the function in question, the contrast with the heat equation may be expressed as follows: The diffusion of heat tends to smooth out irregularities in the initial temperature distribution, but in wave motion, any initial roughness simply propagates without dying out.

We can obtain a positive result by imposing more differentiability hypotheses on f and g. If we assume that not only f and g but also the first two derivatives of f and the first derivative of g are continuous and piecewise smooth, and that not only f and g but also f'' vanishes at the endpoints (so that its odd periodic extension is continuous there), then Theorems 8.26 and 8.29 imply that $\sum n^2 |b_n| < \infty$ and $\sum n^2 |B_n| < \infty$, which guarantees the absolute and uniform convergence of (8.38). This is also enough to guarantee that the formal differentiation of (8.37) that led to the formula for the B_n's is valid.

However, these additional assumptions are rather unnatural from a physical point of view. The obvious model for a plucked string, for example, is to take f to be a piecewise linear function as in Figure 8.8. It is easy to calculate the

FIGURE 8.8: A model for a plucked string.

coefficients b_n explicitly for such an f (Exercise 4), and they turn out to decay exactly like n^{-2}. The series (8.37) therefore converges nicely, and we may expect it to provide a good description of the physical vibration of the string. On the other hand, the twice-differentiated series (8.38) does not converge at all, so it is hard to say in what sense (8.37) satisfies the wave equation. The resolution of this paradox is to expand our vision of what a solution of a differential equation ought to be and to develop a notion of "weak solution" that will encompass examples such as this one. But this is a more advanced topic; see, for example, Folland [6, §9.5].

Taking for granted that the series (8.37) really is the solution of the boundary value problem (8.36), we say a few words about its physical interpretation. Think of the string as being a producer of musical notes such as a guitar string. The nth term in the series (8.37), as a function of t, is a pure sine wave with frequency $n\pi c/l$, which represents a musical tone at a pure, definite pitch. The series (8.37) therefore shows how the sound produced by the string can be resolved into a superposition of these pure pitches. Typically, the coefficients b_n and B_n decrease as n increases, so that the largest contribution comes from the first term, $n = 1$. This is the "fundamental" pitch, and the higher n's are the "overtones" that give the note its particular tone quality.

Related Problems. The heat flow and vibration problems (8.34) and (8.36) can be modified by changing the boundary conditions; this leads to models of other interesting physical processes. Here are a few examples:

1. The boundary value problem

$$\frac{\partial u}{\partial t} = k\frac{\partial^2 u}{\partial x^2}, \qquad u(x,0) = f(x), \qquad u(0,t) = u(l,t) = 0$$

models the flow of heat in a rod that occupies the interval $0 \le x \le l$ when both ends are held at temperature zero — by immersing them in ice water, for instance. (Note that the heat equation doesn't care where the zero point of the temperature scale is located; if u is a solution, so is $u + c$ for any constant c. Of course, this means that the validity of the heat equation as a model for actual thermodynamic processes has its limitations, as absolute zero exists physically.) The method of solution is exactly the same as for the insulated problem (8.34), except that the boundary conditions for $\varphi(x)$ are $\varphi(0) = \varphi(l) = 0$. Thus, as in the vibrating string problem, we obtain $\varphi(x) = \sin(n\pi x/l)$, and the solution is given by

$$u(x,t) = \sum_1^\infty \exp\left(\frac{-n^2\pi^2 kt}{l^2}\right) \sin\frac{n\pi x}{l},$$

where $\sum b_n \sin(n\pi x/l)$ is the Fourier sine series of $f(x)$.

2. The boundary value problem

$$\frac{\partial^2 u}{\partial t^2} = c^2 \frac{\partial^2 u}{\partial x^2},$$

$$u(x,0) = f(x), \qquad \frac{\partial u}{\partial t}(x,0) = g(x), \qquad \frac{\partial u}{\partial x}(0,t) = \frac{\partial u}{\partial x}(l,t) = 0$$

models the vibration of air in a cylindrical pipe occupying the interval $0 \le x \le l$ that is open at both ends. (Examples: flutes and some organ pipes.) Here $u(x,t)$ represents the longitudinal displacement of the air at position x and time t. The boundary conditions $\partial_x u(0,t) = \partial_x u(l,t) = 0$ come from the fact that the change in air pressure due to the displacement u is proportional to $\partial_x u$, and the air pressure at both ends must remain equal to the ambient air pressure. Again, the solution is very similar to (8.37) except that it involves cosines instead of sines in x:

$$u(x,t) = \tfrac{1}{2}(a_0 + A_0 t) + \sum_1^\infty \left(a_n \cos\frac{n\pi ct}{l} + A_n \sin\frac{n\pi ct}{l} \right) \cos\frac{n\pi x}{l},$$

where $\tfrac{1}{2}a_0 + \sum_1^\infty a_n \cos(n\pi x/l)$ and $\tfrac{1}{2}A_0 + \sum_1^\infty (n\pi c A_n/l) \cos(n\pi x/l)$ are the Fourier cosine series of f and g, respectively. (The term $\tfrac{1}{2}(a_0 + A_0 t)$ represents a flow of air down the tube with constant velocity, of no importance for the vibrations.) As with the vibrating string, the vibrations of the pipe are a superposition of vibrations at the definite frequencies $n\pi c/l$ ($n = 1, 2, 3, \ldots$).

3. We can also mix the two types of boundary conditions we have been considering: for the heat equation,

$$\frac{\partial u}{\partial t} = k\frac{\partial^2 u}{\partial x^2}, \qquad u(x,0) = f(x), \qquad u(0,t) = \frac{\partial u}{\partial x}(l,t) = 0,$$

or the wave equation,

$$\frac{\partial^2 u}{\partial t^2} = c^2 \frac{\partial^2 u}{\partial x^2},$$

$$u(x,0) = f(x), \qquad \frac{\partial u}{\partial t}(x,0) = g(x), \qquad u(0,t) = \frac{\partial u}{\partial x}(l,t) = 0.$$

The first of these models heat flow in a rod where one end is held at temperature zero and the other is insulated; the second models vibrations of air in cylindrical pipes where one end is closed and the other is open, such as clarinets and some organ pipes. In both of them, separation of variables leads to the ordinary differential equation $\varphi''(x) = -\alpha\varphi(x)$ with boundary conditions $\varphi(0) = \varphi'(l) = 0$. The general solution of the differential equation is $\varphi(x) = C_1 \cos\sqrt{\alpha}x + C_2 \sin\sqrt{\alpha}x$; the condition $\varphi(0) = 0$ forces C_1 to vanish, and then the condition $\varphi'(l) = 0$ forces $\sqrt{\alpha}$ to be of the form $(n - \frac{1}{2})\pi/l$ with n a positive integer. We are therefore led to try to expand the initial functions in a series of the form

$$f(x) = \sum_1^\infty a_n \sin(n - \tfrac{1}{2})\frac{\pi x}{l}.$$

This can indeed be done; the technique for reducing this problem to one of ordinary Fourier sine series is outlined in Exercise 4 of §8.4.

It is interesting to note that the resulting frequencies for the vibrating pipe are $(n - \frac{1}{2})\pi c/l$ $(n = 1, 2, 3, \ldots)$. In particular, the fundamental frequency for a pipe closed at one end and open at the other, namely $\frac{1}{2}\pi c/l$, is half as great as for a pipe of equal length that is open at both ends. Moreover, only the odd-numbered multiples of this fundamental frequency occur as "harmonics" for half-open pipes, whereas all integer multiples occur for open pipes; as a result, the two kinds of pipes produce notes of different tone qualities.

4. Clearly there are many other variations to be played on this theme — different boundary conditions, other differential equations, and so on. A few further examples are outlined in the exercises, and we shall indicate a more general framework in which such problems can be studied in the next section.

EXERCISES

1. A rod 100 cm long is insulated along its length and at both ends. Suppose that its initial temperature is $u(x, 0) = x$ (x in cm, u in °C, t in sec, $0 \le x \le 100$), and that its diffusivity coefficient k is 1.1 cm²/sec (about right if the rod is made of copper).

   a. Find the temperature $u(x, t)$ for $t > 0$. (For the relevant Fourier series, see Example 1 of §8.4.)

   b. Show that the first three nonvanishing terms of the series (including the constant term) give the temperature accurately to within 1° when $t = 60$ (one minute after starting). What are $u(0, 60)$, $u(10, 60)$, and $u(40, 60)$ to the nearest 1°? (*Hint:* $\sum_1^\infty (2n - 1)^{-2} = \pi^2/8$, so $\sum_3^\infty (2n - 1)^{-2} = (\pi^2/8) - 1 - \frac{1}{9} \approx 0.123$.)

   c. Show that $u(x, t)$ is within $1°$ of its equilibrium value of $50°$ for all x when $t \geq 3600$ (i.e., after one hour). (Don't work too hard; crude estimates are enough.)

2. Find the temperature function $u(\theta, t)$ $(t > 0)$ for a rod bent into the shape of a circular hoop, given the initial temperature $u(\theta, 0) = f(\theta)$. (Here θ denotes the angular coordinate on the circle, and the boundary conditions for a straight rod are replaced by the requirement that u should be a 2π-periodic function of θ.)

3. As we found in §5.6, the inhomogeneous heat equation $\partial_t u = k\partial_x^2 u + G$ can be used to model heat flow in a rod when the total amount of heat energy is not constant; here G is a function of x and t, with units of degrees per unit time, that accounts for the addition or subtraction of heat from the rod. Let us solve the initial value problem with constant-temperature boundary conditions,

$$\partial_t u = k\partial_x^2 u + G, \qquad u(x, 0) = f(x), \qquad u(0, t) = u(l, t) = 0,$$

making appropriate assumptions on f and G so that Fourier expansions are valid. Motivated by the solution (8.35) for the special case $G \equiv 0$, we expand everything in a Fourier sine series. That is, for each t we write $G(x, t) = \sum_1^\infty \beta_n(t) \sin(n\pi x/l)$, and we try to find a solution in the form $u(x, t) = \sum_1^\infty b_n(t) \sin(n\pi x/l)$, where the coefficients $b_n(t)$ are to be determined. Plug this into the equation $\partial_t u = k\partial_x^2 u + G$ to obtain an ordinary differential equation for each $b_n(t)$, with initial condition determined by the requirement that $\sum_1^\infty b_n(0) \sin(n\pi x/l)$ should be the Fourier sine series of $f(x)$. Then solve these ordinary differential equations to obtain u. What conditions on f and G will guarantee the validity of these calculations?

4. Consider a vibrating string occupying the interval $[0, l]$. Suppose the string is plucked at $x = b$ $(0 < b < l)$ so that its initial displacement $u(x, 0)$ is mx/b for $0 \leq x \leq b$ and $m(l - x)/(l - b)$ for $b \leq x \leq l$ (that is, $u(x, 0)$ is linear on $[0, b]$ and on $[b, l]$, and equal to m at $x = b$), and its initial velocity $\partial_t u(x, 0)$ is zero. (*Note:* For this to be a realistic model of a plucked string, we should have $l \gg m$.)

   a. Find the Fourier series for $u(x, t)$ for $t > 0$. (The result of Exercise 2 of §8.3 can be used.)

   b. Compute the coefficients $b_1, \ldots, b_5$ of the first five terms (notation as in (8.37)) numerically when $b = (0.4)l$ and when $b = (0.1)l$. Observe that the higher frequencies contribute a lot more to $u(x, t)$ when $b = (0.1)l$ than when $b = (0.4)l$. (Musically: Plucking a string nearer the end gives a note with more "harmonics.")

5. The model for a vibrating string given by the wave equation is unrealistic because it predicts that the vibration will continue forever without dying out. Real strings, however, are not perfectly elastic, so the vibrational energy is gradually dissipated. A better model is obtained by the following modification of the wave equation:

$$\partial_t^2 u = c^2 \partial_x^2 u - 2\delta \partial_t u,$$

where δ is a small positive constant. (The left side is the acceleration, and the terms on the right are the effects of the elastic restoring force and the damping force that tends to slow the motion down. The factor of 2 is just for convenience.) Find the general solution of this differential equation subject to the boundary conditions $u(0, t) = u(l, t) = 0$ by modifying the method used in the text for the ordinary wave equation. Assume that $\delta < \pi c/l$. You should find that the solutions decay exponentially in time and that the frequencies decrease as the damping constant δ increases.

Exercises 6 and 7 concern the **Dirichlet problem** for a bounded open set $S \subset \mathbb{R}^2$: Given a function f on the boundary ∂S, find a solution of Laplace's equation $\partial_x^2 u + \partial_y^2 u = 0$ on S such that $u = f$ on ∂S. (A physical interpretation: Find the steady-state distribution of heat in S when the temperature on the boundary is given.)

6. Consider the Dirichlet problem for a rectangle:

$$\partial_x^2 u + \partial_y^2 u = 0 \text{ for } 0 < x < l, \ 0 < y < L;$$
$$u(x, 0) = f_1(x), \quad u(x, L) = f_2(x), \quad u(0, y) = g_1(y), \quad u(l, y) = g_2(y).$$

   a. Suppose we can solve this problem in the two special cases $g_1 = g_2 = 0$ and $f_1 = f_2 = 0$. How can the solutions u_1 and u_2 for these cases be combined to yield the solution for the general case?

   b. Henceforth we assume that $g_1 = g_2 = 0$ (the case $f_1 = f_2 = 0$ is similar). Use separation of variables to find solutions of Laplace's equation satisfying $u(0, y) = u(l, y) = 0$ in the form $u(x, y) = \varphi(x)\psi(y)$; then use Fourier techniques to find the (infinite) linear combination of these solutions that satisfies $u(x, 0) = f_1(x)$ and $u(x, L) = f_2(x)$. (*Hint:* The general solution of $\psi'' - c^2\psi = 0$ can be written in the form $\psi(y) = a \sinh cy + b \sinh c(L - y)$. [Why?] This form of the solution is more convenient than the more obvious $a \sinh cy + b \cosh cy$.)

7. Consider the Dirichlet problem for the unit disc:

$$\partial_x^2 u + \partial_y^2 u = 0 \text{ for } x^2 + y^2 < 1, \quad u(\cos\theta, \sin\theta) = f(\theta).$$

If we think of u as a function of the polar coordinates (r, θ) rather than the Cartesian coordinates (x, y), by Proposition 2.51 this becomes

$$r^2 \partial_r^2 u + r \partial_r u + \partial_\theta^2 u = 0 \text{ for } r < 1, \quad u(1, \theta) = f(\theta).$$

a. Use separation of variables to find solutions of this differential equation in the form $u(r, \theta) = \varphi(r)\psi(\theta)$. Keep in mind that the solutions must be 2π-periodic functions of θ and that they must be smooth at the origin, where $r = 0$ and θ is undefined. (*Hint:* The general solution of the Euler equation $r^2 \varphi'' + r\varphi' - c^2 \varphi = 0$ is $\varphi(r) = ar^c + br^{-c}$ if $c \neq 0$, $a + b \log r$ if $c = 0$.) Then use Fourier techniques to find the (infinite) linear combination of these solutions that satisfies $u(1, \theta) = f(\theta)$.

b. You should find that $u(r, \theta)$ equals $A_r f(\theta)$, the Abel approximant to f defined by (8.19). Use (8.20) and (8.22) to derive the **Poisson integral formula** for the solution:

$$u(r, \theta) = \frac{1}{2\pi} \int_{-\pi}^{\pi} \frac{1 - r^2}{1 + r^2 - 2r \cos \varphi} f(\theta - \varphi) \, d\varphi.$$

8.6   The Infinite-Dimensional Geometry of Fourier Series

In this section we shall re-examine the notion of Fourier series in the light of a profound analogy with certain ideas from vector algebra. We begin with a quick review of the latter.

When expressed in algebraic terms, the concepts of Euclidean geometry in n dimensions are based on the vector-space structure of $\mathbb{R}^n$ (that is, the operations of vector addition and scalar multiplication), together with the dot product or inner product $\mathbf{a} \cdot \mathbf{b}$, in terms of which we can define lengths ($|\mathbf{a}| = (\mathbf{a} \cdot \mathbf{a})^{1/2}$) and angles (the angle from $\mathbf{a}$ to $\mathbf{b}$ is $\arccos(\mathbf{a} \cdot \mathbf{b}/|\mathbf{a}|\,|\mathbf{b}|)$). The "natural" coordinate systems for this geometry are the ones arising from an *orthonormal basis* for $\mathbb{R}^n$, that is, a basis $\mathbf{u}_1, \ldots, \mathbf{u}_n$ such that $\mathbf{u}_j \cdot \mathbf{u}_k$ equals 0 for $j \neq k$ and 1 for $j = k$. The formula for expressing an arbitrary vector $\mathbf{x}$ in terms of such a basis is given very simply in terms of inner products:

$$\mathbf{x} = \sum_1^n c_j \mathbf{u}_j, \quad c_j = \mathbf{x} \cdot \mathbf{u}_j.$$

(The formula for c_j results from taking the inner product of both sides of the equation $\mathbf{x} = \sum_1^n c_k \mathbf{u}_k$ with $\mathbf{u}_j$ to yield $\mathbf{x} \cdot \mathbf{u}_j = \sum_1^n c_k \mathbf{u}_k \cdot \mathbf{u}_j = c_j$.)

Similar ideas underlie the study of *complex n-dimensional vectors*. The main difference is that, since the absolute value $|z|$ of a complex number z is given by $(z\bar{z})^{1/2}$ rather than $(z^2)^{1/2}$, the appropriate definition of inner product is

$$(8.39) \qquad \langle \mathbf{a}, \mathbf{b} \rangle = \sum_{1}^{n} a_j \bar{b}_j \qquad (\mathbf{a}, \mathbf{b} \in \mathbb{C}^n).$$

(Recall that the conjugate $\bar{z}$ of a complex number $z = x + iy$ $(x, y \in \mathbb{R})$ is defined to be $x - iy$. The notation $\mathbf{a} \cdot \mathbf{b}$ is also used for the complex inner product, but we introduce the new notation $\langle \mathbf{a}, \mathbf{b} \rangle$ to avoid confusion with the real case and to prepare for further developments.) Thus $\langle \mathbf{a}, \mathbf{b} \rangle$ is a linear function of $\mathbf{a}$ but a conjugate-linear function of $\mathbf{b}$ (meaning that $\langle \mathbf{a}, c\mathbf{b} \rangle$ equals $\bar{c} \langle \mathbf{a}, \mathbf{b} \rangle$ rather than $c \langle \mathbf{a}, \mathbf{b} \rangle$), and $\langle \mathbf{b}, \mathbf{a} \rangle = \overline{\langle \mathbf{a}, \mathbf{b} \rangle}$. The magnitude or norm of the vector $\mathbf{a}$ is still given by $|\mathbf{a}| = \langle \mathbf{a}, \mathbf{a} \rangle^{1/2}$, and we still call two vectors $\mathbf{a}$ and $\mathbf{b}$ *orthogonal* if $\langle \mathbf{a}, \mathbf{b} \rangle = 0$. As in the real case, a basis $\mathbf{u}_1, \ldots, \mathbf{u}_n$ for $\mathbb{C}^n$ is *orthonormal* if $\langle \mathbf{u}_j, \mathbf{u}_k \rangle$ is 0 if $j \neq k$ and 1 if $j = k$. The expansion formula for a vector $\mathbf{x} \in \mathbb{C}^n$ with respect to an orthonormal basis is exactly the same:

$$\mathbf{x} = \sum_{1}^{n} c_j \mathbf{u}_j, \qquad c_j = \langle \mathbf{x}, \mathbf{u}_j \rangle.$$

If the basis $\{\mathbf{u}_j\}$ is orthogonal ($\langle \mathbf{u}_j, \mathbf{u}_k \rangle = 0$ for $j \neq k$) but not normalized ($\|\mathbf{u}_j\|$ not necessarily equal to 1), the formula becomes

$$(8.40) \qquad \mathbf{x} = \sum_{1}^{n} c_j \mathbf{u}_j, \qquad c_j = \frac{\langle \mathbf{x}, \mathbf{u}_j \rangle}{\|\mathbf{u}_j\|^2}.$$

Now we are ready to make the conceptual leap from the discrete and finite-dimensional to the continuous and infinite dimensional. Suppose we are studying functions on an interval $[a, b]$ — let us say, piecewise continuous, complex-valued ones. We regard such a function f as a "vector" whose "components" are the *values $f(x)$* as x ranges over $[a, b]$. We define the **inner product** of two functions f and g just as in (8.39) except that the sum is replaced by an integral:

$$(8.41) \qquad \langle f, g \rangle = \int_a^b f(x) \overline{g(x)} \, dx.$$

Further, we define the **norm** of a function f to be

$$\|f\| = \langle f, f \rangle^{1/2} = \left[\int_a^b |f(x)|^2 \, dx \right]^{1/2},$$

and we define two functions f and g to be **orthogonal** on $[a, b]$ if $\langle f, g \rangle = 0$. A sequence of functions $\{\varphi_n\}$ is called **orthogonal** if $\langle \varphi_m, \varphi_n \rangle = 0$ for $m \neq n$, and **orthonormal** if, in addition, $\|\varphi_n\| = 1$ for all n.

For example, take the interval $[a, b]$ to be $[-\pi, \pi]$, and define $e_n(x) = e^{inx}$. Then, since $\overline{e^{inx}} = e^{-inx}$, by (8.5) we have

$$\langle e_m, e_n \rangle = \int_{-\pi}^{\pi} e^{i(m-n)x} \, dx = \begin{cases} 2\pi & \text{if } m = n, \\ 0 & \text{otherwise.} \end{cases}$$

Thus $\{e_n\}_{-\infty}^{\infty}$ is an orthogonal set; the corresponding orthonormal set is $\{\varphi_n\}_{-\infty}^{\infty}$ where $\varphi_n = (2\pi)^{-1/2} e_n$. The formula for the Fourier series of a function f,

$$f = \sum_{-\infty}^{\infty} c_n e_n, \qquad c_n = \frac{1}{2\pi} \int_{-\pi}^{\pi} f(x) e^{-inx} \, dx = \frac{\langle f, e_n \rangle}{\|e_n\|^2},$$

is an exact analogue of the formula (8.40) for the expansion of a vector in terms of an orthogonal basis!

A similar interpretation holds for Fourier cosine and sine series. To wit, it is easy to verify (Exercise 1) that $\{\cos n\pi x/l\}_0^{\infty}$ and $\{\sin n\pi x/l\}_1^{\infty}$ are orthogonal sets on the interval $[0, l]$, and that the formulas for the Fourier cosine and sine coefficients of a function f on $[0, l]$ are analogous to (8.40).

There are some unanswered questions here, however. The inner product $\langle f, g \rangle$ makes sense when f and g are piecewise continuous on $[a, b]$, but we have proved the validity of Fourier expansions only for piecewise smooth functions. So, what is the "right" class of functions to consider here? Can we make sense out of Fourier series for functions that may not be piecewise smooth?

The key insight is that *pointwise convergence is the wrong notion of convergence in this situation.* Instead, we should use a notion of convergence that arises from the geometry of the inner product. That is, we think of the set

$PC(a, b) = $ set of all piecewise continuous complex-valued functions on $[a, b]$

as an "infinite-dimensional Euclidean space" with the notions of length and angle given by the inner product (8.41). The "distance" between two functions is to be interpreted as the norm of their difference,

$$\text{Distance from } f \text{ to } g = \|f - g\| = \left[\int_a^b |f(x) - g(x)|^2 \, dx \right]^{1/2},$$

and the corresponding notion of convergence is that

$$f_k \to f \quad \Longleftrightarrow \quad \|f_k - f\| \to 0, \text{ i.e., } \int_a^b |f_k(x) - f(x)|^2 \, dx \to 0.$$

This notion of convergence is called **convergence in norm** or **mean-square convergence**.

Note. If the distance $\|f - g\|$ between two piecewise continuous functions is zero, it does not follow that f is identically equal to g, but only that $f(x) = g(x)$ for all except perhaps finitely many values of x. In this setting, it is appropriate not to worry about this technicality and to think of two functions as being the same when they differ only at finitely many points. This issue already arose in connection with the behavior of the Fourier series of f at points where f is discontinuous (cf. Corollary 8.18).

Mean-square convergence is rather different from pointwise convergence, and neither one implies the other. For example, let us take $[a, b] = [-1, 1]$. If

$$f_k(x) = \begin{cases} k & \text{if } 0 < x < 1/k, \\ 0 & \text{otherwise,} \end{cases}$$

then $f_k \to 0$ pointwise but $\|f_k\| = (\int_0^{1/k} k^2 \, dx)^{1/2} = \sqrt{k} \to \infty$. On the other hand, if

$$g_k(x) = \begin{cases} 1 & \text{if } -1/k < x < 1/k, \\ 0 & \text{otherwise,} \end{cases}$$

then $\|g_k\| = (\int_{-1/k}^{1/k} dx)^{1/2} = \sqrt{2/k} \to 0$, but $g_k(0) = 1 \nrightarrow 0$. (By replacing the interval $(-1/k, 1/k)$ here by an interval I_k whose length tends to 0 but whose midpoint oscillates back and forth within the interval $[-1, 1]$ as $k \to \infty$, one can construct examples of sequences $\{g_k\}$ that converge in norm but do not converge at any point.) However, for uniform rather than pointwise convergence there is something to say.

8.42 Proposition. *If $f_k \to f$ uniformly on $[a, b]$, then $f_k \to f$ in norm on $[a, b]$.*

Proof. If $f_k \to f$ uniformly, there is a sequence $\{C_k\}$ of constants such that $|f_k(x) - f(x)| \leq C_k$ for all $x \in [a, b]$ and $C_k \to 0$, so

$$\int_a^b |f_k(x) - f(x)|^2 \, dx \leq (b - a)C_k^2 \to 0.$$

$\square$

More generally, $f_k \to f$ in norm provided that $f_k \to f$ pointwise and there is a constant C such that $|f_k(x)| \leq C$ for all k and all $x \in [a, b]$; this follows from the bounded convergence theorem (4.52).

The introduction of norm convergence is justified by the fact that *the Fourier series of any piecewise continuous function f on $[-\pi, \pi]$ converges in norm to f.* This is a substantial result, but there is more to be said before we state a formal theorem.

The space $PC(a, b)$ of piecewise continuous functions on $[a, b]$ fails to be a good infinite-dimensional analogue of Euclidean space in one crucial respect: it is not *complete*. That is, if $\{f_k\}$ is a sequence in $PC(a, b)$ such that $\|f_j - f_k\| \to 0$ as $j, k \to \infty$, there may not be a function $f \in PC(a, b)$ such that $\|f_k - f\| \to 0$. For example, with $[a, b] = [0, 1]$, let

$$f_k(x) = \begin{cases} x^{-1/4} & \text{if } x > 1/k, \\ 0 & \text{otherwise.} \end{cases}$$

It is easily verified that $\|f_j - f_k\|^2 = 2|j^{-1/2} - k^{-1/2}| \to 0$ as $j, k \to \infty$. However, the function to which the f_k's are converging is clearly $f(x) = x^{-1/4}$ $(x > 0)$, which does not belong to $PC(0, 1)$ because it blows up at 0. Thus, to fill in the "holes" in $PC(a, b)$ one will have to deal with unbounded functions and improper integrals. But even this is not enough; with more cleverness one can construct examples where the limiting function f is not (Riemann) integrable on any subinterval of $[a, b]$.

What is needed here is the Lebesgue integral, which handles integrals of unbounded and discontinuous functions more capably (see §4.8). The appropriate "completion" of the space $PC(a, b)$ is the space of **square-integrable** functions,

$$L^2(a, b) = \left\{ f : f \text{ is Lebesgue measurable on } [a, b] \text{ and } \int_a^b |f(x)|^2 \, dx < \infty \right\},$$

where the integral is a Lebesgue integral. (The name "L^2" is pronounced "L-two"; the L is in honor of Lebesgue and the 2 refers to the exponent in $|f(x)|^2$.)

We can now state the general convergence theorem for Fourier series.

8.43 Theorem. *Let $e_n(\theta) = e^{in\theta}$.*

  a. If $f \in L^2(-\pi, \pi)$, the Fourier series

$$\sum_{-\infty}^{\infty} c_n e_n, \qquad c_n = \frac{1}{2\pi} \int_{-\pi}^{\pi} f(\theta) e^{-in\theta} \, d\theta,$$

  converges in norm to f, that is,

$$\lim_{N \to \infty} \int_{-\pi}^{\pi} \left| f(\theta) - \sum_{-N}^{N} c_n e^{in\theta} \right|^2 \, d\theta = 0.$$

b. *Bessel's inequality is an equality: For any $f \in L^2(-\pi, \pi)$,*

$$\sum_{-\infty}^{\infty} |c_n|^2 = \frac{1}{2\pi} \int_{-\pi}^{\pi} |f(\theta)|^2 \, d\theta.$$

c. *If $\{c_n\}_{-\infty}^{\infty}$ is any sequence of complex numbers such that $\sum_{-\infty}^{\infty} |c_n|^2$ converges, then the series $\sum_{-\infty}^{\infty} c_n e_n$ converges in norm to a function in $L^2(-\pi, \pi)$.*

Proof. A full proof of Theorem 8.43 is beyond the scope of this book. (One may be found in Jones [9, p. 325] or Rudin [18, pp. 328ff.].) However, the idea is as follows. If f is continuous and piecewise smooth, we know that its Fourier series converges uniformly (Theorem 8.29) and hence in norm, so (a) is valid for such f. We then obtain the result for arbitrary $f \in L^2(-\pi, \pi)$ by a limiting argument that involves proving that any function in $L^2(-\pi, \pi)$ is the limit in norm of a sequence of continuous, piecewise smooth functions. (A partial result in this direction is indicated in Exercise 7.) (b) follows easily because, as we showed in the proof of Bessel's inequality,

$$\frac{1}{2\pi} \int_{-\pi}^{\pi} |f(\theta)|^2 \, d\theta - \sum_{-N}^{N} |c_n|^2 = \frac{1}{2\pi} \int_{-\pi}^{\pi} \left| f(\theta) - \sum_{-N}^{N} c_n e^{in\theta} \right|^2 d\theta,$$

and the integral on the right tends to zero as $N \to \infty$ since the series converges in norm to f. (c) follows from (b) and the completeness of $L^2(-\pi, \pi)$. Indeed, by (b),

$$\int_{-\pi}^{\pi} \left| \sum_{M \le |n| \le N} c_n e^{in\theta} \right|^2 d\theta = 2\pi \sum_{M \le |n| \le N} |c_n|^2,$$

so the partial sums of the series $\sum c_n e_n$ are Cauchy in norm; by completeness, the series converges in norm. $\square$

Theorem 8.43 says that $\{e^{inx}\}_{-\infty}^{\infty}$ is an **orthogonal basis** for $L^2(-\pi, \pi)$, that is, an orthogonal set with the property that every function in $L^2(-\pi, \pi)$ can be expanded uniquely as a norm-convergent series of scalar multiples of functions in the set. Likewise, $\{\cos nx\}_0^{\infty}$ and $\{\sin nx\}_1^{\infty}$ are orthogonal bases for $L^2(0, \pi)$; see Exercises 1 and 2.

The equality in Theorem 8.43b,

(8.44) $$\sum_{-\infty}^{\infty} |c_n|^2 = \frac{1}{2\pi} \int_{-\pi}^{\pi} |f(\theta)|^2 \, d\theta,$$

is known as **Parseval's identity**; it is the infinite-dimensional analogue of the Pythagorean theorem for finite-dimensional vectors, if we think of f as an infinite-dimensional vector and the c_n's as the components of this vector with respect to the orthogonal basis $\{e_n\}$. The factor of 2π is there because $\|e_n\|^2 = 2\pi$.

As an illustration of the use of Parseval's identity, we give another derivation of the formula $\sum_1^\infty n^{-2} = \pi^2/6$. (The first one was in Example 2 of §8.2.) Let f be the sawtooth wave function ($f(\theta) = \theta$ for $|\theta| < \pi$). We calculated in §8.1 that its Fourier coefficients are given by $c_0 = 0$ and $c_n = (-1)^{n+1}/in$ for $n \neq 0$. Therefore,

$$\sum_1^\infty \frac{1}{n^2} = \frac{1}{2}\left[\sum_1^\infty \frac{1}{n^2} + \sum_{-\infty}^{-1} \frac{1}{n^2}\right] = \frac{1}{2}\sum_{-\infty}^\infty |c_n|^2 = \frac{1}{4\pi}\int_{-\pi}^\pi \theta^2\,d\theta = \frac{\pi^2}{6}.$$

Parseval's identity easily yields the following generalization of itself, which is often useful:

8.45 Corollary. *If $f, g \in L^2(\pi, \pi)$ have the Fourier series $\sum c_n e_n$ and $\sum \gamma_n e_n$, then*

(8.46)
$$\sum_{-\infty}^\infty c_n\overline{\gamma}_n = \frac{1}{2\pi}\int_{-\pi}^\pi f(\theta)\overline{g(\theta)}\,d\theta.$$

Proof. We apply (8.44) to the functions f, g, and $f + g$:

$$\sum (|c_n|^2 + 2\,\mathrm{Re}\,c_n\overline{\gamma}_n + |\gamma_n|^2) = \sum |c_n + \gamma_n|^2$$
$$= \frac{1}{2\pi}\int_{-\pi}^\pi |f(\theta) + g(\theta)|^2\,d\theta = \frac{1}{2\pi}\int_{-\pi}^\pi (|f(\theta)|^2 + 2\,\mathrm{Re}\,f(\theta)\overline{g(\theta)} + |g(\theta)|^2)\,d\theta$$
$$= \sum |c_n|^2 + \frac{1}{\pi}\,\mathrm{Re}\int_{-\pi}^\pi f(\theta)\overline{g(\theta)}\,d\theta + \sum |\gamma_n|^2.$$

It follows that $\mathrm{Re}\sum c_n\overline{\gamma}_n = \mathrm{Re}(1/2\pi)\int_{-\pi}^\pi f(\theta)\overline{g(\theta)}\,d\theta$. The same calculation, with f replaced by if, shows that the imaginary parts are also equal.   □

The Fourier bases $\{e^{inx}\}_{-\infty}^\infty$, $\{\cos nx\}_0^\infty$ and $\{\sin nx\}_1^\infty$ play a special role among all the orthogonal bases for $L^2(-\pi, \pi)$ and $L^2(0, \pi)$ because these functions are *eigenfunctions* for the differential operators d/dx and d^2/dx^2. To explain this in more detail, we recall that an *eigenvector* for a linear transformation T on $\mathbb{R}^n$ or $\mathbb{C}^n$ is a nonzero vector $\mathbf{x}$ such that $T\mathbf{x} = \lambda\mathbf{x}$ for some scalar λ. (See Appendix A, (A.56)–(A.58)). In our situation, the "vectors" are functions in $L^2(-\pi, \pi)$ or $L^2(0, \pi)$, and the linear transformation in question is d/dx or d^2/dx^2, defined not on the whole L^2 space but on a suitable subspace of functions that possess the requisite derivatives *and* satisfy certain boundary conditions. Indeed, we have

$$\frac{d}{dx}e^{inx} = ine^{inx}, \qquad \frac{d^2}{dx^2}\cos nx = -n^2 \cos nx, \qquad \frac{d^2}{dx^2}\sin nx = -n^2 \sin nx.$$

The functions e^{inx} are precisely the eigenfunctions of d/dx on $[-\pi, \pi]$ that satisfy the periodicity condition $f(-\pi) = f(\pi)$, and the functions $\cos nx$ and $\sin nx$ are precisely the eigenfunctions of d^2/dx^2 on $[0, \pi]$ that satisfy the boundary conditions $f'(0) = f'(\pi) = 0$ and $f(0) = f(\pi) = 0$, respectively. The Fourier expansion of a function therefore provides the analogue of the spectral theorem (A.58) for these fundamental differential operators, with all the resulting simplifications that one expects when one finds an orthonormal eigenbasis for a matrix.

For example, we can rederive the solution (8.35) of the insulated heat flow problem (8.34) as follows. To solve the heat equation $\partial_t u = k\partial_x^2 u$ subject to the boundary conditions $\partial_x u(0, t) = \partial_x u(l, t) = 0$, we take u to be the sum of a series of eigenfunctions of ∂_x^2 satisfying these boundary conditions:

$$u(x, t) = \sum_{0}^{\infty} \alpha_n(t) \cos \frac{n\pi x}{l}.$$

Plugging this into the heat equation turns the *partial* differential equation $\partial_t u = k\partial_x^2 u$ into the *ordinary* differential equations $\alpha_n'(t) = -k(n\pi/l)^2 \alpha_n(t)$ for the coefficients. The latter are easily solved to yield $\alpha_n(t) = a_n e^{-k(n\pi/l)^2 t}$ and hence the solution (8.35).

There is an extensive theory of eigenfunction expansions associated to boundary value problems. Many such expansions yield interesting orthogonal bases for L^2 spaces. Others, in which there is a "continuous spectrum" instead of (or in addition to) a "discrete spectrum," involve integrals instead of (or in addition to) infinite series. A great deal of interesting mathematics has arisen from these ideas, and its ramifications spread far beyond the problems for which it was originally devised. An introduction to this subject can be found, for example, in Folland [6].

EXERCISES

1. Show that $\{\cos nx\}_0^\infty$ and $\{\sin nx\}_1^\infty$ are orthogonal sets of functions on $[0, \pi]$. What are the norms of these functions?

2. Deduce from Theorem 8.43 that if $f \in L^2(0, \pi)$, the Fourier cosine and sine series of f both converge to f in norm.

3. Determine the constants a, b, and c so that the functions

$$f_0(x) = 1, \qquad f_1(x) = x + a, \qquad f_2(x) = x^2 + bx + c$$

form an orthogonal set on $[0, 1]$.

4. Suppose $\{\varphi_n\}_1^\infty$ is an orthonormal set of functions on $[0, l]$, and let φ_n^+ and φ_n^- be the even and odd extensions of φ_n to $[-l, l]$. Show that $\{\varphi_n^+/\sqrt{2}\}_1^\infty \cup \{\varphi_n^-/\sqrt{2}\}_1^\infty$ is an orthonormal set on $[-l, l]$.

5. Suppose $\{\varphi_n\}_1^\infty$ is an orthonormal set of functions on $[a, b]$. Given $c > 0$ and $d \in \mathbb{R}$, let $\psi_n(x) = \sqrt{c}\,\varphi_n(cx + d)$. Show that $\{\psi_n\}_1^\infty$ is an orthonormal set on $[(a - d)/c, (b - d)/c]$.

6. Suppose $\{\varphi_n\}_1^\infty$ is an orthonormal set of functions on $[0, 1]$, and let $\psi_n(x) = \sqrt{2x}\,\varphi_n(x^2)$. Show that $\{\psi_n\}_1^\infty$ is also an orthonormal set on $[0, 1]$.

7. Show that any piecewise continuous function on $[a, b]$ is the limit in norm of a sequence of continuous functions on $[a, b]$ by the argument suggested by the following picture:

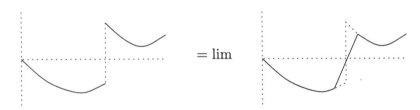

8. Show that in terms of the cosine and sine coefficients a_n and b_n defined by (8.7), Parseval's identity takes the form

$$\int_{-\pi}^{\pi} |f(\theta)|^2 \, d\theta = \frac{\pi}{2}|a_0|^2 + \pi \sum_{1}^{\infty} \left(|a_n|^2 + |b_n|^2 \right).$$

9. Evaluate the following series by applying Parseval's identity, in the form given in Exercise 8, to certain of the Fourier series found in the exercises of §8.1 and §8.3. (Remember that the constant term is $\frac{1}{2}a_0$, not a_0.)

a. $\displaystyle\sum_{1}^{\infty} \frac{1}{n^4}$

b. $\displaystyle\sum_{1}^{\infty} \frac{1}{(2n-1)^6}$

c. $\displaystyle\sum_{1}^{\infty} \frac{1}{n^8}$

d. $\sum_1^\infty \dfrac{\sin^2 na}{n^2}$ (First assume that $0 < a < \pi$, then deduce the general re-

sult.)

10. Suppose that f is 2π-periodic, real-valued, and of class C^1. Show that f' is orthogonal to f on $[-\pi, \pi]$ in two ways: (i) directly from the fact that $2ff' = (f^2)'$, and (ii) by expanding f in a Fourier series and using (8.46). (*Hint:* When f is real we have $c_{-n} = \bar{c}_n$; why?)

8.7   The Isoperimetric Inequality

We conclude this chapter by using Fourier analysis together with Green's theorem (thereby joining two of the main threads of this book) to show that among all simple closed curves in the plane with a given length, the circle is the one that encloses the greatest area.

First, a few preliminaries. Suppose $\mathbf{g} : [a, b] \to \mathbb{R}^2$ is a continuous, piece-wise smooth parametrized curve in the plane. (Thus, the components of $\mathbf{g}$ are continuous, piecewise smooth functions on $[a, b]$; $\mathbf{g}'(t)$ is defined except perhaps at finitely many points, and we make the usual nondegeneracy assumption that $\mathbf{g}'(t) \neq \mathbf{0}$.) The arc-length function $s = \varphi(t) = \int_a^t |\mathbf{g}'(u)| \, du$ is a continuous, piecewise smooth, strictly increasing function on $[a, b]$. It therefore has an inverse function, $t = \varphi^{-1}(s)$, with the same properties, defined on the interval $[0, L]$ where $L = \varphi(b)$ is the total length of the curve. We can then reparametrize the curve by $\mathbf{h}(s) = \mathbf{g}(\varphi^{-1}(s))$, $s \in [0, L]$; we then say that the curve is *parametrized by arc length*. In this parametrization, the speed $|\mathbf{h}'(s)|$ is identically equal to 1 (except at isolated points where it is undefined):

(8.47) $$|\mathbf{h}'(s)| = |\mathbf{g}'(\varphi^{-1}(s))[\varphi^{-1}]'(s)| = \frac{|\mathbf{g}'(t)|}{\varphi'(t)} = 1.$$

Now, suppose in addition that our curve is a *simple closed* curve; this means that, for $0 \leq s_1 < s_2 \leq L$, $\mathbf{h}(s_1) = \mathbf{h}(s_2)$ only when $s_1 = 0$ and $s_2 = L$. We can then extend the function $\mathbf{h}$ from $[0, L]$ to $\mathbb{R}$ by requiring it to be L-periodic; this extension is still continuous and piecewise smooth. (Indeed, this is the natural way to think of a simple closed curve. We think of $\theta = 2\pi s/L$ as the angular coordinate on a circle; then $\mathbf{h}(s)$ traces out the curve as θ goes once around the circle.)

Finally, we observe that we can identify $\mathbb{R}^2$ with the complex plane $\mathbb{C}$ and the vector-valued function $\mathbf{h} = (h_1, h_2)$ with the complex-valued function $\zeta = h_1 + ih_2$. The "velocity" $\mathbf{h}'(s)$ then turns into $\zeta'(s)$, and the condition (8.47) becomes $|\zeta'(s)| \equiv 1$.

Now we are ready to state our theorem:

8.48 Theorem (The Isoperimetric Inequality). *Suppose that C is a piecewise smooth, simple closed curve in the plane. Let L be the length of C and A the area of the region enclosed by C. Then $A \leq L^2/4\pi$, with equality if and only if C is a circle of radius $L/2\pi$.*

Proof. We identify the plane with $\mathbb{C}$. Dilating the plane by a factor of r, $z \to rz$, has the effect of multiplying the length of a curve by r and the area of a region by r^2, so it is enough to consider the case $L = 2\pi$, for which the conclusion is that $A \leq \pi$. By the preceding remarks, then, we can assume that C is given by $z = \zeta(s)$, where ζ is a continuous, piecewise smooth, 2π-periodic, complex-valued function on $\mathbb{R}$, and $|\zeta'(s)| \equiv 1$ (except at isolated points where $\zeta'(s)$ is undefined). We expand ζ in a Fourier series:

$$\zeta(s) = \sum_{-\infty}^{\infty} c_n e^{ins}.$$

Since ζ is continuous and piecewise smooth, the nth Fourier coefficient of ζ' is inc_n, by Theorem 8.26. Since $|\zeta'(s)| \equiv 1$, Parseval's identity implies that

$$(8.49) \qquad 1 = \frac{1}{2\pi} \int_{-\pi}^{\pi} |\zeta'(s)|^2 \, ds = \sum_{-\infty}^{\infty} n^2 |c_n|^2.$$

On the other hand, by Green's theorem (see Example 3 in §5.2), the area of the region enclosed by C is

$$A = \left| \frac{1}{2} \int_C x \, dy - y \, dx \right|.$$

(The absolute value is there because we do not specify whether C is positively or negatively oriented.) Moreover,

$$x \, dy - y \, dx = \mathrm{Im}\big[(x - iy)(dx + i \, dy)\big] = \mathrm{Im}\, \overline{z} \, dz,$$

so

$$A = \left| \tfrac{1}{2} \mathrm{Im} \int_C \overline{z} \, dz \right| = \tfrac{1}{2} \left| \mathrm{Im} \int_{-\pi}^{\pi} \overline{\zeta(s)} \zeta'(s) \, ds \right|.$$

Thus, by the general form (8.46) of Parseval's identity,

$$A = \pi \left| \mathrm{Im} \sum_{-\infty}^{\infty} \overline{c_n} \, inc_n \right| = \pi \left| \sum_{-\infty}^{\infty} n |c_n|^2 \right|.$$

Comparing this with (8.49) yields the desired upper bound for A:

$$A = \pi \left| \sum_{-\infty}^{\infty} n |c_n|^2 \right| \leq \pi \sum_{-\infty}^{\infty} |n| \, |c_n|^2 \leq \pi \sum_{-\infty}^{\infty} n^2 |c_n|^2 = \pi.$$

Moreover, the second inequality is strict unless $c_n = 0$ for $|n| > 1$. In that case, the first inequality becomes

$$\left| \, |c_1|^2 - |c_{-1}|^2 \right| \leq |c_1|^2 + |c_{-1}|^2,$$

which is strict unless either c_1 or c_{-1} vanishes. Thus $A < \pi$ unless $\zeta(s) = c_0 + c_1 e^{is}$ or $\zeta(s) = c_0 + c_{-1} e^{-is}$, both of which describe a circle centered at c_0, traversed counterclockwise or clockwise, respectively. (In either case the radius is 1 since $|c_{\pm 1}| = |\zeta'(s)| = 1$.) $\qquad\qquad\square$

Appendix A

SUMMARY OF LINEAR ALGEBRA

This appendix consists of a brief summary of the definitions and results from linear algebra that are needed in the text (and a little more). Brief indications of proofs are given where it is easy to do so, but lack of any proof does not necessarily mean that a statement is supposed to be obvious. More complete treatments can be found in texts on linear algebra such as Anton [1] and Lay [16].

A.1   Vectors

Most of the basic terminology concerning n-dimensional vectors is contained in §1.1; we introduce a few more items here.

(A.1)   If $\mathbf{x}_1, \ldots, \mathbf{x}_k$ are vectors in $\mathbb{R}^n$, any vector of the form

$$c\mathbf{x}_1 + c_2\mathbf{x}_2 + \cdots + c_k\mathbf{x}_k \qquad (c_1, \ldots, c_k \in \mathbb{R})$$

is called a **linear combination** of $\mathbf{x}_1, \ldots, \mathbf{x}_k$. The set of all linear combinations of $\mathbf{x}_1, \ldots, \mathbf{x}_k$ is called the **linear span** of $\mathbf{x}_1, \ldots, \mathbf{x}_k$.

Geometrically, the linear span of a single nonzero vector $\mathbf{x}$ (that is, the set of all scalar multiples of $\mathbf{x}$) is the straight line through $\mathbf{x}$ and the origin. The linear span of a pair of nonzero vectors $\mathbf{x}$ and $\mathbf{y}$ is the plane containing $\mathbf{x}$, $\mathbf{y}$, and the origin *unless* $\mathbf{y}$ is a scalar multiple of $\mathbf{x}$, in which case it is just the line through $\mathbf{x}$ and the origin.

(A.2)   For $1 \leq j \leq n$, we define $\mathbf{e}_j$ to be the vector in $\mathbb{R}^n$ whose jth component is 1 and whose other components are all 0:

$$\mathbf{e}_1 = (1, 0, 0, \ldots, 0), \ \mathbf{e}_2 = (0, 1, 0, \ldots, 0), \ \ldots, \ \mathbf{e}_n = (0, 0, 0, \ldots, 1).$$

We call $\mathbf{e}_1, \dots, \mathbf{e}_n$ the **standard basis vectors** for $\mathbb{R}^n$. (When $n = 3$, the common notation is $\mathbf{i}, \mathbf{j}, \mathbf{k}$ rather than $\mathbf{e}_1, \mathbf{e}_2, \mathbf{e}_3$.) Every vector $\mathbf{x} \in \mathbb{R}^n$ can be written uniquely as a linear combination of the standard basis vectors:

$$(x_1, x_2, \dots, x_n) = x_1 \mathbf{e}_1 + x_2 \mathbf{e}_2 + \cdots + x_n \mathbf{e}_n.$$

A.2   Linear Maps and Matrices

(A.3)   Let m and n be positive integers. A map $A : \mathbb{R}^n \to \mathbb{R}^m$ is called **linear** if it preserves the vector operations of addition and scalar multiplication:

(A.4) $A(\mathbf{x} + \mathbf{y}) = A(\mathbf{x}) + A(\mathbf{y}), \quad A(c\mathbf{x}) = cA(\mathbf{x}) \qquad (\mathbf{x}, \mathbf{y} \in \mathbb{R}^n, \ c \in \mathbb{R}).$

(A.5)   In elementary mathematics, a "linear function" of the real variable x is something of the form $f(x) = ax + b$. As a mapping from $\mathbb{R}^1$ to $\mathbb{R}^1$, such a function is linear in the sense just defined only when $b = 0$. More generally, mappings from $\mathbb{R}^n$ to $\mathbb{R}^m$ of the form $\mathbf{f}(\mathbf{x}) = A(\mathbf{x}) + \mathbf{b}$, where A satisfies (A.4), are called "linear" in some contexts, as in Chapters 2 and 3 when we speak of the "linear approximation" to a differentiable map. However, within the subject of linear algebra, and in particular throughout this appendix, "linear" is always meant in the strict sense (A.4), and the term **affine** is used for the more general notion. The feature that immediately distinguishes linear maps in the strict sense among the affine ones is that they satisfy $A(\mathbf{0}) = \mathbf{0}$.

(A.6)   If A is linear, we have

$$A\left(\sum_1^n x_j \mathbf{e}_j \right) = \sum_1^n x_j A(\mathbf{e}_j),$$

so A is completely determined by its values on the standard basis vectors. Let us denote the jth component of $A(\mathbf{e}_k)$ by A_{jk}:

$$A(\mathbf{e}_1) = (A_{11}, A_{21}, \dots, A_{m1}), \ \dots, \ A(\mathbf{e}_n) = (A_{1n}, A_{2n}, \dots, A_{mn}).$$

Then for any $\mathbf{x} \in \mathbb{R}^n$ we have

(A.7) $A(\mathbf{x}) = \mathbf{y}, \text{ where } y_j = \displaystyle\sum_{k=1}^n A_{jk} x_k.$

Thus A can be completely described by the $m \cdot n$ numbers A_{jk}.

(A.8) Such a collection $(A_{jk}) = \{A_{jk} : 1 \leq j \leq m, \ 1 \leq k \leq n\}$ of $m \cdot n$ real numbers is called an $m \times n$ **matrix**. It is pictured as a rectangular array, with the first index j labeling the *rows* of the array and the second index k labeling the *columns*:

$$\begin{pmatrix} A_{11} & A_{12} & \cdots & A_{1n} \\ \vdots & \vdots & \ddots & \vdots \\ A_{m1} & A_{m2} & \cdots & A_{mn} \end{pmatrix}.$$

(More precisely, such an array is a *real* $m \times n$ matrix. One can also consider matrices whose entries are other sorts of algebraic objects, such as complex numbers or polynomials.) The formula (A.7) defines a one-to-one correspondence between linear maps from $\mathbb{R}^n$ to $\mathbb{R}^m$ and $m \times n$ matrices. Henceforth we shall use the same letter A to denote either a linear map or its associated matrix; the meaning will be clear from the context.

(A.9) Linear maps from $\mathbb{R}^n$ to $\mathbb{R}^m$ can be added to one another and multiplied by scalars:

$$(A + B)(\mathbf{x}) = A(\mathbf{x}) + B(\mathbf{x}), \qquad (cA)(\mathbf{x}) = c(A(\mathbf{x})).$$

On the level of matrices, this is just addition and multiplication in each entry — that is, vector addition and scalar multiplication, if we think of $m \times n$ matrices as mn-dimensional vectors.

(A.10) Suppose that $A : \mathbb{R}^n \to \mathbb{R}^m$ and $B : \mathbb{R}^m \to \mathbb{R}^l$ are linear maps. We can then consider their composition $B \circ A : \mathbb{R}^n \to \mathbb{R}^l$, and it is easy to check that $B \circ A$ is again linear. It is customary in linear algebra to denote this composition simply by BA, and we do so henceforth.

Given $\mathbf{x} \in \mathbb{R}^n$, let $\mathbf{y} = A(\mathbf{x})$ and $\mathbf{z} = B(\mathbf{y})$. On the one hand, we have

$$z_i = \sum_{j=1}^{m} B_{ij} y_j = \sum_{j=1}^{m} \sum_{k=1}^{n} B_{ij} A_{jk} x_k,$$

and on the other,

$$z_i = \sum_{k=1}^{n} (BA)_{ik} x_k.$$

It follows that the matrix BA is obtained from the matrices B and A by the formula

$$(BA)_{ik} = \sum_{j=1}^{m} B_{ij} A_{jk}.$$

In general, if B is an $l \times m$ matrix and A is an $m \times n$ matrix, the $l \times n$ matrix BA defined by this formula is called the **product** of the matrices B and A.

(A.11)   It is important to note that the product BA is defined *only* if the number of columns in B is the same as the number of rows in A, that is, if the length of a row in B is equal to the length of a column in A. It is also important to note that matrix multiplication is not commutative: In general, $BA \neq AB$, even when both products are defined. However, matrix multiplication is associative; that is, $(CB)A = C(BA)$ for any A, B, C such that all products in question are defined. It also distributes over addition in the obvious way: $C(A + B) = CA + CB$ and $(A + B)D = AD + BD$.

(A.12)   Let I be the identity mapping on $\mathbb{R}^n$, $I(\mathbf{x}) = \mathbf{x}$ for all $\mathbf{x} \in \mathbb{R}^n$. The corresponding matrix is called the $n \times n$ **identity matrix** and is denoted by I or by I_n if the size needs to be specified. It is the matrix whose columns are the standard basis vectors $\mathbf{e}_1, \ldots, \mathbf{e}_n$, that is, the matrix whose entries I_{jk} are equal to 1 when $j = k$ and 0 when $j \neq k$. If A is any $m \times n$ matrix, we have $I_m A = A$ and $AI_n = A$. This is obvious since the composition of any map A with the identity map is just A; it is also easy to verify from the definition of matrix products in (A.10).

(A.13)   Let $A : \mathbb{R}^n \to \mathbb{R}^n$ be a linear map. If there is another linear map $B : \mathbb{R}^n \to \mathbb{R}^n$ such that $AB(\mathbf{x}) = BA(\mathbf{x}) = \mathbf{x}$ for all $\mathbf{x} \in \mathbb{R}^n$ (that is, in terms of matrices, $AB = BA = I_n$), then A (or its associated matrix) is called **invertible** or **nonsingular**, and B is called the **inverse** of A and is denoted by A^{-1}. It is easy to verify that if A_1 and A_2 are both invertible, then so is their product $A_1 A_2$, and $(A_1 A_2)^{-1} = A_2^{-1} A_1^{-1}$. We shall say more about invertibility in (A.50)–(A.55).

(A.14)   Vectors in $\mathbb{R}^n$ can be thought of as $n \times 1$ matrices (called **column vectors**) or as $1 \times n$ matrices (called **row vectors**), and scalars can be thought of as 1×1 matrices. With these identifications, we can reinterpret some of the preceding formulas:

- If $A : \mathbb{R}^n \to \mathbb{R}^m$ and $\mathbf{x} \in \mathbb{R}^n$, then by (A.7), $A(\mathbf{x})$ is the matrix product $A\mathbf{x}$, where $\mathbf{x}$ and $A(\mathbf{x})$ are considered as column vectors. For this reason, we (almost) always think of vectors as column vectors when we perform matrix calculations with linear maps. Moreover, we shall henceforth write $A\mathbf{x}$ in preference to $A(\mathbf{x})$.

- Let B be an $l \times m$ matrix and A an $m \times n$ matrix; then the rows of B and the columns of A can both be considered as vectors in $\mathbb{R}^m$. The (ik)th entry of the product matrix BA is the dot product of the ith row of B with the kth column of A.

(A.15)   The **transpose** or **adjoint** of an $m \times n$ matrix A is the $n \times m$ matrix A^* defined by $(A^*)_{jk} = A_{kj}$. (Many people denote A^* by A^t or A^T.) Thus, the rows of A^* are the columns of A and vice versa. As linear maps, A and A^* are related through the dot product:

$$(A.16) \qquad \qquad \mathbf{x} \cdot A\mathbf{y} = A^*\mathbf{x} \cdot \mathbf{y},$$

since both sides are equal to the double sum $\sum_{j=1}^{m} \sum_{k=1}^{n} x_j A_{jk} y_k$. It is easy to check that $(AB)^* = B^* A^*$.

A.3 Row Operations and Echelon Forms

(A.17)   In high school algebra one learns techniques for solving systems of linear equations that involve multiplying equations by scalars, adding one equation to another one, and so forth. When systematized and translated into matrix language, these methods amount to performing "row operations" on matrices. The three types of **elementary row operations** on a matrix are defined as follows. Let A be an $m \times n$ matrix, and let $\mathbf{r}_1, \ldots, \mathbf{r}_m$ be the rows of A (considered as vectors in $\mathbb{R}^n$).

    i. Multiply one row by a nonzero scalar. (That is, for some j, replace $\mathbf{r}_j$ by $c\mathbf{r}_j$ with $c \neq 0$, and leave all the other rows unchanged.)

    ii. Add a scalar multiple of one row to another row. (That is, for some $j \neq k$, replace $\mathbf{r}_j$ by $\mathbf{r}_j + c\mathbf{r}_k$, and leave all the other rows unchanged.)

    iii. Interchange two rows. (That is, for some $j \neq k$, replace $\mathbf{r}_j$ by $\mathbf{r}_k$ and $\mathbf{r}_k$ by $\mathbf{r}_j$, and leave all other rows unchanged.)

(A.18)   For each elementary row operation, the matrix obtained by performing that operation on the identity matrix I_m is called the corresponding **elementary matrix**. For example, the entries of the elementary matrix corresponding to the operation (ii) are 1 on the main diagonal, c in the (jk)th slot, and 0 elsewhere. We leave it as an easy exercise for the reader to verify that *performing an elementary row operation on a matrix A is the same as multiplying A on the left by the corresponding elementary matrix.*

(A.19)   It is important to note that *the elementary row operations, and their associated matrices, are all invertible, and their inverses are operations of the same types.* Indeed, the inverses of the operations

$$\mathbf{r}_j \to c\mathbf{r}_j, \qquad \mathbf{r}_j \to \mathbf{r}_j + c\mathbf{r}_k, \qquad \mathbf{r}_j \leftrightarrow \mathbf{r}_k$$

are

$$\mathbf{r}_j \to c^{-1}\mathbf{r}_j, \qquad \mathbf{r}_j \to \mathbf{r}_j - c\mathbf{r}_k, \qquad \mathbf{r}_j \leftrightarrow \mathbf{r}_k.$$

(A.20)   Row operations can be used to transform a matrix into certain standard forms that are useful for many purposes. The definitions are as follows. A matrix is said to be in **echelon form** if the following conditions are satisfied:

- In every nonzero row (that is, every row in which at least one entry is non-zero), the first nonzero entry is equal to 1.

- If the jth and kth rows are nonzero, and $j < k$, the initial 1 in row j is to the left of the initial 1 in row k.

- The zero rows (if any) are below all of the nonzero rows.

The following matrices are in echelon form:

$$\begin{pmatrix} 1 & 3 & 5 \\ 0 & 1 & 0 \end{pmatrix}, \qquad \begin{pmatrix} 1 & 2 \\ 0 & 1 \\ 0 & 0 \end{pmatrix}, \qquad \begin{pmatrix} 1 & 4 & 0 \\ 0 & 1 & 1 \\ 0 & 0 & 1 \end{pmatrix}.$$

A matrix is said to be in **reduced echelon form** if it is in echelon form, and in addition,

- The entries above and below the initial 1's in the nonzero rows are all 0.

The matrices displayed above are not in reduced echelon form, but the following matrices are:

$$\begin{pmatrix} 1 & 0 & -5 \\ 0 & 1 & -3 \end{pmatrix}, \qquad \begin{pmatrix} 1 & 4 \\ 0 & 0 \end{pmatrix}, \qquad \begin{pmatrix} 1 & 7 & 0 \\ 0 & 0 & 1 \\ 0 & 0 & 0 \end{pmatrix}.$$

(A.21)   Suppose A is a *square* matrix (say, $n \times n$) in echelon form, and suppose A has no zero rows. The first nonzero entry in each row is a 1, and these initial 1's occur successively farther to the right. Since the n initial 1's must occur in n different columns, the only possibility is that the initial 1 in the jth row occurs precisely in the jth column. In other words, the entries of A on the main diagonal are all equal to 1, and below the main diagonal they are all equal to 0. If A is in *reduced* echelon form, then all the entries above the main diagonal must also be 0. In short, *the only $n \times n$ matrix in reduced echelon form with no zero rows is the identity matrix I_n.*

(A.22)   The simplest algorithm for turning a given $m \times n$ matrix A into one in echelon form by elementary row operations, known as **row reduction** or **Gaussian elimination**, can be described as follows:

1. If necessary, interchange the first row with another row so that the leftmost nonzero column has a nonzero entry in the first row.

2. Multiply the first row by the reciprocal of its first nonzero entry (thus turning the first nonzero entry into a 1).

3. Add multiples of the first row to the rows below so as to make the entries below the initial 1 in the first row equal to 0.

4. Set the first row aside and apply steps 1–3 to the submatrix obtained by omitting the first row. Repeat this process until no nonzero rows remain.

Once this is done, the matrix can be further transformed into one in reduced echelon form as follows:

5. Add multiples of each nonzero row to the rows above so as to make the entries above the initial 1's equal to 0.

(A.23)   All of the ideas in this section have analogues for columns in place of rows. That is, we have the *elementary column operations* (multiply a column by a nonzero scalar, add a multiple of one column to another one, interchange two columns), which are implemented by multiplying a matrix on the *right* by the corresponding elementary matrix. They can be used to transform a matrix into one in *column-echelon form* or *reduced column-echelon form*, whose definitions are the obvious modifications of the ones given above for (row-)echelon forms.

A.4   Determinants

(A.24)   The **determinant** is a function that assigns to each *square* matrix A a certain number $\det A$. For 2×2 and 3×3 matrices, the determinant is given by

(A.25)
$$\det \begin{pmatrix} a & b \\ c & d \end{pmatrix} = ad - bc,$$

(A.26)
$$\det \begin{pmatrix} a & b & c \\ d & e & f \\ g & h & i \end{pmatrix} = a(ei - fh) - b(di - fg) + c(dh - eg).$$

For larger matrices, the explicit formula for the determinant is quite a mess. However, this formula is of little use; the important things about determinants are the properties they possess, which lead to more efficient ways of computing them. The following seven items constitute a list of the most fundamental properties of determinants. In them, A and B denote $n \times n$ matrices.

(A.27) $\det I_n = 1$.

(A.28) $\det(AB) = (\det A)(\det B)$.

(A.29)   For each j, $\det A$ is a linear function of the jth row of A when the other rows are kept fixed. (Thus, for example, when $j = 1$,

$$\det \begin{pmatrix} a\mathbf{r}_1' + b\mathbf{r}_1'' \\ \mathbf{r}_2 \\ \vdots \end{pmatrix} = a \det \begin{pmatrix} \mathbf{r}_1' \\ \mathbf{r}_2 \\ \vdots \end{pmatrix} + b \det \begin{pmatrix} \mathbf{r}_1'' \\ \mathbf{r}_2 \\ \vdots \end{pmatrix},$$

where the $\mathbf{r}_j$'s denote row vectors.) In particular, if A has a zero row, $\det A = 0$.

(A.30)   (*Behavior under elementary row operations*)

- If one row of A is multiplied by c and the other rows are left unchanged, $\det A$ is multiplied by c.

- If a multiple of the kth row of A is added to the jth row and the other rows are left unchanged, $\det A$ is unchanged.

- If two rows of A are interchanged, $\det A$ is multiplied by -1.

(A.31)   Let M^{jk} denote the $(n-1) \times (n-1)$ matrix obtained by deleting the jth row and kth column of A. Then, for each j,

$$\det A = \sum_{k=1}^{n} (-1)^{j+k} A_{jk} \det M^{jk}.$$

This formula is called the **cofactor expansion** of $\det A$ along the jth row. (For example, in view of equation (A.25), equation (A.26) gives the cofactor expansion of the determinant of a 3×3 matrix along its first row.)

(A.32) $\det(A^*) = \det A$. Consequently, properties (A.29) and (A.30) remain valid if "row" is replaced by "column," and we can sum over j instead of k in the cofactor expansion.

(A.33)   (*How to compute determinants*) The cofactor expansion reduces $n \times n$ determinants to determinants of smaller size and so can be used recursively to compute a determinant. However, for large matrices it is much more efficient to use row operations. That is, to compute $\det A$, row-reduce A and keep track of what happens to the determinant as each row operation is performed, according to the rules in (A.30). At the end, we have a matrix in reduced echelon form, which (by (A.21)) is either the identity matrix (whose determinant is 1, by (A.26)) or a matrix with a zero row (whose determinant is 0, by (A.29)).

A.5 Linear Independence

(A.34)   The vectors $\mathbf{x}_1, \ldots, \mathbf{x}_k \in \mathbb{R}^n$ are said to be **linearly dependent** if they satisfy a nontrivial linear equation, that is, if there are scalars $c_1, \ldots, c_k$, not all zero, such that

$$c_1 \mathbf{x}_1 + c_2 \mathbf{x}_2 + \cdots + c_k \mathbf{x}_k = \mathbf{0}.$$

If $c_j \neq 0$, this equation can be solved for $\mathbf{x}_j$:

$$\mathbf{x}_j = -c_j^{-1} \big[c_1 \mathbf{x}_1 + \cdots + c_{j-1} \mathbf{x}_{j-1} + c_{j+1} \mathbf{x}_{j+1} + \cdots + c_k \mathbf{x}_k \big]$$

Hence, $\mathbf{x}_1, \ldots, \mathbf{x}_k$ *are linearly dependent if and only if one of them is a linear combination of the others.* If $\mathbf{x}_1, \ldots, \mathbf{x}_k$ are not linearly dependent, they are said to be **linearly independent**. That is, *linear independence of* $\mathbf{x}_1, \ldots, \mathbf{x}_k$ *means that*

$$c_1 \mathbf{x}_1 + \cdots + c_k \mathbf{x}_k = \mathbf{0} \; \textit{only when } c_1 = \cdots = c_k = 0.$$

In the case $k = 2$, linear independence of $\mathbf{x}_1$ and $\mathbf{x}_2$ means simply that $\mathbf{x}_1$ and $\mathbf{x}_2$ are not scalar multiples of one another.

(A.35)   If A is a matrix in echelon form, then the nonzero rows $\mathbf{r}_1, \ldots, \mathbf{r}_k$ of A are linearly independent. Indeed, suppose that $\sum_1^k c_j \mathbf{r}_j = \mathbf{0}$. In the column in which the initial 1 in the first row appears, the entries in all the other rows are 0; hence, the entry of $\sum_1^k c_j \mathbf{r}_j$ in this column is c_1, and so $c_1 = 0$. That being the case, we have $\sum_2^k c_j \mathbf{r}_j = \mathbf{0}$; the same argument now shows that $c_2 = 0$, and so forth.

(A.36)   A set of vectors $\mathbf{x}_1, \ldots, \mathbf{x}_k$ is called **orthonormal** if they are mutually orthogonal and have unit norm:

$$\mathbf{x}_i \cdot \mathbf{x}_j = 0 \text{ for } i \neq j, \text{ and } |\mathbf{x}_j| = 1 \text{ for all } j.$$

(For example, the standard basis vectors $\mathbf{e}_j$ for $\mathbb{R}^n$ are orthonormal.) If $\mathbf{x}_1, \ldots, \mathbf{x}_k$ are orthonormal, then they are linearly independent. Indeed, suppose $\sum_1^k c_j \mathbf{x}_j = \mathbf{0}$; we wish to show that $c_j = 0$ for all j. To see that a particular coefficient c_i is zero, take the dot product of both sides of the equation $\mathbf{0} = \sum_{j=1}^k c_j \mathbf{x}_j$ with $\mathbf{x}_i$. All of the dot products $\mathbf{x}_j \cdot \mathbf{x}_i$ vanish except for $j = i$, so we obtain $0 = c_i |\mathbf{x}_i|^2 = c_i$.

(A.37)   In general, to determine whether $\mathbf{x}_1, \ldots, \mathbf{x}_k \in \mathbb{R}^n$ are linearly independent, we can regard them as the rows of a $k \times n$ matrix and perform a row reduction. The rows of the resulting echelon matrix are linear combinations of the original rows $\mathbf{x}_j$. If they are all nonzero, then they are linearly independent by (A.35), and so are $\mathbf{x}_1, \ldots, \mathbf{x}_k$. But if there is a zero row, then $\mathbf{x}_1, \ldots, \mathbf{x}_k$ are linearly dependent, because that row is a nontrivial linear combination of the $\mathbf{x}_j$'s.

A.6   Subspaces; Dimension; Rank

(A.38)   A **vector subspace** of $\mathbb{R}^n$, or just a **subspace** for short, is a subset $\mathcal{X}$ of $\mathbb{R}^n$ such that
   i. if $\mathbf{x}, \mathbf{y} \in \mathcal{X}$ then $\mathbf{x} + \mathbf{y} \in \mathcal{X}$, and
   ii. if $\mathbf{x} \in \mathcal{X}$ and $c \in \mathbb{R}$ then $c\mathbf{x} \in \mathcal{X}$.
Subspaces are closed under taking linear combinations; that is, if $\mathbf{x}_1, \ldots, \mathbf{x}_k \in \mathcal{X}$ and $c_1, \ldots, c_k \in \mathbb{R}$, then $c_1\mathbf{x}_1 + \cdots + c_k\mathbf{x}_k \in \mathcal{X}$. The largest subspace of $\mathbb{R}^n$ is $\mathbb{R}^n$ itself, and the smallest one is the trivial subspace consisting of the single element $\mathbf{0}$. When $n > 1$, there are also subspaces of intermediate size. For example, when $n = 2$, the intermediate subspaces are the lines through the origin; when $n = 3$, they are the lines and planes through the origin.

(A.39)   The linear span of any set of vectors in $\mathbb{R}^n$ is easily seen to be a subspace of $\mathbb{R}^n$.

(A.40)   Let $\mathcal{X}$ be a subspace of $\mathbb{R}^n$. A set of vectors in $\mathcal{X}$ is called a **basis** for $\mathcal{X}$ if it is linearly independent and its linear span is $\mathcal{X}$. For example, the standard basis vectors $\mathbf{e}_1, \ldots, \mathbf{e}_n$ for $\mathbb{R}^n$ are a basis for $\mathbb{R}^n$ in this sense. One can show that any two bases for $\mathcal{X}$ have the same number of elements; that number is called the **dimension** of $\mathcal{X}$ and is denoted by $\dim \mathcal{X}$. The dimension of $\mathbb{R}^n$ itself is n, and we define the dimension of the trivial subspace $\{\mathbf{0}\}$ to be 0; the dimension of any other subspace is an integer strictly between 0 and n.

   If $\mathbf{x}_1, \ldots, \mathbf{x}_k$ is a basis for $\mathcal{X}$, then any element of $\mathcal{X}$ can be written in one and only one way as $\sum_1^j c_j\mathbf{x}_j$. Thus, the dimension of $\mathcal{X}$ is the number of real parameters (namely, the coefficients c_j) that are needed to specify an element of $\mathcal{X}$.

(A.41)   Let $\mathcal{X}$ be a subspace of $\mathbb{R}^n$. Its **orthogonal complement** $\mathcal{X}^\perp$ is the set of all vectors that are orthogonal to every vector in $\mathcal{X}$:

$$\mathcal{X}^\perp = \{\mathbf{x} \in \mathbb{R}^n : \mathbf{x} \cdot \mathbf{y} = 0 \text{ for all } \mathbf{y} \in \mathcal{X}\}.$$

It is easy to verify that $\mathcal{X}^\perp$ is also a subspace. For example, in $\mathbb{R}^3$, the orthogonal complement of a plane through the origin is the line through the origin perpendicular to it, and vice versa; the orthogonal complement of $\mathbb{R}^3$ is $\{\mathbf{0}\}$, and vice versa.

The complementary relations between the dimensions of $\mathcal{X}$ and $\mathcal{X}^\perp$ in this example persists in higher dimensions:

(A.42)   *For any subspace $\mathcal{X} \subset \mathbb{R}^n$, $\dim \mathcal{X} + \dim \mathcal{X}^\perp = n$.*

(A.43)   Let $A : \mathbb{R}^n \to \mathbb{R}^m$ be a linear map. There are two subspaces (one of $\mathbb{R}^n$ and one of $\mathbb{R}^m$) naturally associated to A: its **nullspace**

$$\mathcal{N}(A) = \{\mathbf{x} \in \mathbb{R}^n : A\mathbf{x} = \mathbf{0}\},$$

and its **range**

$$\mathcal{R}(A) = \{\mathbf{y} \in \mathbb{R}^m : \mathbf{y} = A\mathbf{x} \text{ for some } \mathbf{x} \in \mathbb{R}^n\}.$$

It is an easy exercise to check that $\mathcal{N}(A)$ and $\mathcal{R}(A)$ are indeed subspaces. If we think of A as an $m \times n$ matrix, $\mathcal{R}(A)$ is the linear span of the columns of A, because these columns are the vectors obtained by applying A to the standard basis vectors for $\mathbb{R}^n$. Hence, $\mathcal{R}(A)$ is sometimes called the **column space** of A.

(A.44)   We can also consider the nullspace and range of the transpose $A^* : \mathbb{R}^m \to \mathbb{R}^n$. The range $\mathcal{R}(A^*)$ is the linear span of the columns of A^*, which are the rows of A; hence $\mathcal{R}(A^*)$ is sometimes called the **row space** of A. The spaces $\mathcal{N}(A)$, $\mathcal{R}(A)$, $\mathcal{N}(A^*)$, and $\mathcal{R}(A^*)$ are related as follows:

(A.45) $\mathcal{N}(A^*) = \mathcal{R}(A)^\perp;$ $\mathcal{N}(A) = \mathcal{R}(A^*)^\perp.$

This follows easily from the relation (A.16). Indeed, $\mathbf{x} \in \mathcal{R}(A)^\perp \iff \mathbf{x} \cdot A\mathbf{y} = 0$ for all $\mathbf{y} \iff A^*\mathbf{x} \cdot \mathbf{y} = 0$ for all $\mathbf{y} \iff A^*\mathbf{x} = 0 \iff \mathbf{x} \in \mathcal{N}(A^*)$, and likewise with A and A^* switched.

(A.46)   The fundamental identity concerning dimensions is the following:

(A.47)   *For any linear map $A : \mathbb{R}^n \to \mathbb{R}^m$, $\dim \mathcal{N}(A) + \dim \mathcal{R}(A) = n$.*

The intuitive reason behind this identity is simple. An element of $\mathbb{R}^n$ has n degrees of freedom (one can vary any of its n components). The elements of the nullspace $\mathcal{N}(A)$ are all mapped by A to the single vector $\mathbf{0}$, resulting in a loss of $\dim \mathcal{N}(A)$ degrees of freedom and leaving $n - \dim \mathcal{N}(A)$ degrees of freedom for the range $\mathcal{R}(A)$.

(A.48)   From the preceding results, we obtain one more important relation:

(A.49)   *For any linear map $A : \mathbb{R}^n \to \mathbb{R}^m$, $\dim \mathcal{R}(A) = \dim \mathcal{R}(A^*)$.*

Indeed, by (A.42), (A.45), and (A.47),

$$\dim \mathcal{R}(A) = n - \dim \mathcal{N}(A) = n - \dim \mathcal{R}(A^*)^\perp = \dim \mathcal{R}(A^*).$$

The common dimension of $\mathcal{R}(A)$ and $\mathcal{R}(A^*)$ is called the **rank** of A.

A.7   Invertibility

(A.50)   We recall from the introduction to Chapter 1 that that a mapping $f : X \to Y$ from a set X to another set Y is *invertible* if there is another mapping $g : Y \to X$ such that $g(f(x)) = x$ for all $x \in X$ and $f(g(y)) = y$ for all $y \in Y$, and that f is invertible if and only if f maps X onto Y and f is one-to-one.

(A.51)   Now let $A : \mathbb{R}^n \to \mathbb{R}^m$ be a linear map. We first observe that A is one-to-one if and only if $N(A) = \{\mathbf{0}\}$, for $A\mathbf{x} = A\mathbf{y}$ if and only if $\mathbf{x} - \mathbf{y} \in N(A)$. In particular, if $m < n$, then by (A.47) we have $\dim N(A) = n - \dim R(A) \geq n - m > 0$, so A cannot be one-to-one. On the other hand, if $m > n$, then by (A.47) again, $\dim R(A) \leq n < m$, so $R(A)$ cannot be all of $\mathbb{R}^m$. Hence, A can be invertible in the sense of (A.50) only when $n = m$; in this case, it is not hard to check that the inverse of A (if it exists) is again a linear map. Thus, for linear maps the definition of invertibility in (A.50) agrees with the one in (A.13).

(A.52)   *For a linear map $A : \mathbb{R}^n \to \mathbb{R}^n$, the following conditions are all equivalent:*

  a. A is invertible.
  b. $R(A) = \mathbb{R}^n$.
  c. $N(A) = \{\mathbf{0}\}$.
  d. $R(A^*) = \mathbb{R}^n$.
  e. $N(A^*) = \{\mathbf{0}\}$.
  f. The columns of the matrix A are linearly independent.
  g. The rows of the matrix A are linearly independent.
  h. $\det A \neq 0$.
  i. The matrix A is a product of elementary matrices.

(A.53)   Let us prove (A.52). First, (a) is equivalent to the conjunction of (b) and (c) by the discussion in (A.50–A.51). (b) and (c) are equivalent to each other by (A.47), as are (d) and (e), and (b) and (d) are equivalent by (A.49). (f) is equivalent to (c), for if $\mathbf{c}_j = A\mathbf{e}_j$ is the jth column of A, we have $\sum_{j=1}^n a_j \mathbf{c}_j = \mathbf{0}$ if and only if $\sum a_j \mathbf{e}_j \in N(A)$; similarly, (g) is equivalent to (e).

Next, we can perform elementary row operations on A to turn A into a matrix B in reduced echelon form; since performing row operations does not change the row space of a matrix, we have $R(A^*) = R(B^*)$. But by (A.21) and (A.33), either $B = I$, in which case $\det A \neq 0$ and $R(A^*) = R(I) = \mathbb{R}^n$; or B contains at least one zero row, in which case $\det A = 0$ and $\dim R(A^*) = \dim R(B^*) < n$; thus (h) is equivalent to (d).

We have shown that (a)–(h) are all equivalent. Finally, we observed in (A.19) that every elementary matrix is invertible, and hence so is every product of elementary matrices. Conversely, if A is invertible, let $B = A^{-1}$. Then B is invertible also, so B can be row-reduced to the identity matrix; that is, there is a product E of elementary matrices such that $EB = I$. But $E = E(BA) = (EB)A = A$, so A is a product of elementary matrices. Thus (a) is equivalent to (i).

(A.54) **(Cramer's Rule)** If A is invertible and $\mathbf{b} \in \mathbb{R}^n$, the vector $\mathbf{x} = A^{-1}\mathbf{b}$ is given by $x_j = (\det B^j)/(\det A)$, where B^j is the matrix obtained from A by replacing its jth column with the column vector $\mathbf{b}$. This is not a computationally efficient way of solving $A\mathbf{x} = \mathbf{b}$ when n is large, but the fact that the solution can be expressed as a quotient of determinants is theoretically important.

(A.55) In particular, computing A^{-1} amounts to solving $A\mathbf{x}_j = \mathbf{e}_j$ for $j = 1, \dots, n$, where the $\mathbf{e}_j$'s are the standard basis vectors: The solutions $\mathbf{x}_j$ are the columns of A^{-1}. It follows that *the entries of A^{-1} are rational functions of the entries of A whose common denominator is* $\det A$.

A.8 Eigenvectors and Eigenvalues

(A.56) Let A be an $n \times n$ matrix. A nonzero vector $\mathbf{x} \in \mathbb{R}^n$ is called an **eigenvector** for A if there is a scalar $\lambda \in \mathbb{R}$ such that $A\mathbf{x} = \lambda\mathbf{x}$; in this case, λ is called the **eigenvalue** of A for the vector $\mathbf{x}$. The equation $A\mathbf{x} = \lambda\mathbf{x}$ can be rewritten as $(A - \lambda I)\mathbf{x} = 0$; hence, λ is an eigenvalue of A (that is, there is a nonzero $\mathbf{x}$ such that $A\mathbf{x} = \lambda\mathbf{x}$) if and only if $\mathcal{N}(A - \lambda I) \neq \{0\}$. By (A.52), this condition is equivalent to $\det(A - \lambda I) = 0$. It is easy to see that $\det(A - \lambda I)$ is a polynomial of degree n in λ, called the **characteristic polynomial** of A, and the eigenvalues of A are precisely the roots of this polynomial.

(A.57) The analysis of a matrix A is greatly facilitated if there is an **eigenbasis** for A, that is, a basis $\mathbf{b}_1, \dots, \mathbf{b}_n$ of $\mathbb{R}^n$ consisting of eigenvectors for A. Indeed, suppose $A\mathbf{b}_j = \lambda_j\mathbf{b}_j$. Any $\mathbf{x} \in \mathbb{R}^n$ can be written as a linear combination of the $\mathbf{b}_j$'s, say $\mathbf{x} = \sum_{j=1}^n c_j\mathbf{b}_j$, and then $A\mathbf{x} = \sum_{j=1}^n \lambda_j c_j\mathbf{b}_j$. In other words, once the basis $\mathbf{b}_1, \dots, \mathbf{b}_n$ is known, the action of A is completely determined by the n numbers λ_j rather than the n^2 numbers A_{jk}.

(A.58) Not all matrices have eigenbases. (In fact, some matrices have no eigenvalues at all, as long as we allow only real numbers. The situation changes dramatically if we consider complex matrices and complex eigenvalues, but even then A may not have an eigenbasis when the characteristic polynomial has multiple roots.) However, there is an important class of matrices that do have eigenbases.

The $n \times n$ matrix A is called **symmetric** if $A = A^*$, that is, if $A_{jk} = A_{kj}$ for all j and k. One can show that *every symmetric matrix has an orthonormal eigenbasis.* This is one of the major results of linear algebra, known as the **spectral theorem** or **principal axis theorem**.

Appendix B

SOME TECHNICAL PROOFS

B.1   The Heine-Borel Theorem

B.1 Theorem. *If S is a subset of $\mathbb{R}^n$, the following are equivalent:*

a. *S is compact.*

b. *If $\mathcal{U}$ is any covering of S by open sets, there is a finite subcollection of $\mathcal{U}$ that still forms a covering of S.*

Proof. If S is not compact, by the Bolzano-Weierstrass theorem there is a sequence $\{\mathbf{x}_k\}$ in S, no subsequence of which converges to any point of S. This means that for each $\mathbf{x} \in S$ there is an open ball $D_{\mathbf{x}}$ centered at $\mathbf{x}$ that contains $\mathbf{x}_k$ for at most finitely many values of k (Exercise 7, §1.5). The collection $\mathcal{U} = \{D_{\mathbf{x}} : \mathbf{x} \in S\}$ is then an open cover of S. Any finite subcollection can contain at most finitely many of the $\mathbf{x}_k$'s and hence cannot cover all of S.

Conversely, suppose S is compact. Since S is bounded, it is contained in some closed rectangular box

$$B_0 = [a_1, b_1] \times [a_2, b_2] \times \cdots \times [a_n, b_n]$$
$$= \{\mathbf{x} : a_1 \leq x_1 \leq b_1,\ a_2 \leq x_2 \leq b_2, \ldots, a_n \leq x_n \leq b_n\}.$$

By bisecting the intervals $[a_j, b_j]$, we can write B_0 as the union of 2^n boxes whose side lengths are half as big as those of B_0; we denote this collection of boxes by $\mathcal{B}_1$. By bisecting the sides of each box in $\mathcal{B}_1$, we can write B_0 as the union of 2^{2n} boxes whose side length are $\frac{1}{4}$ as big as those of B_0; we denote this collection of boxes by $\mathcal{B}_2$. Continuing inductively, for each positive integer k we can write B_0 as the union of 2^{kn} boxes whose side lengths are 2^{-k} times as big as those of B_0, and we denote this collection of boxes by $\mathcal{B}_k$.

Now suppose $\mathcal{U}$ is a covering of S by open sets. We claim that *there is an integer k such that each box in $\mathcal{B}_k$ that intersects S is included in one of the open sets in $\mathcal{U}$.* Once we know this, we are done. There are finitely many (in fact, 2^{kn}) boxes in $\mathcal{B}_k$; let $B_1, \ldots, B_m$ be the ones that intersect S. Each B_j is included in some $U_j \in \mathcal{U}$; the sets $B_1, \ldots, B_m$ cover S, and hence so do $U_1, \ldots, U_m$.

It remains to prove the claim. Suppose, to the contrary, that for each k there is a box $B_k \in \mathcal{B}_k$ containing a point $\mathbf{x}_k \in S$ but not included in any set in $\mathcal{U}$. By the Bolzano-Weierstrass theorem, by passing to a subsequence we may assume that $\{\mathbf{x}_k\}$ converges to some point $\mathbf{x} \in S$. This $\mathbf{x}$ is contained in some open set U in the collection $\mathcal{U}$. Since U is open, there is a positive number ϵ such that every point $\mathbf{y}$ with $|\mathbf{y} - \mathbf{x}| < \epsilon$ is contained in U. Now pick k large enough so that $|\mathbf{x}_k - \mathbf{x}| < \frac{1}{2}\epsilon$ and also $2^{-k}[\sum_1^n (b_j - a_j)^2]^{1/2} < \frac{1}{2}\epsilon$. The latter condition implies that the distance between any two points of the box B_k is less than $\frac{1}{2}\epsilon$. Thus, if $\mathbf{y} \in B_k$, then

$$|\mathbf{y} - \mathbf{x}| \le |\mathbf{y} - \mathbf{x}_k| + |\mathbf{x}_k - \mathbf{x}| < \tfrac{1}{2}\epsilon + \tfrac{1}{2}\epsilon = \epsilon.$$

But this means that $B_k \subset U$, contrary to assumption. This contradiction completes the proof. $\qquad\qquad\square$

B.2   The Implicit Function Theorem

B.2 Theorem. *Let $\mathbf{F}(\mathbf{x}, \mathbf{y})$ be an $\mathbb{R}^k$-valued function of class C^1 on some neighborhood of a point $(\mathbf{a}, \mathbf{b}) \in \mathbb{R}^{n+k}$, and let $B_{ij} = (\partial F_i / \partial y_j)(\mathbf{a}, \mathbf{b})$. Suppose that $\mathbf{F}(\mathbf{a}, \mathbf{b}) = \mathbf{0}$ and $\det B \ne 0$. Then for some positive numbers r_0, r_1, the following conclusions are valid.*

a. *For each $\mathbf{x}$ in the ball $|\mathbf{x} - \mathbf{a}| < r_0$ there is a unique $\mathbf{y}$ such that $|\mathbf{y} - \mathbf{b}| < r_1$ and $\mathbf{F}(\mathbf{x}, \mathbf{y}) = \mathbf{0}$. We denote this $\mathbf{y}$ by $\mathbf{f}(\mathbf{x})$; in particular, $\mathbf{f}(\mathbf{a}) = \mathbf{b}$.*

b. *The function $\mathbf{f}$ thus defined for $|\mathbf{x} - \mathbf{a}| < r_0$ is of class C^1, and its partial derivatives $\partial_{x_j}\mathbf{f}$ can be computed by differentiating the equations $\mathbf{F}(\mathbf{x}, \mathbf{f}(\mathbf{x})) = \mathbf{0}$ with respect to x_j and solving the resulting linear system of equations for $\partial_{x_j} f_1, \ldots, \partial_{x_j} f_k$.*

Proof. The proof proceeds by induction on k. The case $k = 1$ is the implicit function theorem for a single equation, proved in §3.1. We assume that the result is valid when the number of equations is $1, 2, \ldots, k - 1$ and deduce it when the number of equations is k.

Let M^{ij} denote the $(k - 1) \times (k - 1)$ matrix obtained by deleting the ith row and the jth column from the matrix B. By the cofactor expansion along the last row (see (A.31) in Appendix A),

(B.3)
$$\det B = (-1)^{k+1} B_{k1} \det M^{k1} + (-1)^{k+2} B_{k2} \det M^{k2} + \cdots + B_{kk} \det M^{kk}.$$

Since $\det B \neq 0$ by assumption, at least one term in this sum must be nonzero. By reordering the variables if necessary, we can assume that the last term is nonzero, so $\det M^{kk} \neq 0$.

Now, M^{kk} is the matrix of partial derivatives of $F_1, \ldots, F_{k-1}$ with respect to the variables $y_1, \ldots, y_{k-1}$, evaluated at $(\mathbf{a}, \mathbf{b})$, so by inductive hypothesis, the $k-1$ equations

$$F_1(\mathbf{x}, \mathbf{y}) = F_2(\mathbf{x}, \mathbf{y}) = \cdots = F_{k-1}(\mathbf{x}, \mathbf{y}) = 0$$

determine $y_1, \ldots, y_{k-1}$ as C^1 functions of $x_1, \ldots, x_n$ *and* y_k in some neighborhood of $(\mathbf{a}, \mathbf{b})$:

$$y_j = g_j(\mathbf{x}, y_k) \qquad (j \leq k - 1).$$

Let G be the function of $x_1, \ldots, x_n, y_k$ obtained by substituting the g_j's for the y_j's in the last function F_k:

$$G(\mathbf{x}, y_k) = F_k(\mathbf{x}, \mathbf{g}(\mathbf{x}, y_k), y_k).$$

We wish to use the implicit function theorem for a single equation to solve the equation $G(\mathbf{x}, y_k) = 0$ for y_k as a C^1 function of $\mathbf{x}$, say $y_k = f_k(\mathbf{x})$. Then for $j < k$ we will have $y_j = f_j(\mathbf{x})$ where $f_j(\mathbf{x}) = g_j(\mathbf{x}, f_k(\mathbf{x}))$, and the proof will be complete. (The method for computing the partial derivatives of $\mathbf{f}$ stated in (b) is just implicit differentiation, as discussed in §2.5.)

Our task is to verify that the hypothesis of the implicit function theorem, namely $\partial_{y_k} G(\mathbf{a}, b_k) \neq 0$, is satisfied. To do this we need the chain rule, some facts about determinants, and perseverance. To begin with,

$$\frac{\partial G}{\partial y_k} = \sum_{j=1}^{k-1} \frac{\partial F_k}{\partial y_j} \frac{\partial g_j}{\partial y_k} + \frac{\partial F_k}{\partial y_k},$$

so setting $(\mathbf{x}, \mathbf{y}) = (\mathbf{a}, \mathbf{b})$ gives

(B.4) $$\frac{\partial G}{\partial y_k}(\mathbf{a}, b_k) = \sum_{j=1}^{k-1} B_{kj} \frac{\partial g_j}{\partial y_k}(\mathbf{a}, b_k) + B_{kk}.$$

To evaluate $\partial g_j / \partial y_k$, we differentiate the equations $F_i(\mathbf{x}, \mathbf{g}(\mathbf{x}, y_k), y_k) = 0$ for $i < k$, obtaining

$$\sum_{j=1}^{k-1} \frac{\partial F_i}{\partial y_j} \frac{\partial g_j}{\partial y_k} + \frac{\partial F_i}{\partial y_k} = 0 \qquad (i < k),$$

which at $(\mathbf{x}, \mathbf{y}) = (\mathbf{a}, \mathbf{b})$ becomes

(B.5) $$\sum_{j=1}^{k-1} B_{ij} \frac{\partial g_j}{\partial y_k}(\mathbf{a}, b_k) = -B_{ik} \qquad (i < k).$$

These $k-1$ equations can be solved for the desired quantities $(\partial g_j / \partial y_k)(\mathbf{a}, b_k)$ by Cramer's rule (see (A.54) in Appendix A). The coefficient matrix in (B.5), $(B_{ij})_{i,j=1}^{k-1}$, is what we called M^{kk} above, and the matrix obtained by replacing its jth column by the numbers $-B_{ik}$ on the right of (B.5) is

$$\begin{pmatrix} B_{11} & \cdots & -B_{1k} & \cdots & B_{1(k-1)} \\ \vdots & & \vdots & & \vdots \\ B_{(k-1)1} & \cdots & -B_{(k-1)k} & \cdots & B_{(k-1)(k-1)} \end{pmatrix}.$$

But this is just the matrix M^{kj} obtained by deleting the kth row and the jth column from B except that the column involving the B_{ik}'s has been multiplied by -1 and moved from the last slot to the jth slot. The determinant of this matrix is therefore $(-1)^{k-j} \det M^{kj}$ — one factor of -1 because of the minus signs on the column of B_{ik}'s, and $k - j - 1$ more factors of -1 from interchanging that column with the succeeding $k - j - 1$ columns to move it back to its rightful place on the right end. In short, the application of Cramer's rule to the system (B.5) yields

$$\frac{\partial g_j}{\partial y_k}(\mathbf{a}, b_k) = (-1)^{k-j} \frac{\det M^{jk}}{\det M^{kk}}.$$

Now we are done. Substitute this result back into (B.4), noting that $(-1)^{-j} = (-1)^j$, and recall (B.3):

$$\frac{\partial G}{\partial y_k}(\mathbf{a}, b_k) = \sum_{j=1}^{k-1} (-1)^{j+k} B_{kj} \frac{\det M^{kj}}{\det M^{kk}} + B_{kk}$$

$$= \frac{\sum_{j=1}^{k} (-1)^{j+k} B_{kj} \det M^{kj}}{\det M^{kk}} = \frac{\det B}{\det M^{kk}}.$$

Since $\det B \neq 0$ by assumption, this completes the verification that $\partial_{y_k} G(\mathbf{a}, b_k) \neq 0$ and hence the proof of the theorem. $\qquad\square$

B.3 Approximation by Riemann Sums

The subject of this section is Proposition 4.16 and its generalization to multiple integrals.

B.6 Lemma. *Suppose f is an integrable function on $[a, b]$ and $|f(x)| \leq C$ for $x \in [a, b]$. Let $P = \{x_0, \ldots, x_J\}$ be a partition of $[a, b]$ such that $\max_j (x_j - x_{j-1}) < \delta$, and let P' be another partition obtained by adding N extra points to P. Then $S_P f < S_{P'} f + 2CN\delta$ and $s_P f > s_{P'} f - 2CN\delta$.*

Proof. We consider the upper sums $S_P f$ and $S_{P'} f$; the argument for the lower sums is similar. If no extra point is added in the interval (x_{j-1}, x_j) in passing from P to P', both sums contain the term $M_j (x_j - x_{j-1})$, where M_j is the supremum of f on $[x_{j-1}, x_j]$. If extra points are added, the term $M_j (x_j - x_{j-1})$ in $S_P f$ is replaced by a sum of similar terms corresponding to subintervals of $[x_{j-1}, x_j]$. Both $M_j (x_j - x_{j-1})$ and the latter sum are bounded in absolute value by $C(x_j - x_{j-1}) < C\delta$, so their difference is bounded by $2C\delta$. The total change from $S_P f$ to $S_{P'} f$ is the sum of these differences, of which there are at most N, so it is less than $2CN\delta$. $\square$

Remark. The conclusion of this lemma is significant only when $N\delta \ll 1$, and hence when N is much less than the number J of subdivision points of P (since $J\delta > b - a$).

B.7 Theorem. *Suppose f is integrable on $[a, b]$. Given $\epsilon > 0$, there exists $\delta > 0$ such that if $P = \{x_0, \ldots, x_J\}$ is any partition of $[a, b]$ satisfying*

$$\max_{1 \leq j \leq J} (x_j - x_{j-1}) < \delta,$$

any Riemann sum $\sum_1^J f(t_j)(x_j - x_{j-1})$ associated to P differs from $\int_a^b f(x)\,dx$ by at most ϵ.

Proof. It is enough to prove the result for the lower and upper sums $s_P f$ and $S_P f$, as all other Riemann sums lie in between these two. Pick a partition Q of $[a, b]$ such that $S_Q f < \int_a^b f(x)\,dx + \frac{1}{2}\epsilon$ and $s_Q f > \int_a^b f(x)\,dx - \frac{1}{2}\epsilon$. Let N be the number of subdivision points in Q, and let C be an upper bound for $|f|$ on $[a, b]$; we claim that any $\delta < \epsilon/4NC$ will do the job. Indeed, suppose $P = \{x_0, \ldots, x_J\}$ satisfies $\max_j (x_j - x_{j-1}) < \delta$. Then the partition $P \cup Q$ is obtained by adding at most N points to P (namely, the points of Q that are not already in P). By Lemma B.6 and Lemma 4.3,

$$S_P f < S_{P \cup Q} f + 2NC\delta < S_{P \cup Q} f + \tfrac{1}{2}\epsilon \leq S_Q f + \tfrac{1}{2}\epsilon \leq \int_a^b f(x)\,dx + \epsilon,$$

and likewise $s_P f > \int_a^b f(x)\,dx - \epsilon$. Since $s_P f \leq \int f(x)\,dx \leq S_P f$, the proof is complete. $\square$

In the next two sections we shall need the generalization of Theorem B.7 to multiple integrals. The idea is exactly the same, but the notation is more complicated. We give the precise statement of the result but leave the adaptation of the one-dimensional proof to the reader.

B.8 Theorem. *Suppose f is integrable on the rectangular box $B = [a_1, b_1] \times \cdots \times [a_n, b_n]$. Given $\epsilon > 0$, there exists $\delta > 0$ such that if*

$$P = \left\{ x_{10}, \ldots, x_{1J_1}; x_{20}, \ldots, x_{2J_2}; \ldots; x_{n0}, \ldots x_{nJ_n} \right\}$$

is any partition of B satisfying

$$\max_{1 \leq i \leq n} \max_{1 \leq j \leq J_i} (x_{ij} - x_{i(j-1)}) < \delta,$$

any Riemann sum for f associated to B differs from $\int \cdots \int_B f(\mathbf{x}) \, d^n\mathbf{x}$ by at most ϵ.

B.4   Double Integrals and Iterated Integrals

B.9 Theorem. *Let $R = [a, b] \times [c, d]$, and let f be an integrable function on R. Suppose that, for each $y \in [c, d]$, the function f_y defined by $f_y(x) = f(x, y)$ is integrable on $[a, b]$, and the function $g(y) = \int_a^b f(x, y) \, dx$ is integrable on $[c, d]$. Then*

$$\iint_R f \, dA = \int_c^d \left[\int_a^b f(x, y) \, dx \right] dy.$$

Proof. Let $P_{JK} = \{x_0, \ldots, x_J; y_0, \ldots, y_K\}$ be the partition of R obtained by subdividing $[a, b]$ and $[c, d]$, respectively, into J and K equal subintervals of length $\Delta x = (b - a)/J$ and $\Delta y = (d - c)/K$. Given $\epsilon > 0$, there is an integer N such that

(B.10)
$$\left| \iint_R f \, dA - \sum_{j=1}^J \sum_{k=1}^K f(x_j, y_k) \, \Delta x \, \Delta y \right| < \frac{\epsilon}{3}$$

provided that $J \geq N$ and $K \geq N$, and also

(B.11)
$$\left| \int_c^d \left[\int_a^b f(x, y) \, dx \right] dy - \sum_{k=1}^K \int_a^b f(x, y_k) \, dx \, \Delta y \right| < \frac{\epsilon}{3}$$

provided that $K \geq N$. (For (B.10) we are applying Theorem B.8 to the function f, and for (B.11) we are applying Theorem B.7 to the function $g(y) = \int_a^b f(x, y) \, dx$.)

Let us fix K to be equal to N; then the points y_k are also fixed. By Theorem B.7 again, we can choose J large enough so that

$$\left| \int_a^b f(x, y_k)\, dx - \sum_{j=1}^J f(x_j, y_k)\, \Delta x \right| < \frac{\epsilon}{3(d - c)}$$

for all $k = 1, \ldots, K$. Then

$$\left| \sum_{j=1}^J \sum_{k=1}^K f(x_j, y_k)\, \Delta x\, \Delta y - \sum_{k=1}^K \int_a^b f(x, y_k)\, dx\, \Delta y \right|$$

$$\leq \sum_{k=1}^K \left| \sum_{j=1}^J f(x_j, y_k)\, \Delta x - \int_a^b f(x, y_k)\, dx \right| \Delta y < \frac{K \epsilon \Delta y}{3(d - c)} = \frac{\epsilon}{3}.$$

Therefore, by (B.10),

$$\left| \iint_R f\, dA - \sum_{k=1}^K \int_a^b f(x, y_k)\, dx\, \Delta y \right| < \frac{2\epsilon}{3},$$

and hence by (B.11),

$$\left| \iint_R f\, dA - \int_c^d \left[\int_a^b f(x, y)\, dx \right] dy \right| < \epsilon.$$

Since ϵ is arbitrary, the double integral and the iterated integral must be equal. $\quad\square$

B.5   Change of Variables for Multiple Integrals

The object of this section is to show that measurability and the zero-content property are preserved under invertible C^1 transformations, and to prove Theorem 4.37. The arguments are rather difficult, and we must begin by developing some tools.

For the calculations in this section, it will be convenient to measure the magnitude of a vector $\mathbf{x} \in \mathbb{R}^n$ not by the Euclidean norm $|\mathbf{x}|$ but by the "max-norm"

$$\|\mathbf{x}\| = \max\left(|x_1|, |x_2|, \ldots, |x_n| \right).$$

As we observed in (1.3), the norms $|\mathbf{x}|$ and $\|\mathbf{x}\|$ are comparable to each other in the sense that $\|\mathbf{x}\| \leq |\mathbf{x}| \leq \sqrt{n}\|\mathbf{x}\|$. The max-norm shares the following basic properties with the Euclidean norm:

$$\|\mathbf{x} + \mathbf{y}\| \leq \|\mathbf{x}\| + \|\mathbf{y}\|, \qquad \|c\mathbf{x}\| = |c|\,\|\mathbf{x}\|, \qquad \|\mathbf{x}\| = 0 \iff \mathbf{x} = \mathbf{0}.$$

However, the set

$$Q(r, \mathbf{x}) = \{\mathbf{y} : \|\mathbf{y} - \mathbf{x}\| < r\}$$

is not the ball of radius r about $\mathbf{x}$ but rather the open *cube* (or square, if $n = 2$) of side length $2r$ centered at $\mathbf{x}$.

Suppose $A : \mathbb{R}^n \to \mathbb{R}^m$ is a linear map with associated matrix (A_{jk}). For any $\mathbf{x} \in \mathbb{R}^n$ we have

$$\|A\mathbf{x}\| = \max_{j=1}^{m} \left| \sum_{k=1}^{n} A_{jk} x_k \right| \leq \left(\max_{j=1}^{m} \sum_{k=1}^{n} |A_{jk}| \right) \|\mathbf{x}\|.$$

Hence, if we define

(B.12) $$\|A\| = \max_{j=1}^{m} \sum_{k=1}^{n} |A_{jk}|,$$

we have

$$\|A\mathbf{x}\| \leq \|A\| \, \|\mathbf{x}\|.$$

We shall need the variant of Theorem 2.88 that pertains to the norms just defined, and an extension of it to nonconvex sets:

B.13 Lemma. *Suppose $\mathbf{F}$ is a differentiable map from a convex set $W \subset \mathbb{R}^n$ into $\mathbb{R}^m$, and suppose that $\|D\mathbf{F}(\mathbf{x})\| \leq M$ for all $\mathbf{x} \in W$ (where $\|D\mathbf{F}(\mathbf{x})\|$ is defined by (B.12)). Then*

$$\|\mathbf{F}(\mathbf{x}) - \mathbf{F}(\mathbf{y})\| \leq M\|\mathbf{x} - \mathbf{y}\| \text{ for all } \mathbf{x}, \mathbf{y} \in W.$$

Proof. Let $\mathbf{F} = (F_1, \ldots, F_m)$. By the mean value theorem (2.39), for each j there is a point $\mathbf{c}$ on the line segment between $\mathbf{x}$ and $\mathbf{y}$ such that

$$F_j(\mathbf{x}) - F_j(\mathbf{y}) = \nabla F_j(\mathbf{c}) \cdot (\mathbf{x} - \mathbf{y}) = \sum_{k=1}^{n} (\partial_k F_j(\mathbf{c}))(x_k - y_k).$$

But then

$$|F_j(\mathbf{x}) - F_j(\mathbf{y})| \leq \sum_{k=1}^{n} |\partial_k F_j(\mathbf{c})| \, \|\mathbf{x} - \mathbf{y}\| \leq \|D\mathbf{F}(\mathbf{c})\| \, \|\mathbf{x} - \mathbf{y}\| \leq M\|\mathbf{x} - \mathbf{y}\|.$$

Taking the maximum over j, we obtain the desired result. $\square$

B.14 Lemma. *Suppose $\mathbf{F}$ is a map of class C^1 from an open set $U \subset \mathbb{R}^n$ into $\mathbb{R}^m$. For any compact set $R \subset U$ there is a constant C such that*

$$\|\mathbf{F}(\mathbf{x}) - \mathbf{F}(\mathbf{y})\| \leq C\|\mathbf{x} - \mathbf{y}\| \text{ for all } \mathbf{x}, \mathbf{y} \in R.$$

Proof. Since U is open, for each $\mathbf{x} \in R$ there is a positive number r such that the cube $Q(2r, \mathbf{x})$ is contained in U. By the Heine-Borel theorem, R is covered by finitely many of the cubes $Q(r, \mathbf{x})$ with side length half as large, say $R \subset \bigcup_{j=1}^{J} Q(r_j, \mathbf{x}_j)$. Let r_0 be the smallest of the numbers $r_1, \ldots, r_J$. Moreover, let C_1 and C_2 be the maximum values of $\|D\mathbf{F}(\mathbf{x})\|$ and $\|\mathbf{F}(\mathbf{x})\|$ as $\mathbf{x}$ ranges over R. (These maxima exist since R is compact and $\|D\mathbf{F}(\mathbf{x})\|$ and $\|\mathbf{F}(\mathbf{x})\|$ are continuous functions of $\mathbf{x} \in R$.)

Now suppose $\mathbf{x}, \mathbf{y} \in R$; then either $\|\mathbf{x} - \mathbf{y}\| < r_0$ or $\|\mathbf{x} - \mathbf{y}\| \geq r_0$. In the first case, both $\mathbf{x}$ and $\mathbf{y}$ lie in one of the cubes $Q(2r_j, \mathbf{x}_j)$. (Indeed, $\mathbf{x}$ lies in one of the cubes $Q(r_j, \mathbf{x}_j)$ since they cover R, and then $\mathbf{y} \in Q(r_j + r_0, \mathbf{x}_j)$.) Since $Q(2r_j, \mathbf{x}_j)$ is convex, we can apply Lemma B.9 to conclude that $\|D\mathbf{F}(\mathbf{x}) - D\mathbf{F}(\mathbf{y})\| \leq C_1\|\mathbf{x} - \mathbf{y}\|$. In the second case, we simply have

$$\|\mathbf{F}(\mathbf{x}) - \mathbf{F}(\mathbf{y})\| \leq \|\mathbf{F}(\mathbf{x})\| + \|\mathbf{F}(\mathbf{y})\| \leq 2C_2 \leq \frac{2C_2}{r_0}\|\mathbf{x} - \mathbf{y}\|.$$

Hence we can take $C = \max(C_1, 2C_2/r_0)$. $\qquad\square$

Before proceeding, we need to make one more observation: In developing the theory of integration one uses (n-dimensional) rectangles in a number of places; *it is enough to use cubes instead.* First, in defining the integral of an integrable function over a measurable set S, we can enclose S in a cube Q and restrict attention to the approximating sums obtained by partitioning Q evenly into smaller subcubes; these sums converge to the integral by Theorem B.8. Second, in showing that a set has zero content, we consider coverings of a set S by finite unions of rectangles whose total volume is small. We can enlarge each rectangle by an arbitrarily small amount to obtain one whose vertices have rational coordinates, and the latter rectangle can be subdivided into cubes of side length $1/d$ where d is the least common denominator of its side lengths.

Now we are ready to address the central issues of this section. *For the rest of this section,* $\mathbf{G}$ *will denote a one-to-one transformation of class* C^1 *from an open set* $U \subset \mathbb{R}^n$ *onto another open set* $V \subset \mathbb{R}^n$ *whose derivative* $D\mathbf{G}(\mathbf{u})$ *is invertible for all* $\mathbf{u} \in U$. By the inverse mapping theorem, $\mathbf{G}^{-1} : V \to U$ also has the same properties. *Moreover, we denote the n-dimensional volume of a measurable set* $S \subset \mathbb{R}^n$ *by* $V^n(S)$. (Thus, if S is a cube of side length r, we have $V^n(S) = r^n$.)

B.15. Theorem. *Suppose* $K \subset U$ *is a compact set with zero content. Then* $\mathbf{G}(K)$ *also has zero content.*

Proof. First, since U is open, for each $\mathbf{u} \in K$ there is a cube centered at $\mathbf{x}$ whose vertices have rational coordinates and whose closure lies in U. Since K is compact, finitely many of these cubes cover K; thus, $K \subset R^{\text{int}}$ where R is a finite union of closed cubes contained in U. Let C be the constant in Lemma B.14, with R being the set we have just defined.

Since K has zero content, for any $\epsilon > 0$ there is a finite collection of cubes $\{Q(r_j, \mathbf{x}_j)\}$ such that $K \subset \bigcup Q(r_j, \mathbf{x}_j)$ and $\sum V^n(Q(r_j, \mathbf{x}_j)) = \sum r_j^n < \epsilon/C^n$, and these cubes can be taken to be subsets of R. (See the remarks following Lemma B.14.) By Lemma B.14, $\mathbf{G}(Q(r_j, \mathbf{x}_j)) \subset Q(Cr_j, \mathbf{G}(\mathbf{x}_j))$. Thus $\mathbf{G}(K)$ is contained in the union of the cubes $Q(Cr_j, \mathbf{G}(\mathbf{x}_j))$, and the sum of their volumes is $\sum (Cr_j)^n = C^n \sum r_j^n < \epsilon$. It follows that $\mathbf{G}(K)$ has zero content. $\qquad \square$

B.16 Corollary. *Suppose T is a measurable set with $\overline{T} \subset U$. Then $\mathbf{G}(T)$ is also measurable.*

Proof. First we observe that T is bounded (because it is measurable), so its boundary ∂T is compact. Moreover, $\mathbf{G}(\partial T) = \partial(\mathbf{G}(T))$. (This is an easy consequence of the fact that $\mathbf{G}$ and $\mathbf{G}^{-1}$ are both continuous; the proof is left as an exercise to the reader.) Now, measurability means that the boundary ∂T is a set of zero content. (In particular, ∂T is bounded, and hence compact since it is closed.) By Theorem B.15, $\partial(\mathbf{G}(T)) = \mathbf{G}(\partial T)$ has zero content, so $\mathbf{G}(T)$ is measurable. $\quad \square$

B.17 Corollary. *If $f : V \to \mathbb{R}$ is continuous except possibly on a compact set of zero content, then the same is true of $f \circ \mathbf{G} : U \to \mathbb{R}$.*

Proof. Suppose f is continuous on $V \setminus K$, where $K \subset V$ is compact and has zero content. Since $\mathbf{G}$ is continuous, $f \circ \mathbf{G}$ is continuous on $U \setminus \mathbf{G}^{-1}(K)$. Since $\mathbf{G}^{-1}$ is continuous, $\mathbf{G}^{-1}(K)$ is compact (by Theorem 1.22) and has zero content (by Theorem B.15). $\qquad \square$

We now present a sequence of lemmas leading up to the main change-of-variable theorem. The heart of the argument is Lemma B.21.

If S and T are subsets of $\mathbb{R}^n$, the **distance** from S to T is defined to be

$$d(S, T) = \inf \{ |\mathbf{x} - \mathbf{y}| : \mathbf{x} \in S, \ \mathbf{y} \in T \}.$$

B.18 Lemma. *Suppose that S and T are disjoint closed subsets of $\mathbb{R}^n$ and S is compact. Then $d(S, T) > 0$.*

Proof. If the assertion is false, there exist sequences $\{\mathbf{x}_j\}$ in S and $\{\mathbf{y}_j\}$ in T such that $|\mathbf{x}_j - \mathbf{y}_j| \to 0$. Since S is compact, by passing to a subsequence we may assume that $\mathbf{x}_j$ converges to a point $\mathbf{x} \in S$. But then $\mathbf{y}_j \to \mathbf{x}$ also, so $\mathbf{x} \in T$ since T is closed. This is impossible since $S \cap T = \varnothing$. $\qquad \square$

B.19 Lemma. *Suppose $Q \subset U$ is a closed cube. For any invertible linear map $A : \mathbb{R}^n \to \mathbb{R}^n$,*

$$V^n(\mathbf{G}(Q)) \le |\det A| \left(\sup_{\mathbf{u} \in Q} \|A^{-1}D\mathbf{G}(\mathbf{u})\| \right)^n V^n(Q).$$

Proof. Let $C = \sup_{\mathbf{u} \in Q} \|A^{-1}D\mathbf{G}(\mathbf{u})\|$ (which is finite since Q is compact), and notice that $A^{-1}D\mathbf{G}(\mathbf{u}) = D(A^{-1} \circ \mathbf{G})(\mathbf{u})$ since A^{-1} is linear. We apply Lemma B.9 to the map $\mathbf{F} = A^{-1} \circ \mathbf{G}$ on the set $W = Q$ to see that $A^{-1}(\mathbf{G}(Q))$ is contained in a cube Q' whose side length is C times the side length of Q, and whose volume is therefore C^n times that of Q. Hence, by Theorem 4.35,

$$|\det A|^{-1}V^n(\mathbf{G}(Q)) = V^n(A^{-1}(\mathbf{G}(Q))) \le V^n(Q') = C^n V^n(Q),$$

as claimed. $\qquad\square$

B.20 Lemma. *Let R be a compact subset of U. For any $\epsilon > 0$ there is a $\delta > 0$ such that*

$$\left| \|D\mathbf{G}(\mathbf{u})^{-1}D\mathbf{G}(\mathbf{v})\| - 1 \right| < \epsilon \text{ and } \left| |\det D\mathbf{G}(\mathbf{u})|^{-1}|\det D\mathbf{G}(\mathbf{v})| - 1 \right| < \epsilon$$

whenever $\mathbf{u}, \mathbf{v} \in R$ and $\|\mathbf{u} - \mathbf{v}\| < \delta$.

Proof. By (A.55) in Appendix A, the entries of the matrix $D\mathbf{G}(\mathbf{u})^{-1}D\mathbf{G}(\mathbf{v})$ vary continuously as $\mathbf{u}, \mathbf{v}$ vary over R, so the functions $\varphi(\mathbf{u}, \mathbf{v}) = \|D\mathbf{G}(\mathbf{u})^{-1}D\mathbf{G}(\mathbf{v})\|$ and $\psi(\mathbf{u}, \mathbf{v}) = |\det D\mathbf{G}(\mathbf{u})|^{-1}|\det D\mathbf{G}(\mathbf{v})|$ are continuous on $R \times R$. Moreover, $\varphi(\mathbf{u}, \mathbf{u}) = \psi(\mathbf{u}, \mathbf{u}) = 1$ for all $\mathbf{u} \in R$. (It follows easily from the definition (B.12) that $\|I\| = 1$.) Since $R \times R$ is compact, φ and ψ are uniformly continuous (Theorem 1.33). Hence, for any $\epsilon > 0$ there is a $\delta > 0$ such that $|\varphi(\mathbf{u}, \mathbf{v}) - \varphi(\mathbf{u}', \mathbf{v}')| < \epsilon$ whenever $\|\mathbf{u} - \mathbf{u}'\| + \|\mathbf{v} - \mathbf{v}'\| < \delta$, and likewise for ψ. Taking $\mathbf{u}' = \mathbf{v}' = \mathbf{u}$, we obtain the desired conclusions. $\qquad\square$

For the remainder of this section, we denote n-dimensional integrals by a single integral sign $\int$ rather than $\int \cdots \int$.

B.21 Lemma. *Let T be a measurable set such that $\overline{T} \subset U$. Then*

(B.22) $$V^n(\mathbf{G}(T)) \le \int_T |\det D\mathbf{G}| \, dV^n.$$

Proof. Since ∂T and $\partial(\mathbf{G}(T))$ have zero content (Corollary B.16), the quantities on either side of (B.22) are unchanged if we replace T by $\overline{T}$. Hence we may as well assume that $T = \overline{T}$ is compact.

We shall prove (B.22) by approximating the quantities on either side by finite sums corresponding to a grid of small cubes. In detail, the process is as follows. Pick a closed cube Q_0 such that $T \subset Q_0$, and denote the side length of Q_0 by l. By partitioning the sides of Q_0 into M equal pieces, we obtain a partition of Q_0 into M^n equal subcubes of side length l/M; denote this collection of closed cubes by $\mathcal{Q}_M$. Since distance from T to the complement of U is strictly positive by Lemma B.18, all of the cubes in $\mathcal{Q}_M$ that intersect T will be contained in U provided M is sufficiently large, say $M \geq M_0$. For each $M \geq M_0$, let R_M be the union of those cubes in $\mathcal{Q}_M$ that intersect T. Then R_M is a compact set such that $T \subset R_M \subset U$, and $V^n(R_M) \to V^n(T)$ as $M \to \infty$.

Now, let $\epsilon > 0$ be given. We choose $\delta > 0$ as in Lemma B.20, and we then pick $M \geq M_0$ large enough so that $l/M < \delta$ and $V^n(R_M) < V^n(T) + \epsilon$.

Let $Q_1, \ldots, Q_K$ be the cubes in $\mathcal{Q}_M$ that intersect T, so that $R_M = \bigcup_{k=1}^{K} Q_k$, and let $\mathbf{x}_k$ be the center of Q_k. Since $l/M < \delta$, Lemma B.20 applies whenever $\mathbf{u} \in Q_k$ and $\mathbf{v} = \mathbf{x}_k$. Thus, by Lemma B.19, with $A = D\mathbf{G}(\mathbf{x}_k)$,

$$V^n(\mathbf{G}(Q_k)) \leq |\det D\mathbf{G}(\mathbf{x}_k)|(1 + \epsilon)^n V^n(Q_k),$$

so

$$V^n(\mathbf{G}(T)) \leq \sum_{k=1}^{K} V^n(\mathbf{G}(Q_k)) < (1 + \epsilon)^n \sum_{k=1}^{K} |\det D\mathbf{G}(\mathbf{x}_k)| V^n(Q_k).$$

On the other hand, by Lemma B.20 again,

$$|\det D\mathbf{G}(\mathbf{u})| > (1 - \epsilon)|\det D\mathbf{G}(\mathbf{x}_j)| \quad \text{for all } \mathbf{u} \in Q_j,$$

so

$$|\det D\mathbf{G}(\mathbf{x}_k)| V^n(Q_k) = \int_{Q_k} |\det D\mathbf{G}(\mathbf{x}_k)| \, d^n\mathbf{u} < \frac{1}{1 - \epsilon} \int_{Q_k} |\det D\mathbf{G}(\mathbf{u})| \, d^n\mathbf{u}.$$

In short,

$$V^n(\mathbf{G}(T)) \leq \frac{(1 + \epsilon)^n}{1 - \epsilon} \sum_{k=1}^{K} \int_{Q_k} |\det D\mathbf{G}| \, dV^n = \frac{(1 + \epsilon)^n}{1 - \epsilon} \int_{R_M} |\det D\mathbf{G}| \, dV^n.$$

Finally, let C be the maximum of $|\det D\mathbf{G}(\mathbf{u})|$ as $\mathbf{u}$ ranges over the compact set R_{M_0}. Then

$$\left| \int_{R_M} |\det D\mathbf{G}|\, dV^n - \int_T |\det D\mathbf{G}|\, dV^n \right| = \int_{R_M \setminus T} |\det D\mathbf{G}|\, dV^n$$
$$\leq C[V^n(R_M) - V^n(T)] < C\epsilon.$$

Therefore,

$$V^n(\mathbf{G}(T)) \leq \frac{(1+\epsilon)^n}{1-\epsilon} \int_T |\det D\mathbf{G}|\, dV^n + C\frac{(1+\epsilon)^n}{1-\epsilon}\epsilon.$$

Since ϵ is arbitrary and C is independent of ϵ, (B.22) follows. $\qquad\square$

B.23 Lemma. *Let T be a measurable set such that $\overline{T} \subset U$, and let f be a bounded nonnegative function on $\mathbf{G}(T)$ that is continuous except perhaps on a set of zero content. Then*

$$\int_{\mathbf{G}(T)} f(\mathbf{x})\, d^n\mathbf{x} \leq \int_T f(\mathbf{G}(\mathbf{u}))|\det D\mathbf{G}(\mathbf{u})|\, d^n\mathbf{u}.$$

Proof. Consider a lower Riemann sum for $\int_{\mathbf{G}(T)} f$:

$$s_P f = \sum_{j=1}^{J} m_j V^n(Q_j),$$

where the Q_j's are cubes with disjoint interiors contained in $\mathbf{G}(T)$ and $m_j = \inf_{\mathbf{x} \in Q_j} f(\mathbf{x})$. (The hypothesis $f \geq 0$ is needed so that the cubes Q_j satisfy $Q_j \subset \mathbf{G}(T)$, not just $Q_j \cap \mathbf{G}(T) \neq \varnothing$.) By Theorem B.15 and Corollary B.17 (applied to $\mathbf{G}^{-1}$), the sets $\mathbf{G}^{-1}(Q_j)$ are measurable and overlap only in sets of zero content. By Lemma B.21, then, we have

$$s_P f = \sum m_j V^n(Q_j)$$
$$\leq \sum m_j \int_{\mathbf{G}^{-1}(Q_j)} |\det D\mathbf{G}|\, dV^n \leq \sum \int_{\mathbf{G}^{-1}(Q_j)} (f \circ \mathbf{G})|\det D\mathbf{G}|\, dV^n$$
$$= \int_{\bigcup \mathbf{G}^{-1}(Q_j)} (f \circ \mathbf{G})|\det \mathbf{G}|\, dV^n \leq \int_T (f \circ \mathbf{G})|\det D\mathbf{G}|\, dV^n.$$

(For the last inequality we used the fact that $\bigcup \mathbf{G}^{-1}(Q_j) \subset T$ and the assumption that $f \geq 0$.) Taking the supremum over all lower Riemann sums $s_P f$, we obtain the desired conclusion. $\qquad\square$

At last we come to the main result, for which we restate the hypotheses in full. We assume that $f : \mathbf{G}(T) \to \mathbb{R}$ is bounded and continuous except on a set of zero content (and hence is integrable on $\mathbf{G}(T)$); by Corollary B.17, this implies that $f \circ \mathbf{G} : T \to \mathbb{R}$ is also bounded and continuous except on a set of zero content (and hence is integrable on T). It is actually enough to assume that f is integrable on $\mathbf{G}(T)$, but then an additional argument would be necessary to establish the integrability of $f \circ \mathbf{G}$.

B.24 Theorem. *Let $\mathbf{G}$ be a one-to-one transformation of class C^1 from an open set $U \subset \mathbb{R}^n$ onto another open set $V \subset \mathbb{R}^n$ whose derivative $D\mathbf{G}(\mathbf{u})$ is invertible for all $\mathbf{u} \in U$. Let T be a measurable set such that $\overline{T} \subset U$, and let f be a bounded function on $\mathbf{G}(T)$ that is continuous except perhaps on a set of zero content. Then*

$$\int_{\mathbf{G}(T)} f(\mathbf{x})\, d^n\mathbf{x} = \int_T f(\mathbf{G}(\mathbf{u}))|\det D\mathbf{G}(\mathbf{u})|\, d^n\mathbf{u}.$$

Proof. It suffices to show that each of these integrals is less than or equal to the other one. For $f \geq 0$, Lemma B.23 proves one of these inequalities, and the reverse inequality follows by applying Lemma B.23 to the inverse transformation. More precisely, if in Lemma B.23 we replace T by $\mathbf{G}(T)$, $\mathbf{G}$ by $\mathbf{G}^{-1}$, and f by $(f \circ \mathbf{G})|\det D\mathbf{G}|$, we obtain

$$\int_T f(\mathbf{G}(\mathbf{u}))|\det D\mathbf{G}(\mathbf{u})|\, d^n\mathbf{u}$$

$$\leq \int_{\mathbf{G}(T)} f(\mathbf{G}(\mathbf{G}^{-1}(\mathbf{x})))|\det D\mathbf{G}(\mathbf{G}^{-1}(\mathbf{x}))||\det D\mathbf{G}^{-1}(\mathbf{x})|\, d^n\mathbf{x}.$$

But by the chain rule (2.86), the matrices $D\mathbf{G}(\mathbf{G}^{-1}(\mathbf{x}))$ and $D\mathbf{G}^{-1}(\mathbf{x})$ are inverses of each other, so their determinants are reciprocals of each other; hence, the integral on the right is simply $\int_{\mathbf{G}(T)} f(\mathbf{x})\, d^n\mathbf{x}$. Thus the theorem is proved for the case $f \geq 0$. The general case follows by writing $f = (f + C) - C$ where $C \geq 0$ is sufficiently large that $f + C \geq 0$ on T. The argument just given applies to $f + C$ and to the constant functon C; subtracting the results yields the theorem. $\qquad\square$

B.6   Improper Multiple Integrals

In this section we denote multiple integrals by a single integral sign $\int$ rather than $\int \cdots \int$.

B.25 Theorem. *Let S be an open set in $\mathbb{R}^n$, and let f be a nonnegative function on S that is integrable over every compact subset of S. Let $\{U_j\}$ and $\{\tilde{U}_j\}$ be sequences of compact subsets of S such that*

$$U_1 \subset U_2 \subset U_3 \subset \cdots, \quad \tilde{U}_1 \subset \tilde{U}_2 \subset \tilde{U}_3 \subset \cdots, \quad and \quad \bigcup_1^\infty U_j^{\text{int}} = S = \bigcup_1^\infty \tilde{U}_j^{\text{int}}.$$

Then

$$\lim_{j \to \infty} \int_{U_j} f \, dV^n = \lim_{j \to \infty} \int_{\tilde{U}_j} f \, dV^n,$$

where the limits may be finite or $+\infty$.

Proof. The limits in question exist by the monotone sequence theorem, because $\int_{U_j} f \, dV^n$ and $\int_{\tilde{U}_j} f \, dV^n$ increase with j. Let $I = \lim_{j \to \infty} \int_{U_j} f \, dV^n$, and let c be any number less than I. We then have $\int_{U_j} f \, dV^n > c$ when j is sufficiently large, say $j \geq J$. Now $U_J \subset S = \bigcup_1^\infty \tilde{U}_j^{\text{int}}$, so by the Heine-Borel theorem, for some finite K we have $U_J \subset \bigcup_1^K \tilde{U}_j^{\text{int}} \subset \tilde{U}_K$. But then, for $j \geq K$,

$$\int_{\tilde{U}_j} f \, dV^n \geq \int_{\tilde{U}_K} f \, dV^n \geq \int_{U_J} f \, dV^n > c.$$

Since c is an arbitrary number less than I, it follows that $\lim \int_{\tilde{U}_j} f \, dV^n \geq I = \lim \int_{U_j} f \, dV^n$. The same argument works with the roles of the U_j's and $\tilde{U}_j$'s switched, so the two limits are actually equal. $\qquad\square$

B.7 Green's Theorem and the Divergence Theorem

The object of this section is to show how to prove Green's theorem and its analogues in higher dimensions for general C^1 domains. For this purpose we need to develop a technical tool, the notion of partitions of unity, that has many uses in advanced analysis.

B.26 Lemma. *For any rectangular box $B = [a_1, b_1] \times \cdots \times [a_n, b_n]$ in $\mathbb{R}^n$, there is a C^∞ function f on $\mathbb{R}^n$ such that $f(\mathbf{x}) > 0$ for $\mathbf{x} \in B^{\text{int}}$ and $f(\mathbf{x}) = 0$ otherwise.*

Proof. In the case $n = 1$, $B = [a, b]$, we can take f to be

$$f_a^b(x) = \begin{cases} e^{1/(x-a)(x-b)} & \text{if } a < x < b, \\ 0 & \text{otherwise.} \end{cases}$$

(Note that the exponent $1/(x - a)(x - b)$ is negative for $a < x < b$ and tends to $-\infty$ as $x \to a+$ or $x \to b-$.) An argument like that in Exercise 9, §2.1, shows that f and all its derivatives vanish as $x \to a+$ or $x \to b-$, so f is C^∞ even at a and b. For the n-dimensional case, then, the function

$$f(\mathbf{x}) = f_{a_1}^{b_1}(x_1) f_{a_2}^{b_2}(x_2) \cdots f_{a_n}^{b_n}(x_n)$$

does the job.   □

If f is a function on $\mathbb{R}^n$, the **support** of f, denoted by $\mathrm{supp}(f)$, is the closure of the set of all points $\mathbf{x}$ such that $f(\mathbf{x}) \neq 0$; in other words, it is the smallest closed set outside of which f vanishes.

B.27 Theorem. *Suppose $K \subset \mathbb{R}^n$ is compact and $U_1, \ldots, U_J$ are open sets such that $K \subset \bigcup_1^J U_j$. Then there exists a finite collection $\{\varphi_m\}_1^M$ of C^∞ functions such that*

a. *the support of each φ_m is compact and contained in one of the sets U_j, and*
b. *$\sum_1^M \varphi_m(\mathbf{x}) = 1$ for all $\mathbf{x} \in K$.*

Proof. The starting point is a fact we demonstrated in the course of proving Theorem B.1: *There is a grid $\mathcal{B}$ of closed rectangular boxes such that each box in $\mathcal{B}$ that intersects K is contained in one of the sets U_j.* Let $B_1, \ldots, B_M$ be the boxes in $\mathcal{B}$ that intersect K, and let $B_{M+1}, \ldots, B_N$ be the additional boxes in $\mathcal{B}$ that intersect at least one of $B_1, \ldots, B_M$. (Thus, $\bigcup_1^M B_m$ is a compact set contained in U whose interior contains K, and $\bigcup_1^N B_m$ is obtained by adding one additional layer of boxes around the boundary of $\bigcup_1^M B_m$.)

For $1 \leq m \leq M$, the box B_m is contained in one of the U_j's, say $U_{j(m)}$; let $c_m = d(B_m, U_{j(m)}^c)$. (Here $d(S, T)$ is the distance from S to T, defined before Lemma B.18.) On the other hand, for $M < m \leq N$ we have $B_m \cap K = \varnothing$; let $c_m = d(B_m, K)$. The numbers c_m are all positive by Lemma B.18. Let η be the smallest of the side lengths of the B_m's, let

$$\delta = \frac{1}{2\sqrt{n}} \min(c_1, \ldots, c_N, \eta),$$

and for $1 \leq m \leq N$ let $\widetilde{B}_m$ be the closed box with the same center as B_m whose side lengths are larger than those of B_m by the amount δ. Then the boxes $\widetilde{B}_m$ have the following properties: First, each point of B_m is in the interior of $\widetilde{B}_m$. Second, since $\delta < \frac{1}{2}\eta$, for $m \leq M$ each point of $\widetilde{B}_m$ is in the interior of one of the $\widetilde{B}_l$'s. (It is the points on the boundary of $\widetilde{B}_m$ that are at issue here, and it may happen that $l > M$.) Third, if $\mathbf{x} \in \widetilde{B}_m$, there is a point $\mathbf{y} \in B_m$ such that $|x_j - y_j| \leq \delta$ for all

j, and hence $|\mathbf{x} - \mathbf{y}| \leq \delta\sqrt{n}$. Since $\delta\sqrt{n} < \frac{1}{2}c_m$, it follows that $\widetilde{B}_m \subset U_{j(m)}$ for $m \leq M$ and $\widetilde{B}_m \cap K = \varnothing$ for $m > M$.

Now, for $1 \leq m \leq N$, choose a C^∞ function ψ_m such that $\psi_m > 0$ on $\widetilde{B}_m$ and $\psi_m = 0$ on $\widetilde{B}_m^c$, according to Lemma B.26, and let

$$\varphi_m = \frac{\psi_m}{\sum_{l=1}^N \psi_l} \qquad (1 \leq m \leq M),$$

with the understanding that $\varphi_m = 0$ outside $\widetilde{B}_m$. Since the sum in the denominator is strictly positive on $\bigcup_1^N \widetilde{B}_l^{\mathrm{int}}$, an open set that includes $\widetilde{B}_m$, the function φ_m is C^∞ and $\mathrm{supp}(\varphi_m) = \widetilde{B}_m \subset U_{j(m)}$. Finally, for $l > M$ we have $\widetilde{B}_l \cap K = \varnothing$ and hence $\psi_l = 0$ on K; therefore, for $\mathbf{x} \in K$,

$$\sum_{m=1}^M \varphi_m(\mathbf{x}) = \frac{\sum_{m=1}^M \psi_m(\mathbf{x})}{\sum_{l=1}^N \psi_l(\mathbf{x})} = \frac{\sum_{m=1}^M \psi_m(\mathbf{x})}{\sum_{l=1}^M \psi_l(\mathbf{x})} = 1,$$

so the φ_m's have all the desired properties. $\qquad\square$

The collection of functions $\{\varphi_m\}$ in Theorem B.27 is called a **partition of unity on K subordinate to the covering** $\{U_j\}$.

We are now ready to prove Green's theorem for general regions with smooth boundary. Afterwards, we shall indicate how to extend the proof to regions with piecewise smooth boundary.

B.28 Theorem. *Suppose S is a compact region in $\mathbb{R}^2$ whose boundary ∂S is a finite union of simple closed curves of class C^1, equipped with the positive orientation. Suppose also that P and Q are C^1 functions on S. Then*

$$\int_{\partial S} P\, dx + Q\, dy = \iint_S \left(\frac{\partial Q}{\partial x} - \frac{\partial P}{\partial y} \right) dA.$$

Proof. The starting point is the special case of Green's theorem, proved in §5.2, in which S is x-simple and y-simple. (What we actually need here is the case where S is a rectangle with sides parallel to the axes.) In contrast to the method used in §5.2 to handle more general regions, instead of cutting up the *region* into simple pieces, we shall use a partition of unity to cut up the *integrand* into pieces that are easily analyzed by a change of variables.

By Theorem 3.13, for every point $\mathbf{x} \in \partial S$ there is an open disc D centered at $\mathbf{x}$ such that the portion of ∂S within D is the graph of a C^1 function, either $y = f(x)$ or $x = f(y)$. By the Heine-Borel theorem, we can select finitely many of these

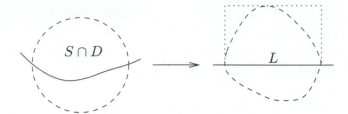

FIGURE B.1: The transformation $(x, y) \rightarrow (u, v)$. The disc D is the indicated by the dashed circle on the left; the rectangle R to which Green's theorem is to be applied is dotted on the right.

discs, say $D_1, \ldots, D_J$, so that $\partial S \subset \bigcup_1^J D_j$. Then $D_1, \ldots, D_J$, and S^{int} form an open covering of S.

By Theorem B.27 we can choose a partition of unity $\{\varphi_m\}_1^M$ on S subordinate to this covering. Then $P = \sum_1^m \varphi_m P$ and $Q = \sum_1^m \varphi_m Q$ on S, and $\varphi_m P$ and $\varphi_m Q$ are still of class C^1, so it suffices to prove the theorem with P and Q replaced by $\varphi_m P$ and $\varphi_m Q$ for $m = 1, \ldots, M$. In short, *it is enough to prove the theorem when* $\operatorname{supp}(P)$ *and* $\operatorname{supp}(Q)$ *are either (a) contained in* S^{int} *or (b) contained in a disc* D *such that* $D \cap \partial S$ *is the graph of a* C^1 *function.*

In case (a), P and Q both vanish on ∂S, so $\int_{\partial S} P \, dx + Q \, dy = 0$. Also, P and Q remain C^1 if we extend them to be zero outside of S. But then we can apply Green's theorem on any rectangle R that includes S to conclude that

$$\iint_S \left(\frac{\partial Q}{\partial x} - \frac{\partial P}{\partial y} \right) dA = \iint_R \left(\frac{\partial Q}{\partial x} - \frac{\partial P}{\partial y} \right) dA = \int_{\partial R} P \, dx + Q \, dy = 0.$$

Thus the theorem is true in case (a).

Case (b) is the more interesting one. Suppose, to be definite, that P and Q are supported in D, where $\partial S \cap D$ is a portion of the graph of a C^1 function $y = f(x)$. (The case $x = f(y)$ is similar.) We define a change of variables $(x, y) = \mathbf{G}(u, v)$ on D by

$$x = u, \quad y = v + f(u); \qquad \text{that is,} \qquad u = x, \quad v = y - f(x).$$

The transformation $\mathbf{G}^{-1}(x, y) = (u, v)$ maps D to a bounded region in the uv-plane, $\partial S \cap D$ to a line segment L in the u-axis, and $S \cap [\operatorname{supp}(P) \cup \operatorname{supp}(Q)]$ to a bounded region T in either the upper or the lower half-plane. More precisely, T will be in the upper half-plane if S lies above the graph $y = f(x)$ and in the lower half-plane if S lies below; thus, the relative orientations of T and L are the same as those of S and ∂S. See Figure B.1.

Let R be a rectangle in the uv-plane, one of whose sides is the segment L, that includes T. Then the functions $\widetilde{P} = P \circ \mathbf{G}$ and $\widetilde{Q} = Q \circ \mathbf{G}$ are C^1 functions on R that vanish on the three sides of R other than L.

Now, $dx = du$ and $dy = f'(u)\, du + dv$, so

$$\int_{\partial S} P\, dx + Q\, dy = \int_L \widetilde{P}\, du + \widetilde{Q}[f'(u)\, du + dv],$$

where L is oriented as a portion of ∂R. Since $\widetilde{P}$ and $\widetilde{Q}$ vanish on the other sides of R, we can apply Green's theorem on R to conclude that

$$\int_{\partial S} P\, dx + Q\, dy = \int_{\partial R} [\widetilde{P}(u,v) + \widetilde{Q}(u,v) f'(u)]\, du + \widetilde{Q}(u,v)\, dv$$

(B.29)
$$= \iint_R \left(\frac{\partial \widetilde{Q}}{\partial u} - \frac{\partial \widetilde{P}}{\partial v} - \frac{\partial \widetilde{Q}}{\partial v} f'(u) \right) du\, dv.$$

But by the chain rule,

$$\frac{\partial \widetilde{Q}}{\partial u} - \frac{\partial \widetilde{Q}}{\partial v} f'(u) = \frac{\partial Q}{\partial x}\Big|_{(x,y)=\mathbf{G}(u,v)}; \qquad \frac{\partial \widetilde{P}}{\partial v} = \frac{\partial P}{\partial y}\Big|_{(x,y)=\mathbf{G}(u,v)}.$$

Also,

$$\frac{\partial(x,y)}{\partial(u,v)} = \det D\mathbf{G} = \det \begin{pmatrix} 1 & 0 \\ f'(u) & 1 \end{pmatrix} = 1,$$

so $du\, dv = dx\, dy$ by Theorem B.24. It follows that the double integral (B.29) is equal to $\iint_S (\partial_x Q - \partial_y P)\, dx\, dy$, which completes the proof. $\qquad \square$

Let us indicate how this argument can be extended to region S with piecewise smooth boundary. Recall from §5.1 that "piecewise smooth" means that ∂S consists of curves that are smooth except at finitely many points, where they have "corners," i.e., where the direction of the curve changes abruptly. If $\mathbf{x}_0$ is such a point, there is a small disc D centered at $\mathbf{x}_0$ such that $\partial S \cap D$ is the union of portions of two smooth curves that intersect at $\mathbf{x}_0$. By Theorem 3.13, by shrinking D if necessary we may assume that these curves are the loci of equations $F(\mathbf{x}_0) = 0$ and $G(\mathbf{x}_0) = 0$, where $\nabla F \neq \mathbf{0}$ and $\nabla G \neq \mathbf{0}$ on D. We shall assume that $\nabla F(\mathbf{x}_0)$ and $\nabla G(\mathbf{x}_0)$ are linearly independent. (The exceptional case where they are not — that is, where the two curves are tangent at $\mathbf{x}_0$ and the region has a sharp "cusp" rather than a "corner" at $\mathbf{x}_0$ — must be handled by an additional limiting argument, in which S is approximated by regions with smooth boundaries.) Then, by

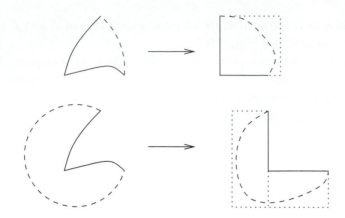

FIGURE B.2: Transformation of a region with a corner. The rectangles to which Green's theorem is to be applied are dotted on the right.

the inverse mapping theorem, by shrinking D yet further we may assume that the transformation $u = F(x, y)$, $v = G(x, y)$ has a C^1 inverse on D.

Now, as in the proof of Theorem B.28, we can cover ∂S by finitely many discs $D_1, \ldots D_J$ such that $\partial S \cap D_j$ is the graph of a smooth function, together with finitely many discs $D_{J+1}, \ldots D_K$ centered at the corners and satisfying the conditions of the preceding paragraph. By using of a partition of unity subordinate to the covering $\{D_1, \ldots, D_K, S^{\text{int}}\}$ of S, we reduce to the case where P and Q are supported in one of these discs. The discs D_j of the first kind ($j \leq J$) are handled as before. For the ones centered at a corner, we use the change of variables $u = F(x, y)$, $v = G(x, y)$ described above to reduce to the case where the boundary consists of a segment of the u-axis and a segment of the v-axis that meet at the origin. (This change of variables is not as simple as the one we used before, so the calculations are more complicated, but the idea is the same.) If S occupies the "inside" of the corner, the calculation boils down to Green's theorem on a rectangle as before; if S occupies the "outside," it boils down to Green's theorem for two rectangles; see Figure B.2.

Finally, we prove the divergence theorem for general regions with C^1 boundary. The argument can be extended to handle regions with piecewise smooth boundary in a manner similar to that in the preceding paragraphs.

B.30 Theorem. *Suppose R is a compact region in $\mathbb{R}^3$ with piecewise smooth boundary ∂R, oriented so that the positive normal points out of R. Suppose also* *that* $\mathbf{F}$ *is a vector field of class* C^1 *on R. Then*

$$\iint_{\partial R} \mathbf{F} \cdot \mathbf{n}\, dA = \iiint_{R} \operatorname{div} \mathbf{F}\, dV.$$

Proof. The proof is very similar to that of Theorem B.28, so we shall omit many details. By using a partition of unity, we reduce the problem to proving the theorem when $\operatorname{supp}(\mathbf{F}) \subset R^{\mathrm{int}}$ or when $\operatorname{supp}(\mathbf{F}) \subset B$ where B is a ball such that $\partial R \cap B$ is the graph of a C^1 function, say $z = \varphi(x, y)$. In the first case, the integrals $\iint_{\partial R} \mathbf{F} \cdot \mathbf{n}\, dA$ and $\iiint_{R} \operatorname{div} \mathbf{F}\, dV$ both vanish, as in Theorem B.28. In the second case, we introduce a change of variables on B, $(x, y, z) = \mathbf{G}(u, v, w)$, defined by

$$x = u, \; y = v, \; z = w - \varphi(u, v); \quad \text{that is,} \quad u = x, \; v = y, \; w = z - \varphi(x, y),$$

and set $\widetilde{\mathbf{F}} = \mathbf{F} \circ \mathbf{G}$. The set corresponding to $\partial R \cap B$ in uvw-space is a region S in the uv-plane. Let Q be a rectangular box in uvw-space that includes $\mathbf{G}^{-1}(R \cap \operatorname{supp}(\mathbf{F}))$, one face of which is a rectangle in the uv-plane that includes S. (S and Q correspond to L and R in the proof of Theorem B.28.)

We parametrize the surface $z = \varphi(x, y)$ by $(u, v) \to (u, v, \varphi(u, v))$ to see that

$$\iint_{\partial R} \mathbf{F} \cdot \mathbf{n}\, dA = \pm \iint_{S} \left[-(\partial_u \varphi)\widetilde{F}_1 - (\partial_v \varphi)\widetilde{F}_2 + \widetilde{F}_3 \right] dA = \iint_{\partial Q} \mathbf{H} \cdot \mathbf{n}\, dA,$$

where

$$\mathbf{H} = \left[-(\partial_u \varphi)\widetilde{F}_1 - (\partial_v \varphi)\widetilde{F}_2 + \widetilde{F}_3 \right]\mathbf{k}.$$

Here the $\pm$ is $+$ or $-$ depending on whether R (resp. Q) lies below or above the surface $z = \varphi(x, y)$ (resp. the uv-plane), that is, on whether the outward normal to Q on S is $+\mathbf{k}$ or $-\mathbf{k}$; the last equality holds because $\widetilde{\mathbf{F}}$ vanishes on $\partial Q \setminus S$. In the vector field $\mathbf{H}$, the functions $\widetilde{F}_j$ depend on (u, v, w), but φ depends only on (u, v). By the divergence theorem for the box Q (proved in §5.5), then,

$$(\text{B.31}) \qquad \iint_{\partial R} \mathbf{F} \cdot \mathbf{n}\, dA = \iiint_{Q} \operatorname{div} \mathbf{H}\, dV$$

$$= \iiint_{Q} \left[-\frac{\partial \varphi}{\partial u}\frac{\partial \widetilde{F}_1}{\partial w} - \frac{\partial \varphi}{\partial v}\frac{\partial \widetilde{F}_2}{\partial w} + \frac{\partial \widetilde{F}_3}{\partial w} \right] dV.$$

Now, $\widetilde{\mathbf{F}}(u, v, w) = \mathbf{F}(u, v, w + \varphi(u, v))$, so by the chain rule,

$$\frac{\partial \widetilde{F}_1}{\partial u} = \frac{\widetilde{\partial F_1}}{\partial x} + \frac{\widetilde{\partial F_1}}{\partial z}\frac{\partial \varphi}{\partial u}, \qquad \frac{\partial \widetilde{F}_2}{\partial v} = \frac{\widetilde{\partial F_2}}{\partial y} + \frac{\widetilde{\partial F_2}}{\partial z}\frac{\partial \varphi}{\partial v},$$

and

$$\frac{\partial \widetilde{F_j}}{\partial w} = \frac{\partial \widetilde{F_j}}{\partial z} \text{ for } j = 1, 2, 3,$$

where the tildes continue to denote composition with $\mathbf{G}$. Substituting these formulas into (B.31), we obtain

(B.32) $$\iint_{\partial R} \mathbf{F} \cdot \mathbf{n} \, dA = \iiint_{Q} \left[\frac{\partial \widetilde{F_1}}{\partial x} - \frac{\partial \widetilde{F_1}}{\partial u} + \frac{\partial \widetilde{F_2}}{\partial y} - \frac{\partial \widetilde{F_2}}{\partial v} + \frac{\partial \widetilde{F_3}}{\partial z} \right] dV.$$

We are almost done. On the one hand, by integrating first with respect to u or v, we see that

$$\iiint_{Q} \frac{\partial \widetilde{F_1}}{\partial u} \, dV = \iiint_{Q} \frac{\partial \widetilde{F_2}}{\partial v} \, dV = 0,$$

because F_1 and F_2 vanish on the vertical faces of Q. On the other hand, the transformation $\mathbf{G}$ is volume-preserving,

$$\frac{\partial(x, y, z)}{\partial(u, v, w)} = 1,$$

so by Theorem B.24,

$$\iiint_{Q} \left[\frac{\partial \widetilde{F_1}}{\partial x} + \frac{\partial \widetilde{F_2}}{\partial y} + \frac{\partial \widetilde{F_3}}{\partial z} \right] dV = \iiint_{Q} \widetilde{\operatorname{div} \mathbf{F}} \, dV = \iiint_{R} \operatorname{div} \mathbf{F} \, dV.$$

Therefore, (B.32) reduces to the desired result:

$$\iint_{\partial R} \mathbf{F} \cdot \mathbf{n} \, dA = \iiint_{R} \operatorname{div} \mathbf{F} \, dV.$$

In conclusion, we remark that these calculations appear more natural if the argument is recast in the language of differential forms as described in §5.9. $\qquad\square$

Answers to Selected Exercises

CHAPTER 1

Section 1.1

1. $\|\mathbf{x}\| = 2\sqrt{3}$, $\|\mathbf{y}\| = 3$, $\theta = 5\pi/6$.

Section 1.2

1. (a) Not open or closed; $\partial S = \{(0,0)\} \cup \{(x,y) : x^2 + y^2 = 4\}$.
   (b) Closed; $\partial S = \{(x,0) : 0 \le x \le 1\} \cup \{(x, x^2 - x) : 0 \le x \le 1\}$.
   (c) Open; $\overline{S} = \{(x,y) : x \ge 1,\ y \ge 1,\ \text{and } x + y \ge 1\}$.
   (d) Closed; $S^{\text{int}} = \varnothing$.

Section 1.3

3. $f(0, y) = y$.
5. Discontinuous only at $(0, 0)$.
7. Continuous at every irrational.

Section 1.4

1. (a) $1/\sqrt{2}$.   (b) 0.   (c) Diverges.
2. Any $K \ge (19/\epsilon) + 5$ will work.
3. $\lim x_k = 0$.

Section 1.5

1. (a) $\sup S = 1$, $\inf S = -1$.
   (b) $\sup S = 2$, $\inf S = -1$.
   (c) $\sup S = \infty$, $\inf S = \pi/4$.
5. $\lim x_k = 2$.

CHAPTER 2

Section 2.2

1. (a) $\nabla f(x, y) = (2xy + \pi y \cos \pi xy, \; x^2 + \pi x \cos \pi xy)$;
   $[\nabla f(1, -2)] \cdot (\frac{3}{5}, \frac{4}{5}) = -\frac{1}{5}(8 + 2\pi)$.
2. (a) $df = e^{x-y+3z}[(2x + x^2) \, dx - x^2 \, dy + 3x^2 \, dz]$;
   $f(1.1, 1.2, -0.1) - f(1, 1, 0) \approx -0.2$.
3. (a) $dz = 0.036$. (b) z.

Section 2.3

1. (Derivatives of f, g, and h are to be evaluated at the same points as f, g, and h
   themselves.) (a) $dw/dt = f_1(g_1 h' + g_2) + f_2 h' + f_3$.
   (b) $\partial_x w = f_1 + f_2 g_1 + f_3 h_1, \; \partial_y w = f_2 g_2, \; \partial_z w = f_3 h_2$.
   (c) $dw/dx = f'(g_1 + g_2 h')$.
2. (a) $\partial_x w = 2 f_1 + (\sin 3y) f_2 + 4x^3 f_3, \; \partial_y w = -2y f_1 + (3x \cos 3y) f_2$　(f_1 and
   f_2 evaluated at $(2x - y^2, \; x \sin 3y, \; x^4)$).
   (c) $\partial_x w = 2(\partial_2 f)/(f^2 + 1), \; \partial_y w = (2y \partial_1 f - \partial_2 f)/(f^2 + 1)$　(f and its
   derivatives evaluated at $(y^2, \; 2x - y)$).
6. (a) $z = 4x - 3y - 6$.
   (b) $2x + 4y - 6z = 12$.

Section 2.5

1. (a) $\partial z/\partial x = (1 - 3yz)/(3xy - 3z^2), \quad \partial z/\partial y = (2y - 3xz)/(3xy - 3z^2)$.
3. $dz/dt = (2yz + 5y^4 t + zte^{yz})/(10y^4 z^3 - 4z^4 e^{yz} - y^2 e^{yz} - yt^2)$.
4. $2x, \quad 2x + 6xz^2$.
5. $(\partial V/\partial h)|_r = \pi r^2, \quad (\partial V/\partial h)|_S = \pi r^2 - 2\pi r^2 h/(2r + h), \quad (\partial V/\partial S)|_r = r/2, \quad (\partial S/\partial V)|_r = 2/r$.

Section 2.6

2. $r \sin \theta \cos \theta (f_{yy} - f_{xx}) + r(\cos^2 \theta - \sin^2 \theta) f_{xy} - (\sin \theta) f_x + (\cos \theta) f_y$.
3. (a) $\partial_x^2 w = 4 f_{11} + 4(\sin 3y) f_{12} + 16x^3 f_{13} + (\sin^2 3y) f_{22} + 8x^3 (\sin 3y) f_{23} + 16x^6 f_{33} + 12x^2 f_3, \quad \partial_x \partial_y w = -4y f_{11} + (3x \cos 3y - 2y \sin 3y) f_{12} - 8x^3 y f_{13} + 3x(\sin 3y \cos 3y) f_{22} + 12x^4 (\cos 3y) f_{23} + 3(\cos 3y) f_2$.

Section 2.7

1. (b) $1/24$.
2. (a) $P_{1,3}(h) = h - \frac{1}{2} h^2 + \frac{1}{3} h^3, C = 4$.
   (b) $P_{1,3}(h) = 1 + \frac{1}{2} h - \frac{1}{8} h^2 + \frac{1}{16} h^3, C = 5 \cdot 2^{-7/2}$. (*Note:* These C's come
   from Lagrange's formula and may not be optimal.)

4. 0.747.

5. (a) $x^2 + xy - \frac{1}{6}(x^4 + 3x^3y + 3x^2y^2 + xy^3)$.
   (b) $1 + xy - \frac{1}{2}(x^4 + x^2y^2 + y^4)$.

6. $P_{(3,1),3}(h, k) = 2 + h + 3k + hk + \frac{1}{2}(\pi^2 - 3)k^2 - \frac{1}{2}hk^2 + k^3$.

7. $P_{(1,2,1),3}(h, k, l) = 3 + 4h + k + l + 2h^2 + 2hk + h^2k$.

Section 2.8

1. (a) $(0, -2)$ and $(0, 1)$ minima, $(0, 0)$ saddle.
   (b) $(\pm 1, \sqrt{2})$ minima, $(0, -\sqrt{2})$ maximum, $(\pm 1, -\sqrt{2})$ and $(0, \sqrt{2})$ saddles.
   (c) $(1, \pm 1)$ and $(0, 0)$ saddles, $(\frac{2}{3}, 0)$ minimum.
   (e) $(0, 0)$ minimum, $(\pm 1, 0)$ maxima, $(0, \pm 1)$ saddles.
   (f) $((a^2/b)^{1/3}, (b^2/a)^{1/3})$, a minimum if a and b have the same sign and a maximum otherwise.

Section 2.9

1. $\min = -\frac{1}{2}$, $\max = 4$.
2. $\min = -4$, $\max = \frac{16}{5}$.
3. $\min = (308 - 62\sqrt{31})/27$, $\max = 2/3\sqrt{3}$.
4. $\min = -\frac{85}{3}$, $\max = 56$.
5. $A^2/(1 + b^2 + c^2)$.
6. $\min = 0$, $\max = 2/e$.
7. $\min = -2/e$, $\max = 1/e$.
8. $3(12)^{1/3}$
9. $\min = 1$, $\max = 3$.
11. $(\sqrt{a} + \sqrt{b} + \sqrt{c})^2$.
12. $3V^{1/3}$.
13. $(\frac{1}{2}, \frac{1}{2}, 0)$.
14. $V_{\max} = A^{3/2}/6\sqrt{3}$.
15. $(2, 0, 2)$.
16. a^2b^2.

Section 2.10

1. $\partial(u, v)/\partial(x, y) = 3xy^2z^2 - yz^3 + 24y^3$, $\partial(u, v)/\partial(x, z) = -y^2z^2 - 6xy^3z$,
   $\partial(u, v)/\partial(y, z) = xyz^2 + 8y^2 - 12x^2y^2z$.

2. $\partial(u, v)/\partial(x, y) = 3x - 18y$, $\partial(v, w)/\partial(x, y) = -6x^2 - 18y^2$,
   $\partial(u, w)/\partial(x, y) = -12x - 6y$.

3. (b) $\begin{pmatrix} -15 & -20 \\ 3 & 4 \\ 2 & 4 \end{pmatrix}$.

4. (b) $\begin{pmatrix} 8 & 6 & -21 \\ 18 & 10 & -43 \end{pmatrix}$.

CHAPTER 3

Section 3.1

3. y yes; z no.
5. $\partial_2 F(0,0) \neq 0$ and $\partial_1 F(0,0) \neq -1$.
6. Can solve for x and y or y and z.
7. Can solve for any pair.
9. Yes.

Section 3.2

1. (a), (c), (f) are smooth curves.
3. (a), (c), (e) are smooth curves.

Section 3.3

1. (a) Plane.
   (b) Elliptic cone.
   (c) Hyperboloid of revolution.
   (d) Paraboloid of revolution.
2. (a) $2x - y - z = 3$. (b) $x - y = 3$.
3. (a) One possibility: $\mathbf{f}(u,v) = (u \cos v, \, u \sin v, \, f(u))$ $(a < u < b, \, |v| \leq \pi)$.
4. (a) One possibility: $\mathbf{f}(t) = (1 + t, \, \frac{1}{3} + t, \, \frac{8}{3} + t)$.
5. (a) One possibility: $\mathbf{f}(t) = \frac{1}{2}(1 + \cos t, \, \sqrt{2} \sin t, \, 1 - \cos t)$. (b) One possibility: $\mathbf{f}(t) = \frac{1}{2}(1 + t, \, -\sqrt{2}, \, 1 - t)$.

Section 3.4

1. (a) $\det D\mathbf{f} = e^{2x}$; $x = \frac{1}{2} \log(u^2 + v^2)$; y is given up to multiples of 2π by $\arctan(v/u)$ when $u > 0$, $\frac{1}{2}\pi - \arctan(u/v)$ when $v > 0$, $\pi + \arctan(v/u)$ when $u < 0$, $\frac{3}{2}\pi - \arctan(u/v)$ when $v < 0$.
2. (a) $(x, y) = \frac{1}{3}(2v - u, \, v - 2u)$.
4. (d) $\mathbf{g}(u,v) = \frac{1}{2}(u - \sqrt{u^2 + 4v}, \, -u - \sqrt{u^2 - 4v})$.

Section 3.5

1. One relation for (a), (c), and (e); two relations for (d).

CHAPTER 4

Section 4.3

1. (a) $\frac{4}{5}$. (b) $\frac{32}{35}(5 - \sqrt{2})$.
2. $\frac{1}{20}$.
3. (a) $\int_{-2}^{0} \int_{4x}^{x^3} f(x,y)\,dy\,dx$, $\int_{-8}^{0} \int_{y^{1/3}}^{y/4} f(x,y)\,dx\,dy$.
   (b) $\int_{0}^{2} \int_{x/3}^{x} f(x,y)\,dy\,dx + \int_{2}^{3} \int_{x/3}^{4-x} f(x,y)\,dy\,dx$,
   $\int_{0}^{1} \int_{y}^{3y} f(x,y)\,dx\,dy + \int_{1}^{2} \int_{y}^{4-y} f(x,y)\,dx\,dy$.
4. (a) $\int_{0}^{1} \int_{y^3}^{y^{1/2}} f(x,y)\,dx\,dy$.
   (b) $\int_{-1}^{0} \int_{-x}^{1} f(x,y)\,dy\,dx + \int_{0}^{2} \int_{x/2}^{1} f(x,y)\,dy\,dx$.
5. (a) $\frac{5}{8}e^6 - \frac{33}{8}e^2$. (b) $\frac{1}{3}(\sin 2 - \sin 1)$. (c) $\frac{1}{2}e^2 - e$.
6. $\int_{0}^{1} f(y)\sqrt{y/2}\,dy + \int_{1}^{2} f(y)(\sqrt{y/2} - y + 1)\,dy$.
8. (a) $\int_{-1}^{1} \int_{-\sqrt{1-x^2}}^{\sqrt{1-x^2}} \int_{x^2+y^2}^{1} f\,dz\,dy\,dx$.
   (b) $\int_{-1}^{1} \int_{x^2}^{1} \int_{-\sqrt{z-x^2}}^{\sqrt{z-x^2}} f\,dy\,dz\,dx$.
   (c) $\int_{0}^{1} \int_{-\sqrt{z}}^{\sqrt{z}} \int_{-\sqrt{z-y^2}}^{\sqrt{z-y^2}} f\,dx\,dy\,dz$.
9. (b) $\int_{0}^{1} \int_{0}^{\sqrt{1-x}} \int_{0}^{y} f\,dz\,dy\,dx$.
   (c) $\int_{0}^{1} \int_{0}^{\sqrt{1-x}} \int_{z}^{\sqrt{1-x}} f\,dy\,dz\,dx$.
10. $\frac{1}{4}(a,b,c)$.
11. mass $= 8$, center of mass $= (1, \frac{4}{3}, \frac{4}{3})$.
12. $-\frac{126}{5}$.

Section 4.4

1. $3\pi/2$.
2. $(\frac{1}{\pi}, 0, \frac{3}{4})$.
3. $4\pi(\frac{8}{3} - \sqrt{3})$.
4. $2\pi - \frac{32}{9}$.
5. $\frac{1}{2}\pi c R^2 h^2$.
6. $5\pi/3$.
7. $\pi c R^4/3$.
8. $(\frac{3}{8}, \frac{3}{8}, \frac{3}{8})$.
9. $\frac{1}{14}(55, -5)$.
10. $\frac{4}{81}$.
11. $\pi/3\sqrt{3}$.
12. $A = \frac{3}{2}\log 4$, $\quad \bar{x} = \frac{14}{9\log 4}$, $\quad \bar{y} = \frac{28}{9\log 4}$.
13. 3.

14. 3.
15. $\frac{1}{2}\pi^2 R^4$.

Section 4.5

2. (a) $\dfrac{1}{x}\log\left(\dfrac{1+ex}{1+x}\right)$.
   (b) $(2x)^{-1}(5\cos x^5 - \cos x)$.
   (c) $x^{-1}(2e^{3x^2} - e^x)$.

Section 4.6

1. (a) Converges. (b) Diverges. (c) Converges. (d) Converges. (e) Diverges.
2. (a) Converges. (b) Diverges. (c) Converges. (d) Converges. (e) Diverges.
3. (a) Converges. (b) Diverges. (c) Converges. (d) Diverges. (e) Converges.
   (f) Diverges.
4. (b) $p > 1$.
5. (b) $p > 1$.
10. $-\frac{1}{2}\log 3$.

Section 4.7

2. (a) Diverges.   (b) $\frac{1}{4}\pi$.   (c) $2\pi/3$.   (d) $\frac{1}{2}\sqrt{\pi}$.   (e) Diverges.

CHAPTER 5

Section 5.1

1. (a) $2\pi\sqrt{a^2 + b^2}$.   (b) $\frac{14}{3}$.   (c) e^2.   (d) 24.
2. (a) $4aE(\sqrt{1 - (b/a)^2})$.   (b) $2^{3/2}E(2^{-1/2})$.
3. $\left(0,\ (2 + \sinh 2)/4\sinh 1)\right)$.
4. $\frac{1}{3}[(1 + 4\pi^2)^{3/2} - 1]$.
5. (a) 1.   (b) $\frac{23}{21}$.   (c) -2π.   (d) $\frac{9856}{45}$.
6. (a) $\frac{1}{2}(1 - e^{-1}) + (2/\pi)$.   (b) $-\pi$.   (c) $-\frac{4}{3}$.

Section 5.2

1. (c) 12.   (d) 0.
2. $\frac{15}{2}\pi$.
3. The circle $x^2 + y^2 = 1$.
4. $3\pi R^2$.

Section 5.3

1. $\frac{2}{3}\pi[(1 + a^2)^{3/2} - 1]$.
2. $\frac{1}{6}\pi[(1 + 4a^2)^{3/2} - 1]$.
3. $4\pi^2 ab$.
4. $2\pi a^2 + \dfrac{2\pi ab^2}{\sqrt{a^2 - b^2}} \log\left(\dfrac{a + \sqrt{a^2 - b^2}}{b}\right)$ if $a > b$,

   $2\pi a^2 + \dfrac{2\pi ab^2}{\sqrt{b^2 - a^2}} \arcsin\left(\dfrac{\sqrt{b^2 - a^2}}{b}\right)$ if $b > a$.
5. $(0, 0, \frac{1}{2})$.
6. $20\pi/3$.
7. 0.
8. (a) $-\frac{17}{9}$. (b) 0. (c) 2. (d) $\pi(b^2 - a^2)$. (e) $\pi(2^{5/2} - \frac{7}{2})$.

Section 5.4

1. (a) $\operatorname{curl} \mathbf{F} = x\mathbf{i} - y\mathbf{j} + (y - y^2)\mathbf{k}$, $\operatorname{div} \mathbf{F} = x + y^2$.
   (b) $\operatorname{curl} \mathbf{F} = \mathbf{0}$, $\operatorname{div} \mathbf{F} = -x(y^2 + z^2)\sin yz$.
   (c) $\operatorname{curl} \mathbf{F} = (1 - 4xy)\mathbf{i} - (x^2 - 3z^2)\mathbf{j} + 4yz\mathbf{k}$, $\operatorname{div} \mathbf{F} = 0$.
2. (a) 0. (b) $2x - 24yz$. (c) $a(a + n - 2)|\mathbf{x}|^{a-2}$. (d) 0.

Section 5.5

1. (c) $3a^4$. (d) $4\pi(a^2 b^2 + b^2 c^2 + a^2 c^2)/3abc$. (e) $3A$.
2. $4\pi a^5$.
6. (a) $-\mathbf{x}/|\mathbf{x}|^3$.

Section 5.6

3. (a) $2\rho(x\mathbf{i} + y\mathbf{j})/(x^2 + y^2)$.

Section 5.7

1. 2π.
2. $-\pi a^2/\sqrt{2}$.
4. 0.
5. 0.
7. $5 + 3\pi(r^2 - 1)$.

Section 5.8

1. (a) $x^2 y + \frac{1}{3}x^3 - \frac{1}{3}y^3 + C$.
   (b) Not a gradient.
   (c) $e^{2x}\sin y - 3xy + 5x + C$.

(d) $xyz + \cos xy + \sin yz + C$.

(e) Not a gradient.

(f) $x^2 y + (y + 2) \log z + C$.

(g) $\frac{1}{2}(xw + yz)^2 - e^{2y+z} + \cos zw + C$.

2. (a) Not a curl.

(b) $\frac{1}{2}xz^2\mathbf{i} - (xyz - \frac{1}{2}x^2 - \frac{1}{2}z^2)\mathbf{j} + \nabla f$.

(c) $(5yz + z^2)\mathbf{i} + (6xz - x\int_0^z e^{-x^2 t^2}\, dt)\mathbf{j} + \nabla f$.

CHAPTER 6

Section 6.1

1. (a) $-1 - 2^{-1/3} < x < -1 + 2^{-1/3}$; $(2x + 2)/[1 - 2(x + 1)^3]$.

   (b) $x < -\sqrt{2}$ or $x > \sqrt{2}$; $10/(x^2 - 2)$.

   (c) $x > 0$; $\frac{1}{2}(1 + x^{-1})$.

   (d) $e^{-1} < x < e$; $\log x/(1 - \log x)$.

2. (a) Diverges.   (b) 1.   (c) Diverges.   (d) Diverges.

Section 6.2

1. Converges.
2. Converges.
3. Diverges.
4. Converges.
5. Diverges.
6. Converges.
7. Diverges.
8. Diverges.
9. Converges.
10. Converges.
11. Diverges.
12. Converges.
13. Converges.
14. Diverges.
15. Diverges.
16. Converges.
17. Converges.
18. Diverges.
21. $p > 1$.

Section 6.4

1. Converges absolutely for $-3 \leq x \leq -1$.
2. Converges absolutely for $0 < x < 1$.
3. Converges absolutely for all x.
4. Converges absolutely for $-5 < x < 5$, conditionally for $x = -5$.
5. Converges absolutely for $2 < x < 6$, conditionally for $x = 6$.
6. Converges absolutely for $x > 0$, conditionally for $x = 0$.
7. Converges absolutely for $4 < x < 8$.
8. Converges absolutely for $-2 < x < 0$, conditionally for $x = -2$ and $x = 0$.
9. Converges absolutely for $-\frac{3}{2} < x < \frac{3}{2}$, conditionally for $x = -\frac{3}{2}$.
10. Converges conditionally.
11. Converges conditionally.
12. Diverges.
13. Converges absolutely.
14. Converges conditionally.
18. Converges when $|x| < 1$ and $\theta \in \mathbb{R}$, when $x = 1$ and $\theta \neq 2k\pi$, or when $x = -1$ and $\theta \neq (2k+1)\pi$.

CHAPTER 7

Section 7.1

1. (a) Uniform convergence on $[0, 1 - \delta]$ $(\delta > 0)$.
   (b) Uniform convergence on $[\delta, 1]$ $(\delta > 0)$.
   (c) Uniform convergence on $[0, \frac{1}{2}\pi - \delta]$ and $[\frac{1}{2}\pi + \delta, \pi]$ $(\delta > 0)$.
   (d) Uniform convergence on $\mathbb{R}$.
   (e) Uniform convergence on $[\delta, \infty)$ $(\delta > 0)$.
   (f) Uniform convergence on $[0, b]$ $(b < \infty)$.
   (g) Uniform convergence on $[0, 1 - \delta]$ and $[1 + \delta, \infty)$ $(\delta > 0)$.
2. (a) Uniform convergence on $[\delta, \infty)$ $(\delta > 0)$.
   (b) Uniform convergence on $[-1, 1]$.
   (c) Uniform convergence on $[-2 + \delta, 2 - \delta]$ $(\delta > 0)$.
   (d) Uniform convergence on $\mathbb{R}$.
   (e) Uniform convergence on $\mathbb{R}$.
   (f) Uniform convergence on $[1 + \delta, \infty)$ $(\delta > 0)$.

Section 7.3

5. (a) $\displaystyle\sum_{0}^{\infty} \frac{(-1)^n x^{2n+1}}{n!(2n+1)}, \quad x \in \mathbb{R}.$

(b) $\displaystyle\sum_{0}^{\infty} \frac{(-1)^n x^{4n+1}}{(2n)!(4n+1)}$, $x \in \mathbb{R}$.

(c) $\displaystyle\sum_{1}^{\infty} \frac{(-1)^{n-1}(2x)^n}{n^2}$, $|x| \le \frac{1}{2}$.

10. (a) $e^x + x^{-1}(1 - e^x)$.

    (b) $\int_0^x t^{-2}(1 - \cos t)\,dt$.

    (c) $x^{-1} \int_0^x t^{-1}(e^t - 1)\,dt$.

    (d) $\cos x - x \sin x$.

Section 7.5

4. $\dfrac{1 \cdot 3 \cdots (2n-3)}{2 \cdot 4 \cdots (2n-2)} \dfrac{\pi}{2x^{(2n-1)/2}}$.

Section 7.6

3. (a) $\frac{3}{8}\sqrt{\pi}$.   (b) $\frac{1}{2}\sqrt{\pi/27}$.   (c) $\frac{3}{16}\sqrt{\pi}$.

5. $\Gamma((a+1)/b)\Gamma(c+1)/b\Gamma(c+1+(a+1)/b)$.

7. $\dfrac{1 \cdot 3 \cdots (k-1)}{2 \cdot 4 \cdots k} \dfrac{\pi}{2}$ if k is even, $\dfrac{2 \cdot 4 \cdots (k-1)}{3 \cdot 5 \cdots k}$ if k is odd (and $k > 1$).

10. (a) Diverges.   (b) Converges.

CHAPTER 8

Section 8.1

1. $\dfrac{4}{\pi} \displaystyle\sum_{1}^{\infty} \dfrac{\sin(2m-1)\theta}{2m-1}$.

2. $\frac{1}{2} - \frac{1}{2}\cos 2\theta$.

3. $\dfrac{2}{\pi} - \dfrac{4}{\pi} \displaystyle\sum_{1}^{\infty} \dfrac{\cos 2m\theta}{4m^2 - 1}$.

4. $\dfrac{\pi^2}{3} + 4 \displaystyle\sum_{1}^{\infty} \dfrac{(-1)^n}{n^2} \cos n\theta$.

5. $\dfrac{\sinh b\pi}{\pi} \displaystyle\sum_{-\infty}^{\infty} \dfrac{(-1)^n}{b - in} e^{in\theta}$.

6. $\dfrac{8}{\pi} \displaystyle\sum_{1}^{\infty} \dfrac{\sin(2m-1)\theta}{(2m-1)^3}$.

7. $\dfrac{2}{a(\pi - a)} \displaystyle\sum_{1}^{\infty} \dfrac{\sin na}{n} \cos n\theta$.

8. $\dfrac{1}{2\pi} + \dfrac{2}{\pi} \displaystyle\sum_{1}^{\infty} \dfrac{1 - \cos na}{n^2 a^2} \cos n\theta.$

Section 8.2

1. (a) $\displaystyle\sum_{1}^{\infty} \dfrac{\sin 2n\theta}{n}.$

    (b) $1 - \dfrac{2}{\pi} \displaystyle\sum_{1}^{\infty} \dfrac{\sin 2n\theta}{n}.$

2. $\dfrac{\pi^2}{3} + 4 \displaystyle\sum_{1}^{\infty} \dfrac{(-1)^n}{n^2} (\cos \tfrac{1}{4}n\pi \cos n\theta + \sin \tfrac{1}{4}n\pi \sin n\theta).$

3. (a) $\dfrac{1}{2} + \dfrac{2}{\pi} \displaystyle\sum_{1}^{\infty} \dfrac{\sin(2m - 1)\theta}{2m - 1}.$

    (b) $\dfrac{1}{\pi} - \dfrac{2}{\pi} \displaystyle\sum_{1}^{\infty} \dfrac{\cos 2m\theta}{4m^2 - 1} + \dfrac{1}{2} \sin \theta.$

    (c) $\dfrac{1}{2\pi} + \dfrac{1}{\pi} \displaystyle\sum_{1}^{\infty} \dfrac{\sin na}{na} \cos n\theta.$

    (d) $\dfrac{2 \sinh \pi}{\pi} \displaystyle\sum_{1}^{\infty} \dfrac{(-1)^{n-1} n}{n^2 + 1} \sin n\theta.$

4. (a) $\tfrac{1}{2}, \quad \tfrac{1}{4}(\pi - 2).$
    (b) $\tfrac{1}{6}\pi^2, \quad \tfrac{1}{12}\pi^2.$
    (c) $(\pi b \operatorname{csch} \pi b - 1)/2b^2, \quad (\pi b \coth \pi b - 1)/2b^2.$
    (d) $\tfrac{1}{32}\pi^3.$

Section 8.3

2. (b) $\dfrac{2}{a(\pi - a)} \displaystyle\sum_{1}^{\infty} \dfrac{\sin na}{n^2} \sin n\theta.$

6. (a) $k = 6.$
    (b) $k = \infty.$
    (c) $k = 0$, i.e., the function is merely continuous. (It is known to be nowhere differentiable.)

Section 8.4

1. (a) $1;\quad \dfrac{4}{\pi} \displaystyle\sum_{1}^{\infty} \dfrac{\sin(2m - 1)\theta}{2m - 1}.$

    (b) $\dfrac{2}{\pi} - \dfrac{4}{\pi} \displaystyle\sum_{1}^{\infty} \dfrac{\cos 2m\theta}{4m^2 - 1};\quad \sin \theta.$

(c) $\dfrac{\pi^2}{3} + 4 \displaystyle\sum_1^\infty \dfrac{(-1)^n}{n^2} \cos n\theta$; $2\pi \displaystyle\sum_1^\infty \dfrac{(-1)^{n+1}}{n} \sin n\theta - \dfrac{8}{\pi} \displaystyle\sum_1^\infty \dfrac{\sin(2m-1)\theta}{(2m-1)^3}$.

(d) $\dfrac{\pi}{4} - \dfrac{2}{\pi} \displaystyle\sum_1^\infty \dfrac{\cos(4m-2)\theta}{(2m-1)^2}$; $\dfrac{4}{\pi} \displaystyle\sum_1^\infty (-1)^{m+1} \dfrac{\sin(2m-1)\theta}{(2m-1)^2}$.

2. (a) $\dfrac{4}{\pi} \displaystyle\sum_1^\infty \dfrac{\sin(2m-1)\pi x}{2m-1}$.

(b) $\dfrac{4}{\pi} \displaystyle\sum_1^\infty \dfrac{(-1)^{m+1} \cos(\frac{1}{2}m - \frac{1}{4})\pi x}{2m-1}$.

(c) $\dfrac{8l^2}{\pi^3} \displaystyle\sum_1^\infty \dfrac{\sin(2m-1)\pi x/l}{(2m-1)^3}$.

(d) $(e-1) \displaystyle\sum_{-\infty}^\infty \dfrac{e^{2\pi i n x}}{1 - 2\pi i n}$.

Section 8.5

1. (a) $u(x,t) = 50 - \dfrac{400}{\pi^2} \displaystyle\sum_1^\infty \dfrac{1}{(2m-1)^2} e^{-(0.00011)(2m-1)^2 \pi^2 t} \cos \dfrac{(2m-1)\pi x}{100}$.

2. $u(x,t) = \sum_{-\infty}^\infty c_n \exp[in\theta - n^2 kt]$ where $f(\theta) = \sum_{-\infty}^\infty c_n e^{in\theta}$.

3. $b_n(t) = \exp(-n^2\pi^2 kt/l^2) \left[b_n(0) + \int_0^t \beta_n(s) \exp(n^2\pi^2 ks/l^2)\, ds \right]$.

4. (a) $u(x,t) = \dfrac{2l^2 m}{\pi^2 b(l-b)} \displaystyle\sum_1^\infty \dfrac{1}{n^2} \sin \dfrac{n\pi b}{l} \sin \dfrac{n\pi x}{l} \cos \dfrac{n\pi c t}{l}$.

5. $u(x,t) = \sum_1^\infty e^{-\delta t}(b_n \cos \omega_n t + B_n \sin \omega_n t) \sin(n\pi x/l)$,
   where $\omega_n^2 = (n\pi c/l)^2 - \delta^2$.

6. (b) $u(x,y) = \displaystyle\sum_1^\infty \left[a_n^1 \dfrac{\sinh(n\pi(L-y)/l)}{\sinh(n\pi L/l)} + a_n^2 \dfrac{\sinh(n\pi y/l)}{\sinh(n\pi L/l)} \right] \sin(n\pi x/l)$,
   where $f_1(x) = \sum_1^\infty a_n^1 \sin(n\pi x/l)$ and $f_2(x) = \sum_1^\infty a_n^2 \sin(n\pi x/l)$.

Section 8.6

3. $a = -\frac{1}{2},\quad b = -1,\quad c = \frac{1}{6}$.

9. (a) $\pi^4/90$.

(b) $\pi^6/960$.

(c) $\pi^8/9450$.

(d) $\frac{1}{2}a(\pi - a)$ if $0 \le a \le \pi$, π-periodic as a function of a.

Bibliography

[1] H. Anton, *Elementary Linear Algebra* (7th ed.), John Wiley, New York, 1994.

[2] R. G. Bartle, Return to the Riemann integral, *Amer. Math. Monthly* **103** (1996), 625–632.

[3] H. S. Bear, *A Primer of Lebesgue Integration*, Academic Press, San Diego, 1995.

[4] G. Birkhoff and S. Mac Lane, *A Survey of Modern Algebra* (5th ed.), A K Peters, Wellesley, MA, 1997.

[5] J. D. DePree and C. W. Swartz, *Introduction to Real Analysis*, John Wiley, New York, 1988.

[6] G. B. Folland, *Fourier Analysis and its Applications*, Brooks/Cole, Pacific Grove, CA, 1992.

[7] J. H. Hubbard and B. B. Hubbard, *Vector Calculus, Linear Algebra, and Differential Forms*, Prentice-Hall, Upper Saddle River, NJ, 1999.

[8] T. W. Hungerford, *Abstract Algebra: an Introduction* (2nd ed.), Saunders College Publishing, Fort Worth, 1997.

[9] B. F. Jones, *Lebesgue Integration on Euclidean Space*, Jones and Bartlett, Boston, 1993.

[10] D. W. Kammler, *A First Course in Fourier Analysis*, Prentice Hall, Upper Saddle River, NJ, 2000.

[11] T. W. Körner, *Fourier Analysis*, Cambridge University Press, Cambridge, UK, 1988.

[12] S. G. Krantz, *Real Analysis and Foundations*, CRC Press, Boca Raton, FL, 1991.

[13] J. C. Lagarias, The $3x + 1$ problem and its generalizations, *Amer. Math. Monthly* **92** (1985), 3–23.

[14] P. D. Lax, Change of variables in multiple integrals, *Amer. Math. Monthly* **106** (1999), 497–501.

[15] P. D. Lax, Change of variables in multiple integrals II, *Amer. Math. Monthly* **108** (2001), 115–119.

[16] D. C. Lay, *Linear Algebra and its Applications* (2nd ed.), Addison-Wesley, Reading, MA, 1997.

[17] J. W. Lewin, A truly elementary approach to the bounded convergence theorem, *Amer. Math. Monthly* **93** (1986), 395–397.

[18] W. Rudin, *Principles of Mathematical Analysis* (3rd ed.), McGraw-Hill, New York, 1976.

[19] S. H. Weintraub, *Differential Forms: A Complement to Vector Calculus*, Academic Press, San Diego, 1997.

Index